Seascape Ecology

Seascape Ecology

Edited by Simon J. Pittman
NOAA Biogeography Branch, Silver Spring, USA
and Marine Institute, Plymouth University, Plymouth, UK

Registered Offices
John Wiley & Sons, Inc., 111 River Street, Hoboken, NJ 07030, USA
John Wiley & Sons Ltd, The Atrium, Southern Gate, Chichester, West Sussex, PO19 8SQ, UK

Editorial Office
9600 Garsington Road, Oxford, OX4 2DQ, UK

For details of our global editorial offices, customer services, and more information about Wiley products visit us at www.wiley.com.

Wiley also publishes its books in a variety of electronic formats and by print-on-demand. Some content that appears in standard print versions of this book may not be available in other formats.

Library of Congress Cataloging-in-Publication Data

Names: Pittman, Simon J., editor.
Title: Seascape ecology / edited by Dr. Simon J. Pittman.
Description: Hoboken, NJ : John Wiley & Sons, 2018. | Includes
    bibliographical references and index. |
Identifiers: LCCN 2017036860 (print) | LCCN 2017044138 (ebook) | ISBN
    9781119084457 (pdf) | ISBN 9781119084440 (epub) | ISBN 9781119084433 (pbk.
    : alk. paper)
Subjects: LCSH: Marine ecology. | Landscape ecology.
Classification: LCC QH541.5.S3 (ebook) | LCC QH541.5.S3 S36 2017 (print) |
    DDC 577.7–dc23
LC record available at https://lccn.loc.gov/2017036860

Cover Design: Wiley
Cover Image: (Shoal of Surgeonfish and reef) © Jason Edwards/Gettyimages;
(Offshore windfarm and service boat) © Monty Rakusen;
(Sea water surface) © Andrey_Kuzmin/Shutterstock;
(Structure molecule) © pro500/Shutterstock;
(Wireframe grid) © DmitriyRazinkov/Shutterstock; (Square grid) Wiley

Set in 10/12pt WarnockPro by SPi Global, Chennai, India

Contents

Contributors

Diego Alvarez-Berastegui
Balearic Islands Coastal Observing and
Forecasting System (SOCIB)
Palma de Mallorca
Balearic Islands
Spain

Donna-marie Audas
Great Barrier Reef Marine Park Authority
Townsville
Australia

Edward B. Barbier
Department of Economics
School of Global Environmental
Sustainability
Colorado State University
Fort Collins, CO
United States

Christoffer Boström
Åbo Akademi University
Faculty of Science and Engineering
Environmental and Marine Biology
Turku
Finland

Mark H. Carr
Department of Ecology and Evolutionary
Biology
University of California Santa Cruz
Santa Cruz, CA
United States

Samantha Coccia-Schillo
NatureServe
Arlington, VA
United States

Bryan Costa
Biogeography Branch
National Centers for Coastal Ocean
Science
National Oceanic and Atmospheric
Administration
Silver Spring
Maryland, MD
United States

Benjamin Davis
Estuary and Tidal Wetland Ecosystems
Research Group
School of Marine and Tropical Biology
James Cook University
Townsville
Queensland, QLD
Australia

Jennifer A. Dijkstra
Center for Coastal and Ocean Mapping
University of New Hampshire
Jere A. Chase Ocean Engineering Lab
Durham, NH
United States

Clare Embling
Marine Biology and Ecology Research
Centre
School of Marine Science and
Engineering
Plymouth University
Plymouth
United Kingdom

Kim A. Falinski
The Nature Conservancy of Hawai'i
Honolulu
O'ahu, HI
United States

Ben L. Gilby
School of Science and Engineering
University of the Sunshine Coast
Maroochydore
Queensland, QLD
Australia

Paul Groves
Great Barrier Reef Marine Park Authority
Townsville
Queensland, QLD
Australia

Kevin A. Hovel
Department of Biology
San Diego State University
San Diego, CA
United States

Chantal M. Huijbers
Australian Rivers Institute
Coast and Estuaries, and School of
Environment
Griffith University
Gold Coast
Queensland, QLD
Australia

Simon Ingram
Marine Biology and Ecology Research
Centre
School of Marine Science and
Engineering
Plymouth University
Plymouth
United Kingdom

Emma L. Jackson
School of Medical and Applied Sciences
Central Queensland University
Gladstone
Queensland, QLD
Australia

Johnathan Kool
National Earth and Marine Observations
Branch
Geoscience Australia
Canberra
Australia

Chris A. Lepczyk
School of Forestry and Wildlife Science
Auburn University
AL
United States

Ivan Nagelkerken
Southern Seas Ecology Laboratories
School of Biological Sciences and The
Environment Institute
The University of Adelaide
Adelaide, SA
Australia

Andrew D. Olds
School of Science and Engineering
University of the Sunshine Coast
Maroochydore
Queensland, QLD
Australia

Kirsten L. L. Oleson
Department of Natural Resources and
Environmental Management
University of Hawai'i Mānoa
Hawai'i
United States

Camille Parrain
Littoral, Environment and Societies
Université de La Rochelle
La Rochelle
France

Simon J. Pittman
Biogeography Branch
National Centers for Coastal Ocean
Sciences
National Oceanic and Atmospheric
Administration
Silver Spring, MD
United States

and

Marine Institute
Plymouth University
Plymouth
United Kingdom

Helen M. Regan
Department of Biology
University of California Riverside
Riverside, CA
United States

Rolando O. Santos Corujo
Southeast Environmental Research
Center
Department of Earth and Environment
Florida International University
Miami, FL
United States

Steven Saul
College of Integrative Sciences and Arts
Arizona State University
Polytechnic Campus
Wanner Hall
Mesa, AZ
United States

Kylie L. Scales
School of Science and Engineering
University of the Sunshine Coast
Maroochydore
Queensland, QLD
Australia

formerly at

Environmental Research Division
Southwest Fisheries Science Center
National Oceanic and Atmospheric
Administration
Monterey, CA
United States

Thomas A. Schlacher
School of Science and Engineering
University of the Sunshine Coast
Maroochydore
Queensland, QLD
Australia

David C. Schneider
Department of Ocean Sciences
Memorial University
St John's
Canada

Charles A. Simenstad
School of Aquatic and Fishery Sciences
University of Washington
Seattle, WA
United States

Lida Teneva
Conservation International
Honolulu
Hawai'i
USA

Eric A. Treml
School of BioSciences
The University of Melbourne
Victoria
Australia

Dean L. Urban
Nicholas School of the Environment
Duke University
Durham, NC
United States

Mary A. Young
Centre for Integrative Ecology
Department of Life and Environmental
Sciences
Deakin University
Warrnambool
Victoria, VIC
Australia

Brian K. Walker
GIS and Spatial Ecology Lab
Nova Southeastern University
Halmos College of Natural Sciences and
Oceanography
GIS and Spatial Ecology Lab
Dania Beach, FL
United States

Lisa M. Wedding
Center for Ocean Solutions
Stanford University
Monterey, CA
United States

John A. Wiens
Department of Forest Ecosystems and
Society
Oregon State University
Corvallis, OR
United States

Jianguo Wu
Landscape Ecology and Sustainability Lab
School of Life Sciences and School of
Sustainability
Arizona State University
Tempe, AZ
United States

and

Center for Human-Environment System
Sustainability (CHESS)
Beijing Normal University
Beijing
China

Foreword

From Seascapes to Landscapes and Back Again

When I entered ecology from mathematics in the late 1960s, I was fortunate to be at Cornell, where colleagues like Robert Whittaker and Gene Likens were among the leaders in building linkages from individuals to populations, to communities and to ecosystems, and expanding the spatial scale of study to what would now be called landscapes. A transformational moment for the field of landscape ecology came with the Allerton Park meeting in 1983, organized by visionary pioneers Paul Risser, James Karr and Richard Forman. Among the topics addressed at that workshop were organismic influences on fluxes, the relationship between process and pattern, and the effects of landscape heterogeneity on the spread of disturbance. These major research themes have continued to shape the subject to this day.

But the ideas that helped shape landscape ecology were being developed in parallel in the oceans; and indeed there was and is a healthy coevolution between landscape ecology and seascape ecology, though it took a while for the latter to get its name. My earliest introduction to the subject was through my interest in spatial ecology and the development with Robert Paine of a patch-dynamic, mosaic approach to the intertidal; but our work was influenced heavily by A. S. Watt's foundational paper on forest communities; by the development of the theory of island biogeography by Robert MacArthur, E. O. Wilson and Daniel Simberloff; and by Richard Levins' introduction of metapopulation theory. The seeds of landscape and seascape ecology alike thus came from multiple complementary sources, from the land and sea and from the interplay of empiricism and theory. The emergence of seascape ecology today, therefore, as beautifully captured in this book, is not just the reapplication of ideas borrowed from landscape ecology; it is the natural continuation of developments with roots on land and in the oceans.

For me, a key development, six years before Allerton, was the NATO School on Spatial Pattern in Plankton Communities, organized by John Steele in 1977 with the assistance of Akira Okubo, Trevor Platt, Gunnar Kullenberg and Bruno Battaglia. Though the focus there was the oceans, Steele was prescient enough to make sure that terrestrial perspectives were also represented. The new insights I acquired at that meeting helped shape my research for decades, and were well captured in the edited book that emerged; those themes are central both to seascape ecology and to landscape ecology, and quite naturally permeate the current book. John Steele was much more than the chief organizer of the meeting. He was a synthesizer of ideas, emphasizing the importance of viewing systems across scales, and of looking for commonalities in the study of landscapes and seascapes. In a very real sense, the current book owes much to his seminal efforts.

This book represents an important contribution to the literature, and hopefully will provide a touchstone for the field, much as the Allerton report did for landscape ecology. It has roots in the NATO volume edited by Steele, in the recognition that management of the oceans must acknowledge their nature as systems, and in the large-marine-ecosystem paradigm long advocated by Kenneth Sherman. The focus on seascapes raises a range of fundamental questions, inspired by these earlier works:

How does the scale of study influence the description of a system and how do fundamental descriptors like the variance of population densities vary across scales? These issues are addressed head on in Chapters 1 and 2, and are pervasive throughout the book. Unification of the subject must also recognize, for example, that physical factors largely shape patterning and species distributions on broad scales and that biological factors become increasingly important on finer scales. Thanks to the support of the Gordon and Betty Moore Foundation and the Simons Foundation, as well as traditional funding sources such as the National Science Foundation, great advances have been made in the past decade in understanding microbial processes at fine scales. Attention has now turned, including by many of the same researchers, to ask how what has been learned scales up to the seascape level. This certainly will be one of the most exciting areas of investigation in the continuing development of seascape ecology.

Can we develop an adequate statistical mechanics of ecological interactions, explaining how macroscopic patterns emerge from interactions at much lower levels of organization? Regularities like the Sheldon spectrum and Redfield ratios, for all their variation, provide striking examples. The issue of pattern, quite naturally, is the first topic discussed in this book, and forms the focus of Part I; complementarily, pattern and process are addressed in Part II. In a sense, these sections address the key issues in tying work on the oceans into a coherent discipline, by providing the linkages across scales. Efforts like those of Michael Follows, Penny Chisholm and their collaborators have led in building a framework that embeds ecological interactions within an evolutionary context, and complementary efforts by numerous groups to create trait-based approaches to marine ecosystems are transforming our understanding of the oceans.

Can we characterise regime shifts in large marine ecosystems and identify early warning indicators? John Steele pioneered this notion for the oceans and it is receiving great current interest across a range of systems. Are there early warning signals of impending regime shifts? What are the features of systems that confer robustness and resilience, and how can we utilize this knowledge to manage fisheries and other dimensions of the oceans?

How do concepts like patch dynamics (Chapter 6), originally developed in terrestrial and intertidal ecosystems, extend to the marine environment? The oceans may seem more well-mixed than terrestrial environments, but physical and biological factors alike lead to coherent structures and aggregations that create great heterogeneities. These themes are addressed throughout this book but especially in Part I on pattern and Part III on connectivity.

How do concepts of public goods apply to organisms in the ocean and to phenomena like chelation and nitrogen fixation? Striking examples include extracellular macromolecules that bind iron, polysaccharides, and enzymes, as well as fixed nitrogen; information itself is a public good and is central to understanding the collective motions of organisms from plankton to tuna and whales. Evolution is mediated most strongly at the level of the individual, but individual decisions and actions – from bacteria to fishermen – have emergent consequences at the level of whole systems. So, too, evolutionary changes at the level of organisms and populations have signatures at larger scales of organization. There is considerable theory, largely from terrestrial populations, of how and when co-operation emerges, and it will be exciting to see how that theory can be extended to seascapes.

Part IV of this book addresses the question of how we can manage the oceans sustainably, recognizing both their nature as systems, and the interplay with human societies. Climate change and other anthropogenic influences have had and are having major influences, for example on fish populations and associated fisheries. As Malin Pinsky, his collaborators and others have shown, fish population ranges are shifting in response to climate; this will also be reflected to some extent in where fishermen ply their trades. Managing the oceans from the perspective of coupled human-environmental complex adaptive systems is an essential next step. As Jane Lubchenco and collaborators have documented recently, rights-based approaches hold great potential for dealing with problems of the Commons but a comprehensive theory of management will have many dimensions, integrating rights-based approaches, marine reserves and protected areas, and more conventional management methods.

These are among the questions that will occupy our attention for decades to come and for which this book will provide a valuable point of departure. The book not only identifies a fascinating set of perspectives but also takes steps towards integrating them into a coherent discipline of seascapes, bringing theory and empirical work together with advances in remote sensing and geospatial techniques. In the spirit of John Steele, one of the most attractive features is the creation of a nexus of common interest for scientists and practitioners from multiple backgrounds and perspectives. The editor, Simon Pittman, and the authors are to be congratulated for this foundational effort.

Simon Levin
Princeton University

Preface

The impetus for this book is fuelled by the recognition that a growing number of marine scientists, geographers and marine managers around the world are increasingly asking questions about the marine environment that are best addressed with a landscape ecology perspective, yet few are aware that it exists. There is a substantial gap in the academic knowledge base, which if left untreated will leave a new generation of researchers without the benefits that the more cohesive and well-established marine science subjects provide. That emerging subject is seascape ecology – the application of landscape ecology to the seas.

This book includes a collection of contributions from researchers who are active at the frontiers of seascape ecology and are beginning to refer to themselves as seascape ecologists.

Their unique contributions encompass the concepts, methodology and tools and techniques that demonstrate the importance of a seascape ecology approach in ecology and marine management. The book captures the state-of-the-science in seascape ecology and provides guidance for future research priorities. We also feature a special epilogue, which presents the perspectives of three eminent landscape ecologists, who have been and continue to be instrumental in the development of modern landscape ecology. In fact, this final chapter serves equally well as a prologue for the book!

Many of us drawn to seascape ecology have an affinity with maps, spatial patterning and scale awareness. Today, as a society, we are immersed in spatial information. I have long held a fascination with pattern and scale in nature, perhaps from childhood, where my parents encouraged me to be mindful of plants, animals and ecological relationships in our small urban garden in South London and the many places that we were fortunate to visit around the world. At school, I was captivated by geography, particularly plate tectonics and the linkages between distant flora and fauna which once shared the same land mass. My first scientific research projects, however, involved identifying and counting dinoflagellate cyst assemblages in Scottish sea lochs followed by a research scholarship with Howard Platt in the Nematode Research Unit of the Natural History Museum in London, where I filtered sediment samples and identified nematodes and other meiofauna to assess impacts from a gas refinery in the Thames estuary. Long hours sat at the microscope not only indelibly etched the fine microstructures of these tiny but extremely important organisms into my memory but also helped me to decide that microscopic was not the resolution at which I wanted to focus my career as an ecologist.

My interest in applying landscape ecology to the sea was realised with my doctoral research and the challenge of mapping ecologically meaningful habitats for highly mobile fishes in Queensland, Australia. Fishes in nearshore areas of Moreton Bay were undergoing extensive (>1 km) daily tidal excursions across spatially complex seafloor connecting subtidal seagrasses with intertidal seagrasses and mangroves. It became clear that to understand fish-habitat relationships required an approach capable of considering structure at a range of spatial scales relevant to the way that animals used space over time. To do this needed remote sensing, geospatial tools, and a different conceptual framework than anything I had considered before.

Since my doctoral research at the University of Queensland, where I had the unconventional but greatly influential, experience of working alongside a terrestrial landscape ecologist studying koalas and a remote sensing expert mapping coral reefs from space, I have been fascinated and excited by landscape ecology. Had I not left the familiar mindspace of marine zoology to venture out through the more holistic geographical sciences, and so fusing the two conceptual frameworks, I would probably not be writing this now. Although they had little interest in fish ecology, I owe my doctoral committee my gratitude for introducing me to the landscape ecology literature, where my new academic heroes became the likes of John Wiens, Kevin McGarigal (creator of FRAGSTATS software), Lenore Fahrig, Jingle Wu, Dean Urban and others.

My interest in applying landscape ecology to the marine environment led to my joining NOAA's Biogeography Branch, a globally rare team of applied marine spatial ecologists and technicians in the United States working intensively at the interface between geospatial science and management. It was here that I learnt how seascape ecology can be applied to address complex marine management problems and this book owes a lot to the technical expertise, proactive approach and boldness for innovative thinking demonstrated by the Biogeography Branch of the National Centers for Coastal Ocean Science.

Although the draft contents for a book on seascape ecology had been languishing (but never forgotten) in my 'Papers in Progress' folder since 2007 when I first brainstormed the project with Professor John Wiens, the then Chief Scientist of The Nature Conservancy, it was not until I was invited to speak on the topic at the Plymouth Marine Lab lunchtime seminar programme in 2013 that the vision crystalised. My presentation titled 'Taking landscape ecology into the sea' was seen by a commissioning editor for John Wiley & Sons who encouraged me to submit a proposal for a book, which then underwent academic peer review. After approval, the project slowly began to take shape, yet for the most part I had to fit it around my day job and a busy family life.

Fortunately, the time spent living and working as a marine ecologist in Australia, the United States of America, the Caribbean and in Europe enabled me to form a network of like-minded and specialist marine spatial scientists who provided an immense and valuable shortlist of potential contributors. I was delighted and relieved by the enthusiasm with which these individuals accepted the challenge and in the end we had 16 chapters from 36 authors based in Australia, the United States and various European countries. Thank you authors for accepting my guidance and my suggestions, which, I know, for many of you, pushed you a little deeper into the landscape ecology way of thinking than you had been before. I hope the process was interesting and will influence your future research directions as we consolidate and build a new discipline in the marine sciences.

We are hopeful that the individual parts of the book stimulate interest, provide useful specific guidance, and act as a catalyst for future studies that will reveal exciting new ecological insights. Furthermore, we hope that the book in its entirety generates sufficient synergy to support the continued evolution of seascape ecology as a distinct discipline.

We are grateful for the focused time provided by reviewers which included several of the chapter authors, as well as the following external reviewers: Siân Rees (Plymouth University), Carrie Kappel (National Center for Ecological Analysis and Synthesis, University of California Santa Barbara), Charlotte Berkström (Department of Ecology, Environment and Plant Sciences, Stockholm University), Ron Kneib (RTK Consulting Services and University of Georgia), Nicola Foster (Plymouth University), Katherine Moseley (Seascape Analytics Ltd), Dave Whitall (NOAA National Centers for Coastal Ocean Science), Patrick Crist (NatureServe), and Manuel Hidalgo (Balearic Islands Oceanographic Centre).

Simon J. Pittman
Plymouth, UK

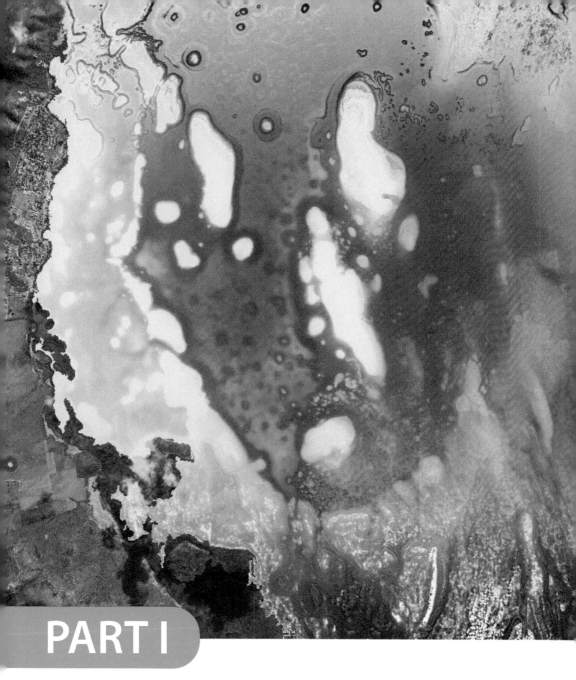

PART I

Spatial Patterning in the Sea

1

Introducing Seascape Ecology

Simon J. Pittman

Here, on the edge of what we know, in contact with the ocean of the unknown, shines the mystery and beauty of the world. And it's breathtaking.
Carlo Rovelli, *Sette brevi lezioni di fisica* (2014)

1.1 Introduction

A conceptual shift is under way in the way we perceive, understand and interact with the marine environment. The once widespread view of the sea as a vast featureless expanse of water providing unlimited resources is now giving way to a deeper more ecologically meaningful perspective. In recent years we have started to recognise and study the sea as a highly interconnected system exhibiting complex spatiotemporal patterning and previously unanticipated vulnerability to human activities. A primary catalyst for this change in worldview has been the technological advancement and proliferation of space-, air- and water-based ocean-sensing systems, together with increased sophistication in geospatial tools and mathematical simulation models. These technologies have allowed us to collect, integrate, analyse and visualise vast quantities of marine data that have revealed unimaginable structural complexity and interconnectedness across the seafloor, sea surface and throughout the water column. Whilst these patterns are visually captivating, it is their ecological implications that are scientifically intriguing and most relevant to society. However, significant knowledge gaps still exist in our understanding of how structural patterning in the sea, across multiple spatial and temporal scales, influences marine species distributions, biodiversity patterns, ecosystem services and human wellbeing. The breadth and depth of our existing knowledge has been constrained not only by technological limitations and data availability but also by the dominant philosophical and methodological scientific frameworks in marine science that have resulted in a preponderance of nonspatial, single scale and reductionist approaches in marine ecology. This has limited the ability of marine science to support holistic management strategies such as ecosystem-based management, where the human dimensions are integral to understanding the functioning of the system. In a rapidly changing world, where environmental patterns are being modified by human activity and the ocean economy is expanding and diversifying, society urgently requires a deeper and more holistic understanding of the linkages between seascape patterns and ecological processes to inform effective marine stewardship.

The scientific discipline of landscape ecology offers an appropriately holistic and interdisciplinary spatially explicit framework to address complex ecological questions. Although landscape ecology was once considered esoteric in science, its conceptual and analytical frameworks now permeate many areas of ecological research offering important new ecological insights (Turner 2005). In contrast, spatially explicit studies of seascape patterning are still relatively rare and although seascape ecology, the marine equivalent of landscape ecology, is on the verge of entering mainstream marine ecology, the level of familiarity among marine scientists is still comparable to that reported by terrestrial landscape ecologists in the 1980s, whereby 'ideas were new and were received with a mixture of scepticism and excitement' (Turner 2005).

Regardless, this new direction of scientific enquiry, with a focus on interpretation of spatial patterning, is not isolated to ecology but is part of a broader technological shift often referred to as the geospatial revolution. Our global society is undergoing a spatial information revolution fuelled by rapid innovations emerging from spatial computing, such as online maps, geoportals, location-based public services and a proliferation of augmented reality applications for all ages (Downs 2014; Shekhar *et al.* 2015). The geospatial revolution is also influencing the curriculum in schools and universities (Coulter 2014). This new wave of technological innovation and open access to geospatial data, which allows us to construct detailed and dynamic multidimensional digital representations of the global system, is the inspiration for the vision of Digital Earth as highlighted by Al Gore (Gore 1998) and others (Craglia *et al.* 2012). Consequently, a transformative shift is also underway in marine ecology. A new generation of marine ecologists, known as seascape ecologists, are bringing enhanced spatial awareness to ecological thinking, together with the tools to work with 'big data', a holistic perspective and a desire to ask new types of applied research questions. Yet, despite our best efforts to acquire and make accessible vast datasets that capture in detail the multidimensional patterning of the oceans, we still know surprisingly little about the ecological consequences of spatial patterning, including the implications for people and society.

In this introduction to the first book on seascape ecology, I introduce this emerging discipline and its relationship to landscape ecology and then touch on several key topics central to the conceptual and operational framework of seascape ecology.

1.2 Landscape Ecology and the Emergence of Seascape Ecology

Seascape ecology, the application of landscape ecology concepts to the marine environment, has been slowly emerging since the 1970s (Box 1.1) (Sousa 1979; Paine & Levin 1981; Walsh 1985; Steele 1989; Jones & Andrew 1992), yielding new ecological insights and showing growing potential to support the development of ecologically meaningful science-based management practices (Boström *et al.* 2011; Pittman *et al.* 2011). Seascape ecology, which draws heavily from conceptual and analytical frameworks developed in landscape ecology, focuses on understanding the causes and ecological consequences of the complex and dynamic spatial patterning that exists in marine environments (Robbins & Bell 1994; Pittman *et al.* 2011; see also Chapter 16 in this book). Landscape ecology shares some common ground with the broader subject of spatial ecology, which is also interested in spatial heterogeneity, but landscape ecology is defined by a set of concepts and techniques that are widely recognised as a specialisation in ecology (see foundation papers in Wiens *et al.* 1987) warranting an identity as a distinct discipline within ecology.

Influenced by a fusion of geography and ecology, the European roots of landscape ecology can be traced back to the early twentieth century and perhaps even to the Prussian geographer and explorer Alexander von Humboldt (1769–1859) through his contributions in *Essai sur la géographie des plantes* (von Humboldt & Bonpland 1807) and his holistic perspective of the universe outlined in *Kosmos* (published between 1845 and 1862). In 1939, Carl Troll, a German biogeographer, first introduced the term 'landschaftsökologie' (landscape ecology) to describe the study of landscape patterns mapped from aerial photography and then in 1971 defined it as 'the study of the main complex causal relationships between the life communities and their environment' that 'are expressed regionally in a definite distribution pattern' (Troll 1971). Troll's focus was primarily terrestrial but his studies of landscape formations also included detailed mapping of the spatial patterning and hydrological characteristics of mangrove vegetation, Gezeitenwälder, in southeast Asia (Troll 1939).

Modern landscape ecology, whose practitioners are primarily interested in the geometry of patterns across the landscape and revealing the empirical relationships between pattern, ecological processes and environmental change, has developed a unique set of concepts and analytical tools, which combined with a holistic and interdisciplinary perspective have made valuable contributions to the way we understand and manage terrestrial environments (Forman 1995; Gutzwiller 2002; Bissonette & Storch 2002; Turner 2005; see also Chapter 12 in this book). Key focal research themes in landscape ecology include: linkages between spatial pattern and ecological process; importance of spatial and temporal scales; spatial heterogeneity effects on fluxes and disturbance; changing spatial patterns; and the development of a framework for natural resource management (Risser *et al.* 1984; Wiens & Moss 2005). A central tenet in landscape ecology is that patch context matters, where local conditions are influenced by attributes of the surroundings (for definitions of a patch see Chapter 6 in this book). For instance, the physical arrangement of objects in space and their location relative to other things influence how they function (Bell *et al.* 1991). With this perspective, landscape ecologists will typically ask different questions focused at different scales than other scientists, such as: 'How do landscape patterns influence the way that animals find food, evade predators and interact with competitors? What are the ecological consequences of patches with different sizes, quality, spatial arrangement and diversity across the landscape? At what scale(s) is structure most influential? How does human activity alter the structure and function of landscapes?'

Although the majority of focus in landscape ecology has been in terrestrial systems, some crossover into aquatic systems has occurred. For freshwater systems, for example, landscape ecology, in the form of landscape limnology, has now influenced the way we study rivers and lakes (Schlosser 1991; Poff 1997; Wiens 2002). For marine systems, the application of landscape ecology came about through a recognition that many of the concepts developed in the theory of island biogeography (MacArthur & Wilson 1967) and the study of patch dynamics (precursors to modern landscape ecology) could be applicable to a range of marine environments from plankton patches

(Steele 1978) to patch reefs (Molles 1978), intertidal mussel beds (Paine & Levin 1981) and seagrasses meadows (McNeill & Fairweather 1993). Typically, the spatial scale of early investigations into patchiness in the sea was limited by either the logistical challenges of conducting manipulative field experiments or the lack of instrumentation to measure dynamic structural patterns at broader spatial and temporal scales. In the 1970s, the ease of access to shallow water coral reef ecosystems (seagrasses, coral reefs, mangroves) encouraged intensive observations of marine animal movements leading researchers to consider the effect of the spatial arrangement of patches on functional connectivity (Ogden & Zieman 1977; Ogden & Gladfelter 1983; Birkeland 1985; Parrish 1989). Although these earlier studies referred to concepts associated with landscape ecology, they were not spatially explicit and did not focus on quantifying the geometry of seascape patterning.

Progress in the ecological understanding of spatial patterning was not confined to shallow seafloor environments. For the open ocean, advances in ocean observing systems since the 1970s have allowed us to map, classify and track dynamic spatial structure in the form of eddies, surface roughness, currents, runoff plumes, ice, temperature fronts and plankton patches (Scales *et al*. 2014; Kavanaugh *et al*. 2016; see also Chapter 3 in this book). Subsurface structures too, such as internal waves, thermoclines, haloclines, boundary layers and stratification resulting in distinct layering of organisms, is increasingly being mapped and modelled in multiple dimensions (Ryan *et al*. 2005; Kavanaugh *et al*. 2016). For example, ecologically important layers of dense phytoplankton have been discovered in stratified coastal waters, parallel to thermoclines, where little turbulent mixing occurs, creating biological structure in the water column that can form a horizontal sheet of productivity extending for many kilometres (Ryan *et al*. 2008).

With the integration of satellite data, field sampling, animal telemetry and geospatial modelling, studies of pelagic seascapes have demonstrated that spatial patterning can help explain ecological processes such as animal movement patterns and foraging behaviour and that dynamic geometric features of the ocean can be both persistent and predictable across ocean space (Bakun 1996; Alvarez-Berastegui *et al*. 2014; Hidalgo *et al*. 2016; see also Chapter 3 in this book). Three decades ago, John Steele drew attention to the importance of spatial and temporal patterning in the open ocean and introduced the term 'ocean landscapes' to encapsulate his observations of scale dependent structure (Steele 1989). Now researchers under the banner of pelagic seascape ecology are realising his vision of a landscape ecology for the seas (see also Chapter 3 in this book).

1.3   What is a Seascape?

Here, seascapes are defined as spatially heterogeneous and dynamic spaces that can be delineated at a wide range of scales in time and space. In seascape ecology, spatial structure is typically represented as a two-dimensional construct (*e.g.*, benthic habitat map) applying the patch-matrix or patch-mosaic models (*i.e.*, a collection of internally homogeneous patches) (section 1.3.1), or as continuously varying two or three-dimensional surfaces (section 1.3.2), sometimes referred to as spatial gradients (Wedding *et al*. 2011; see also Chapter 2 in this book). The choice of spatial model must be linked to ecological theory through an appropriate conceptual framework, which will be driven by the question(s) being addressed. Furthermore, the way in which seascape structure is represented will be fundamental to the types of analyses conducted, the tools required and to the subsequent understanding of ecological patterns and processes. With regard to scaling seascapes, one approach is to select spatial and temporal scales to be ecologically meaningful to the organism's movements or other processes of interest (Wiens 1989; Pittman & McAlpine 2003; see also Chapter 7 in this book). Clearly, a seascape for an amphipod will be very different in size and structure than the seascape for a blue whale. With regard to seascapes defined by sampling units, a 1 m^2 quadrat can be a valid seascape sample unit (SSU), just as can a 1 km^2 analytical window in a geographical information system. The wide diversity of possible focal scales in marine ecology means that the term seascape cannot be used as an indication of scale, or a level of organisation. That is, we must avoid statements such as: 'a seascape-scale study was conducted' or 'at the seascape level', which are ambiguous regarding spatial and temporal scale and approach. Instead, a specific and numeric scale, or scale range, must be provided, even if approximate, when referring to the grain (minimum resolution of data – *e.g.*, sample unit area, pixel size) and extent of a study (size of the study area). Defining scales is especially important because different patterns emerge at different scales of investigation (Wiens 1989). This problem of semantics was discussed in detail by Allen (1998) who issued a caution supported by a well-reasoned argument against the unsubstantiated use of the term 'landscape scale' and 'landscape level', a problem that has also been propagated in application to marine environments.

Although often represented as static maps, seascape structure is dynamic at a range of temporal scales adding additional challenges in ecological applications. Figure 1.1 depicts a hypothetical ocean space (a seascape cuboid) illustrating examples of seascape patterns exhibiting very different temporal dynamics – *e.g.*, ephemeral plankton patches (hours or days) to relatively stable seafloor morphology (thousands or millions of year). The range of structures include morphological features that exist as continuous spatial gradients in physical, chemical, geological and biological variables (*i.e.*, seafloor terrain morphology, salinity gradients, runoff plumes); biological patches (*i.e.*, plankton patches, benthic patch mosaic, within-patch structures); subsurface laminar and stratified features – thermoclines, chemoclines (*e.g.*, haloclines), internal waves. Fronts and eddies are key dynamic physical features important to many marine organisms. Fronts that persistent are usually associated with a convergence zone, where mixing of two water masses of different densities results in a body of sinking water at the interface between them (Acha *et al.* 2015). For example, river plume convergence zones occur where less dense low salinity plume waters float over denser oceanic surface waters. Where waters mix, plankton concentrate leading to the well-documented proliferation of consumers from multiple trophic levels (Bakun 1996). Convergence zones between water masses are often visible as surface drift lines (Figure 1.1). In addition to plankton, detached seaweeds, logs and flotsam accumulate at these fronts providing physical structure for settlement and refuge from predation and food. Eddy motions in the ocean yield a rich seascape of configurations where mechanisms of enrichment, concentration and retention can result in rapidly enhanced biological productivity (Bakun 1996).

Figure 1.1 Seascape showing spatial structure in the sea: **A.** Runoff plume; **B.** Temperature front; **C.** Eddies with entrained phytoplankton; **D.** Thermal front; **E.** Salinity gradients; **F.** Surface roughness; **G.** Plankton patches; **H.** Thin horizontal layer of plankton; **I.** Internal wave; **J.** Thermocline; **K.** Seafloor terrain morphology from bathymetry (three dimensional); **L.** Benthic habitat map representing patch-mosaic patterns (two dimensional); **M.** Geological features (canyons and seamounts); **N.** Within-patch structure (biological assemblages); **O.** Surficial sediment and geological strata. *Source*: Artwork created in partnership with The Agency at Plymouth College of Art, United Kingdom.

1.3.1   The Patch-Matrix and Patch-Mosaic Models of Seascape Structure

The *patch-matrix* model has its origin in island biogeography (MacArthur & Wilson 1967) and typically depicts the environment as a binary seascape composed of focal habitat patches or 'islands' surrounded by an inhospitable matrix *i.e.*, habitat and non-habitat (Forman 1985). This model has found great utility in studies of fragmentation and metapopulation biology and has been applied more widely in marine applications of landscape ecology (particularly seagrass-dominated environments) than any other single habitat model (McNeill & Fairweather 1993; Robbins & Bell 1994; Irlandi *et al.* 1995; Hovel & Regan 2008). Characteristic variables include patch size, perimeter to area ratio and patch isolation (reviewed by Boström *et al.* in Chapter 5 in this book). The patch-matrix model, however, has been found to be too simplistic for many species-environment relationships, particularly for highly mobile multihabitat organisms and where the matrix is heterogeneous and dynamic (Bender & Fahrig 2005). In contrast, the *patch-mosaic* model represents structural heterogeneity as a collection of patch types, without any particular focal patch but where the composition and spatial arrangement (or spatial configuration) of the mosaic as a whole influences ecological function (Wiens *et al.* 1993). Composition usually refers to the amount and variety of patch types in the seascape and spatial arrangement to the physical distribution of patches in space (Dunning *et al.* 1992; Wiens *et al.* 1993). The underlying premise is that the composition and spatial arrangement of patches influence ecological processes in ways that would be different if the composition and arrangement were different (Wiens *et al.* 1993; Wiens 1995). At a fundamental level, the composition and spatial arrangement of the seascape determines the amount, variety and distribution of resources that organisms must locate and utilise (Senft *et al.* 1987). Patch structure can be quantified using an extensive range of pattern metrics (McGarigal & Marks 1994; Gustafson 1998; McGarigal & Cushman 2002), for most of which the functional significance for marine species remains to be determined (Robbins & Bell 1994; Grober-Dunsmore *et al.* 2009; Wedding *et al.* 2011).

Benthic habitat maps are cartographic products that are commonly represented as patch mosaics (see Chapter 2 in this book). When a seascape is represented as a patch-mosaic (or patch-matrix), discrete homogeneous patches, usually with sharp boundaries, are used to depict major discontinuities in horizontal structure. Fuzzy boundaries are also possible, yet they are rarely used for marine environments (Wang & Hall 1996). The patch mosaic model has proved effective by virtue of its simplifying organisational framework that facilitates experimental design, analyses and management consistent with conventional tools and methodologies (McGarigal & Cushman 2002). In fact, many of the concepts and analytical techniques in landscape ecology are based on the assumption of the ecological importance of patch mosaic structure, usually in the form of categorical habitat maps. In the process of mapmaking, however, measured heterogeneity is usually subsumed in order to produce a greatly simplified model of reality, resulting inevitably in all patches within a single class, or habitat type, exhibiting equal homogeneity (example of LiDAR versus benthic habitat map). Hence, the choice of spatial and thematic resolution must be considered very carefully because the loss of relevant within-patch heterogeneity may confound analyses. Selection of thematic resolution and spatial resolution is particularly pertinent to species-environment studies making use of maps (Kendall *et al.* 2011; Lecours *et al.* 2015; see also Chapter 4 in this book).

The applicability of a patch-mosaic model to the benthic structure of patchy seascapes, such as some coral reef ecosystems, is usually relatively clear since coral reefs, mangroves and seagrasses often exist as discrete components of an interconnected mosaic of patches. It is still not yet clear, however, if the patch-matrix and patch-mosaic models are as useful for characterising the fluid driven patterning observed in pelagic marine environments as they have been for terrestrial landscapes and benthic environments (Kavanaugh *et al.* 2016; Manderson 2016). There is potential for application, however, because biological distributions, such as plankton, can be highly patchy, albeit spatially dynamic at finer temporal scales than seafloor terrain (Hardy 1936; Bainbridge 1957; Steele 1978; Martin 2003). Intensive field sampling of zooplankton patches has revealed a complex spatial mosaic of patches and aggregations of patches, each dominated by a single species, with some patterns exhibiting fractal properties of self-similarity across spatial scales (Tsuda 1995; Seuront 2009). Much of this structural heterogeneity is driven by the interaction between water movements and benthic topography. Both horizontal and vertical water movements create a dynamic mosaic of patterns and patch structure, with elevated productivity often occurring at the boundaries or edges of patches (Powell & Okubo 1994; Haury *et al.* 2000; Martin 2003; Yen *et al.* 2004).

Box 1.1 Semantics: Seascapes or marine landscapes?

Keyword searches of the scientific literature using Google Scholar for the period 1960 to 2016 inclusive revealed an increase in the use of the terms 'seascape ecology' and 'seascape' since the turn of the twenty-first century (Figure 1.2a), with the highest prevalence occurring after 2010. Outside of art and poetry, the term seascape was rarely used before the 1970s. The use of the term 'marine landscape' has also been frequently applied to describe benthic structure for shallow water inshore areas but has given way to 'seascape' in recent years (Figure 1.2b). In 1991, with reference to the use of 'landscape', John J. Magnuson wrote that 'all aquatic ecologists worth their salt would change this category to seascape … [because] … three-quarters of the earth is covered by the seas, and the continents are embedded in the seascape at the global scale' (Magnuson 1991). In the literature, these descriptors have rarely been defined explicitly and rarely was the spatial structure of the seascape quantified before 2000. In this book, we advocate the continued use of the descriptor 'seascape' for both benthic and pelagic environments instead of 'landscape', partly due to the inappropriateness of the term 'landscape' when referring to heterogeneous areas of water surface or water column and also because differences in marine systems will likely require the development of new concepts and analytical approaches that could be unique to the marine environment *i.e.*, a seascape approach. Throughout the book we use descriptors such as seagrass seascapes, pelagic seascapes, benthic seascapes dependent on the primary focus of the research.

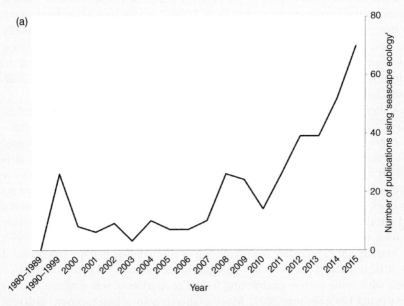

Figure 1.2 (a) Number of publications using 'seascape ecology' over time. (b) Comparison of terms used over time.

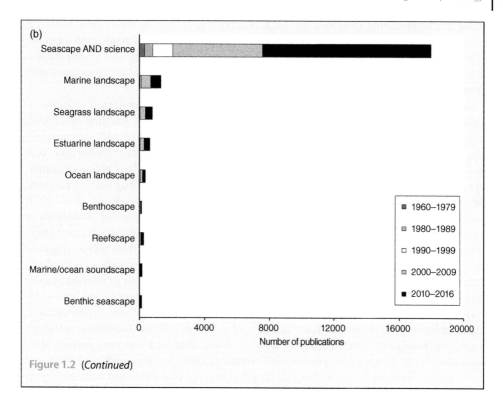

Figure 1.2 (*Continued*)

1.3.2   The Spatial Gradient Model of Seascape Structure

In addition to patchiness, the marine environment exhibits spatial variability in the form of continuous multidimensional gradients. For example, hydrodynamic interactions with coastline geomorphology, riverine influences and seafloor topography create a wide range of spatial gradients in marine environmental conditions (*e.g.,* salinity, wave action, depth, temperature, nutrients). Recently, spatial gradients or continuously varying surfaces have been used successfully to represent structure in terrestrial landscape ecology studies (McGarigal & Cushman 2002; Fischer & Lindenmayer 2006; McGarigal *et al.* 2009) and in certain situations the gradient perspective may also have greater utility in studies of marine organisms and marine ecosystems. The spatial gradient model represents seascape structure as a continuously varying surface without discrete patch boundaries, although discontinuities and boundary ecotones may still be represented. The recent resurgence of interest in gradient models including the 'variegation model' (McIntyre & Barrett 1992) and the 'continuum model' derived from continuum theory (Manning *et al.* 2004) is an attempt to broaden the perception of spatial heterogeneity in landscape ecology and ultimately to recognise species-specific responses and to better integrate ecological process-based variables. The continuum model recognises that a range of ecological processes may affect habitat suitability for different species through time, in a spatially continuous and potentially complex way. The premise is that species are likely to respond individualistically to the gradual changes taking place along spatial gradients in resources, such as food and refuge, or other environmental conditions (Austin & Smith 1989; Fischer & Lindenmayer 2006). An additional benefit of the gradient model is that it retains the captured heterogeneity and avoids subjectivity associated with boundary delineation and thematic designation associated with thematic habitat maps.

High-resolution bathymetric data and associated backscatter collected using techniques such as airborne laser altimetry (*e.g.*, Light Detection and Ranging or LiDAR) and sonar (*e.g.*, multibeam or side-scan sonar) can be used to construct continuously varying surfaces exhibiting complex vertical, horizontal and compositional structure (see Chapter 2 in this book). A wide range of surface metrics, such as geomorphometrics, are available to quantify continuously varying surface patterning such as rugosity, slope and curvature (Wilson *et al.* 2007; Pittman *et al.* 2009; Wedding *et al.* 2011; Lecours *et al.* 2016). In the absence of continuous sampling, interpolated surfaces or predictive surfaces can be modelled using point data from intensive georeferenced field samples. In many cases, surrogate or indirect environmental variables will be required to represent some of the more difficult to quantify direct variables such as prey abundance. For example, to study leatherback turtle movement behaviour in relation to their prey, Witt *et al.* (2007) used spatial interpolation to construct monthly and long-term seasonal means of the spatial distributions of gelatinous organisms (coelenterates, siphonophores and Thaliacea) from 50 years of zooplankton trawl samples. Although spatial patterning was not quantified explicitly, it was observed that prey patches were coincident with distinct bathymetric and oceanographic features such as eddies and tracking data indicated that turtles had targeted regions that exhibited seasonally persistent prey over decadal timescales. Sea surface topography may also be analysed in a continuous spatial framework. For instance, patterns in sea surface topography may represent hydrodynamic structures caused by changes in pressure, wind-generated waves, currents and seafloor topography. Ocean altimetry, ocean colour and temperature data collected by space-borne sensors (*i.e.*, TOPEX-Poseidon, SEAWIFS, MODIS) have been used to create maps of eddies and fronts that have then been used to locate and explain aggregation areas and movement pathways for marine megafauna such as cetaceans, seabirds and tuna (Gottschalk *et al.* 2005; Ballance *et al.* 2006; Palacios *et al.* 2006).

1.3.3   Combining Spatial Gradients and Patch Mosaics

Exploratory seascape studies focusing on linking species to their environment will likely use a suite of environmental layers that include combinations of benthic habitat maps, continuous bathymetry and spatial gradients representing various properties of the water column (*e.g.*, turbidity, temperature, salinity, stratification, plankton, *etc.*). Both continuous spatial gradients and discrete patch mosaics may together explain more of the variability in a response variable than either type used in isolation. Some species may respond more distinctly to emergent properties of structural variability that are best represented by patch mosaics, whilst others may respond to gradual changes along gradients, although threshold effects may also occur requiring a pluralistic approach (Price *et al.* 2009). The response to either patchiness or gradients, or even gradients of patchiness, or patchy gradients! is likely to be scale dependent. For instance, Sims *et al.* (2006) used spatially intensive and long-term plankton survey data to construct spatial surfaces of prey abundance (referred to as 'prey landscapes') to explain movement behaviour of satellite tracked basking sharks (*Cetorhinus maximus*). The authors found that individual sharks responded to seasonally persistent across-shelf gradients in zooplankton abundance at broad scales (50–500 km), as well as complex patchiness of zooplankton prey at finer spatial scales (<10 km) (Sims & Quayle 1998). In the open ocean, water column attributes can be spatially modelled to reveal a complex three-dimensional representation of structure (Ryan *et al.* 2005) that integrates a patch mosaic perspective, as well as varying intensity within individual patches.

In many regions, nearshore environments such as coral reef ecosystems are characterised by a complex environmental structure with inshore-offshore and alongshore spatial gradients in variables such as turbidity, salinity, depth, exposure, as well as structural patchiness in the distribution of benthic habitat types at multiple spatial scales. Pittman *et al.* (2007) used a suite of spatial explanatory variables comprising of both continuous (bathymetric complexity) and patch mosaic variables (areal extent of benthic habitat types, habitat richness) quantified at multiple spatial scales to examine fish abundance and diversity in Caribbean coral reef ecosystems. In estuarine environments, Martino & Able (2003) showed that fish assemblage structure was influenced by broad-scale environmental gradients at scales of tens of kilometres and species-specific responses to habitat structure, predation and competition at finer spatial scales. Hierarchical response scales are not well known for many marine species as studies are usually conducted at a single scale and at relatively fine spatial scales that are often even finer than the daily movements of the organisms of interest (Pittman & McAlpine 2003).

1.3.4   Chemical Seascapes and Ocean Soundscapes

In seascape ecology, focus has primarily been on linking species to the spatial pattern of readily observable structure in the seascape (*e.g.*, seafloor morphology, seagrass patch size and edges, *etc.*), but focus is now also turning to the challenge of understanding the spatial and temporal patterning of more visually obscure but crucially important dynamic patterns and processes across seascapes. Many examples now exist that demonstrate the importance of chemical odours and sound in generating temporal and spatial structure, which has often been described as analogous to the patterning of a landscape, but yet is considerably more challenging to accurately map and measure in the sea. Spatial and temporal heterogeneity of chemical and auditory variables has important implications for the functional ecology of organisms and can influence their survival. For example, seabirds, fish, invertebrates and mammals use chemical cues and / or sound to locate suitable habitat patches for foraging, settlement or locating mates and avoiding predators (Atema 1996; Moore & Crimaldi 2004; Gardiner *et al.* 2015). *In situ* tracking of larval fish swimming behaviour has found that reef fish larvae use olfactory cues (Paris *et al.* 2013) and auditory cues (Leis & Carson-Ewart 1997) to locate coral reefs from several kilometres offshore, resulting in fish changing their swimming speed and direction to track gradients to settlement habitat. Dimethylsulfonioprionate (DMSP) produced by phytoplankton provides foraging cues for several species of marine invertebrates, fish, birds and mammals and plays an important role in shaping planktonic food webs (Nevitt 1999; Seymour *et al.* 2010; Brooker & Dixson 2016). High concentrations of DMSP are typically associated with physical hydrodynamic features such as fronts, gyres, eddies, river outflows, coastal shelf waters and upwelling zones.

If we are able to map the chemical seascape, then should we represent the patterns as patch mosaics, gradients or both. If chemicals concentrate in distinct patches (sources and sinks) then the patch-mosaic construct could be appropriate, with organisms responding to patch size, shape, corridors and edges and the spatial configuration and composition of the chemical seascape as a whole. Alternatively, we know that chemicals and sound permeate / propagate through water and often form complex spatial gradients in concentration or magnitude, which organisms from bacteria to whales respond to in a variety of ways. Many exciting research challenges will emerge with the need to map, quantify and characterise these and other dynamic properties of the seascape and new questions will arise that will no doubt require new sensing technologies and novel ways of thinking about seascape geometry. In the field of marine soundscape ecology, passive acoustic monitoring is being used to map variability in the dynamic patterning of biological noise (biophonic sound) in nearshore temperate environments (McWilliam 2016) and coral reefs (Nedelec *et al.* 2015). Soundscape ecology has developed novel diversity and complexity metrics to analyse recorded biophonic sound and these can be linked to different seascape patterns and ecological conditions as has been achieved on land (Pijanowski *et al.* 2011; Fuller *et al.* 2015).

Interestingly, both noise and water chemistry in the sea are influenced by human activity, which creates and modifies the chemical seascape and soundscape. Runoff from land, river plumes, outflows, acidification and changes to habitat patterning and biological communities all influence the chemistry of the seascape. Likewise, the sea is becoming a noisier place with more vessels, coastal development and many other sources of noise that will have considerable impact on the marine soundscape (Slabbekoorn *et al.* 2010). As a result, the mapping of chemical patterning and acoustic patterning and investigation of its consequences for ecological processes at a range of scales is fast becoming an active research area with potential for many new questions requiring novel technological innovation through interdisciplinary research linking physics, chemistry, biology and geosciences. Studies that can link compositional characteristics of the soundscape and chemical seascape with other structural and functional characteristics, including human activities, will move us closer to a more holistic ecological understanding of the seascape.

1.4   Why Scale Matters in Seascape Ecology

Scale, the spatial or temporal dimensions of a phenomenon, is central to seascape ecology and the topic permeates all of the chapters in this book. Seascape ecology acknowledges that decisions made for scaling ecological studies influence our perspective and ultimately our understanding of ecological patterns and processes (see Chapter 4 in this book). Yet, despite the fact that, historically, marine ecologists have played a significant role in communicating the importance of scale in ecology (*e.g.,* Stommel 1963; Steele 1978; Dayton & Tegner 1984; Schneider 2001), scale selection and scale-dependency is still rarely dealt with explicitly in ecological studies of marine systems. In seascape ecology, however, the issue of scale and scale awareness is crucial to guiding scale selection and evaluating scale effects. Whether creating habitat maps, or designing a sampling strategy, at almost every stage of the decision-making process, careful consideration must be given to the effects of spatial and temporal scale. This is because the sequence of choices and tradeoffs related to scale selection are likely to influence the results and ecological interpretation of results, a topic handled eloquently by the geographer Vernon Meetenmeyer (Meetenmeyer 1989). In a terrestrial system, Wiens *et al.* (1987) showed that species-environment associations can change from strongly positive to strongly negative with a change in the scale of analysis. In a marine system, Pittman *et al.* (2004) designed a multiscale study that showed that both within-patch structural variability (including vertical plant structure) and the composition and configuration of the surrounding seascape were important explanatory variables for spatial patterns of fish and crustaceans. Schneider, in Chapter 4 in this book, highlights some of the consequences of what he refers to as 'scale-inadequate' science for public policy.

Levin (1992) considered the linkage between pattern and scale to be the central problem in ecology. When quantifying spatial pattern for studies of organism-seascape relationships, seascape ecologists must recognise that structure mapped at one scale may be functionally meaningful for one organism while irrelevant to another. For example, the way that an amphipod perceives, experiences and responds to the surrounding seascape will be different from a shark (Pittman & McAlpine 2003). Furthermore, this organism-seascape relationship will change throughout an organism's life, with juveniles responding differently than adults. Furthermore, seascape ecology recognises that individuals, species and communities will respond to seascape patterning across a hierarchy of spatial scales. So how do we select spatial scales for field investigations? Movement behaviour is one approach that is increasingly feasible with advances in remote sensing. The rationale is that knowing an organism's actual activity space, or ecological neighbourhood (*sensu* Addicott *et al.* 1987) over time allows us to anchor the scale selection to a specific set of functionally meaningful spatiotemporal scales that can form the focal scale in a hierarchical approach (Pittman & McAlpine 2003; see also Chapter 7 in this book). As such, the size of a seascape will depend on the research questions and the ecology of the organism or process of interest. In landscape ecology, this approach to scaling is referred to as the organism-based perspective (Wiens 1989). For many marine species, however, we do not have sufficient knowledge of movement patterns to select a meaningful focal scale(s) and therefore, an exploratory multiscale approach to measurement of environmental patterning and analysis of organism-seascape relationships is probably more appropriate than selecting a single arbitrary scale.

This book has generally avoided the comparison of differences between marine and terrestrial ecosystems, in part because they are covered elsewhere in the literature (Steele 1985; Carr *et al.* 2003; Stergiou *et al.* 2005; Manderson 2016). Here we deviate from divisive comparison between realms primarily to highlight a difference in the scale selection for ecological studies on land versus the sea. To examine if terrestrial and marine applications of landscape ecology differ in focal scales and scientific approach, a random 20 published landscape ecology studies from both marine (n = 20) and terrestrial (n = 20) environments were reviewed. In marked contrast to terrestrial landscape ecology studies, where studies have been conducted at relatively broad spatial scales (75% >1 km^2), with few experimental studies on ecological processes (5%), the majority (65%) of marine studies had focused on field experimentation to investigate mechanisms such as faunal growth, predation and movement focused at considerably finer spatial scales (75% < 1 km^2) (see also Chapter 5 in this book). This scale difference relates to differences in constraints to working in marine versus terrestrial systems, conventional scale selection and to the limited availability of data to support questions directed at broader scale patterns and processes. Furthermore, our review indicated that approximately 65% of marine studies have been conducted at a single spatial scale, with the surrounding spatial patterning considered in only 35% of studies (*i.e.,* dominated by patch-centred studies). Studies of the impact of human activity on spatial patterning were very rare in these marine studies but relatively common in the terrestrial studies. Additional examination revealed that spatial scales were usually selected without *a priori* information on the scales most likely to be ecologically meaningful to the species or processes of interest. In many of these studies, especially fine-scale experimental studies that have focused on highly mobile fish and crustaceans, measures of physical structure may not have coincided with the structure that was most functionally meaningful to the species of interest, thereby confounding results. Careful consideration of spatial and temporal scale effects and, where appropriate, the application of a hierarchical multiscale framework is one way that a landscape ecology perspective can help to avoid misinformation in marine ecology. The topic of *ecological scale* is dealt with comprehensively in the book of that same name (Peterson & Parker 1998) and by Schneider in Chapter 4 in this book.

1.5   Seascape Ecology can Inform Marine Stewardship

Evidence that human activity is capable of dramatically modifying the spatial structure of the environment is increasing and thus a better understanding of the specific causes and ecological consequences of change to spatial structure in estuaries, bays, seas and the open oceans, across multiple spatial and temporal scales, is critical to the development of effective ecosystem-based management strategies (see Chapters 12 and 13 in this book). Gaining a better understanding of interlinked socio-ecological spatial structure and processes represents one of the most important future challenges for both marine ecologists and environmental managers. Of considerable and immediate societal importance is the potential for seascape ecology to provide ecological informa-tion at operationally relevant scales for environmental management, thus bridging the information gap that has so often hindered the successful application of results from conventional ecological investigations. For example, seascape ecology can help identify and prioritise ocean spaces for conservation actions based on an understanding of the functional implications of seascape patterning. A good example is the design and placement of MPAs and networks of MPAs to optimise socio-ecological performance. Chapter 14 in this book presents a methodological framework for MPA network design and evaluation based on a landscape ecology approach and the application of spatial pattern metrics. In some instances, the surrounding seascape may exert a greater influence on differences between patches than the within-patch composition and therefore comparative studies too would benefit from considering the surrounding seascape at the sampling design stage to avoid potentially spurious conclusions, a step that is particularly important in effects studies, where results are likely to influence management decisions.

Marine spatial planning (MSP) is rapidly emerging as a viable approach for com-prehensive and efficient management of coastal and marine environments around the world (Caldow *et al.* 2015). Implementing MSP, however, is a considerable challenge for marine stewardship agencies, because data on spatially heterogeneous and dynamic socio-ecological systems are extremely complex with insufficient guidance on appropriate analyses available in the marine sciences. Marine spatial planning typically requires spatial information on the spatial patterning of seascapes such as identifying essential fish habitat, diversity and abundance hotspots (and 'cold spots' of low diversity or abundance) and 'blue corridors' to incorporate ecological connectivity into the planning process. Consequently, marine managers are increasingly framing questions through the lens of landscape ecology and asking similar questions of their jurisdictions as their terrestrial counterparts have done. Which seascapes support high biodiversity and optimal functional connectivity and maintain valuable ecosystem services? How should we modify seascape patterning to optimise restoration strategies and to adapt better to the effects of environmental change?

Information on the spatial patterns of connectivity through key life stages (*e.g.*, larval, postsettlement juvenile and adult), as well as for predicting resilience or vulnerability to extinction across metapopulations (*e.g.*, seascape genetics) is fast becoming a standard request in management planning and can often be the primary driver for ecological studies focused on modelling marine connectivity. This book features three chapters (Chapters 7, 9 and 10) dealing with different aspects of seascape connectivity, which describe a wide range of analytical techniques for studying structural and functional aspects of connectivity. The chapters present evidence that connectivity influences the outcomes of marine protected area function and is required knowledge for the design of functionally efficient networks of protected areas; making such information of intense interest to marine stewardship agencies.

In addressing these types of questions (and those presented throughout this book), spatial and temporal scale selection will need to be redefined, the design of sampling schemes may need re-evaluation and the capture of appropriate environmental features and statistical examination of ecological relationships is likely to require a different suite of tools than those currently familiar to most marine ecologists. As marine scientists begin to embrace landscape ecology, it is becoming apparent that while many of the concepts and analytical techniques are equally applicable to some marine environments, bringing landscape ecology to the sea will almost certainly require the development of unique seascape concepts and tools. To advance seascape ecology requires that we go beyond just observing spatial patterning, to actively engage in its systematic quantification across multiple scales and ultimately linking spatial patterning to ecological processes and environmental change.

Furthermore, we will need to integrate knowledge across the land-sea interface. In coastal systems, where landscapes are most heavily modified, the biophysical characteristics of the sea are inextricably linked to the land and it is now abundantly clear that landscape structure is one of the most important factors influencing nutrient and organic matter runoff to the sea. In the 1970s, ecologists Gene Likens and Herbert Bormann called for a more holistic approach in addressing problems associated with runoff with an explicit need to consider landscape mosaics. The authors wrote 'legislation which ignores the biospheric perspective, or the complexity of the landscape mosaic, is ultimately naïve' (Likens & Bormann 1974). In Chapter 11 in this book, Oleson *et al.* explore how landscape ecology concepts and spatial modelling techniques can improve our understanding of the consequences of terrestrial landscape patterning on seascape condition.

1.6   Conclusions and Future Directions

The new generation of ecologists embracing big data and the geospatial revolution are fast becoming equipped with the tools to conduct the sophisticated spatial analyses now required to address many of the complex spatial questions presented in seascape ecology. Instead of blaming complex spatial heterogeneity for surprising results, or condemning ecological theory for oversimplifying spatial variation, as many ecologists have done in the past (Kareiva 1994), seascape ecologists place spatial heterogeneity firmly at the front and centre of their approach. The recent global surge in interest in marine spatial planning should fuel the evolution of conceptual and operational approaches in seascape ecology. From an academic perspective, however, the immediate focus of seascape ecology should be to determine which theoretical constructs, analytical techniques and structural patterns or features from landscape ecology are relevant to understanding the relationships between marine organisms and their environment. More specifically, we now also need to strengthen the evidence for causal linkages between seascape geometry and ecological processes, including the impact of changing patterns on predator-prey dynamics, animal movement pathways, foraging behaviour and individual growth rates. From an evolutionary perspective, the spatial patterning of the environment has been perceived as a dynamic templet with which organisms interact and tactics and strategies have evolved (Southwood 1977). Studies that link the evolution of tactics and strategies with specific seascape configuration and composition have potential to help us understand the consequences of seascape change. For example, species response to change in seascape structure will be different for a seascape generalist (adapted to thrive in a wide range of seascape types) than it would for a seascape specialist (adapted to thrive in a narrow range of seascape types). Although these adaptations exist along a continuum, some generalities may exist that will allow us to identify threshold effects and predict future community composition in response to seascape change. For example, studies in terrestrial systems have reported functional homogenisation in bird and insect communities, across a wide range of landscapes, in response to disturbance (Clavel *et al.* 2011).

Seascape ecology helps define the relevance of the surroundings. Without such an approach, important pieces of the ecological puzzle will be missing. In a world where spatial data is a core component of decision making throughout society, seascape ecologists have an academically rewarding challenge ahead and great potential to change the way we perceive and manage the marine environment. One has only to look at the rise of terrestrial landscape ecology since the late 1980s to obtain a sense of what lies ahead. To date, however, studies that have applied landscape ecology concepts and tools to the marine environment are geographically sparse, with distinct patches of activity but with varying levels of influence from the parent conceptual framework provided by landscape ecology and sometimes with ambiguous use of terminology. Yet, as demonstrated by this book, connectivity does exist and great potential exists for a more cohesive approach through the development of conceptual and analytical frameworks that are appropriate to a wide range of marine ecosystems. In developing this book, we the collective of authors, are strengthening that connectivity and the foundations of the discipline. An investment in the foundational framework is essential to the effective uptake of a new field of study and to ensure that applications in ecology have the appropriate context. This book aims to help define, consolidate, evaluate and guide the future of seascape ecology. We hope to inspire the new generations of geospatially enabled marine scientists to adopt and evolve the science of seascape ecology and for the research community to continue testing the applicability of existing concepts from landscape ecology and to help develop seascape-specific concepts and tools. We encourage you to read the landscape ecology literature and engage with the landscape ecology community to explore opportunities with which to learn from one another.

References

Acha EM, Piola A, Iribarne O, Mianzan H (2015) Ecological Processes at Marine Fronts: Oases in the Ocean. Springer, Mar del Plata.

Addicott JF, Aho JM, Antolin MF, Padilla DK, Richardson JS, Soluk DA (1987) Ecological neighbourhoods: scaling environmental patterns. Oikos 1: 340–346.

Allen TFH (1998) The landscape 'level' is dead: persuading the family to take off the respirator. In Peterson DL, Parker VT (eds) Ecological scale: Theory and applications. Columbia University Press, New York, pp. 35–54.

Alvarez-Berastegui D, Ciannelli L, Aparicio-Gonzalez A, Reglero P, Hidalgo M, López-Jurado JL, Tintoré J, Alemany F (2014) Spatial scale, means and gradients of hydrographic variables define pelagic seascapes of bluefin and bullet tuna spawning distribution. PloS One 9(10): e109338.

Atema (1996) Eddy chemotaxis and odor landscapes: exploration of nature with animal sensors. Biological Bulletin 191(1): 129–138.

Austin MP, Smith TM (1989) A new model for the continuum concept. Vegetatio 83: 35–47.

Bainbridge R (1957) The size, shape and density of marine phytoplankton concentrations. Biological Reviews 32(1): 91–115.

Bakun A (1996) Patterns in the Ocean: Ocean Processes and Marine Population Dynamics. University of California Sea Grant, San Diego, CA, United States in cooperation with Centro de Investigaciones Biológicas de Noroeste, La Paz, Baja California Sur, Mexico.

Ballance LT, Pitman RL, Fiedler PC (2006) Oceanographic influences on seabirds and cetaceans of the eastern tropical Pacific: a review. Progress in Oceanography 69(2): 360–390.

Bell S, McCoy ED, Mushinsky HR (eds) (1991) Habitat Structure: The Physical Arrangement of Objects in Space. Springer Science & Business Media, Tampa, FL.

Bender DJ, Fahrig L (2005) Matrix structure obscures the relationship between interpatch movement and patch size and isolation. Ecology 86(4): 1023–1033.

Birkeland C (1985) Ecological interactions between mangroves, seagrass beds, and coral reefs. In Birkeland C & Grosenbaugh D (eds) Ecological Interactions between Tropical Coastal Ecosystems. UNEP Regional Seas Reports and Studies 73. United Nations Environment Programme, Nairobi, Kenya, pp. 1–26.

Bissonette JA, Storch I (eds) (2002) Landscape Ecology and Resource Management: Linking Theory with Practice. Island Press, Washington, DC.

Boström C, Pittman SJ, Simenstad C, Kneib RT (2011) Seascape ecology of coastal biogenic habitats: advances, gaps, and challenges. Marine Ecology Progress Series 427: 191–217.

Brooker RM, Dixson DL (2016) Assessing the role of olfactory cues in the early life history of coral reef fish: Current methods and future directions. In Schulte B, Goodwin T, Ferkin M (eds) Chemical Signals in Vertebrates. Springer, Basel, pp. 17–31.

Caldow C, Monaco ME, Pittman SJ, Kendall MS, Goedeke TL, Menza C, Kinlan BP, Costa BM (2015) Biogeographic assessments: a framework for information synthesis in marine spatial planning. Marine Policy 51: 423–432.

Carr MH, Neigel JE, Estes JA, Andelman S, Warner RR, Largier JL (2003) Comparing marine and terrestrial ecosystems: implications for the design of coastal marine reserves. Ecological Applications 1: S90–107.

Clavel J, Julliard R, Devictor V (2011) Worldwide decline of specialist species: toward a global functional homogenization? Frontiers in Ecology and the Environment 9(4): 222–228.

Coulter B (2014) Moving out of flatland: Toward effective practice in geospatial inquiry. In MaKinster J, Trautmann N, Barnett M (eds) Teaching Science and Investigating Environmental Issues with Geospatial Technology: Designing Effective Professional Development for Teachers. Springer, Dordrecht, pp. 287–302.

Craglia M, de Bie K, Jackson D, Pesaresi M, Remetey-Fülöpp G, Wang C, Annoni A, Bian L, Campbell F, Ehlers M, van Genderen J (2012) Digital Earth 2020: towards the vision for the next decade. International Journal of Digital Earth 5(1): 4–21.

Dayton PK, Tegner MJ (1984) The importance of scale in community ecology: a kelp forest example with terrestrial analogs. In Price PW, Slobodchikoff CN & Gaud WS (eds) A New Ecology. Novel Approaches to Interactive Systems. John Wiley & Sons, New York, pp. 457–481.

Downs RM (2014) Coming of age in the geospatial revolution: The geographic self re-defined. Human Development 57(1): 35–57.

Dunning JB, Danielson BJ, Pulliam HR (1992) Ecological processes that affect populations in complex landscapes. Oikos 1: 169–175.

Fischer J, B Lindenmayer D (2006) Beyond fragmentation: the continuum model for fauna research and conservation in human-modified landscapes. Oikos 112(2): 473–480.

Forman RT (1995) Land Mosaics: The Ecology of Landscapes and Regions. Island Press, Washington, DC.

Fuller S, Axel AC, Tucker D, Gage SH (2015) Connecting soundscape to landscape: Which acoustic index best describes landscape configuration? Ecological Indicators 58: 207–215.

Gardiner JM, Whitney NM, Hueter RE (2015) Smells like home: the role of olfactory cues in the homing behaviour of blacktip sharks, *Carcharhinus limbatus*. Integrative and Comparative Biology 55(3): 495–506.

Gore A (1998) The digital earth: Understanding our planet in the 21st century. Australian Surveyor 43(2): 89–91.

Gottschalk T, Huettmann F, Ehlers M (2005) Thirty years of analysing and modelling avian habitat relationships using satellite imagery data: a review. International Journal of Remote Sensing 26(12): 2631–2656.

Grober-Dunsmore R, Pittman SJ, Caldow C, Kendall MS, Fraser TK (2009) A landscape ecology approach for the study of ecological connectivity across tropical marine seascapes. In Nagelkerken I (ed.) Ecological Connectivity among Coral Reef Ecosystems. Springer, New York, NY, pp. 493–529.

Gustafson EJ (1998) Quantifying landscape spatial pattern: what is the state of the art? Ecosystems 1(2): 143–156.

Gutzwiller KJ (2002) Applying landscape ecology in biological conservation: principles, constraints, and prospects. In Gutzwiller K (ed.) Applying Landscape Ecology in

Biological Conservation. Springer Science & Business Media, New York, NY, pp. 481–495.

Hardy SA (1936) Observations on the Uneven Distribution of Oceanic Plankton. Cambridge University Press, Cambridge.

Haury L, Fey C, Newland C, Genin A (2000) Zooplankton distribution around four eastern North Pacific seamounts. Progress in Oceanography 45(1): 69–105.

Hidalgo M, Secor DH, Browman HI (2016) Observing and managing seascapes: linking synoptic oceanography, ecological processes, and geospatial modelling. ICES Journal of Marine Science 73(7): 1831–1838.

Hovel KA, Regan HM (2008) Using an individual-based model to examine the roles of habitat fragmentation and behaviour on predator-prey relationships in seagrass landscapes. Landscape Ecology 23(1): 75–89.

Irlandi EA, Ambrose Jr WG, Orlando BA (1995) Landscape ecology and the marine environment: how spatial configuration of seagrass habitat influences growth and survival of the bay scallop. Oikos 1: 307–313.

Jones GP, Andrew NL (1992) Temperate reefs and the scope of seascape ecology. In Battershill CN, Schiel DR, Jones GP, Creese RG, MacDiarmid AB (eds) 2nd International Temperate Reef Symposium (7–10 January 1992). NIWA Marine, Auckland, pp. 63–76.

Kareiva P (1994) Space: The final frontier for ecological theory. Ecology 75: 1.

Kavanaugh MT, Oliver MJ, Chavez FP, Letelier RM, Muller-Karger FE, Doney SC (2016) Seascapes as a new vernacular for pelagic ocean monitoring, management and conservation. ICES Journal of Marine Science 73(7): 1839–1850.

Kendall MS, Miller TJ, Pittman SJ (2011) Patterns of scale-dependency and the influence of map resolution on the seascape ecology of reef fish. Marine Ecology Progress Series 427: 259–274.

Lecours V, Devillers R, Schneider DC, Lucieer VL, Brown CJ, Edinger EN (2015) Spatial scale and geographic context in benthic habitat mapping: review and future directions. Marine Ecology Progress Series 535: 259–284.

Lecours V, Dolan MFJ, Micallef A, Lucieer V (2016) A review of marine geomorphometry, the quantitative study of the seafloor. Hydrology and Earth System Sciences 20(8): 3207–3244.

Leis JM, Carson-Ewart BM (1997) In situ swimming speeds of the late pelagic larvae of some Indo-Pacific coral-reef fishes. Marine Ecology Progress Series 159: 165–174.

Levin (1992) The problem of pattern and scale in ecology: The Robert H. MacArthur award lecture. Ecology 73(6): 1943–1967.

Likens GE, Bormann FH (1974) Linkages between terrestrial and aquatic ecosystems. BioScience 24(8): 447–456.

MacArthur RH, Wilson EO (1967) The Theory of Island Biogeography. Monographs in Population Biology. Princeton University Press, Princeton, NJ.

Magnuson JJ (1991) Fish and fisheries ecology. Ecological Applications 1(1): 13–26.

Manderson JP (2016) Seascapes are not landscapes: an analysis performed using Bernhard Riemann's rules. ICES Journal of Marine Science 73(7): 1831–1838.

Manning AD, Lindenmayer DB, Nix HA (2004) Continua and Umwelt: novel perspectives on viewing landscapes. Oikos 104(3): 621–628.

Martin AP (2003) Phytoplankton patchiness: the role of lateral stirring and mixing. Progress in Oceanography 57(2): 125–174.

Martino EJ, Able KW (2003) Fish assemblages across the marine to low salinity transition zone of a temperate estuary. Estuarine, Coastal and Shelf Science 56(5): 969–987.

McGarigal K, Cushman SA (2002) Comparative evaluation of experimental approaches to the study of habitat fragmentation effects. Ecological Applications 12(2): 335–345.

McGarigal K, Marks BF (eds) (1994) FRAGSTATS. Spatial pattern analysis program for quantifying landscape structure. Forest Science Department, Oregon State University, Corvalis, OR.

McGarigal K, Tagil S, Cushman SA (2009) Surface metrics: an alternative to patch metrics for the quantification of landscape structure. Landscape Ecology 24(3): 433–450.

McIntyre S, Barrett GW (1992) Habitat variegation, an alternative to fragmentation. Conservation Biology 6(1): 146–147.

McNeill SE, Fairweather PG (1993) Single large or several small marine reserves? An experimental approach with seagrass fauna. Journal of Biogeography 1: 429–440.

McWilliam J (2016) Spatial patterns of inshore marine soundscapes. In Popper AN, Hawkins A (eds) The Effects of Noise on Aquatic Life II. Springer, New York, NY, pp. 697–703.

Meentemeyer V (1989) Geographical perspectives of space, time, and scale. Landscape Ecology 3(3–4): 163–173.

Molles MC Jr (1978) Fish species diversity on model and natural reef patches: experimental insular biogeography. Ecological Monographs 48: 289–305.

Moore P, Crimaldi J (2004) Odor landscapes and animal behaviour: tracking odor plumes in different physical worlds. Journal of Marine Systems 49(1): 55–64.

Nedelec SL, Simpson SD, Holderied M, Radford AN, Lecellier G, Radford C, Lecchini D (2015) Soundscapes and living communities in coral reefs: temporal and spatial variation. Marine Ecology Progress Series 524: 125–135.

Nevitt G (1999) Foraging by seabirds on an olfactory landscape. American Scientist 87(1): 46–53.

Ogden JC, Gladfelter E H (eds) (1983) Coral Reefs, Seagrass Beds, and Mangroves: Their Interaction in the Coastal Zones of the Caribbean. UNESCO Reports in Marine Science. UNESCO, Montevideo.

Ogden JC, Zieman JC (1977) Ecological aspects of coral reef-seagrass bed contacts in the Caribbean. Proceedings of the Third International Coral Reef Symposium 1: 377–382.

Paine RT, Levin SA (1981) Intertidal landscapes: disturbance and the dynamics of pattern. Ecological Monographs 51(2): 145–178.

Palacios DM, Bograd SJ, Foley DG, Schwing FB (2006) Oceanographic characteristics of biological hot spots in the North Pacific: a remote sensing perspective. Deep Sea Research Part II: Topical Studies in Oceanography 53(3): 250–269.

Paris CB, Atema J, Irisson J-O, Kingsford M, Gerlach G, Guigand CM (2013) Reef odor: A wake up call for navigation in reef fish larvae. PLoS One 8(8): e72808.

Parrish JD (1989) Fish communities of interacting shallow-water habitats in tropical oceanic regions. Marine Ecology Progress Series 58: 143–160.

Peterson DL, Parker VT (1998) Ecological scale: Theory and applications. Columbia University Press, New York, NY.

Pijanowski BC, Villanueva-Rivera LJ, Dumyahn SL, Farina A, Krause BL, Napoletano BM, Gage SH, Pieretti N (2011). Soundscape ecology: the science of sound in the landscape. BioScience 61(3): 203–216.

Pittman SJ, Christensen JD, Caldow C, Menza C, Monaco ME (2007) Predictive mapping of fish species richness across shallow-water seascapes in the Caribbean. Ecological Modelling 204(1): 9–21.

Pittman SJ, Costa BM, Battista TA (2009) Using lidar bathymetry and boosted regression trees to predict the diversity and abundance of fish and corals. Journal of Coastal Research 25: 27–38.

Pittman SJ, Kneib RT, Simenstad CA (2011) Practicing coastal seascape ecology. Marine Ecology Progress Series 427: 187–190.

Pittman SJ, McAlpine CA (2003) Movements of marine fish and decapod crustaceans: Process, theory and application. Advances in Marine Biology 44: 205–294.

Pittman SJ, McAlpine CA, Pittman KM (2004) Linking fish and prawns to their environment: a hierarchical landscape approach. Marine Ecology Progress Series 283: 233–254.

Poff NL (1997) Landscape filters and species traits: towards mechanistic understanding and prediction in stream ecology. Journal of the North American Benthological Society 1: 391–409.

Powell TM, Okubo A (1994) Turbulence, diffusion and patchiness in the sea. Philosophical Transactions of the Royal Society of London B: Biological Sciences 343(1303): 11–18.

Price B, McAlpine CA, Kutt AS, Phinn SR, Pullar DV, Ludwig JA (2009) Continuum or discrete patch landscape models for savanna birds? Towards a pluralistic approach. Ecography 32(5): 745–756.

Risser PG, Karr JR and Forman RTT (1984) Landscape ecology: directions and approaches. Special Publication 2, Illinois Natural History Survey, Champaign, Illinois.

Robbins BD, Bell SS (1994) Seagrass landscapes: a terrestrial approach to the marine subtidal environment. Trends in Ecology and Evolution 9(8): 301–314.

Rovelli C (2014) Sette brevi lezioni di fisica. Piccola Biblioteca 666, Adelphi.

Ryan JP, Chavez FP, Bellingham JG (2005) Physical-biological coupling in Monterey Bay, California: topographic influences on phytoplankton ecology. Marine Ecology Progress Series 287: 23–32.

Ryan JP, McManus MA, Paduan JD, Chavez FP (2008) Phytoplankton thin layers caused by shear in frontal zones of a coastal upwelling system. Marine Ecology Progress Series 354: 21–34.

Scales KL, Miller PI, Hawkes LA, Ingram SN, Sims DW, Votier SC (2014) On the front line: frontal zones as priority at-sea conservation areas for mobile marine vertebrates. Journal of Applied Ecology 51(6): 1575–1583.

Schlosser IJ (1991) Stream fish ecology: a landscape perspective. BioScience 41(10): 704–712.

Schneider DC (2001) The rise of the concept of scale in ecology: The concept of scale is evolving from verbal expression to quantitative expression. BioScience 51(7): 545–553.

Senft RL, Coughenour MB, Bailey DW, Rittenhouse LR, Sala OE, Swift DM (1987) Large herbivore foraging and ecological hierarchies. BioScience 37(11): 789–799.

Seuront L (2009) Fractals and multifractals in ecology and aquatic science. CRC Press, Boca Raton, FL.

Seymour JR, Simó R, Ahmed T, Stocker R (2010) Chemoattraction to dimethylsulfoniopropionate throughout the marine microbial food web. Science 329(5989): 342–345.

Shekhar S, Feiner S, Aref WG (2015) From GPS and virtual globes to spatial computing – 2020. Geoinformatica 19(4): 799–832.

Sims DW, Quayle VA (1998) Selective foraging behaviour of basking sharks on zooplankton in a small-scale front. Nature 393(6684): 460–464.

Sims DW, Witt MJ, Richardson AJ, Southall EJ, Metcalfe JD (2006) Encounter success of free-ranging marine predator movements across a dynamic prey landscape. Proceedings of the Royal Society of London B: Biological Sciences 273(1591): 1195–1201.

Slabbekoorn H, Bouton N, van Opzeeland I, Coers A, ten Cate C, Popper AN (2010) A noisy spring: the impact of globally rising underwater sound levels on fish. Trends in Ecology and Evolution 25(7): 419–427.

Sousa WP (1979) Disturbance in marine intertidal boulder fields: the nonequilibrium maintenance of species diversity. Ecology 60(6): 1225–1239.

Southwood TR (1977) Habitat, the templet for ecological strategies? The Journal of Animal Ecology 1: 337–365.

Steele JH (1978) Spatial Pattern in Plankton Communities. Plenum Press, New York, NY.

Steele JH (1985) A comparison of terrestrial and marine ecological systems. Nature 313: 355–358.

Steele JH (1989) The ocean 'landscape'. Landscape Ecology 3(3–4): 185–192.

Stergiou KI, Browman HI, Cole JJ, Halley JM, Paine RT, Raffaelli D, Solan M, Webb TJ, Stenseth NC, Mysterud A, Durant JM (2005) Bridging the gap between aquatic and terrestrial ecology. Marine Ecology Progress Series 304: 271–307.

Stommel H (1963) Varieties of oceanographic experience. Science 139(3555): 572–576.

Troll C (1939) Luftbildplan und ökologische bodenforschung. Zeitschraft der Gesellschaft für Erdkunde Zu Berlin 7/8: 241–298.

Troll C (1971) Landscape ecology (Geoecology) and biogeocenology: A terminology study. Geoforum 8(71): 43–46.

Tsuda A (1995) Fractal distribution of an oceanic copepod, *Neocalanus cristatus*, in the subarctic Pacific. Journal of Oceanography 51(3): 261–266.

Turner MG (2005) Landscape ecology: what is the state of the science? Annual Review of Ecology, Evolution, and Systematics 31: 319–344.

von Humboldt A, Bonpland A (1807) Essai sur la géographie des plantes, accompagné d'un tableau physique des régions equinoxiales. Fr. Schoell Libraire, Paris.

Walsh WJ (1985) Reef fish community dynamics on small artificial reefs: the influence of isolation, habitat structure, and biogeography. Bulletin of Marine Science 36(2): 357–376.

Wang F, Hall GB (1996) Fuzzy representation of geographical boundaries in GIS. International Journal of Geographical Information Systems 10(5): 573–590.

Wedding LM, Lepczyk CA, Pittman SJ, Friedlander AM, Jorgensen S (2011) Quantifying seascape structure: extending terrestrial spatial pattern metrics to the marine realm. Marine Ecology Progress Series 427: 219–232.

Wiens JA (1989) Spatial scaling in ecology. Functional Ecology 3: 385–397.

Wiens JA (1995) Landscape mosaics and ecological theory. In Wiens JA (ed.) Mosaic landscapes and ecological processes. Springer, Dordrecht, pp. 1–26.

Wiens JA (2002) Riverine landscapes: taking landscape ecology into the water. Freshwater Biology 47(4): 501–515.

Wiens JA, Chr N, Van Horne B, Ims RA (1993) Ecological mechanisms and landscape ecology. Oikos 1: 369–380.

Wiens J, Moss M (eds) (2005) Issues and perspectives in landscape ecology. Cambridge University Press, Cambridge.

Wiens JA, Rotenberry JT, Van Horne B (1987) Habitat occupancy patterns of North American shrubsteppe birds: the effects of spatial scale. Oikos 1: 132–147.

Wilson MF, O'Connell B, Brown C, Guinan JC, Grehan AJ (2007) Multiscale terrain analysis of multibeam bathymetry data for habitat mapping on the continental slope. Marine Geodesy 30(1–2): 3–5.

Witt MJ, Broderick AC, Johns DJ, Martin C, Penrose R, Hoogmoed MS, Godley BJ (2007) Prey landscapes help identify potential foraging habitats for leatherback turtles in the NE Atlantic. Marine Ecology Progress Series 337: 231–243.

Yen PP, Sydeman WJ, Hyrenbach KD (2004) Marine bird and cetacean associations with bathymetric habitats and shallow-water topographies: implications for trophic transfer and conservation. Journal of Marine Systems 50(1): 79–99.

2

Mapping and Quantifying Seascape Patterns

Bryan Costa, Brian K. Walker and Jennifer A. Dijkstra

2.1 Introduction

Distributions of marine organisms are influenced by their physical environment (*e.g.,* depth, temperature, currents, light availability, *etc.*) and surrounding ecological processes (*e.g.,* competition, predation, reproduction, recruitment, *etc.*). The interplay of these processes drives the composition, configuration and complexity of patterning in the seascape (Grober-Dunsmore *et al.* 2009; Wedding *et al.* 2011; Walker *et al.* 2012). A fundamental goal of ecological research is to determine the influence of spatial and temporal environmental patterns on ecological processes (Forman & Godron 1986; Levin 1992). Much of this research has been conducted in terrestrial systems with a landscape ecology approach, where the understanding of spatial patterning is central to the research question (Turner 1989). Landscape ecology has provided valuable insights into how changes in land-use patterns affect the organization of, and processes in, ecological communities (Turner 2005). Since the late 1990s, landscape ecology has slowly been adapted and applied to marine systems, primarily shallow-water coastal ecosystems, giving rise to the analogous field of seascape ecology (Robbins & Bell 1994; Pittman *et al.* 2011). Like landscape ecology, seascape ecology seeks to understand the relationship between spatial patterns and ecological processes at a range of spatial and temporal scales (Hinchey *et al.* 2008; Li & Mander 2009).

Here, seascape patterns are broadly defined as the spatial and temporal distribution of physical and biological drivers (on the seafloor and in the water column) that influence species distributions. These drivers are often oceanographic or topographic, but they can include other drivers as well (*e.g.,* atmospheric, anthropogenic). Patterns may be two dimensional (2D), changing geographically (longitude (x) and latitude (y)) or they may be three dimensional (3D), changing geographically (longitude (x) and latitude (y)) as well as with depth (z) and / or time (t). Surface currents are an example of an oceanographic process where different spatial and temporal patterns are visible across the seascape. Currents may be faster at different times of the year (*e.g.,* winter versus summer) or in different locations (*e.g.,* between islands) (Figure 2.1). These current patterns can be modelled and mapped to provide a better understanding of what is potentially influencing the distributions of marine organisms.

The methods used to measure and map marine systems depend on the number of desired dimensions (2D / 3D) and scale of the seascape pattern. Some seascape patterns

Seascape Ecology, First Edition. Edited by Simon J. Pittman.

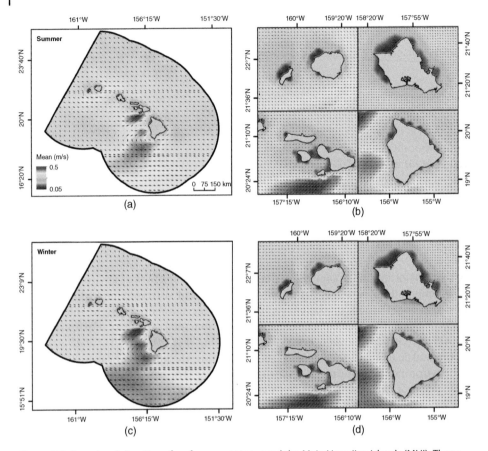

Figure 2.1 Speed and direction of surface currents around the Main Hawaiian islands (MHI). These maps depict the average speed (m/s) and direction (° denoted by arrows) of surface currents in the summer (a, b) and winter (c, d) within the MHI.

can be measured and mapped using sensors mounted on satellites, ships or planes. Others are better measured and mapped *in situ* using underwater sensors or SCUBA divers. Seascape patterns operate at a range of scales, including: (i) patterns that are constant in time but change across space (*i.e.,* spatial patterns); (ii) patterns that are constant in space but change in time (*i.e.,* temporal patterns); or (iii) patterns that change in both space and time (*i.e.,* spatiotemporal patterns) (Levin 1992). The interconnectedness of scale and pattern (Hutchinson 1953; Levin 1992) is one of the main challenges that seascape ecologists face. Consequently, they spend a significant amount of time thinking about the optimal methodologies to map seascape patterns. Advances in mapping technologies have given seascape ecologists the tools and data access to reliably quantify seascape structure at a range of scales in time and space. The resulting maps and images (hereafter referred to as 'seascape maps') are critical to seascape ecology by providing a spatial framework with which to quantify and compare seascape patterning at a range of scales (Wedding *et al.* 2011) and to explore linkages between seascape geometry and marine ecological processes (Walker *et al.* 2009; Walker 2012; Fisco 2016).

Given the importance of maps in seascape ecology, this chapter focuses on the considerations and challenges associated with developing and applying maps to ecological and management questions. These considerations are important because they will affect the spatial patterns that are identified in seascape maps and will influence the direction and strength of relationships among spatial patterns and species distributions. These relationships could be misleading (Li & Wu 2004) if our decisions about seascape maps are not linked to our understanding about what is ecologically relevant (Wedding *et al.* 2011). The relevance of these choices may change based on an organism's taxonomy, mobility and ontogeny (Appeldoorn *et al.* 2003; Grober-Dunsmore *et al.* 2008; Walker *et al.* 2009; Kendall *et al.* 2011). Gaps in our knowledge about species distributions and behaviours (Grober-Dunsmore *et al.* 2009) can make choices about relevant scales or datasets difficult. Limitations associated with available technologies can make some data collection efforts suboptimal or infeasible under certain environmental conditions. The growing number of analytical techniques may make it difficult to determine how best to characterize seascape patterns and apply seascape maps. Combined, these factors make *a priori* decisions related to observing, characterizing and applying seascape maps to ecological or management questions extremely challenging. Working through these many challenges can be daunting for marine ecologists and marine managers, alike. Here, we developed a conceptual framework (Figure 2.2) to help guide this

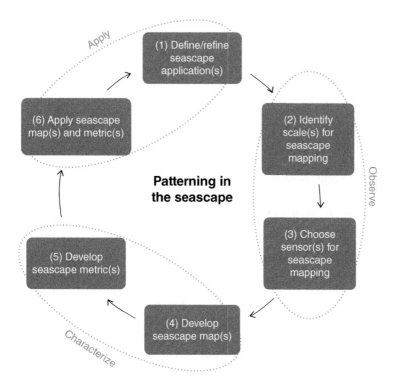

Figure 2.2 Process for investigating seascape patterns. The process for choosing how to observe, characterize and apply seascape patterns to support management applications and / or answer specific ecological questions. This process is cyclical because the way in which seascape patterns and metrics are used will influence how they should be developed and vice versa.

decision making process. The framework describes general considerations associated with observing and characterizing patterning in the seascape, as well as with applying seascape maps to answer questions for specific applications. It was designed with the collection of new remote sensing or *in situ* datasets in mind and is cyclical because the application of seascape maps will influence how they are developed and vice versa. If existing datasets are used, some decisions (*e.g.*, spatial scale) were already made, which may limit how they can be interpreted and applied. While the focus of this chapter is on seascape patterns on the seafloor at broader scales (*i.e.*, tens of square kilometres), many of the techniques and considerations discussed here are also applicable to seascape patterns in the water column and at finer scales (*e.g.*, 1 m^2).

2.2 Defining Seascape Applications

The first step in our framework (Figure 2.2) is to identify the intended application(s), including the ecological question(s) or management issue(s) of interest. These questions or issues are critical to identify at the beginning because they will affect each step in the framework. They will also help guide subsequent choices about which ecological relationships and seascape patterns to focus on and help inform decisions about how to appropriately observe (Steps 2 & 3), characterize (Steps 4 & 5) and apply (Steps 6 & 1) those patterns and relationships via exploratory data mining or hypothesis testing. Some questions and applications may only require one seascape map, while others may require dozens of seascape maps to describe the relevant patterns in the ocean. The majority of questions and applications will fall somewhere in between these extremes. This continuum means that there is no 'standard' or 'correct' way to observe, characterize and apply seascape maps. In many cases, it will differ based on the researcher, application, project, geographic location and available information. It will also be limited by logistic and financial constraints. In these cases, the most defensible way forward is to link decisions about seascape maps to the goals of the project and to root these decisions in what is relevant to the ecology of a species or geographic area.

To better illustrate this point, here is an example describing how a single seascape map is used to optimize biological sampling in the field. The basic goal of many biological surveys is to collect *in situ* samples to understand the abundance and distribution of organisms. It is often preferred to collect as few samples as possible, since *in situ* sampling is often expensive and time consuming. Surveys can be stratified based on habitat preferences to reduce the number of samples needed (Smith *et al.* 2011). Without stratification, large numbers of samples are needed to account for variability in the spatial distributions of organisms, since organisms utilize different physical and biological habitats with varying frequencies. For example, hardbottom habitats are often associated with higher fish diversity, richness and abundance values, while softbottom habitats are associated with lower values (Sale & Douglas 1984; Garcia Charton & Perez Ruzafa 1998; Gratwicke & Speight 2005; Pirtle & Stoner 2010; Dijkstra *et al.* 2012). These ecological relationships are useful to understand when designing a study to measure the spatial distribution, diversity and abundance of fish populations in an area. Since fish population structure varies more on hardbottom (particularly on highly complex hardbottom), more samples could be allocated to highly complex hardbottom habitats and fewer samples to softbottom habitats without impacting the targeted coefficient of variation. These relationships also identify the need for a seascape map depicting the distribution of hard and softbottom habitats across the study area, linking decisions about seascape maps directly to the ecology of fish and the goals of the project.

The second example links *in situ* data and seascape maps to develop species distribution models. The basic goal of species distribution models is to mathematically describe the relationships between environmental drivers and the distribution of species and then to use these relationships to predict the distribution of species in new locations as accurately as possible. Since species often respond to several environmental drivers at multiple scales, many seascape maps are needed to explain and predict species distributions. Without *a priori* knowledge, the number of seascape map / scale combinations can become enormous. However, in many cases, a thorough review of relevant literature can provide clues about important ecological relationships and can guide choices about ecologically relevant scales for seascape maps. For example, recent research suggests that oceanic fronts are influential drivers of cetacean distributions in temperate environments (Miller & Christodoulou 2014; Scales *et al.* 2014; Miller *et al.* 2015). This research suggests that predictive models may explain more variation in cetacean distribution data by including oceanic frontal maps as explanatory predictors. The choice to use oceanic frontal maps also links decisions about seascape maps directly to the ecology of cetaceans and to the goals of the project (see Chapter 3 in this book).

2.3   Identifying Scales for Seascape Mapping

Seascape patterns depend on the scales at which they are observed and analysed (Wiens 1989; Levin 1992). This dependency makes understanding scale critical to seascape ecology. Consequently, the second step in our framework (Figure 2.2) requires thinking critically about ecologically relevant scales. Scale and pattern are important to consider together because they are coupled, *i.e.*, changing the scale will also change the seascape pattern (Hutchinson 1953; Levin 1992; Wu *et al.* 2002; Wu 2004; Wu 2007). Since scale affects patterns, it also influences the maps used to describe these patterns, making seascape maps equally sensitive to changing scale (Li and Wu 2004). In cartography, scale refers to the ratio of the map units to those on the ground and is usually described by the terms 'large' and 'small'. Somewhat counterintuitively, 'small scale', in cartography, refers to a coarser resolution map than 'large scale'. In seascape ecology, the terms 'fine' and 'broad' are more commonly used as relative terms. These terms also suffer from ambiguity, therefore it is important to use quantitative terms (*e.g.*, 100 m^2) when discussing scale to avoid confusion (Meentemeyer 1989; Sayre 2005).

Scale can be described in different ways (Meentemeyer 1989; Sayre 2005). Here, we define scale as both the resolution (*i.e.*, grain) and extent of a dataset or pattern (Wedding *et al.* 2011; Lecours *et al.* 2015). Resolution and extent can vary: (i) spatially, (ii) temporally and (iii) thematically. For our purposes, we refer to these scales collectively as 'data scales'. In addition to data scales, there are three other types of scales to consider, which we will refer collectively to as 'process scales'. These process scales are divided into ecological, observational and analytical scales. Ecological scales (also known as operational scales, Lam & Quattrochi 1992) are the scales at which an ecological pattern or process occurs. Observational scales (also known as measurement scales, Lam & Quattrochi 1992) refer to the resolution and extent at which a physical or biological process or pattern was observed. Analytical scale refers to the scale(s) at which spatial and temporal patterns are characterized (Lechner *et al.* 2012a, Lechner *et al.* 2012b, Lecours *et al.* 2015). Data scales are nested within these process scales, since ecological, observational and analytical scales can all occur at different spatial, temporal and thematic scales. For more discussion on scale, please see Addicott *et al.* 1987; Palumbi 2004; Wu & Li 2006.

Mismatches between ecological, observational and analytical scales can be problematic because they can inhibit detecting certain species-habitat relationships and ecological patterns (Garcia & Oritz-Pulido 2004; Gambi & Danovaro 2006; Purkis *et al.* 2008; Walker *et al.* 2009; Lecours *et al.* 2015) or allow erroneous relationships and patterns to be detected (Li & Wu 2004). Although choosing relevant process scales is critical, choices about scale are often made arbitrarily (Levin 1992) because of technical challenges, financial limitations, time constraints (Meentemeyer 1989) and gaps in our ecological knowledge (Grober-Dunsmore *et al.* 2009). In the absence of a scale-specific question, exploring seascape patterns at multiple scales is often the most efficient approach to fill in key data and information gaps. This exploration can also help explain how ecological relationships change across scales (Pittman & McAlpine 2003; Walker *et al.* 2009). This type of exploratory analysis requires manipulating the analytical scale of the data, including changing the spatial, temporal and / or thematic scales of the datasets before characterizing seascape patterns and creating seascape maps. We discuss some common methods for altering analytical scale in Step 4. While these analyses (and analytical scales) can change, they do not change the scale at which the data were observed. Observational scale is static because it is an inherent part of that dataset (Li & Wu 2004). For example, if the observational resolution of an *in situ* survey is on seagrass patch size, information on shoot density at the within-patch scale may be difficult or impossible to mine from the survey data afterwards. As a result, great care should be taken when choosing observational scales and the sensor, platform, equipment and field data collection design on which they depend. The next section is dedicated to describing the considerations, limitations and challenges associated with these parameters and how they affect our ability to map and sample the seascape.

2.4   Sensor Selection for Seascape Mapping

Step 3 in our framework is the selection of an appropriate sensor type for mapping the seascape. Historically, marine remote sensing was limited by technology, computing and data-storage capabilities of different sensors and platforms. Over the last decade, there have been major advances in the field of marine remote sensing (Craglia *et al.* 2012; Goodman *et al.* 2013; Kachelriess *et al.* 2014). These advances have given researchers and managers the ability to collect large amounts of data describing the marine environment (Brown *et al.* 2011). While these advances are encouraging, no one sensor can map every type of seascape pattern, at multiple observational scales under all environmental conditions. These different limitations make some sensors more suitable for use in certain applications and geographic areas and require integrating multiple sensors when mapping seascape patterns from the shoreline to deep ocean depths. They also underscore the need to think critically about the information that is required for the study and the environmental conditions in the study area before choosing a sensor and

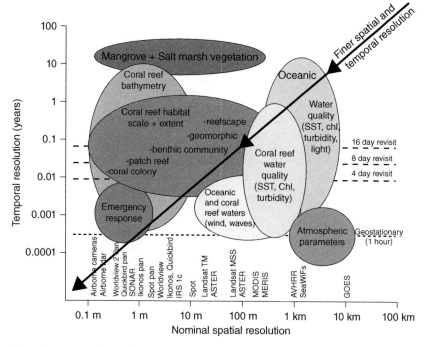

Figure 2.3 Data scales collected by different sensors. Spatial and temporal scales of some commercially available sensors used for mapping seascape patterns. *Source*: Adapted from Phinn *et al.* (2010) and reproduced with permission.

platform. To help guide these decisions, the third step in our framework (Figure 2.2) explains how different sensor and platform combinations collect data (*i.e.*, images) at various observational spatial, temporal and thematic scales (Figure 2.3). The advantages and disadvantages of every sensor is beyond the scope of this section, but more detailed information and references can be found in Horning *et al.* (2010), Wang (2011) and Goodman *et al.* (2013).

2.4.1 Passive and Active Sensors

There are two main groups of sensors: (i) passive and (ii) active (Figure 2.4). Passive sensors map the seascape by receiving and recording energy (*e.g.*, electromagnetic and sound) that reflects off the Earth's surface. They can map a variety of 2D patterns in the seascape, including patterns related to sea surface temperature, chlorophyll-*a* concentrations, turbidity and euphotic depth, dissolved organic material, photosynthetically active radiation, oceanographic upwelling, eddies, fronts, net primary productivity and seafloor habitats. Some passive sensors (called multispectral sensors) can record multiple wavelengths of light (*e.g.*, red, green and blue) in the electromagnetic spectrum, while other passive sensors (called hyperspectral sensors) can record hundreds of wavelengths of light (usually in the visible to infrared range). Both multispectral and hyperspectral sensors are most commonly mounted on satellites or airplanes. The type of platform on which the sensor is mounted will affect the spatial and temporal scale of the image. Generally speaking, cameras mounted on airplanes can collect images with higher spatial

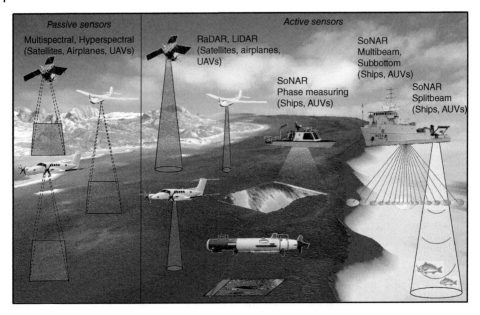

Figure 2.4 Active and passive sensors for mapping seascape patterns. This diagram shows different types and platforms for active and passive sensors. Multispectral, hyperspectral, RaDAR, LiDAR and SoNAR are some of the more commonly used sensors in seascape ecology. Sensor platforms can include satellites, airplanes, ships, UAVs (Unmanned Aerial Vehicles), ASVs (Autonomous Surface Vehicles) and AUVs (Autonomous Underwater Vehicles).

and temporal resolutions, but over smaller spatial and temporal extents. These differences occur because the airplane is closer to the ground, forcing them to image smaller geographic areas with a single pass (*e.g.,* hundreds of meters) than satellites (*e.g.,* hundreds of kilometres). Repeat time (*i.e.,* the amount of time between imagery collects) is another key consideration when choosing a platform. Airplanes can circle back more quickly to reimage an area than satellites in nongeostationary orbits (*i.e.,* orbits where the satellite moves over different parts of the Earth's surface). However, they cannot recollect imagery more quickly in areas covered by geostationary satellites, which are fixed at the same location above the Earth all the time.

Active sensors map the seascape by emitting energy (*e.g.,* radio waves, light or sound) and then receiving and recording the returned energy. They can map a variety of 3D patterns in the seascape, including sea surface heights and seafloor depth (*i.e.,* bathymetry). Radio Detection and Ranging (RaDAR), Light Detection and Ranging (LiDAR), Sound Navigation and Ranging (SoNAR) are examples of active sensors. RaDAR sensors actively emit radio waves to collect information about the surrounding environment, including the distance, direction and speed of patterns in the seascape (*e.g.,* wind, currents, sea surface heights, internal waves, upwelling, surfactants) (Jackson and Apel 2004). RaDAR sensors can be mounted at fixed coastal locations, or on airplanes, ships, satellites and autonomous vehicles. LiDAR sensors actively emit laser light to collect information about seafloor depth (*i.e.,* bathymetry), topography and relative reflectivity (*i.e.,* amount of light returned from the seafloor) (Purkis & Brock, 2013). Similarly, SoNARs actively emit sound to collect information about the seascape (Riegl & Guarin,

2013). Splitbeam SoNARs can use sound to map the abundance, size and distribution of organisms in the water column. Sub-bottom SoNARs can collect information about the depth and thickness of marine sediment layers below the sediment / water interface. Multibeam SoNARs, sidescan SoNARs and phase measuring SoNARs can collect information about seafloor depth, topography and backscatter (*i.e.*, the amount of sound returned from the seafloor).

Similar to passive sensors, the type of active sensor / platform combination will affect the spatial and temporal scales of the images. Mounting sensors at fixed locations (*e.g.*, on land or underwater) often ensures finer temporal scales, but limits the spatial extent of the study area. These types of active sensor / platform combinations are better suited to studies that investigate temporal changes at a specific location. Networks or arrays of these fixed sensors are becoming more common. Synchronizing and using imagery from the entire array can increase the spatial coverage and extent of a project substantially. For example, in southern California, a network of high-frequency RaDARs were used to map patterns in ocean surface currents to better understand the spatial and temporal connectivity of marine protected areas (Zelenke *et al.* 2009). Compared to a fixed location, mounting active sensors on ships, airplanes, satellites or autonomous vehicles will provide imagery at lower temporal resolutions. The tradeoff is that moving platforms can collect imagery over much broader spatial extents than fixed platforms.

The increasing use of AUVs, ASVs and UAVs is changing this paradigm. Autonomous vehicles are transforming our ability to map and monitor the landscape and seascape (Grasmueck *et al.* 2006; Jaramillo & Pawlak 2011; Klemas 2015; Wynn *et al.* 2014; Vincent, Werden & Ditmer 2015). Since these vehicles are autonomous, they can collect data over much longer time periods (and broader geographic extents) than manned vessels. Depending on the vehicle and payload, some autonomous vehicles can stay airborne or at sea for days, weeks or even months (Klemas 2015; Wynn *et al.* 2014). This capability gives them the potential to reduce the tradeoffs between spatial and temporal resolution and extents. It also makes them increasingly attractive and cost-efficient platforms (Wynn *et al.* 2012) to collect imagery in challenging or dangerous environments. This advantage is especially applicable to SoNARs operating in deep water, where AUVs can collect depth and backscatter information at spatial resolutions up to twice that of SoNARs mounted on ships (Murton *et al.* 1992; Scheirer *et al.* 2000). That said, autonomous vehicles are not a panacea for seascape mapping. Flight restrictions on UAVs, for example, prevent them from being used near commercial airline routes and populated areas. There are also privacy concerns associated with their use. Ultimately, there is no replacement for having trained experts in the field, allowing them to be in the environment itself and make real-time decisions during a mission. While there are still questions and challenges surrounding the use of these new technologies, future developments in autonomous vehicles will most likely continue to expand our ability to map and monitor the seascape at finer scales with enhanced hovering, long endurance, extreme depth, or rapid response capabilities (Klemas 2015; Wynn *et al.* 2014).

2.4.2 Environmental Conditions Limiting Passive and Active Sensors

Local environmental conditions also drive and limit our ability to detect and map patterns in the seascape. Different environmental conditions affect some passive and active sensors and platforms (and the images that they acquire) more than others. How these conditions affect the images (generally speaking) depends on whether a sensor is using light, sound or radio waves and whether a sensor is mounted on a satellite, ship, airplane or autonomous vehicle. This makes some sensor and platform combinations better suited and more robust choices to collect images in certain weather and water conditions (Friedlander *et al.* 2011). Multispectral and hyperspectral sensors mounted on satellites or airplanes have several environmental limitations when mapping patterns in the seascape. First, these sensors are affected by the sun's reflection off the water (*i.e.*, sunglint), the intensity of which depends on the time of day and the angle of the sensor. Sunglint is problematic because it obscures the seafloor and reduces the ability to accurately map pattern across the seascape. Second, multispectral and hyperspectral sensors (that record visible light) can only capture imagery during the daytime, since they rely on sunlight reflected from the Earth's surface. Third, most multispectral and hyperspectral sensors cannot collect imagery in deep (>30 m) or turbid water because sunlight quickly attenuates as it travels through the water column. LiDAR sensors have similar depth and turbidity limitations. In deep and turbid areas, light (be it from the sun or from a laser) does not penetrate to the seafloor because the majority is absorbed by water molecules and scattered by material suspended in the water column (Jerlov 1976). In temperate and tropical areas, this often means that multispectral, hyperspectral and LiDAR sensors cannot capture information about the seafloor deeper than ~5 to 10 m and ~25 to 30 m, respectively (Davis 1987; Goodman *et al.* 2013).

In contrast to the above sensors, sound does not attenuate as rapidly as light in the water (Kinsler *et al.* 2000). This ability gives SoNARs the capacity to map seascape patterns in turbid environments and deeper waters (Friedlander *et al.* 2011; Foster *et al.* 2013; Goodman *et al.* 2013). This advantage is critical for research focused in turbid nearshore, mid-depth and / or deep (>30 m) offshore areas. That said, SoNARs are limited in other ways that multispectral, hyperspectral and LiDAR sensors are not. They acquire images at different spatial and thematic scales than optical sensors. They are often less efficient (*i.e.*, map less geographic area in a single pass) in shallower (<30 m) depths. Acquiring data using SoNARs in the very shallow (<5 m) water can also be challenging due to the navigational hazards to ships. These challenges leave data gaps along the shoreline and in intertidal areas. LiDAR, multispectral or hyperspectral sensors can acquire seascape data seamlessly in these very shallow-water (<5 m) and intertidal environments (assuming turbidity, sunglint and cloud conditions are optimal). In areas consistently obscured by turbidity or clouds, RaDAR sensors have the capacity to acquire seascape data in nearshore environments (Jackson & Apel 2004). The tradeoff is that RaDAR sensors collect images at fundamentally different spatial, temporal and thematic scales than other passive and active sensors. This difference makes RaDARs suitable for detecting some seascape patterns (*e.g.*, currents, sea ice, surface and internal waves), but not all seascape patterns that can be detected using other sensors (*e.g.*, benthic habitats, sea surface temperature, chlorophyll-*a* concentrations, *etc.*).

2.5   Representing Patterns in Seascape Maps

Step 4 in our framework (Figure 2.2) describes different analytical methods frequently used to describe seascape patterns and develop maps. Seascapes can be mapped in 2D by passive sensors and in 3D by most active sensors. Patterns seen in these 2D and 3D images can be represented using two different conceptual models (Wedding *et al.* 2011). The 'patch-mosaic model' (Forman 1995) divides the seascape into discrete areas or 'patches' that are distinct from one another. Benthic habitat maps are a common example of the patch-mosaic model, represented as 2D categorical maps (Figure 2.5). The classes in these 2D categorical maps correspond to distinct habitat patches. They can be used to derive seascape metrics describing the composition, configuration and complexity of patches across the mosaic of habitats (Figure 2.5, right). While the patch-mosaic model is useful, its classification does not capture the gradations and true heterogeneity found across the seascape (McGarigal & Cushman 2005). An alternative model, called the 'continuous gradient' concept (McGarigal & Cushman 2005), preserves this environmental gradation and can be used to explore and identify ecologically relevant environmental thresholds and scales.

2.5.1   The Continuous Gradient Concept

Continuous 2D and 3D images are useful for seascape ecology studies (Mellin *et al.* 2012; Tanner *et al.* 2015) because they describe the diffuse boundaries associated with many seascape patterns. Previous studies suggest this gradational approach is a more realistic representation of the marine environment (Brown *et al.* 2011). The gradation allows researchers and managers to investigate and better understand potentially important environmental thresholds (or ranges) that drive the distribution and abundance of species and communities. This shift in thinking has encouraged more seascape ecologists to include continuous variables in their analysis (Ryan *et al.* 2005; Pittman *et al.* 2007; Wedding *et al.* 2008; Pittman *et al.* 2009; Walker 2009; Walker *et al.* 2009; Pittman & Brown 2011; Costa *et al.* 2014). Morphometrics are some of the most commonly used continuous variables in these seascape ecological studies. Morphometrics are variables that describe the vertical structure and morphology of the seafloor. Seafloor structure and morphology are known to be important drivers for many marine organisms from fish to cetaceans (Hyrenbach *et al.* 2000; Pittman & Costa 2010; Graham & Nash 2013; Pittman & Knudby 2014).

Continuous 3D images describing biological, physical and chemical patterns in the water column are also becoming more commonly used in seascape ecology studies (Schick *et al.* 2013; Bauer *et al.* 2016; Pittman *et al.* 2016; Stamoulis *et al.* 2016; Winship *et al.* 2016). Realistic mapping of physical and biological patterns in the water column is an ongoing challenge. Advances in big data management, distribution and modelling now allow greater access to large datasets and enable finer scale predictions using vast quantities of *in situ* oceanographic data (*e.g.*, from the World Ocean Database and the Hybrid Coordinate Ocean Model). These datasets, paired with telemetry data describing animal movements, are a powerful combination for investigating and quantifying the influence of physical patterns in the water column on animal distributions and ecological processes (Block *et al.* 2011).

As images are acquired and archived from around the world, it is becoming increasingly common to have access to multiple types of images that overlap spatially. Overlapping images can be integrated to take advantage of the different sensors and observational scales to enhance our understanding of and ability to map patterns in the seascape (Walker *et al.* 2008). For example, LiDAR images have been integrated with multispectral (Chust *et al.* 2008; Costa *et al.* 2012) and SoNAR images (Costa *et al.* 2012)

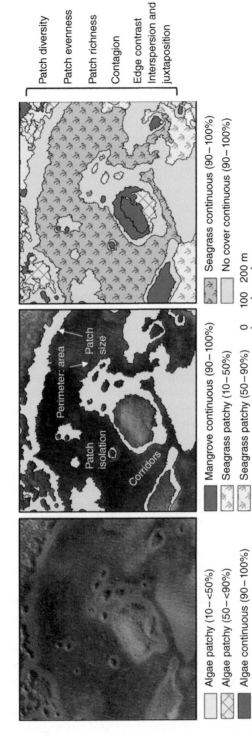

Figure 2.5 Patch-mosaic versus continuous-gradient maps. Examples of how seascape patterns can be depicted as continuous gradients (2D and 3D images) (left) and patch mosaics (2D categorical maps) describing single habitats (middle) or an entire mosaic of habitats (right).

to improve the classification of nearshore benthic habitats. Spatial predictions of deep coral, cetacean and seabird distributions have also been developed by integrating images and data from multiple passive sensors (*i.e.*, MODIS, SeaWiFS), active sensors (*i.e.*, Shoals LiDAR, Multibeam SoNARs and QuickSCAT RaDAR) and 3D ocean circulation models (*i.e.*, HYCOM) (Bauer *et al.* 2016; Pittman *et al.* 2016; Winship *et al.* 2016). However, in each of these studies, the different image types had to be reduced to a common analytical scale before they could be integrated. Caution is required when choosing a method and interpreting ecological results from rescaled data (Li & Wu 2004), since using one technique versus another may lead to slightly different results (Figure 2.6) and conclusions about boundaries, thresholds and patterning in the seascape. For example, spatial rescaling using bilinear interpolation will produce output images with smoother, more continuous looking seascape features than nearest

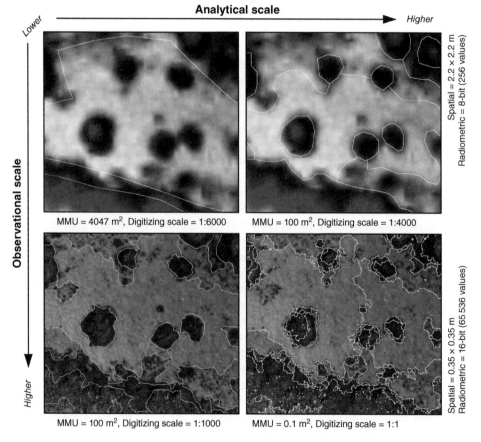

Figure 2.6 The effect of observational and analytical scales on 2D categorical maps. The four map panels show the same geographic location north of St Croix in the US Virgin Islands. The white lines denote distinct habitat patches on the seafloor. The map panels (top to bottom) show how choosing 2 or 3D continuous images with finer spatial and thematic resolutions can increase the number of habitat patches in the seascape. The map panels (left to right) show how choosing a finer resolution analytical scale (*i.e.*, decreasing the MMU size and the delineation scale) can also increase the number of habitat patches in the seascape.

neighbour resampling – see Environmental Systems Research Institute (2009). Images rescaled into annual or mean temporal bins will often have smoother appearances than images rescaled in monthly or majority bins. These differences do not indicate that one approach or statistic is necessarily more appropriate than another. Rather, it suggests that some thought needs to be put into the ecological relevance of one approach and statistic over another.

2.5.2 The Patch-Mosaic Model

The classes in 2D categorical maps correspond to distinct habitat patches in the seascape. Categorical 2D maps are developed from 2D or 3D continuous images. The spatial, temporal and thematic scales of these 2D categorical maps are directly impacted by the type of continuous images used to develop the map (Figure 2.7, Wedding *et al.* 2011). Generally speaking, 2D or 3D images with coarser data scales will produce coarser 2D categorical maps. The spatial scales of 2D categorical maps are defined by both the smallest feature that can be reliably characterized (*i.e.*, its minimum mapping unit or MMU) and the scale at which these features are delineated. In tropical waters, many 2D benthic habitat maps are delineated at scales between 1 : 1000 and 1 : 6000 and have MMUs that are between $100\,m^2$ and $4000\,m^2$ (Kendall *et al.* 2001; Battista *et al.* 2007a and 2007b; Walker *et al.* 2012; Costa *et al.* 2013). Coarser data scales also impact the accuracy of 2D categorical maps. For example, Andréfouët *et al.* (2003) found that benthic habitat map accuracies were reduced by up to 28% when developed from 2D images with lower spatial resolutions.

The spatial and thematic scales of 2D categorical maps are also directly impacted by the analytical approach used to develop the map. Several approaches can be used including: (i) manual digitizing and classification, (ii) unsupervised, pixel-based classification, (iii) supervised, pixel-based classification, (iv) object-based classification and (v) hybrid classification, which mixes these different approaches. Manual digitizing and classification is the process by which habitats are delineated by hand and classified by an expert analyst. The spatial scale at which these habitat patches are classified depends on a predetermined MMU and the map scale at which these features are delineated. The three other approaches are computer based. Unsupervised pixel-based classifications use algorithms to categorize pixels into a predetermined number of classes. These output classes are determined by the computer and do not have ecologically relevant names (*e.g.*, seagrass, coral reef, *etc.*). The supervised pixel-based classification approach is similar to the unsupervised approach, except that the algorithms require a set of ground truthing points to inform the model. These points are collected and classified by an analyst *a priori* using a predetermined classification scheme (Lillesand & Kiefer 2000). For both unsupervised and supervised classification, the spatial resolution of the classified outputs will be the same as the spatial resolution of the input images. Object-based classification methods (also known as segmentation) use algorithms to delineate the edges of habitat patches and then classify each patch using algorithms similar to those implemented by the unsupervised and supervised approaches. The spatial resolution of the classified output depends on both the spatial resolution of the input image(s) and on the input parameters used to segment the image(s). Lastly, hybrid methods blend all these different classification approaches to leverage the advantages and mitigate the disadvantages, of each technique.

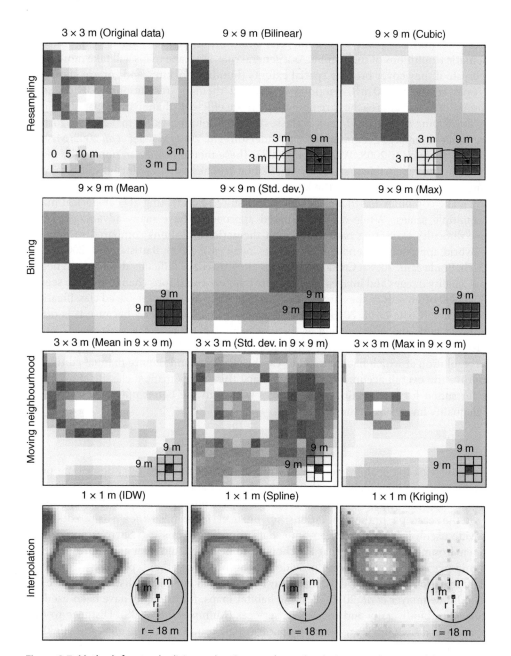

Figure 2.7 Methods for standardizing scales. Commonly used techniques to change and / or standardize the spatial scale of continuous 2D or 3D seascape maps. The neighborhood shape (*e.g.*, square or circle), size (*e.g.*, 3 × 3 m, 9 × 9 m) and statistic (*e.g.*, mean, standard deviation) used can change the patterning seen in the seascape map.

There are a few, key considerations when choosing the most appropriate method to identify and classify seascape patterns. These considerations are: (i) the spatial extent of the area, (ii) the quality of the image(s) and (iii) the required spatial, temporal and thematic resolutions of the output. As a general rule, seascape maps will take more time to create if they cover broader spatial extents (hundreds of km^2) and/or are classified at finer spatial (<100 m^2), temporal (days) and thematic resolutions (>10 classes). Computer-based approaches are potentially more efficient (than manual classification) for developing simple (<10 habitat classes), high resolution (<100 m^2) 2D categorical maps that cover large geographic areas (Anderson *et al.* 2001; Maeder *et al.* 2002; Mishra *et al.* 2006; Lucieer 2008; Weiss *et al.* 2008). These methods are also potentially more objective and repeatable than classifying habitat patches by hand, since the accuracy of hand-made maps depends on the knowledge and skill of the analyst. These advantages may make computer-based approaches better suited for creating 2D categorical maps at multiple scales. While computer-based approaches have many advantages, they have difficulty extracting patterns at high thematic resolutions compared to manual or hybrid approaches (Kendall *et al.* 2001; Coyne *et al.* 2003; Battista *et al.* 2007a and 2007b; Prada *et al.* 2008a; Costa *et al.* 2013). They also have difficulty dealing with noisy, poor quality images and images where environmental conditions change across a scene, which will often reduce the effectiveness and accuracy of computer-based classification approaches (Costa *et al.* 2013). Noise can reduce map accuracies because it is often confused with and misclassified as distinct habitat patches in the seascape. For this reason, poor quality images will often add time to the classification process, reduce the resolution at which habitat patches can be classified and require manual editing to improve the overall accuracy of the final map. This makes manual classification often better suited for interpreting noisy images (although the better option is to acquire high quality images from the start!).

2.5.3 Spatial Surrogates (Proxies)

Often, in seascape ecology, ecologically relevant seascape maps (*e.g.*, describing prey distribution) are unavailable, or they have other limitations that make them of limited utility. These data gaps can create information gaps in our understanding about patterning in the seascape. In these situations, surrogate variables (also known as proxies) can provide useful information when other data sources are missing. Surrogate variables are environmental variables that are thought to be correlated with and representative of, missing data. They can often provide insight into what other types of environmental information and seascape patterns may be driving an organism's distribution (Mellin 2015; Lindenmayer *et al.* 2015). Benthic habitat maps, for example, have been used as proxies for biodiversity and ecosystem services (Mumby *et al.* 2008). Distance to the shoreline and shelf break are becoming increasingly common proxies for studies linking organism distributions to patterns in the seascape (Pittman *et al.* 2009; Pittman & Brown 2011; Costa *et al.* 2014; Stamoulis *et al.* 2016; Pittman *et al.* 2016; Winship *et al.* 2016). Latitude and longitude similarly are being included as geographic surrogates and as a way to account for spatial autocorrelation in ecological data (Mellin *et al.* 2010; Bauer *et al.* 2016; Stamoulis *et al.* 2016; Pittman *et al.* 2016; Winship *et al.* 2016).

While surrogates are often useful, they are by no means a panacea for missing data. They too have limitations and should be used with caution. First, they may be collinear with other environmental variables (beyond the variable that was intended). While there are methods to handle collinearity in datasets, it is often still important to understand these correlations when interpreting results. Second, surrogates may be misleading since their association is indirect and many times removed from the driving ecological mechanism. For example, Kendall *et al.* (2011) found that habitat diversity (which has been used as a surrogate for fish diversity in marine protected area design) is not useful for predicting overall fish and biological diversity in the Caribbean. These results suggest that marine protected areas that used habitat diversity as a surrogate – see NAC (National Research Council) (2001) – for fish diversity may not be as effective at conserving and preserving fish diversity as intended. Clearly, surrogates are only a starting point for understanding how patterns in the seascape drive the distributions of organisms. Additional research and data collections are almost always needed to confirm and more fully explain these ecological relationships. A process for evaluating the quality and feasibility of using proxies is described in Lindenmayer *et al.* (2015).

2.6 Quantifying Seascape Structure

Spatial pattern metrics (sometimes referred to as landscape indices or landscape metrics), commonly applied in landscape ecology, quantitatively describe patches and the composition, spatial configuration and connectivity of seascapes (Gustafson 1998; Uuemaa *et al.* 2009; Wedding *et al.* 2011). Spatial patterns are analysed in numerous ways using algorithms to calculate different characteristics and statistics of a classified landscape. There are four levels of spatial pattern metrics: Cell, patch, class and landscape (McGarigal 2006). The cell is the lowest level of resolution (*e.g.*, pixel in a raster map, MMU in a vector map). The patch is the finest grouping of cells that comprise a map category (*e.g.*, individual polygon). A class is a group of patches of the same type. Landscape is a combination of all classes in the categorized mapped space. To identify redundancy across hundreds of metrics, Cushman *et al.* (2008) used principal components analysis (PCA) applied to metrics in FRAGSTATS software to isolate independent components of landscape structure and groups of components. Significant redundancy was revealed, but Cushman *et al.* (2008) found seven universal metrics at the class and landscape levels (Table 2.1 and 2.2). At the class level, these were edge contrast, patch shape complexity, aggregation, nearest neighbour distance, patch dispersion, large patch dominance and neighbourhood similarity. The landscape-level metrics were contagion / diversity, large patch dominance, interspersion / juxtaposition, edge contrast, patch shape variability, proximity and nearest neighbour distance. Cushman *et al.* (2008) recommended that metrics should be selected based on the research question and known metric behaviour. It is the responsibility of the researcher to understand the behaviour of metrics, some of which are known to exhibit nonlinear behaviour, particularly where data are being interpreted to understand the response of an organism or ecological process (Tischendorf 2001; Li & Wu 2004).

2.6.1 Sensitivity to Scale

Two-dimensional categorical maps and spatial pattern metrics have been shown to be sensitive to changing scales (Prada *et al.* 2008b; Kendall and Miller 2008; Kendall and Miller 2010; Kendall *et al.* 2011). For example, Prada *et al.* (2008b) found that metrics

Table 2.1 Universal class-level metrics.

Component name	Description
Edge contrast	Degree of 'contrast' between the focal class and its neighbourhood, where contrast is user defined and represents the magnitude of difference between classes in one or more attributes.
Patch shape complexity	Shape complexity of patches of the focal class, where shape is defined by perimeter–area relationships.
Aggregation	Degree of aggregation of cells of the focal class, where large, compact clusters of cells of the focal class are considered to be aggregated.
Nearest neighbour distance	Proximity of patches of the focal class, based on the average or area-weighted average distance between nearest neighbours.
Patch dispersion	Spatial dispersion of patches across the landscape, reflecting whether patches of the focal class tend to be uniformly distributed or overdispersed (clumped) based on variability in nearest neighbour distances.
Large patch dominance	Degree of concentration of focal class area in few, large patches with large core areas.
Neighbourhood similarity	Degree of isolation of patches from nearby patches of the same or similar class (*i.e.,* degree of similarity of the neighbourhood surrounding patches of the focal class in terms of patch composition).

Source: Adapted from Cushman *et al.* (2008).

Table 2.2 Universal landscape-level metrics.

Component name	Description
Contagion / diversity	Degree of aggregation of patch types (or the overall clumpiness of the landscape) and the diversity / evenness of patch types. Contagion and diversity are inversely related, clumped landscapes containing large, compact patches and an uneven distribution of area among patch types have high contagion and low diversity.
Large patch dominance	Degree of landscape dominance by large patches.
Interspersion / juxtaposition	Degree of intermixing of patch types.
Edge contrast	Degree of "contrast" among patches, where contrast is user defined and represents the magnitude of difference between classes in one or more attributes.
Patch shape variability	Variability in patch shape complexity, where shape is defined by perimeter–area relationships.
Proximity	Degree of isolation of patches from nearby patches of the same class.
Nearest neighbour distance	Proximity of patches to neighbours of the same class, based on the area-weighted average distance between nearest neighbours.

Source: Adapted from Cushman *et al.* (2008).

(*e.g.,* habitat patch size, shape, distribution, diversity, evenness, cohesion) changed when a 2D benthic habitat map's minimum mapping unit (MMU) was coarsened from 4 m² to 400 m². Similarly, Kendall & Miller (2008) found that rare benthic habitat map categories became rarer (and common ones, more dominant) as the MMU coarsened from 100 m² to 4048 m². They also found that increasing the thematic resolution of the benthic habitat map greatly increased the number of polygons, total edge length and diversity of the map. Subsequent analysis by Kendall *et al.* (2011) found that when these different maps were applied, the spatial and thematic scale of the map directly affected the correlation strength between fish species and seafloor habitats. This suggests that caution should prevail when interpreting results from analyses at a single scale (unless the question is directed at a single, specific scale).

While the sensitivity of benthic patterns to scale has been described in marine and terrestrial ecosystems (Kendall & Miller 2008), the effect of scale on oceanographic seascape patterns has been less well studied. Like benthic patterns, oceanographic patterns (*e.g.,* fronts, eddies, plankton *etc.*) can be thought of as 2D and 3D patches and gradients since they represent distinct physical and biological structural features in the water column and across the sea surface. Also like benthic seascapes, oceanographic patterns affect the distribution of organisms by influencing foraging, reproduction, mortality and predator-prey interactions for many species (Brown *et al.* 2011; Wedding *et al.* 2011). Oceanographic seascape patterns are often more dynamic and have a depth dimension that makes them more challenging to characterize with conventional habitat mapping techniques. However, some oceanographic patterns are persistent in time, particularly those created by interactions with the seafloor or coastline (*e.g.,* fronts, upwelling), which improves our ability to identify, map and quantify these structures over the long term (Hyrenbach *et al.* 2000).

2.7   Applications of Seascape Maps and Spatial Pattern Metrics

Seascape maps and metrics can help bridge the gap between *in situ* data and broader ecological patterns in the marine environment. This connection can help inform many different activities in the seascape from biological survey design to fisheries management and marine spatial planning. Seascape maps are critical for biological survey design. For example, map stratification reduces the number of surveys needed to reach a desired coefficient of variation (Smith *et al.* 2011). This equates to considerable cost savings of periodic assessments through time, once the relationship between the target organism and map categories is understood. Similarly, obtaining the location of every individual organism is often not practical on a broad scale, especially in highly populated areas. Organism distributions can be linked to seascape maps by subsampling populations to determine broad scale patterns (Fisco 2016). If a continuous variable is known at each location (*e.g.,* abundance or percentage cover), then hot spot analyses can be used to determine spatial clustering related to the location and amount of the variable (Getis & Ord 1992; Anselin 1995). These analyses can be constrained to the relevant habitats in the seascape maps.

Multivariate distribution data (*e.g.,* all species abundances by site, the amount of habitat types along a transect) can also be analysed to determine broader scale spatial relationships. Figure 2.8 illustrates an example using seafloor habitat data along cross-shelf transects to determine seascape patterns along the coast in southeast Florida. Multivariate analyses create a matrix of the data at all sites and ranks them relative to each another. A cluster analysis of these data reveals their similarities in a dendrogram. Groups of sites at selected similarities can be characterized by their clusters and evaluated in GIS for spatial patterns. If spatial patterns are evident, sites that are spatially similar can be grouped into new strata that are analysed to determine which variables were responsible for the differences between strata. Figure 2.8 shows clear spatial patterns in the types and amounts of habitats along the coast. For example, the main difference between the Broward-Miami and Biscayne regions (strata) was the presence and absence of seagrass. This is also useful in determining assemblage spatial distributions (Fisco 2016).

Seascape maps are also essential for identifying habitats important for resource managers. For example, Parrish *et al.* (2016), combined underwater photos with LiDAR and multispectral images to examine the distribution of seagrass habitats in Barnegat Bay, New Jersey. The distribution of these habitats may be important for larger organisms, such as predatory fish. Such maps can therefore be used to determine areas of potential habitat suitability for commercially important adult fish and shellfish

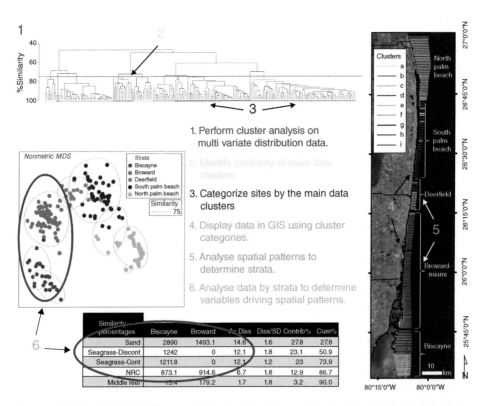

Figure 2.8 Methods for identifying spatial distributions. Six steps involved in analysing multivariate data to determine seascape patterns. The coloured numbers correspond to the coloured list of steps in the middle. The steps go from the top left to the middle right to the bottom left. *Source*: Example data are from Walker (2012).

species and deep-sea coral reef habitat (Kostylev *et al.* 2003; Young *et al.* 2010; Brown *et al.* 2011; Green *et al.* 2011; Huang *et al.* 2011; Monk *et al.* 2011; Pittman & Brown 2011). They also provide valuable information to fishermen on the areas of the seabed that are suitable or unsuitable for trawling (Pirtle *et al.* 2015) and can be used to help manage conflicts among ocean users.

Seascape maps are also critical for marine spatial planning. For example, Costa *et al.* (2015) developed maps predicting the distribution of mesophotic hard corals offshore of the island of Maui in the main Hawaiian islands (MHI). These predictions informed the siting of energy transmission cables between Maui and Oʻahu. They also supported the process to update the Hawaiian islands Humpback Whale National Marine Sanctuary's management plan. Additional seascape mapping work around the MHI (Costa & Kendall 2016) was funded by the Bureau of Ocean Energy Management to support their review of offshore renewable energy lease and development requests. Several hundred maps were developed for this project describing seasonal patterns in the oceanography and distribution of benthic organisms, fishes, sea turtles, marine mammals and seabirds. These seascape maps will also support coastal and ocean management efforts by other local, state and federal agencies working in the MHI.

2.7.1 Understanding Uncertainty in Seascape Maps

Uncertainty can be thought of as a combination of measurement error, processing error and application error, as well as conceptual flaws in pattern analysis (Li & Wu 2007). Uncertainty is introduced when observing, characterizing and applying seascape maps and quantifying spatial patterning and can propagate through all six steps in the seascape mapping and analysis process. Instrument error and sampling bias during acquisition, processing error during the preparation of continuous maps, classification error during the development of categorical maps, and application error introduced when choosing scales and analytical techniques can all combine and interact in unknown ways, potentially leading to inaccurate conclusions about ecological relationships (Wedding *et al.* 2011). The uncertainty at each step in the process may translate into positional or ecological uncertainty across the seascape, influencing the location of optimal boundaries for marine protected areas (Tulloch *et al.* 2013) or sights for developing offshore infrastructure. Currently, there is a need for marine scientists to describe uncertainty in ways that are easily applied by marine managers to make real-world decisions (Guisan *et al.* 2013). Standardizing and automating workflows may make it possible to build these estimates into analyses and make them easy to incorporate into decision making processes. Including estimates of uncertainty would also allow seascape ecologists to explore uncertainty in new and useful ways, including how uncertainty changes in space and time for individual seascape maps and how it propagates through analytical workflows. By better understanding uncertainty, researchers could more efficiently identify and target areas with higher uncertainties for further research and study.

2.8   Conclusions and Future Research Priorities

Conceptually, space, time and organization are central elements to the framework of seascape ecology, since they provide a useful way to identify emergent properties and extract general patterns from complex systems. This chapter described the challenges associated with developing seascape maps to characterize these elements. Seascape maps are important for a variety of applications in seascape ecology. They enhance our ability to sample the seascape efficiently and save money; enable us to bridge the gap between *in situ* and remotely sensed information; inform our understanding about the distribution of species and communities; and support decisions about marine spatial planning and ocean use. While seascape maps are critical for these and other applications, filling fundamental knowledge gaps would improve their ecological relevance. These knowledge gaps also highlight five key research priorities facing the seascape mapping community.

One of the first key questions is whether seascape ecology adheres to fundamental theories in landscape ecology. Originally, landscape ecology theories were developed with the distribution of plants and nutrients in mind. While these theories may provide some insight in marine environments, more research comparing terrestrial landscape ecology and seascape ecology would help us better understand parallels among these fundamentally different ecological systems. Another key question relates to mapping at ecologically relevant spatial and temporal scales. A large part of the existing knowledge in seascape ecology stem from fine scale (<1 m^2) experimental research or single snapshots in time that target a specific hypothesis or question. There is a need to broaden our understanding of ecologically relevant spatial and temporal scales in the context of seascape mapping. Advances in remote sensing provide opportunities to perform experiments in marine ecosystems over broad spatial and temporal scales, since they can acquire data over hundreds of kilometres and with greater frequency. These experiments could be designed to complement the more ubiquitous fine scale experiments to examine questions related to scale and ecological processes. The last key research question and priority focuses on quantifying and mapping the uncertainty associated with seascape maps. Managers and scientists alike are becoming increasingly interested with not only patterning in the seascape, but how confident can they be when making a decision about scarce resources in an increasingly crowded ocean. Focusing on these five research areas would help improve the science of marine mapping and our understanding of patterning in the seascape.

References

Addicott JF, Aho JM, Antolin MF, Padilla DK, Richardson JS, Soluk DA (1987) Ecological neighbourhoods: scaling environmental patterns. Oikos 49: 340–346.

Anderson AJ, Reed TB, Winn CD (2001) Marine sediment classification using sidescan sonar and Geographical Information System software in Kaneohe Bay, Oahu, Hawaii. Oceans Conference Record. MTS / IEEE Conference and Exhibition 4: 2653–2657.

Andréfouët S, Robinson JA, Hu C, Feldman GC, Salvat B, Payri C, Muller-Karger FE (2003) Influence of the spatial resolution of SeaWiFS, Landsat-7, SPOT, and International Space Station data on estimates of landscape parameters of Pacific Ocean atolls. Canadian Journal of Remote Sensing 29: 210–218.

Anselin L (1995) Local indicators of spatial association-LISA. Geographical Analysis 27(2): 93–115.

Appeldoorn RS, Friedlander A, Sladek Nowlis J, Usseglio P, Mitchell-Chui A (2003) Habitat connectivity in reef fish communities and marine reserve design in Old Providence-Santa Catalina, Colombia. Gulf Caribbean Research 14(2): 61–77.

Battista TA, Costa BM, Anderson SM (2007a) Shallow-Water Benthic Habitats of the Republic of Palau. NOAA Technical Memorandum NOS NCCOS 59. NOAA, Silver Spring, MD.

Battista TA, Costa BM, Anderson SM (2007b) Shallow-Water Benthic Habitats of the Main Eight Hawaiian Islands. NOAA Technical Memorandum NOS NCCOS 61. NOAA, Silver Spring, MD.

Bauer L, Poti M, Costa BM, Wagner D, Parrish F, Donovan M, Kinlan B (2016) Benthic habitats and corals. In Costa BM & Kendall MS (eds) Marine Biogeographic Assessment of the Main Hawaiian Islands. OCS Study BOEM 2016-035 and NOAA Technical Memorandum NOS NCCOS 214. Bureau of Ocean Energy Management and National Oceanic and Atmospheric Administration, Washington, DC, pp. 57–136.

Block BA, Jonsen ID, Jorgensen SJ, Winship AJ, Shaffer SA, Bograd SJ, Hazen EL, Foley DG, Breed GA, Harrison AL, Ganong JE, Swithenbank A, Castleton M, Dewar H, Mate BR, Shillinger GL, Schaefer KM, Benson SR, Weise MJ, Henry RW, Costa DP (2011) Tracking apex marine predator movements in a dynamic ocean. Nature 475: 86–90.

Brown CJ, Smith SJ, Lawton P, Anderson JT (2011) Benthic habitat mapping: A review of the progress towards improved understanding of the spatial ecology of the seafloor using acoustic techniques. Estuarine Coastal and Shelf Science 92: 502–520.

Bureau of Ocean Energy Management and National Oceanic and Atmospheric Administration. OCS Study BOEM 2016-035 and NOAA Technical Memorandum NOS NCCOS 214.

Chust G, Galparsoro I, Borja A, Franco J, Uriarte A (2008) Coastal and estuarine habitat mapping, using LIDAR height and intensity and multi-spectral imagery. Estuarine Coastal and Shelf Science 78: 633–643.

Costa BM & Kendall MS (eds) (2016) Marine Biogeographic Assessment of the Main Hawaiian Islands. OCS Study BOEM 2016-035 and NOAA Technical Memorandum NOS NCCOS 214. Bureau of Ocean Energy Management and National Oceanic and Atmospheric Administration, Washington, DC.

Costa BM, Kendall MS, Edwards K, Kagesten G, Battista T (2013) Benthic habitats of fish bay, Coral Bay and the St Thomas East End Reserve. NOAA Technical Memorandum NOS NCCOS 175. NOAA, Silver Spring, MD.

Costa B, Kendall MS, Parrish FA, Rooney J, Boland RC, Chow M, Lecky J, Montgomery A, Spalding H (2015) Identifying suitable locations for mesophotic hard corals offshore of Maui, Hawai'i. PLoS One 10: e0130285.

Costa, BM, Taylor, JC, Battista, TA, Kracker L, Pittman SJ (2014) Mapping reef fish and the seascape: Using acoustics and spatial modelling to guide coastal management. PLoS One 9: e85555.

Costa BM, Tormey S, Battista TA (2012) Benthic habitats of Buck Island Reef National Monument. NOAA Technical Memorandum NOS NCCOS 142. NOAA, Silver Spring, MD.

Coyne MS, Battista TA, Anderson M, Waddell J, Smith W, Jokiel P, Kendall MS, Monaco ME (2003) Benthic habitats of the main Hawaiian Islands. NOAA Technical Memorandum NOS NCCOS CCMA 152. NOAA, Silver Spring, MD.

Craglia M, Bie KD, Jackson D, Pesaresi M, Remetey-Fülöpp G, Wang C, Annoni A, Bian L, Campbell F, Ehlers M, Genderen JV, Goodchild M, Guo H, Lewis A, Simpson R, Skidmore A, Woodgate P (2012) Digital Earth 2020: towards the vision for the next decade. International Journal of Digital Earth 5: 4–21.

Cushman SA, McGarigal K, Neel MC (2008) Parsimony in landscape metrics: Strength, universality, and consistency. Ecological Indicators 8: 691–703.

Davis RA (ed.) (1987) Oceanography: an introduction to the marine environment. William C Brown Pub. Dubuque, IA.

Dijkstra JA, Boudreau J, Dionne M (2012) Species-specific mediation of temperature and community interactions by multiple foundation species. Oikos 121: 646–654.

Environmental Systems Research Institute (2009) Cell size and resampling in analysis. ArcGIS Desktop Help 9.3, Available online: http://webhelp.esri.com/arcgisdesktop/9.3/index.cfm?TopicName=Cell%20size%20and%20resampling%20in%20analysis (accessed 25 May 2017).

Fisco DP (2016) Reef Fish Spatial Distribution and Benthic Habitat Associations on the Southeast Florida Reef Tract. Master of Science thesis. Nova Southeastern University, Dania Beach, FL.

Forman RT, Godron M (1986) Landscape Ecology. John Wiley & Sons, Inc., New York, NY.

Forman, RT (1995) Land Mosaics: The Ecology of Landscapes and Regions. Cambridge University Press, Cambridge.

Foster G, Gleason A, Costa B, Battista T, Taylor C (2013) Acoustic applications. In Goodman JA, Purkis SJ, Phinn S (eds) Coral Reef Remote Sensing: A Guide for Mapping, Monitoring and Management. Springer, New York, NY.

Friedlander AM, Wedding LM, Caselle JE, Costa BM (2011) Integration of remote sensing and in situ ecology for the design and evaluation of marine protected areas: Examples from tropical and temperate ecosystems. In Wang Y (ed.) Remote Sensing of Protected Lands. CRC Press Inc., Boca Raton, FL.

Gambi C, Danovaro R (2006) A multiple-scale analysis of metazoan meiofaunal distribution in the deep Mediterranean Sea. Deep-Sea Research 53: 1117–1134.

Garcia D, Oritz-Pulido R (2004) Patterns of resource tracking by avian frugivores at multiple spatial scales: two case studies on discordance among scales. Ecography 27: 265–268.

Garcia Charton JA, Perez Ruzafa A (1998) Correlation between habitat structure and a rocky reef assemblage in the southwest Mediterranean. Marine Biology 19: 111–128.

Getis A, Ord JK (1992) The Analysis of spatial association by use of distance statistics. Geographical Analysis 24(3): 189–206.

Goodman JA, Purkis SJ, Phinn S (eds) (2013) Coral Reef Remote Sensing: A Guide for Mapping, Monitoring and Management. Springer, New York, NY.

Graham NAJ, Nash KL (2013) The importance of structural complexity in coral reef ecosystems. Coral Reefs 32: 315–326.

Grasmueck M, Eberli G P, Viggiano DA, Correa T, Rathwell G, Luo J (2006) Autonomous underwater vehicle (AUV) mapping reveals coral mound distribution, morphology, and oceanography in deep water of the Straits of Florida. Geophysical Research 33: L23616.

Gratwicke B, Speight MR (2005) The relationship between fish species richness, abundance and habitat complexity in a range of sub-tropical marine habitats. Journal of Fish Biology 66: 650–667.

Green HG, O'Connell VM, Brylinsky CK (2011) Tectonic and glacial related seafloor geomorphology as possible demersal shelf rockfish habitat surrogates – Examples along the Alaskan convergent transform plate boundary. Continental Shelf Research 31: S39–S53.

Grober-Dunsmore R, Frazer TK, Beets JP, Lindberg WJ, Zwick P, Funicelli NA (2008) Influence of landscape structure on reef assemblages. Landscape Ecology 23: 37–53.

Grober-Dunsmore R, Pittman SJ, Caldow C, Kendall MS, Fraser TK (2009) A landscape ecology approach for the study of ecological connectivity across tropical marine seascapes. In Nagelkerken I (ed.) Ecological Connectivity among Coral Reef Ecosystems. Springer, New York, NY.

Guisan A, Tingle R, Baumgartner JB, Naujokaitis-Lewis I, Sutcliffe PR, Tulloch AI, Regan TJ, Brotons L, McDonald-Madden E, Mantyka-Pringle C, Martin TG, Rhodes JR, Maggini R, Setterfield SA, Elith J, Schwartz MW, Wintle BA, Broennimann O, Austin M, Ferrier S, Kearney MR, Possingham HP, Buckley YM (2013) Predicting species distributions for conservation decisions. Ecology Letters 16: 1424–1435.

Gustafson EJ (1998) Quantifying landscape spatial pattern: what is the state of the art? Ecosystems 1(2): 143–156.

Hinchey EK, Nicholson MC, Zajac RN, Irlandi EA (2008) Marine and coastal applications in landscape ecology. Landscape Ecology 23: 1–5.

Horning N, Robinson J, Sterling E, Turner W, Spector S (2010) Remote sensing for ecology and conservation. Oxford University Press, Oxford.

Huang Z, Brooke PB, Harris PT (2011) A new approach to mapping marine benthic habitats using physical environmental data. Continental Shelf Research 31: S4–S16.

Hutchinson GE (1953) The concept of pattern in ecology. Proceedings of the National Academy of Sciences 105: 1–12.

Hyrenbach KD, Forney KA, Dayton PK (2000) Marine protected areas and ocean basin management. Aquatic Conservation: Marine and Freshwater Ecosystems 10: 437–458.

Jackson CR, Apel JR (2004) Synthetic aperture radar marine user's manual. NOAA NESDIS, College Park, MD.

Jaramillo S, Pawlak G (2011) AUV-based bed roughness mapping over a tropical reef. Coral Reefs 30: 11–23.

Jerlov NG (1976) Marine Optics (Vol 14). Elsevier, Amsterdam.

Kachelriess D, Wegmann M, Gollock M, Pettorelli N (2014) The application of remote sensing for marine protected area management. Ecological Indicators 36: 169–177.

Kendall MS, Miller TJ (2008) The influence of spatial and thematic resolution on maps of a coral reef ecosystem. Marine Geodesy 31: 75–102.

Kendall MS, Miller TJ (2010) Relationships among map resolution, fish assemblages, and habitat variables in a coral reef ecosystem. Hydrobiologia 637: 101–119.

Kendall MS, Miller TJ, Pittman SJ (2011) Patterns of scale dependency and the influence of map resolution on the seascape ecology of reef fish. Marine Ecology Progress Series 427: 259–274.

Kendall MS, Monaco ME, Buja KR, Christensen JD, Kruer CR, Finkbeiner M, Warner RA (2001) Methods used to map the benthic habitats of Puerto Rico and the US Virgin Islands. NOAA Biogeography Team, Silver Spring, MD.

Kinsler LE, Frey AR, Coppens AB, Sanders JV (2000) Fundamentals of acoustics (4th ed.). Wiley, New York, NY.

Klemas VV (2015) Coastal and environmental remote sensing from unmanned aerial vehicles: An overview. Journal of Coastal Research 31: 1260–1267.

Kostylev VE, Courtney RC, Robert G, Todd BJ (2003) Stock evaluation of giant scallop (*Placopecten magellanicus*) using high-resolution acoustics for seabed mapping. Fisheries Research 60: 479–492.

Lam NSN, Quattrochi DA (1992) On the issues of scale, resolution, and fractal analysis in the mapping sciences. The Professional Geographer 44: 88–98.

Lechner AM, Langford WT, Bekessy SA, Jones SD (2012a) Are landscape ecologists addressing uncertainty in their remote sensing data? Landscape Ecology 27: 1249–1261.

Lechner AM, Langford WT, Jones SD, Bekessy SA, Gordon A (2012b) Investigating species environment relationships at multiple scales: differentiating between intrinsic scale and the modifiable areal unit problem. Ecological Complex 11: 91–102.

Lecours V, Devillers R, Schneider DC, Lucieer VL, Brown CJ, Edinger EN (2015) Spatial scale and geographic context in benthic habitat mapping: review and future directions. Marine Ecology Progress Series 535: 259–284.

Levin SA (1992) The problem of pattern and scale in ecology. Ecology 73(6): 1943–1967.

Li XZ, Mander U (2009) Future options in landscape ecology: development and research. Progress in Physical Geography 33: 31–48.

Li H, Wu J (2004) Use and misuse of landscape indices. Landscape Ecology 19: 389–399.

Li H, Wu J (2007) Landscape pattern analysis: key issues and challenges. p39–61. In Wu J, Hobbs R (eds) Key Topics in Landscape Ecology. Cambridge University Press, Cambridge.

Lindenmayer D, Pierson J, Barton P, Beger M, Branquinho C, Calhoun A, Caro T, Greig H, Gross J, Heino J, Hunter M (2015) A new framework for selecting environmental surrogates. Science of the Total Environment 538: 1029–1038.

Lillesand TM, Kiefer RW (2000) Remote Sensing and Image Interpretation. 4th edition. John Wiley & Sons, Inc., New York, NY.

Lucieer VL (2008) Object-oriented classification of sidescan sonar data for mapping benthic marine habitats. International Journal of Remote Sensing 29: 905–921.

Maeder J, Narumainai S, Rundquist DC, Perk RL, Schalles J, Hutchins K, Keck J (2002) Classifying and mapping general coral-reef structure using IKONOS Data. Photogrammetric Engineering and Remote Sensing 68: 1297–1305.

McGarigal K (2006) Landscape Pattern Metrics. Encyclopedia of Environmetrics, John Wiley & Sons Ltd, Hoboken, NJ.

McGarigal K, Cushman SA (2005) The gradient concept of landscape structure. In Wiens JA, Moss MR (eds) Issues and Perspectives in Landscape Ecology. Cambridge University Press, Cambridge.

Meentemeyer V (1989) Geographical perspectives of space, time, and scale. Landscape Ecology 3: 163–173.

Mellin C (2015) Abiotic surrogates in support of marine biodiversity conservation. Indicators and Surrogates of Biodiversity and Environmental Change 2: 125–135.

Mellin C, Bradshaw CJA, Meekan MG, Caley MJ (2010) Environmental and spatial predictors of species richness and abundance in coral reef fishes. Global Ecology and Biogeography 19: 212–222.

Mellin C, Parrott L, Andrefouet S, Bradshaw CJA, MacNeil MA, Caley MJ (2012) Multi-scale marine biodiversity patterns inferred efficiently from habitat image processing. Ecological Applications 22: 792–803.

Miller PI, Christodoulou S (2014) Frequent locations of oceanic fronts as an indicator of pelagic diversity: application to marine protected areas and renewables. Marine Policy 45: 318–329.

Miller PI, Scales KL, Ingram SN, Southall EJ, Sims DW (2015) Basking sharks and oceanographic fronts: quantifying associations in the north-east Atlantic. Functional Ecology 29: 1099–1109.

Mishra D, Narumaiani S, Rundquist D, Lawson M (2006) Benthic habitat mapping in tropical marine environments using Quickbird multispectral data. Photogrammetric Engineering and Remote Sensing 72: 1037–1048.

Monk J, Ierodiaconou D, Bellgrove A, Harvey E, Laurenson L (2011) Remotely sensed hydroacoustics and observation data for predicting habitat suitability. Continental Shelf Research 31: S17–S27.

Mumby PJ, Broad K, Brumbaugh DR, Dahlgren C, Harborne AR, Hastings A, Holmes KE, Kappel CV, Micheli F, Sanchirico JN (2008) Coral reef habitats as surrogates of species, ecological functions, and ecosystem services. Conservation Biology 22(4): 941–951.

Murton BJ, Rouse IP, Millard NW, Flewellen CG (1992) Multisensor, deep-towed instrument explores ocean floor. EOS Transactions American Geophysical Union 73: 225–232.

NAC (National Research Council) (2001) Marine protected areas: tools for sustaining ocean ecosystems. National Academy Press, Washington, DC.

Palumbi SR (2004) Marine reserves and ocean neighbourhoods: the spatial scale of marine populations and their management. Annual Review of Environment and Resources 29: 31–68.

Parrish CE, Dijkstra JA, O'Neil-Dunne JPM, McKenna L, Pe'eri S (2016) Post-sandy benthic habitat mapping using new topo bathymetric lidar technology and object-based image classification. Journal of Coastal Research 76(sp1): 200–208.

Phinn S, Roelfsema C, Stumpf RP (2010) Remote sensing: discerning the promise from the reality. In: Dennison WC (ed.) Integrating and applying science: a handbook for effective coastal ecosystem assessment. IAN Press, Cambridge, Maryland.

Pirtle JL, Stoner AW (2010) Red king crab (*Paralithodes camtschaticus*) early post-settlement habitat choice: structure, food and ontogeny. Journal of Experimental Marine Biology and Ecology 393: 130–137.

Pirtle JL, Weber TC, Wilson CD, Rooper CN (2015) Assessment of trawlable and untrawlable seafloor using multibeam-derived metrics. Methods in Oceanography 12: 18–35.

Pittman SJ, Brown KA (2011) Multi-scale approach for predicting fish species distributions across coral reef seascapes. PloS One 6: e20583.

Pittman SJ, Christensen JD, Caldow C, Menza C, Monaco ME (2007) Predictive mapping of fish species richness across shallow-water seascapes in the Caribbean. Ecological Modelling 204: 9–21.

Pittman SJ, Costa BM (2010) Linking cetaceans to their environment: Spatial data acquisition, digital processing and predictive modelling for marine spatial planning in the northwest Atlantic. In Huettmann F, Cushman SA (eds) Spatial Complexity, Informatics and Wildlife Conservation. Springer, Tokyo, Japan.

Pittman SJ, Costa BM, Battista TA (2009) Using LiDAR bathymetry and boosted regression trees to predict the diversity and abundance of fish and corals. Journal of Coastal Research, special issue 53: 27–38.

Pittman SJ, Kneib RT, Simenstad CA (2011) Practicing coastal seascape ecology. Marine Ecology Progress Series 427: 187–190.

Pittman SJ, Knudby A (2014) Spatial predictive mapping of reef fish species and assemblages. In Bortone SA (ed.) Interrelationships between Coral Reefs and Fisheries. CRC Press, Boca Raton, FL.

Pittman SJ, McAlpine CA (2003) Movements of marine fish and decapod crustaceans: process, theory and application. Advances in Marine Biology 44: 205–294.

Pittman SJ, Winship AJ, Poti M, Kinlan BP, Leirness JB, Baird RW, Barlow J, Becker EA, Forney KA, Hill MC, Miller PI, Mobley J, Oleson EM (2016) Marine mammals – cetaceans. In Costa BM, Kendall MS (eds) Marine Biogeographic Assessment of the Main Hawaiian Islands. Bureau of Ocean Energy Management and National Oceanic and Atmospheric Administration. OCS Study BOEM 2016-035 and NOAA Technical Memorandum NOS NCCOS 214. Bureau of Ocean Energy Management and National Oceanic and Atmospheric Administration, Washington, DC, pp. 227–265.

Prada MC, Appeldoorn RS, Rivera JEA (2008a) Improving coral reef habitat mapping of the Puerto Rico insular shelf using side scan sonar. Marine Geodesy 31: 49–73.

Prada MC, Appeldoorn RS, Rivera JA (2008b) The effects of minimum map unit in coral reefs maps generated from high resolution side scan sonar mosaics. Coral Reefs 27: 297–310.

Purkis S, Brock J (2013) LiDAR overview. In Goodman JA, Purkis SJ, Phinn S (eds) Coral Reef Remote Sensing: A Guide for Mapping, Monitoring and Management. Springer, New York, NY.

Purkis S, Graham N, Riegl B (2008) Predictability of reef fish diversity and abundance using remote sensing data in Diego Garcia (Chagos Archipelago). Coral Reefs 27: 167–178.

Riegl B, Guarin H (2013) Acoustic methods overview. In Goodman JA, Purkis SJ, Phinn S (eds) Coral reef remote sensing: A guide for mapping, monitoring and management. Springer, New York, NY.

Robbins BD, Bell SS (1994) Seagrass landscapes: a terrestrial approach to the marine subtidal environment. Trends in Ecology and Evolution 9: 301–304.

Ryan JP, Chavez FP, Bellingham JG (2005) Physical-biological coupling in Monterey Bay, California: topographic influences on phytoplankton ecology. Marine Ecology Progress Series 287: 23–32.

Sale PF, Douglas WA (1984) Temporal variability in the community structure of fish on coral patch reefs and the relation of community structure to reef structure. Ecology 65: 409–422.

Sayre NF (2005) Ecological and geographical scale: parallels and potential for integration. Progress in Human Geography 29: 276–290.

Scales KL, Miller PI, Hawkes LA, Ingram SN, Sims DW, Votier SC (2014) On the front line: frontal zones as priority at-sea conservation areas for mobile marine vertebrates. Journal of Applied Ecology 51: 1575–1583.

Scheirer DS, Fornari DJ, Humphris SE, Lerner S (2000) High-resolution seafloor mapping using the DSL-120 sonar system: Quantitative assessment of sidescan and phase-bathymetry data from the Lucky Strike Segment of the Mid-Atlantic Ridge. Marine Geophysical Researchers 21: 121–142.

Schick RS, Roberts JJ, Eckert SA, Halpin PN, Bailey H, Chai F, Shi L, Clark JS (2013) Pelagic movements of pacific leatherback turtles (*Dermochelys coriacea*) highlight the role of prey and ocean currents. Movement Ecology 1(1): 11.

Smith SG, Ault JS, Bohnsack JA, Harper DE, Luo J, McClellan DB (2011) Multispecies survey design for assessing reef-fish stocks, spatially explicit management performance, and ecosystem condition. Fisheries Research 109: 25–41.

Stamoulis K, Poti M, Delevaux J, Friedlander A, Kendall MS (2016) Chapter 4: Fishes – Reef Fish. In Costa B, Kendall MS (eds) (2016) Marine Biogeographic Assessment of the Main Hawaiian Islands. Bureau of Ocean Energy Management and National Oceanic and Atmospheric Administration, OCS Study BOEM 2016-035 and NOAA Technical Memorandum NOS NCCOS 214. NOAA, Silver Spring, MD.

Tanner JE, Mellin C, Parrott L, Bradshaw CJA (2015) Fine-scale benthic biodiversity patterns inferred from image processing. Ecological Complexity 22: 76–85.

Tischendorf L (2001) Can landscape indices predict ecological processes consistently? Landscape Ecology 16: 235–254.

Tulloch VJ, Possingham HP, Jupiter S, Roelfsema C, Tulloc AI, Klein CJ (2013) Incorporating uncertainty associated with habitat data in marine reserve design. Biological Conservation 162: 41–51.

Turner MG (1989) Landscape ecology: the effect of pattern on process. Annual Review of Ecology and Systematics 20: 171–197.

Turner MG (2005) Landscape ecology: what is the state of the science? Annual Review of Ecology, Evolution and Systematics 36: 319–344.

Uuemaa E, Antrop M, Roosaare J, Marja R, Mander Ü (2009) Landscape metrics and indices: An overview of their use in landscape research. Living Reviews in Landscape Research 3(1): 1–28.

Vincent JB, Werden LK, Ditmer MA (2015) Barriers to adding UAVs to the ecologist's toolbox: Peer-reviewed letter. Frontiers in Ecology and the Environment 13: 74–75.

Walker BK (2008) A Seascape Approach for Predicting Reef Fish Distribution. PhD dissertation, Nova Southeastern University, Fort Lauderdale, FL.

Walker BK (2009) A model framework for predicting reef fish distributions across the seascape using GIS topographic metrics and benthic habitat associations. Proceedings of the 11th International Coral Reef Symposium, Ft. Lauderdale, FL.

Walker BK (2012) Spatial analyses of benthic habitats to define coral reef ecosystem regions and potential biogeographic boundaries along a latitudinal gradient. PLoS One 7: e30466.

Walker BK, Jordan LKB, Spieler RE (2009) Relationship of reef fish assemblages and topographic complexity on Southeastern Florida coral reef habitats. Journal of Coastal Research 53: 39–48.

Walker BK, Larson E, Moulding A, Gilliam D (2012) Small-scale mapping of indeterminate arborescent acroporid coral (*Acropora cervicornis*) patches. Coral Reefs 31: 885–894.

Walker BK, Riegl B, Dodge RE (2008) Mapping coral reef habitats in southeast Florida using a combined technique approach. Journal of Coastal Research 24(5): 1138–1150.

Wang Y (ed.) (2011) Remote Sensing of Protected Lands. CRC Press Inc, Boca Raton, FL.

Wedding LM, Friedlander AM (2008) Determining the influence of seascape structure on coral reef fishes in Hawaii using a geospatial approach. Marine Geodesy 31: 246–266.

Wedding LM, Friedlander AM, McGranaghan M, Yost RS, Monaco ME (2008) Using bathymetric lidar to define nearshore benthic habitat complexity: implications for management of reef fish assemblages in Hawaii. Remote Sensing of Environment 112(11): 4159–4165.

Wedding L, Lepczyk C, Pittman S, Friedlander A, Jorgensen S (2011) Quantifying seascape structure: extending terrestrial spatial pattern metrics to the marine realm. Marine Ecology Progress Series 427: 219–232.

Weiss J, Miller J, Rooney J (2008) Seafloor characterization using multibeam and optical data at French Frigate Shoals, Northwestern Hawaiian Islands. Proceedings of the 11th International Coral Reef Symposium, Fort Lauderdale, FL.

Wiens JA (1989) Spatial scaling in ecology. Functional Ecology 3: 385–397.

Winship AJ, Kinlan BP, Balance LT, Joyce T, Leirness JB, Costa BM, Poti M, Miller PI (2016) Seabirds. In Costa BM, Kendall MS (eds), Marine Biogeographic Assessment of the Main Hawaiian Islands. OCS Study BOEM 2016-035 and NOAA Technical Memorandum NOS NCCOS 214. Bureau of Ocean Energy Management and National Oceanic and Atmospheric Administration, Washington, DC, pp. 283–319 .

Wu J (2004) Effects of changing scale in landscape pattern analysis: scaling relations. Landscape Ecology 19: 125–138.

Wu J (2007) Scale and scaling: a cross-disciplinary perspective. In Wu, J., Hobbs, R. (eds), Key Topics in Landscape Ecology. Cambridge University Press, Cambridge, pp. 115–136.

Wu J, Li H (2006) Concepts of scale and scaling. In Wu J, Jones B, Li H, Loucks OL (eds) Scaling and Uncertainty Analysis in Ecology. Springer, Dordrecht.

Wu J, Shen W, Sun W, Tueller P (2002) Empirical patterns of the effects of changing scale on landscape metrics. Landscape Ecology 17: 761–782.

Wynn RB, Bett BJ, Evans AJ, Griffiths G, Huvenne VAI, Jones AR, Palmer MR, Liu J, Taylor WW (eds) (2012) Integrating Landscape Ecology into Natural Resource Management. Cambridge University Press, Cambridge.

Wynn RB, Huvenne VAI, Le Bas TP, Murton BJ, Connelly DP, Bett BJ, Ruhl HA, Morris KJ, Peakall J, Parsons D., Sumner EJ, Darby SE, Dorrell RM, Hunt JE (2014) Autonomous underwater vehicles (AUVs): their past, present and future contributions to the advancement of marine geoscience. Marine Geology 352: 451–468.

Young MA, Iampietro PJ, Kvitek RG, Garza CD (2010) Multivariate bathymetry-derived generalized linear model accurately predicts rockfish distribution on Cordell Bank, California, United States. Marine Ecology Progress Series 415: 247–261.

Zelenke B, Moline MA, Crawford GB, Garfield N, Jones BH, Largier JL, Paduan JD, Ramp SR, Terrill EJ, Washburn L (2009) Evaluating connectivity between marine protected areas using CODAR high-frequency radar. OCEANS2009 MTS/IEEE Conference, Biloxi, MS.

3

Pelagic Seascapes

Kylie L. Scales, Diego Alvarez-Berastegui, Clare Embling and Simon Ingram

3.1   Introduction

The global ocean covers approximately 71% of the Earth's surface, and by volume represents 97% of the Earth's water. Comprising all of the world's oceans and seas, the global ocean is an inter-connected, highly dynamic fluid system coupled to the atmosphere, which regulates life-sustaining processes on Earth.

The pelagic ocean environment is perhaps the least understood and most difficult to monitor of all biomes on Earth. Physical, biological and chemical processes operating over a continuum of spatial and temporal scales create complex heterogeneity in the distribution of pelagic marine organisms with some regions teeming with life while others are oceanic deserts. Seascape ecology enables us to measure, monitor, analyse and understand the spatial and temporal heterogeneity in the oceans, and to understand the ecological consequences of spatial patterns and structural features.

Seascape ecology embraces the marine environment as a holistic system of systems, constructed of integral parts functioning together across scales and exhibiting nonlinear dynamics. This vision of the 'ocean landscape' was described by John Steele in his important paper published in the journal *Landscape Ecology* (Steele 1989). Steele recognized the importance of understanding biophysical coupling, the consequences of variability in ocean topography, with features such as fronts playing a key role in trophic aggregations, the importance of scale effects in predator-prey interactions and that patchiness is a critical component of the ecosystem.

In addition to the understanding generated by decades of ship-based surveys, recent technological advances have initiated a paradigm shift to the more holistic, ecosystem-based approach in management, which can be supported by seascape ecology (see Chapter 12 this book). Advances in satellite remote sensing techniques made since the 1970s have facilitated the remote and near-constant acquisition of data describing physical and biological conditions at the surface (Wilson *et al.* 2002). Increasingly sophisticated ocean circulation models are reconstructing the physical dynamics of the oceans in virtual form and assimilating real data from satellites and *in situ* sensors (Shchepetkin & McWilliams 2005). Primary productivity and the trophodynamics of lower trophic level organisms are now being incorporated into ecosystem models (Fiechter *et al.* 2015). The technical progress and proliferation in animal tracking and biologging technologies has revealed the previously cryptic

Seascape Ecology, First Edition. Edited by Simon J. Pittman.
© 2018 John Wiley & Sons Ltd. Published 2018 by John Wiley & Sons Ltd.

movements and at-sea behaviours of a diverse set of wide-ranging marine predators (Pittman & McAlpine 2003; Hussey *et al.* 2015; see also Chapter 7 in this book). Autonomous vehicles, gliders and passive acoustic telemetry networks are providing unprecedented insight into pelagic ecosystem dynamics (*e.g.*, Integrated Ocean Observing System, IOOS, http://www.ioos.noaa.gov (accessed 25 May 2017); Marine Biodiversity Observation Network, Muller-Karger *et al.* 2014).

Taken together, these advances in remote sensing and spatial dynamic modelling enable us to apply landscape ecology techniques to the pelagic realm. An emerging area of research is the development and application of pattern metrics that quantify seascape structure, followed by the use of these metrics to explain species distributions and biodiversity patterns in relation to the biophysical environment.

This chapter extends the conceptual framework of seascape ecology into the pelagic realm, drawing on concepts and vocabulary common to landscape ecology and highlighting new applications to open ocean environments. We refer to this emerging discipline as pelagic seascape ecology, with a focus on quantification of spatial patterning in the open oceans, and the study of its ecological consequences. The first section discusses pattern and process in the pelagic realm, from broad-scale biogeographic provinces to finer scale patchiness and patch dynamics. We examine the concept and quantification of boundaries and edges (ecoclines and ecotones) in the pelagic realm and explore structure in the water column. We then discuss geospatial and ecoinformatics approaches used to link physical processes and biological responses in pelagic systems, with specific reference to the at-sea movements of highly mobile marine species such as seabirds, cetaceans, sharks, turtles and pinnipeds. Finally, we outline exciting research frontiers for future exploration of pelagic seascape ecology.

3.2   Pattern and Process in the Pelagic Realm

Although seemingly featureless from our surface-bound perspective, the open oceans are a spatially structured, three-dimensional fluid interplay of distinct water masses and oceanographic features. Structural patterns in oceanic conditions extend across an array of interconnected spatial scales, and fundamentally control the ecological processes that regulate marine biodiversity (see Figure 3.1; see also Bakun 1996; Dickey 2003). This spatial structuring facilitates the application of landscape ecology concepts to the open oceans, as we seek to improve our understanding of pelagic ecosystem functioning.

3.2.1   Broad-scale Biogeographic Provinces

Over ocean-basin scales, spatially explicit patterns of primary productivity are now well established, owing primarily to the measurement of ocean colour from satellites initiated in the 1970s (Figure 3.2). Alongside extensive and repeated ship-board oceanographic surveys such as the Atlantic Meridional Transect (Aiken *et al.* 2000), satellite remote sensing has enabled measurement of physical and biological properties of pelagic systems across ocean basins. In turn, this has allowed for objective classification of regions

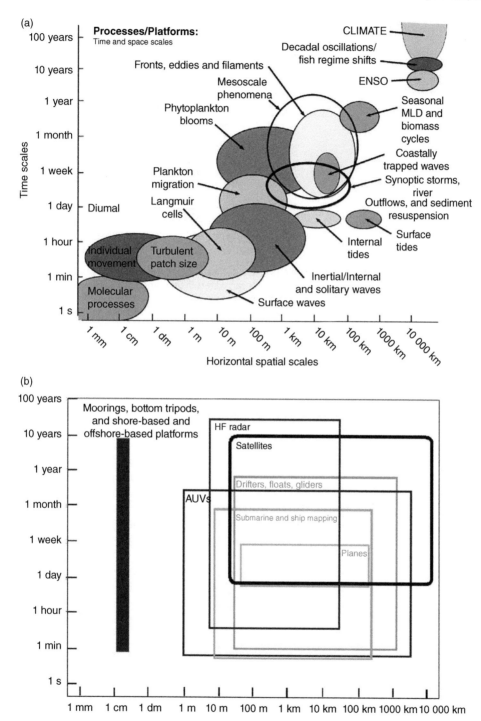

Figure 3.1 Interacting temporal and spatial scales in marine ecosystems. (a) Time-space diagram showing the scale of several physical and biological processes from very short-lived, small-scale phenomena such as molecular processes and individual movements to global-scale, decadal climate signals. (b) Approximate time and horizontal space scales of sampling platforms used in the marine environment. *Source*: Reproduced with permission from Dickey (2003).

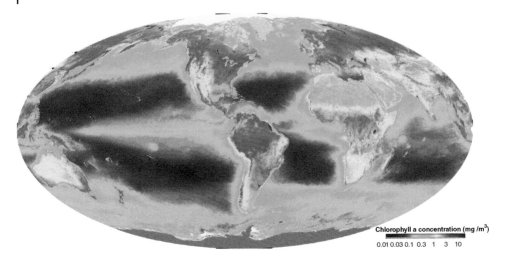

Figure 3.2 Global chlorophyll-*a* climatology, 1997–2000. Mean primary productivity over the global ocean, showing oligotrophic (nutrient-poor) open ocean waters in blue, and productive waters associated with equatorial and eastern boundary upwelling in green. Extremely productive coastal regions with high sediment concentrations such as the outflow from the Amazon delta are highlighted in warm colours. Surface chlorophyll-*a* concentration from ocean colour satellites is a useful metric for identifying and classifying pelagic seascapes over broad scales. Image from NASA Ocean Biology (OB), Sea-viewing Wide Field-of-view Sensor (SeaWiFS) Ocean Color Data, 2014.0 Reprocessing. *Source*: Obtained from NASA OB.DAAC, Greenbelt, MD, http://oceancolor.gsfc.nasa.gov, (accessed 29 May 2017).

of the open oceans into broad-scale *biogeographic provinces* – spatially coherent oceanic regions defined by a set of coincident hydrographic and biotic conditions. As there are no clear hard or fixed boundaries in the pelagic environment, the shape, size and structure of these provinces is defined primarily by the dominant mode of regional circulation. For example, *ocean deserts*, oligotrophic (nutrient-poor) regions where very little primary productivity is observed, are enclosed by the major gyres of the Pacific, Atlantic and Indian Ocean basins (Irwin & Oliver 2009).

Pioneering work in marine remote sensing in the mid-1990s developed the idea of water mass characterization for defining *Longhurst provinces*, with particular focus on the northwest Atlantic (Longhurst *et al.* 1995). More recent work has developed this concept, incorporating biophysical dynamics to move from static, rectilinear province boundaries to classifications that incorporate the shifting, dynamic nature of the oceans, known variously as *dynamic biogeographic provinces* (Platt & Sathyendranath 1999; Devred *et al.* 2007; Oliver & Irwin 2008; Devred *et al.* 2009; Reygondeau *et al.* 2013), *pelagic provinces, marine ecoregions* (Spalding *et al.* 2012) or *seascapes* (Kavanaugh *et al.* 2014). These approaches identify and delineate water masses that differ in temperature, salinity, density, nutrient concentration and associated plankton community structure, identifying distinct broad-scale pelagic seascapes. More recent attempts also include spatial patterning in forage fish distributions (*e.g.*, Briggs & Bowen 2012).

Although not often explicitly stated in these terms, the landscape ecology concepts of *scale, extent, patch* and *grain* are inherent in this seascape classification process. The importance of spatial scale and the *hierarchical patch mosaic* nature of pelagic ecosystems are further reflected in the conceptual organization of such provinces or seascapes into dynamic, fluctuating *superseascapes* at very broad scales and *subseascapes* at finer scales (Kavanaugh *et al.* 2014).

3.2.2 Finer Scale Patchiness and Patch Dynamics

The *patch dynamics* concept, integral to landscape ecology, is the study of spatial patterns and patchiness, the environmental and ecological processes that create these patterns, and the dynamics of fluxes within patches over time (Forman & Godron 1981; Paine & Levin 1981; see also Chapter 6 in this book). In a landscape ecology context, these patches may be features such as forests, wetlands or agricultural land, arranged in a piecewise-continuous harlequin environment, matrix or patch mosaic structure (Turner *et al.* 2001). In the pelagic realm, the biological seascape can also be considered a *patch mosaic* or *harlequin* environment (Steele 1989; Jelinski 2014). In fact, it has long been known that biological distributions can be highly patchy (Hardy 1936; Bainbridge 1957; Stommel 1963; Steele *et al.* 1978). Intensive field sampling of zooplankton patches has revealed a complex spatial mosaic of patches and aggregations of patches, each dominated by a single species, with some patterns exhibiting fractal properties of self-similarity across spatial scales (Tsuda 1995; Tsuda *et al.* 2000).

Much of this structural heterogeneity is driven by the interaction between water movements and benthic topography. Both horizontal and vertical water movements create a dynamic mosaic of patterns and patch structure, with elevated productivity often occurring at the boundaries or edges of patches (Powell & Okubo 1994; Haury *et al.*, 2000; Martin 2003; Yen *et al.* 2004). Few studies, however, have applied landscape ecology concepts and tools to quantify the two- and three-dimensional geometry of patchiness in the ocean. Although the patch-matrix and patch-mosaic models dominate terrestrial landscape ecology and have also been applied widely to benthic marine environments, they are arguably more representative of human modified terrestrial landscapes, with well-defined patch mosaics such as agricultural-urban landscapes, than of the fluid driven patterning observed in pelagic marine environments.

Patterns of primary productivity display extreme heterogeneity over a range of spatial and temporal scales. For example, the distributions of marine phytoplankton lead to distinct and dynamic patchiness across the global ocean, influencing patterns in ocean colour and creating what Adrian Martin refers to as 'the kaleidoscope ocean' (Martin 2005). Physical-biological coupling regulates the patchiness of pelagic seascapes. The extent, structure and properties of *subseascapes* or seascape patches are a direct result of patterns of physical forcing, particularly turbulent flow at the mesoscale (tens to hundreds of kilometres) and sub-mesoscale (~1k m; Martin 2005).

Advances in satellite oceanography in recent decades have revealed the geometric patterning inherent in the structure of phytoplankton blooms, and facilitated understanding of the (sub)mesoscale physical processes leading to this patterning.

Phytoplankton go with the flow – that is, the shape and structure of phytoplankton blooms are a function of the interactions between biological processes, such as proliferation, senescense, death and grazing by zooplankton, and hydrographic processes such as stirring, advection and concentration by (sub)mesoscale turbulence (Martin 2005; Benoit-Bird 2009). Seasonal cycles of primary productivity add an important temporal dimension to the patchiness of pelagic systems, particularly in temperate latitudes (Sathyendranath *et al.* 1995). This complexity in biophysical structuring of the open ocean is dynamic, and difficult to quantify, presenting a challenge for the implementation of pelagic seascape ecology.

3.2.3   Ecoclines and Ecotones in Pelagic Seascapes

In keeping with the *ecocline* and *ecotone* concepts in landscape ecology, oceanic waters are structured over both horizontal and vertical planes into water masses separated by fluid gradients (*ecocline*), or abrupt transitions (*ecotone*) in biophysical properties. Understanding the importance of these interfaces, edges or boundaries in pelagic systems is central to the development and application of seascape ecology.

Ecoclines are defined in this context as continuous gradients in hydrographic properties such as sea surface temperature (SST) or salinity, or in biological properties such as chlorophyll-*a* concentration. Global and ocean-basin scale gradients in SST are surface manifestations of hydrographic processes that fundamentally regulate patterns of marine biodiversity and species richness (Tittensor *et al.* 2010). Gradients in hydrographic properties are linked to broad-scale patterns in phyto- and zooplankton diversity (Barton *et al.* 2013; Irigoien *et al.* 2004). At regional scales, persistent gradients in salinity and chlorophyll-*a* concentration are important to spatial patterning in pelagic biodiversity (Hidalgo *et al.* 2015).

In contrast with terrestrial ecosystems, which frequently display a piecewise-continuous distribution of relatively static patches, the oceans are fluid and so gradients in physical and biological conditions, ecoclines, are ubiquitous throughout pelagic seascapes. *Ecotones* are abrupt transitions in biophysical characteristics of a landscape or seascape, delineating boundaries between patches and often incorporating aspects of the patches on either side, such as hedgerows in agricultural systems (Attrill & Rundle 2002). In terms of pelagic seascape ecology, oceanic fronts are an example of a pelagic ecotone. Fronts are sharp transitions in water mass properties such as temperature, salinity, density, colour, productivity and biotic community structure and are ecologically significant features of pelagic ecosystems over a range of spatiotemporal scales (Bakun 1996).

At ocean-basin scales, the sharp transition in surface chlorophyll between subpolar and subtropical gyres commonly observed in ocean basins is an important ecotone. For example, the Transition Zone Chlorophyll Front (TZCF) in the north Pacific is a major seascape feature and a biological hotspot owing to bottom-up forcing and the manifestation of conditions suitable for foraging and migration for large pelagic predators such as sea turtles and tuna (Polovina *et al.* 2001; Bograd *et al.* 2004; Polovina *et al.* 2015).

Given the relative lack of static seascape features in the open oceans, pelagic ecotones can, in general terms, be considered more dynamic and variable than their terrestrial equivalents. While the interaction of current flows with static bathymetric features, such as continental shelf breaks (Marra *et al.* 1990) and seamounts (Morato *et al.* 2010;

Morato *et al.* 2015), can lead to the manifestation of spatially persistent features such as upwelling zones and fronts (Hyrenbach *et al.* 2000), dynamic hydrographic processes are the primary drivers of ecotone formation in pelagic systems (Acha *et al.* 2004; Manderson 2016). For example, oceanic fronts that manifest at (sub)mesoscales are defining features of pelagic seascapes throughout the global ocean (Fedorov 1986; Bost *et al.* 2009; Scales *et al.* 2014b; Acha *et al.* 2015).

Horizontal stirring and vertical mixing in mesoscale features such as fronts, eddies, jets and filaments are key to nutrient influx to surface layers. Eddies influence the growth, structure and persistence of phytoplankton blooms (Mahadevan *et al.* 2012) and can initiate blooms within otherwise barren ocean deserts (Oschlies 2002). Understanding these (sub)mesoscale processes, considering them as ecotones and mapping them and their propagation through space and time are, therefore, fundamental aspects of pelagic seascape ecology (Figure 3.3; Scales *et al.* 2014b, Miller *et al.* 2015).

Mesoscale fronts can act as hydrographic barriers to the dispersal of low trophic-level marine organisms, separating and, in some cases, enabling interaction between distinct community assemblages (Bakun 2006; Letessier *et al.* 2012; Hidalgo *et al.* 2015; Powell & Ohman 2015). Hydrographic and biological processes that occur at mesoscale and sub-mesoscale (~1 km) fronts are analogous to *edge effects* in landscape ecology, and fronts and eddies themselves as sub-mesoscale ecotones. Localized upwelling and

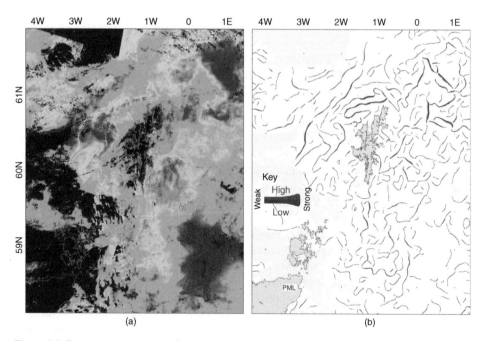

(a) (b)

Figure 3.3 Fronts as ecotones in pelagic seascape ecology. Example of thermal front mapping using satellite SST imagery, to identify potentially important ecotones in pelagic seascapes. Image shows North Sea waters to the north-east of the Scottish mainland. Transitions between water masses of different thermal characteristics are identified and the relative strength of the temperature gradient is shown by line thickness in (b). Red and blue lines trace the warm and cold sides of each thermal front. *Source*: Reproduced with permission from Miller *et al.* (2015).

enhanced vertical mixing, nutrient retention and convergent flow fields can create conditions suitable for development of front-associated phytoplankton blooms, and the physical aggregation of zooplankton and other small, drifting or swimming marine organisms (Le Fevre 1986; Franks 1992; Bakun 1996, 2006).

In addition, the ecological relevance of island effects has also been well-documented. For example, wave conditions and circulation differ on the leeward versus windward side of islands and eddies and gyres at island promontories support foraging hotspots (Hazen *et al.* 2013) and are often associated with the locations of fish spawning aggregations (Kobara *et al.* 2013).

3.2.4   Beneath the Surface: the Vertical Dimension of Pelagic Seascapes

Critically, biophysical conditions in the pelagic realm are highly heterogeneous over three dimensions. Vertical structuring of water column properties is perhaps the key distinguishing feature of pelagic seascapes in comparison to more familiar landscapes in terrestrial biomes. Of course, terrestrial landscapes have three-dimensional structure that fundamentally regulates ecological processes, yet the importance of the vertical dimension is more pronounced in the open oceans. Vertical gradients in physical properties, such as temperature, pressure, light penetration and the concentrations of oxygen, dissolved nutrients and minerals (*e.g.*, nitrate, Iron) are also important *ecoclines* or *ecotones* in pelagic systems (depending on the rate of change with depth), which strongly influence the dynamic distributions of marine organisms and, hence, ecosystem functioning (Benoit-Bird *et al.* 2009). For example, the thermocline, which marks an abrupt drop in temperature at the base of the well-mixed surface layer, is a highly significant ecotone (Spear *et al.* 2001; Pelletier *et al.* 2012). These structures vary in depth and strength through time, ranging from very stable to relatively ephemeral or seasonally present, for example through seasonal stratification in shelf sea systems (Simpson & Sharples 2012).

Light penetration through oceanic waters strongly regulates vertical structuring, separating the water column into distinct zones, which are regarded as vertical patches or *ecotopes* in a seascape ecology context. Sunlight attenuates rapidly in seawater (by 50–100 m light penetration is just 1% of that at the surface – Martin 2005). The epipelagic (also known as euphotic, photic, sunlight zone) is a relatively thin layer, extending to approximately 200 m depth, in which pelagic marine life is largely concentrated. This zone is illuminated and warmed by sufficient incoming sunlight to support photosynthesis, so is where the vast majority of primary production occurs. As light penetration diminishes with depth, community structure changes accordingly through the mesopelagic, bathypelagic and into the abyssopelagic (Bakun 1996).

However, not all primary production occurs in surface layers. Sub-surface chlorophyll maxima (SCM, aka deep chlorophyll maxima, DCM), formed as phytoplankton aggregations capitalize on nitrate flux across the base of the mixed layer (*i.e.*, at the thermocline, Sharples *et al.* 2013), are notable examples of the interplay between three-dimensional physical structure and biological function, particularly in seasonally stratified shelf seas (Embling *et al.* 2012; Scott *et al.* 2013; Macías 2014). Water-column processes such as interactions between internal waves and bottom topography can lead to the development of DCM, or phytoplankton layers at depth (Figure 3.4; Ryan *et al.* 2005). Thin phytoplankton layers can exert a strong bottom-up influence on pelagic

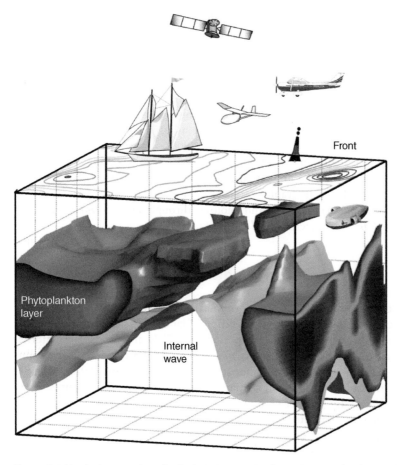

Figure 3.4 Vertical structuring of pelagic seascapes. A schematic example of vertical structuring with depth in pelagic seascapes, from a 3-D biophysical description of Monterey Bay, California. Interactions between thermal fronts at the surface, phytoplankton layers at depth and internal waves – gravity waves that oscillate at depth rather than at the water surface – generate spatial structure in pelagic seascapes. These biophysical coupling processes can be observed and measured using *in situ* or remote sensing technologies, such as shipboard surveys and satellite oceanography. *Source*: Image modified from Ryan *et al.* (2005) by S. J. Pittman.

ecosystem functioning (Benoit-Bird 2009; Benoit-Bird & McManus 2012; Benoit-Bird *et al.* 2013a and b; Greer *et al.* 2013). Internal waves transport huge volumes of heat, nutrients and prey across the ocean and these dynamic structures can help to explain ocean conditions and biological aggregations. Advanced remote sensing techniques using space-borne synthetic aperture radar (SAR) can map and track internal wave velocities as deep as 80 m below the surface (Romeiser & Graber 2015).

Another sub-surface feature of particular relevance to pelagic seascape ecology are Oxygen Minimum Zones (OMZs) – three-dimensional patches with extremely low concentrations of dissolved Oxygen (hypoxia), which is critical for multiple organismal-level processes. While OMZs are a persistent natural feature of the water column in most oceanic regions, expansion of OMZs into previously well-oxygenated regions can affect biological diversity and ecosystem health (Kalvelage *et al.* 2013). The importance of OMZs is likely to increase as hypoxic zones expand and increase in number (Grantham *et al.* 2004; Bograd *et al.* 2008; Deutsch *et al.* 2011). The capacity to map and track these patches, and quantify their effects on ecosystem function, would be invaluable in understanding and predicting change in pelagic systems.

However, only the very surface of the epipelagic zone is measurable from satellites. While surface conditions (*e.g.*, SST) can often serve as useful proxies of important sub-surface structure and process (Oliver & Irwin 2008), it may be that effectively incorporating the vertical dimension of pelagic seascape ecology hinges upon the more widespread application of four-dimensional (latitude, longitude, depth, time), data-assimilative regional oceanographic models (Moore *et al.* 2011), or ideally, a coupling of four dimensional physics and biology through end-to-end ecosystem modelling (Franks *et al.* 2013; Fiechter *et al.* 2015).

3.3   Spatial Pattern Metrics for Pelagic Seascapes

Developing and testing adequate spatial pattern metrics is essential to pelagic seascape ecology. In landscape ecology, the parameterization of landscape structure has been a major focus for more than three decades (Wu & Hobbs 2002). Spatial landscape patterns are described using a wide number of *landscape metrics*, which provide information about multiple parameters of seascape composition and spatial configuration such as edge density, fragmentation, dispersion, spatial heterogeneity, structural connectivity and patch area and diversity (Turner *et al.* 2001; Uuemaa *et al.* 2009; McGarigal *et al.* 2012). The techniques to compute these metrics is based on the *patch mosaic* concept of landscape structure that define the landscape as a mosaic of discrete and well delineated habitat patches (see section 3.3.1). This approach has the advantage of simplifying the design and computation of metrics by applying functions widely extended in GIS software that use cartographies where habitats are classified attending to categories, *i.e.* rivers, forests (McGarigal *et al.* 2012). Oceanic seascapes are typically mapped and represented as continuously varying gridded values (*e.g.*, surface roughness, height, sea surface temperature, chlorophyll concentrations). In landscape ecology, a suite of pattern metrics have been used to measure structural patterns across spatial gradients (McGarigal *et al.* 2009; Wedding *et al.* 2011).

Alongside designing and developing software to compute spatial pattern metrics, landscape ecology researchers have focused on understanding metric inter-relationships and redundancy, as well as the behaviour of metrics when changing the scale of analysis, the area of observation (or *extent)*, or the spatial resolution of input maps (or *grain*; Wu & Hobbs 2002; Cushman *et al.* 2008; see Chapter 2 in this book). It is intuitive that properties defining a landscape are multiple and inter-related, and therefore the combination of multiple different metrics is often necessary in the parameterization of landscape structure. It is challenging yet important to include measures of patchiness into our analyses of pelagic ecosystem functioning, using spatial pattern metrics to quantify seascape structure (Wedding *et al.* 2011).

3.3.1 Patch Mosaic Metrics

The derivation of spatial pattern metrics in landscape ecology set a precedent for seascape ecology. Initial seascape ecology studies focused on the analyses of how benthic coastal habitat structure affects the spatial distribution of marine species (Hinchey *et al.* 2008; Wedding *et al.* 2011). Where benthic habitat maps are represented as discrete, internally homogeneous patches with sharp boundaries, spatial pattern metrics suitable for the *patch mosaic* framework can be applied to quantitatively describe seascape structure (see Chapter 2 in this book). For instance, one could analyse the pattern of habitat patches using spatial pattern metrics such as mean patch size of reefs, the relationships between patch perimeter and patch area, or the number of patches within a unit area (McGarigal *et al.* 2009). These metrics can be applied together to model and investigate fish abundance, presence-absence or other parameters related to population structure. A snapshot of seascape structure in the form of a two-dimensional horizontal surface can be a useful way to address species-environment questions where patch and patch-mosaic features are ecologically meaningful.

The pelagic realm, however, is not configured by static and well-delineated habitat patches that can be easily categorized. It is better represented as a continuous surface over which biophysical parameters change gradually in space and time. The intensity of this spatial variability is what results in the presence of more or less intense ecoclines and ecotones. This is clear when we consider a remotely sensed image showing, for example, mesoscale variability in SST in a dynamic upwelling system (Figure 3.5). Mesoscale features such as filaments, fronts and eddies configure pelagic seascape structure and can be easily recognized from the spatial variability in the continuous parameter (*i.e.*, SST itself).

Continuously varying spatial patterning occurs in many readily available satellite products at a range of spatial and temporal scales. While some surface water patterning is a result of the interactions between hydrodynamics and seafloor topographic features, resulting in spatially and temporally predictable and sometimes persistent biophysical patterns, it is intuitive that the static *patch mosaic* model may not always be appropriate in pelagic systems. This is particularly true for ocean regions characterized by intense (sub)mesoscale variability. This necessitates the consideration of alternative conceptual models that can sufficiently capture the inherent dynamism of pelagic systems.

3.3.2 Surface Model Metrics – Identifying Ecoclines and Ecotones

The input environmental data required for pelagic seascape ecology is mostly obtained from satellite remote sensing or ocean circulation (hydrodynamic) models, as these sources provide spatially continuous and synoptic environmental information that allows for investigation of linkages between oceanographic and ecological processes. Hobday *et al.* (2014) proposed that derived environmental variables, such as metrics of frontal activity, upwelling / downwelling areas or eddy kinetic energy computed from altimetry provide additional information for the analysis of pelagic species ecology. Derived variables can be more ecologically relevant than primary variables for the

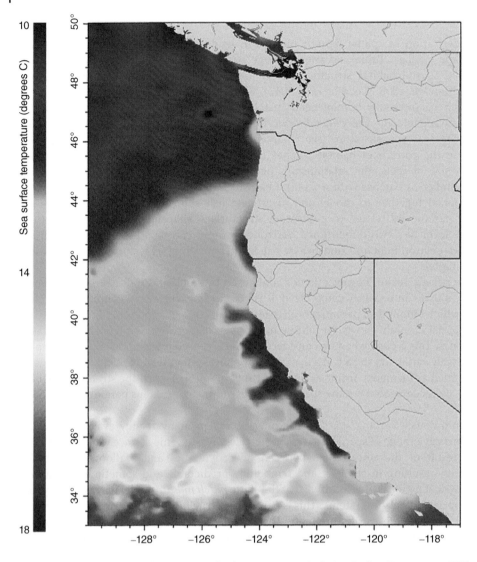

Figure 3.5 Broad-scale spatial structuring of pelagic seascapes. Daily Sea Surface Temperature (SST) image, California Current Large Marine Ecosystem, 1 June 2010. Merged Ultra-high Resolution (MUR) daily SST, generated by the Global High Resolution SST (GHRSST) project and distributed by NOAA Coastwatch http://coastwatch.pfeg.noaa.gov/erddap/griddap/jplMURSST (accessed 29 May 2017).

study of pelagic species (Chassot *et al.* 2011; Hobday & Hartog 2014; Alvarez-Berastegui 2016). Both primary and derived variables are valid pelagic seascape metrics as they express numerically the spatiotemporal dynamics of biophysical processes occurring in the pelagic environment.

Some of the simplest pelagic seascape metrics are those obtained from the calculation of the spatial variance of a hydrographic parameter within a particular area. Spatial gradients calculated directly over SST or chl-*a* imagery have been used in many studies as proxies for frontal areas and correlated with the spatial ecology of various pelagic species (Worm *et al.* 2005; Druon *et al.* 2011; Louzao *et al.* 2011; Mannocci *et al.* 2014; Hidalgo *et al.* 2015). Temporal gradients of primary variables may also contain relevant information about the biophysical dynamics of the pelagic environment. Changes in SST or chl-*a* over a specific period of time may provide appropriate information to parameterize more complex biophysical mechanisms (Druon *et al.* 2011; Alvarez-Berastegui *et al.* 2016).

The spatial scale and temporal lags used for the calculation of seascape metrics are critical to using the surface model, because particular biophysical mechanisms can be identified only when observations are made at appropriate scales. For example, fine-scale nearshore processes may affect recruitment of littoral species (Basterretxea *et al.* 2012). These mechanisms may not be identified if the grain of the input data is too coarse. In other cases, broad frontal zones may drive migration of top predators (Polovina *et al.* 2001; Shillinger *et al.* 2008; Scales *et al.* 2014b) and using a fine-scale analysis may not identify these ocean-basin scale linkages. The relevance of spatial scale in identifying ecological relationships is a critical issue in pelagic seascape ecology (Alvarez-Berastegui *et al.* 2014). Similarly, if time lags are not set appropriately then the mechanisms linking the physical environment with species ecology may not emerge. For example, chl-*a* data from remote sensing provides information about the presence of productive areas, but identification of distinct blooms requires a more refined approach. In the Gulf of Mexico, anomalies of chl-*a* between one day and the mean over the preceding two months identified harmful algal blooms of *Karenia brevis* (Stumpf *et al.* 2003). The time lags used in the computation of anomalies was related to the local enhancement of chl-*a* during observed blooms. An analysis conducted at a different temporal scale would likely have missed these relationships.

More advanced techniques than just computing spatial gradient statistics are available for the identification of frontal areas associated with convergence, divergence, river runoffs or strong eddies. When these processes generate surface ecoclines or ecotones, they may be identified and delineated from satellite data. SST imagery has been the main source of information and various techniques exist to observe these gradients and trace the edges they generate at the surface. The two main approaches to front detection are based on further analysis of spatial gradients of the SST field (Doniol-Valcroze *et al.* 2007) and the frequency distribution of SST within a predefined area (single image edge detection – Cayula & Cornillon 1992). The various techniques available for front detection from SST data propose different solutions to common limitations, such as low signal to noise ratios and cloud cover. More recently, algorithms have been developed that overcome these limitations, facilitate processing and extend the approach to detect fronts in ocean colour fields (Figure 3.3; Belkin & O'Reilly 2009; Miller 2009; Nieto *et al.* 2012).

3.3.3 Lagrangian Approaches

Hydrodynamic models offer a huge number of possibilities to develop metrics characterizing the dynamic nature of the pelagic environment. These sources do not present some of the inherent limitations of remote sensing like data gaps or signal noise and usually provide information at finer spatial and temporal resolution of relevance to individual animal movements (Scales *et al.* 2017). Nevertheless, hydrodynamic models may present bias from *in situ* observation and location of specific oceanographic features and the resolution and accuracy of the data inputs, therefore validation of outputs is an important issue.

Lagrangian approaches to defining pelagic seascape metrics are based upon tracking particles through time as they are moved by fluid motion in hydrodynamic models. This contrasts with the previous approaches where surface SST or chl-*a* data are used as snapshots of biophysical conditions. The application of Lagrangian techniques offers a wide range of possibilities to derive metrics describing dynamic biophysical processes. Satellite altimetry provides information about sea surface height (SSH) from which geostrophic currents can be measured and eddy dynamics tracked (Chelton *et al.* 2011). This information can be quantified using a number of metrics, for example those that describe how geostrophic currents are developing or reveal characteristics of eddies important to biological processes (Bakun 2006). The application of Lyapunov exponents to surface flow fields from satellites or hydrodynamic models allows for additional analysis on the spatial and temporal variability of sea surface currents, providing information such as the position of mesoscale fronts (d'Ovidio *et al.* 2009). Current motion vectors also provide the possibility of modelling how kinetic energy affects larval dispersal and recruitment success (Ruiz *et al.* 2013).

After summarizing the most common metrics applied to describe the dynamic pelagic environment, the main question is how all this information can be applied to the wider concept of pelagic seascape ecology, and how seascape ecology can advance in a way that benefits from advances in landscape ecology. Recent work has expanded the traditional approaches applied in landscape ecology by considering the landscape as a continuous surface instead of a patch mosaic and integrating surface-based metrics with patch mosaic metrics to describe landscape pattern (McGarigal *et al.* 2009). Advances in both seascape and landscape ecology using similar approaches will progress the quantitative analyses of spatial patterning in ecology.

There are still many questions that remain unanswered that will benefit from further research. For example, we have little understanding of how the spatiotemporal scales at which seascapes are modelled affects our ability to link species ecology to oceanographic processes. Furthermore, where multiple metrics are computed, we need to better understand the inter-relationships and redundancy among metrics. These are examples of the information needed to guide metric selection and interpretation, similar to progress made in landscape ecology (McGarigal *et al.* 2009). In all cases, the design and selection of seascape metrics must respond to the parameterization of biophysical processes identified *a priori* on the basis of well-defined hypotheses (Wedding *et al.* 2011). If not, the high number of different metrics available may result in the identification of spurious inference on the ecological relationships between organisms and ocean dynamics.

Spatial proxies or surrogates – measurable variables that can provide information on processes that are unavailable or not directly measurable – can prove useful in pelagic seascape ecology, owing to information gaps in other primary and derived variables (Lindenmayer *et al.* 2015). For example, spatially explicit data describing the distributions of common prey species such as small pelagic fish are often lacking, particularly over scales matching those of animal movement datasets. The use of proxies such as measures of the Finite-Time Lyapunov Exponent (FTLE) – which identifies Lagrangian Coherent Structures such as fronts and eddies (d'Ovidio *et al.* 2009, 2013) – can be informative surrogates for prey distribution, as convergent frontal systems are known to aggregate low trophic level organisms (Le Fevre 1986). Surrogates, however, should be used with care, particularly where mechanistic linkages are unknown (Hazen *et al.* 2013). As outlined above, Lagrangian techniques may be more successful in discovering the mechanisms underlying biological responses to the physical environment than correlative approaches that use the *patch mosaic* approach. Adequate design of pelagic seascape metrics is in itself an objective of seascape ecology and critical to developing ecoinformatics approaches to understanding marine ecosystem functioning.

3.4 Spatial Ecoinformatics in the Pelagic Realm: from Physics to Predators

Ecoinformatics is an emergent discipline focused on the development and application of information technologies relevant to the processing and management of ecological data. Spatial ecoinformatics techniques such as resource selection functions, species distribution modelling, bioclimatic envelope modelling and predictive habitat modelling have revolutionized quantitative ecology in recent years (Matthiopoulos 2003; Elith & Leathwick 2009; Araújo & Peterson 2012). Through application of spatial ecoinformatics to the pelagic realm, pelagic seascape ecology can be considered a 'Big Data' science, particularly where studies operate over broad scales.

The development of pelagic seascape ecology requires an advanced understanding of the mechanisms that underlie the responses of mobile marine organisms to physical conditions. Ecoinformatics provides capacity to quantify and model these mechanisms over a range of spatial and temporal scales. Recent developments in electronic tagging and tracking technologies have revealed many previously cryptic aspects of animal movements at sea (Godley *et al.* 2008; Hazen *et al.* 2012). Combined with advances in remote sensing and marine ecosystem modelling, this has provided unprecedented opportunity to investigate the linkages between the physical environment and animal behaviour in the pelagic realm.

3.4.1 Broad-scale Migrations across Pelagic Seascapes

Many wide-ranging pelagic foragers such as seabirds, cetaceans, pinnipeds, turtles, sharks and large predatory fish undertake migrations over ocean-basin scales through ontogenetic development and annual breeding cycles. These migrations take place over a range of timescales, between areas in which conditions are conducive for certain life history stages or activities such as foraging or breeding. Animals often show fidelity to the same routes and thus these migrations can be predictable in space and time.

For many wide-ranging taxa, the extent and timing of migratory movements are implicitly tied to predictability in the availability of particular conditions or resources within a particular domain. For example, the annual migration of the eastern Pacific grey whale (*Eschrichtius robustus*) between its boreal summer feeding grounds in the Arctic and winter calving grounds off Mexico is one of the most extensive known migrations, yet also highly predictable at the population level (Pike 1962). Comparable broad-scale movements between breeding colonies and spatially predictable overwintering locations are common among seabird populations (Shaffer *et al.* 2006; Guilford *et al.* 2009; Orben *et al.* 2015). The postbreeding dispersal of colonial-breeding seabirds often involves extensive movements across the largely nutrient-poor pelagic seascape, to locate predictably profitable foraging areas (Piatt *et al.* 2006; Weimerskirch 2007).

Identifying preferred habitats of wide-ranging species, and the routes taken between them, is an important step in defining pelagic seascapes owing to the insights such studies can provide regarding the interplay between spatial patterning and ecological processes. From a pelagic seascape ecology perspective, these predictable broad-scale migrations could be informative in defining seascape extent. The paths animals take between preferred habitats can be represented as seascape vertices, or as paths traversing multiple subseascapes or patches within a seascape, depending on the definition of seascape extent. In marine spatial planning, these pathways have been considered as key features termed 'blue corridors' analogous to green corridors in terrestrial conservation planning. In landscape ecology, scaling landscapes based on organism movements or other processes is referred to as an organism-centred or organism-based approach and is typically investigated using environmental variables, quantified at multiple spatial scales (Addicott *et al.* 1987; Wiens & Milne 1989; Pittman & McAlpine 2003; Palumbi 2004). Taking an organism-centred approach is an appropriate corollary to current practice in pelagic seascape ecology.

The geographical ranges of marine species, and movements between breeding and foraging habitats, have been contextualized using an organism-centred approach combined with a patch mosaic model, through determining levels of association with particular biogeographic provinces. Species-specific tracking has revealed the propensity for individual animals to migrate between biogeographic provinces through ontogenetic development or different stages of the annual cycle. For example, electronic tracking of 23 marine species over more than a decade has revealed spatiotemporally predictable associations with the California Current Large Marine Ecosystem (CCLME; Block *et al.* 2011). Similarly, multispecies tracking work in the Southern Ocean has identified regions of particular ecological significance, primarily attributed to accessibility from breeding colonies of penguins, albatrosses and seals (Raymond *et al.* 2015).

Application of the *surface model* – using continuous measures of physical gradients at the sea surface – has revealed the ecological significance of broad-scale gradients in physical properties of the open ocean. Key examples include the North Pacific Transition Zone (NPTZ) including the Transition Zone Chlorophyll Front (TZCF), which is utilized as forage and migration habitat by sea turtles, tuna (Polovina *et al.* 2001, 2015), albatrosses (Kappes *et al.* 2010; Nishizawa *et al.* 2015) and elephant seals (Robinson *et al.* 2012). Similarly, although located in a contrasting ocean region, the Antarctic Polar Frontal Zone (APFZ) is an ecologically significant broad-scale ecocline (Guinet *et al.* 2001; Biuw *et al.* 2007; Bost *et al.* 2009, 2015).

3.4.2 Linking Animal Movements to the Spatial Patterning of Pelagic Seascapes

Over finer spatio-temporal scales, the responses of wide-ranging pelagic foragers to biophysical conditions may be more spontaneous as animals react in real time to the contemporaneous environment. Biophysical processes occurring over (sub-)mesoscales determine the distributions of forage species, such as pelagic fish and squid, and so strongly influence the at-sea movements and behaviour of a wide range of predators (Hazen *et al.* 2013). Over these scales, oceanographic conditions can be highly dynamic and so a *surface model* approach is often more appropriate than a static, *patch mosaic* model. A surface model approach has been used extensively to describe and model the broad to mesoscale habitat preferences of marine predators (*e.g.*, cetaceans, Becker *et al.* 2014; Forney *et al.* 2015; Campbell *et al.* 2015; Correia *et al.* 2015); sea turtles (Benson *et al.* 2007; Hawkes *et al.* 2007; Witt *et al.* 2007; Shillinger *et al.* 2008); and seabirds (Suryan *et al.* 2006; Louzao *et al.* 2011; Nur *et al.* 2011).

A considerable body of work has focused on linking the movements of tagged animals with derived variables to elucidate the influence of sub-mesoscale features such as fronts and eddies on habitat selection. A taxonomically diverse range of marine predators are known to associate with ocean fronts (reviewed in Bost *et al.* 2009; Scales *et al.* 2014b; Acha *et al.* 2015; see also Sims & Quayle 1998; Sims *et al.* 2000; Sims & Southall 2002; Ainley *et al.* 2005; Sabarros *et al.* 2009; Queiroz *et al.* 2012; Lowther *et al.* 2014; Pirotta *et al.* 2014; Scales *et al.* 2014a). Eddies are also now known to be particularly important mesoscale features for foraging marine predators throughout the global ocean (reviewed in Godø *et al.* 2012; see also (Hyrenbach *et al.* 2006; Yen *et al.* 2006; Bailleul *et al.* 2010; Dragon *et al.* 2010; Tew Kai & Marsac 2010; Woodworth *et al.* 2012; Hsu *et al.* 2015).

Satellite altimetry has proven particularly useful for Lagrangian approaches to spatial ecoinformatics. Several progressive studies have found relationships between animal movements and Lagrangian Coherent Structures (LCS; *e.g.*, fronts, filaments, eddies; d'Ovidio *et al.* 2013). For example, Great Frigatebirds (*Fregata minor*) foraging in the Mozambique Channel were found to precisely track LCS to locate prey patches (Tew Kai *et al.* 2009; De Monte *et al.* 2012). Similar techniques have revealed associations between marine mammals, including fin whales, elephant seals, penguins and fur seals and LCS in other pelagic systems (Cotté *et al.* 2007; Cotté *et al.* 2011; d'Ovidio *et al.* 2013; Nordstrom *et al.* 2013; Cotté *et al.* 2015). Similarly, Lagrangian techniques have produced novel insights into the behaviour of southern elephant seals (*Mirounga leonina*), which display quasi-planktonic drifting behaviour during intensive foraging bouts (Della Penna *et al.* 2015).

3.4.3 Incorporating the Vertical Dimension in Spatial Ecoinformatics

The inclusion of the vertical dimension of pelagic seascapes in tracking studies is now generating insight into mechanistic linkages between physical processes and marine predator habitat selection (Jouma'a *et al.* 2016; Scheffer *et al.* 2016). *In situ* studies that simultaneously sample oceanographic conditions, movements and behaviours of detectable predators and characteristics of prey aggregations have contributed much to our understanding of linkages between fine-scale hydrography, forage fish and predator responses (Bertrand *et al.* 2002; Hazen *et al.* 2011; Embling *et al.* 2012; Benoit-Bird *et al.* 2013a; Cox *et al.* 2013; Scott *et al.* 2013). Biophysical properties of the water column have been measured directly using animal-attached sensors (Lawson *et al.* 2015), with corresponding behavioural information simultaneously inferred from three-dimensional movement data (Dewar *et al.* 2011; Hazen *et al.* 2015). Alternatively, ocean models can hindcast likely physical conditions along an animal's movement trajectory, resolving features of the proximate environment in three-dimensions (Woodworth *et al.* 2012; Schick *et al.* 2013). This is a rapidly growing field, with developments now being made in coupled biophysical models including nutrient fluxes, phytoplankton, lower trophic level consumers and, increasingly, marine predators in ecosystem models that use a mechanistic framework to quantify pelagic ecosystem functioning (*e.g.,* ROMS-NEMURO, see Fiechter *et al.* 2015). Sub-surface information is becoming more readily available for integration into studies using spatial ecoinformatics to understand biophysical linkages in the oceans. In turn, pelagic seascape ecology holds much promise for further development of the Integrated Ecosystem Assessment (IEA) approach to marine resource management (Levin *et al.* 2009).

3.5   Conclusions and Future Research Priorities

Developing the pelagic seascape ecology paradigm requires a holistic, interdisciplinary, multi-scale, biogeographic approach to measuring and analysing the properties of oceanic ecosystems. To understand how pelagic ecosystems function, we need to sample the physical and biological properties of the ocean and spatial heterogeneity of these properties, in multiple complementary ways. To do this effectively will require better integration and cross-flow across multiple disciplines including physical oceanography, remote sensing science, ecosystem modelling, fisheries acoustics, fish biology and physiology, fisheries stock assessment science, spatial ecology and species distribution modelling.

A major goal of pelagic seascape ecology is a holistic understanding of the biophysical mechanisms that underlie marine ecosystem functioning. Biological responses to physical conditions occur over a continuum of spatial and temporal scales and with differing spatiotemporal dynamics. Pelagic seascape ecology should seek to identify the relevant scales over which pattern and process are linked, matching animal movements with ecological drivers. This will require more fine-scale synoptic studies that measure multiple physical and biological variables simultaneously, from physical processes through primary and secondary productivity to predator responses (*e.g.,* Embling *et al.* 2012). We also need to understand the life history, ecological niche, habitat preferences and movement ecology of the diverse array of high trophic-level marine vertebrates that inhabit the global ocean (Hays *et al.* 2016). We have much to learn and the emerging fields of biologging (Rutz & Hays 2009) and movement ecology (Nathan *et al.* 2008) hold promise for filling these important knowledge gaps.

Pelagic seascape ecology should seek to provide new concepts and tools to improve spatial ecoinformatics in the oceans. We now have the opportunity to integrate oceanographic dynamics and sub-surface processes into spatial pattern metrics, using the *surface model*, Lagrangian approaches and end-to-end ecosystem models to move towards a mechanistic understanding of biophysical linkages in the ocean. Increasing spatial and temporal resolution in remotely-sensed and modelled environmental data fields is likely to contribute significantly to this process, although a major challenge remains in aligning the spatial and temporal scales over which data are available.

Advances in the use of satellite remote sensing in terrestrial ecology hold promise for pelagic seascape ecology, particularly in modelling the responses of mobile animals to physical conditions (He *et al.* 2015). Developing pelagic seascape metrics is important in spatial ecoinformatics when linking spatial patterning to oceanic ecosystem function and this forms a novel and progressive research arena. The environmental data from which these seascape metrics can be derived are freely available to all via data portals that collate and serve satellite remote sensing data to users (*e.g.*, NOAA CoastWatch, http://coastwatch.noaa.gov/cw_html/index.html, accessed 29 May 2017), allowing for the development of pelagic seascape metrics for regional applications (Kavanaugh *et al.* 2016).

Pelagic seascape ecology has great potential to enhance our capacity to implement ecosystem-based management in the open oceans (Levin *et al.* 2009). Spatial ecoinformatics applied to the pelagic realm can facilitate near real-time management of anthropogenic threat to populations of conservation concern, such as fisheries bycatch (Hobday *et al.* 2013; Lewison *et al.* 2015; Maxwell *et al.* 2015). Monitoring of the marine environment using near-real time environmental data from satellites and models could enable the establishment of dynamic marine protected areas (Wedding *et al.* 2013; Dunn *et al.* 2016) and systematic conservation planning in the high seas, where many protected species spend the majority of their lives and are most at risk (Ban *et al.* 2014).

Finally, pelagic seascape ecology could increase our capacity to understand and predict change in oceanic systems as global climate change intensifies ('ecological forecasting'), particularly where spatial ecoinformatics is coupled with global climate models (Delworth *et al.* 2012). As pelagic seascapes shift and change, and as marine organisms respond to thermal stress and changes in circulation, novel species interactions, ocean acidification and hypoxia (Poloczanska *et al.* 2013), we must urgently develop the tools to detect, understand, monitor and manage these changes.

3.6 Glossary

ecotone   a region of transition between two ecological communities

ecocline   a cline from one ecosystem to another, with a continuous gradient between the two extremes

ecotope   the smallest ecologically distinct landscape feature in a landscape mapping and classification system

ecoinformatics   the science of information in Ecology and Environmental Science, which integrated environmental and information sciences to define entities and natural processes with language common to both humans and computers

Lagrangian   an approach to measurement of physical and biological properties of the ocean that directly incorporates measures of particle flow and fluid dynamics

patch mosaic   an approach to defining spatial heterogeneity within an ecosystem based on the concept that an area of an ecosystem is made up of a mosaic of spatially coherent patches

pelagic   of or relating to the open ocean

pelagic seascape   the three dimensional configuration of the pelagic realm within a specific area or volume, at a specific spatial and temporal scale

pelagic seascape ecology   an ecological discipline focused on understanding mechanistic linkages between the three-dimensional physical structure of pelagic seascapes and ecosystem functioning

pelagic seascape metric   a spatially and temporally explicit, quantitative measure of the physical or biological structure of a pelagic ecosystem

surface model   an approach to defining spatial heterogeneity within an ecosystem that uses continuous measurements of gradients in physical and biological properties rather than defining static, spatially coherent patches

References

Acha EM, Mianzan HW, Guerrero RA, Favero M, Bava J (2004) Marine fronts at the continental shelves of austral South America: physical and ecological processes. Journal of Marine Systems 44: 83–105.

Acha EM, Piola A, Iribarne O, Mianzan H (2015) Ecological Processes at Marine Fronts: Oases in the Ocean. Springer, London.

Addicott JF, Aho JM, Antolin MF, Padilla DK, Richardson JS, Soluk DA (1987) Ecological neighbourhoods: scaling environmental patterns. Oikos: 340–346.

Aiken J, Rees N, Hooker S, Holligan P, Bale A, Robins D, Moore G, Harris R, Pilgrim D (2000) The Atlantic Meridional Transect: overview and synthesis of data. Progress in Oceanography 45: 257–312.

Ainley DG, Spear LB, Tynan CT, Barth JA, Pierce SD, Ford RG, Cowles TJ (2005) Physical and biological variables affecting seabird distributions during the upwelling season of the northern California Current. Deep Sea Research Part II: Topical Studies in Oceanography 52: 123–143.

Alvarez-Berastegui D, Ciannelli L, Aparicio-Gonzalez A, Reglero P, Hidalgo M, López-Jurado JL, Tintoré J, Alemany F (2014) Spatial scale, means and gradients of hydrographic variables define pelagic seascapes of bluefin and bullet tuna spawning distribution. PloS One 9: e109338.

Alvarez-Berastegui D, Hidalgo M, Tugores MP, Reglero P, Aparicio-González A, Ciannelli L, Juza M, Mourre B, Pascual A, López-Jurado JL, García A (2016) Pelagic seascape ecology for operational fisheries oceanography: modelling and predicting spawning distribution of Atlantic bluefin tuna in western Mediterranean. ICES Journal of Marine Science: Journal du Conseil 73(7): 1851–1862.

Araújo MB, Peterson AT (2012) Uses and misuses of bioclimatic envelope modelling. Ecology 93: 1527–1539.

Attrill M, Rundle S (2002) Ecotone or ecocline: ecological boundaries in estuaries. Estuarine, Coastal and Shelf Science 55: 929–936.

Bailleul F, Cotté C, Guinet C (2010) Mesoscale eddies as foraging area of a deep-diving predator, the southern elephant seal. Marine Ecology Progress Series 408: 251–264.

Bainbridge R (1957) The size, shape and density of marine phytoplankton concentrations. Biological Reviews 32(1): 91–115.

Bakun A (1996) Patterns in the Ocean: Ocean Processes and Marine Population Dynamics. California Sea Grant, in cooperation with Centro de Investigaciones Biologicas del Noroeste, La Paz, Mexico.

Bakun A (2006) Fronts and eddies as key structures in the habitat of marine fish larva: opportunity, adaptive response and competitive advantage. Scientia Marina 70: 105–122.

Ban NC, Bax NJ, Gjerde KM, Devillers R, Dunn DC, Dunstan PK, Hobday AJ, Maxwell SM, Kaplan DM, Pressey RL (2014) Systematic conservation planning: a better recipe for managing the high seas for biodiversity conservation and sustainable use. Conservation Letters 7: 41–54.

Barton AD, Pershing AJ, Litchman E, Record NR, Edwards KF, Finkel ZV, Kiørboe T, Ward BA (2013) The biogeography of marine plankton traits. Ecology Letters 16: 522–534.

Basterretxea G, Jordi A, Catalan IA, Sabates A (2012) Model-based assessment of local-scale fish larval connectivity in a network of marine protected areas. Fisheries Oceanography 21: 291–306.

Becker EA, Forney KA, Foley DG, Smith RC, Moore TJ, Barlow J (2014) Predicting seasonal density patterns of California cetaceans based on habitat models. Endangered Species Research 23: 1–22.

Belkin IM, O'Reilly JE (2009) An algorithm for oceanic front detection in chlorophyll and SST satellite imagery. Journal of Marine Systems 78: 319–326.

Benoit-Bird KJ (2009) Dynamic 3-dimensional structure of thin zooplankton layers is impacted by foraging fish. Marine Ecology Progress Series 396: 61–76.

Benoit-Bird KJ, Battaile BC, Heppell SA, Hoover B, Irons D, Jones N, Kuletz KJ, Nordstrom CA, Paredes R, Suryan RM (2013a) Prey patch patterns predict habitat use by top marine predators with diverse foraging strategies. PLoS One 8: e53348.

Benoit-Bird KJ, Cowles TJ, Wingard CE (2009) Edge gradients provide evidence of ecological interactions in planktonic thin layers. Limnology and Oceanography 54: 1382–1392.

Benoit-Bird KJ, McManus MA (2012) Bottom-up regulation of a pelagic community through spatial aggregations. Biology Letters 8: 813–816.

Benoit-Bird KJ, Shroyer EL, McManus MA (2013b) A critical scale in plankton aggregations across coastal ecosystems. Geophysical Research Letters 40: 3968–3974.

Benson SR, Forney KA, Harvey JT, Carretta JV, Dutton PH (2007) Abundance, distribution and habitat of leatherback turtles (*Dermochelys coriacea*) off California, 1990–2003. Fisheries Bulletin 105: 337–347.

Bertrand, A., Josse, E., Bach, P., Gros, P. & Dagorn, L. (2002) Hydrological and trophic characteristics of tuna habitat: consequences on tuna distribution and longline catchability. Canadian Journal of Fisheries and Aquatic Sciences 59: 1002–1013.

Biuw M, Boehme L, Guinet C, Hindell M, Costa D, Charrassin J-B, Roquet F, Bailleul F, Meredith M, Thorpe S (2007) Variations in behaviour and condition of a Southern Ocean top predator in relation to in situ oceanographic conditions. Proceedings of the National Academy of Sciences 104: 13705–13710.

Block BA, Jonsen I, Jorgensen S, Winship A, Shaffer SA, Bograd S, Hazen E, Foley D, Breed G, Harrison A-L (2011) Tracking apex marine predator movements in a dynamic ocean. Nature 475: 86–90.

Bograd SJ, Castro CG, Di Lorenzo E, Palacios DM, Bailey H, Gilly W, Chavez FP (2008) Oxygen declines and the shoaling of the hypoxic boundary in the California Current. Geophysical Research Letters 35(12).

Bograd SJ, Checkley DA, Wooster WS (2003) CalCOFI: a half century of physical, chemical, and biological research in the California Current System. Deep Sea Research Part II: Topical Studies in Oceanography 50: 2349–2353.

Bograd SJ, Foley DG, Schwing FB, Wilson C, Laurs RM, Polovina JJ, Howell EA, Brainard RE (2004) On the seasonal and interannual migrations of the transition zone chlorophyll front. Geophysical Research Letters 31: L17204.

Bost C-A, Cotté C, Bailleul F, Cherel Y, Charrassin J-B, Guinet C, Ainley DG, Weimerskirch H (2009) The importance of oceanographic fronts to marine birds and mammals of the southern oceans. Journal of Marine Systems 78: 363–376.

Bost CA, Cotte C, Terray P, Barbraud C, Bon C, Delord K, Gimenez O, Handrich Y, Naito Y, Guinet C, Weimerskirch H (2015) Large-scale climatic anomalies affect marine predator foraging behaviour and demography. Nature Communications 6: 8220.

Briggs JC, Bowen BW (2012) A realignment of marine biogeographic provinces with particular reference to fish distributions. Journal of Biogeography 39: 12–30.

Campbell GS, Thomas L, Whitaker K, Douglas AB, Calambokidis J, Hildebrand JA (2015) Inter-annual and seasonal trends in cetacean distribution, density and abundance off southern California. Deep Sea Research Part II: Topical Studies in Oceanography 112: 143–157.

Cayula J-F, Cornillon P (1992) Edge detection algorithm for SST images. Journal of Atmospheric and Oceanic Technology 9: 67–80.

Chassot E, Bonhommeau S, Reygondeau G, Nieto K, Polovina JJ, Huret M, Dulvy NK, Demarcq H (2011) Satellite remote sensing for an ecosystem approach to fisheries management. ICES Journal of Marine Science 68(4): 651–666.

Chelton DB, Schlax MG, Samelson RM (2011) Global observations of nonlinear mesoscale eddies. Progress in Oceanography 91: 167–216.

Correia AM, Tepsich P, Rosso M, Caldeira R, Sousa-Pinto I (2015) Cetacean occurrence and spatial distribution: Habitat modelling for offshore waters in the Portuguese EEZ (NE Atlantic). Journal of Marine Systems 143: 73–85.

Cotté C, d'Ovidio F, Chaigneau A, Lévy M, Taupier-Letage I, Mate B, Guinet C (2011) Scale-dependent interactions of Mediterranean whales with marine dynamics. Limnology and Oceanography 56: 219–232.

Cotté C, d'Ovidio F, Dragon A-C, Guinet C, Lévy M (2015) Flexible preference of southern elephant seals for distinct mesoscale features within the Antarctic Circumpolar Current. Progress in Oceanography 131: 46–58.

Cotté C, Park Y-H, Guinet C, Bost C-A (2007) Movements of foraging king penguins through marine mesoscale eddies. Proceedings of the Royal Society of London B: Biological Sciences 274: 2385–2391.

Cox S, Scott B, Camphuysen C (2013) Combined spatial and tidal processes identify links between pelagic prey species and seabirds. Marine Ecology Progress Series 479: 203–221.

Cushman SA, McGarigal K, Neel MC (2008) Parsimony in landscape metrics: strength, universality, and consistency. Ecological Indicators 8: 691–703.

Della Penna A, De Monte S, Kestenare E, Guinet C, d'Ovidio F (2015) Quasi-planktonic behaviour of foraging top marine predators. Scientific Reports 5.

Delworth TL, Rosati A, Anderson W, Adcroft AJ, Balaji V, Benson R, Dixon K, Griffies SM, Lee H-C, Pacanowski RC (2012) Simulated climate and climate change in the GFDL CM2. 5 high-resolution coupled climate model. Journal of Climate 25: 2755–2781.

De Monte S, Cotté C, d'Ovidio F, Lévy M, Le Corre M, Weimerskirch H (2012) Frigatebird behaviour at the ocean–atmosphere interface: integrating animal behaviour with multi-satellite data. Journal of The Royal Society Interface 9: 3351–3358.

Deutsch C, Brix H, Ito T, Frenzel H, Thompson L (2011) Climate-forced variability of ocean hypoxia. Science 333: 336–339.

Devred E, Sathyendranath S, Platt T (2007) Delineation of ecological provinces using ocean colour radiometry. Marine Ecology Progress Series 346: 1–13.

Devred E, Sathyendranath S, Platt T (2009) Decadal changes in ecological provinces of the Northwest Atlantic Ocean revealed by satellite observations. Geophysical Research Letters 36(19): L19607.

Dewar H, Prince ED, Musyl MK, Brill RW, Sepulveda C, Luo J, Foley D, Orbesen ES, Domeier ML, Nasby-Lucas N (2011) Movements and behaviours of swordfish in the Atlantic and Pacific Oceans examined using pop-up satellite archival tags. Fisheries Oceanography 20: 219–241.

Dickey TD (2003) Emerging ocean observations for interdisciplinary data assimilation systems. Journal of Marine Systems 40–41: 5–48.

Doniol-Valcroze T, Berteaux D, Larouche P, Sears R (2007) Influence of thermal fronts on habitat selection by four rorqual whale species in the Gulf of St Lawrence. Marine Ecology Progress Series 335: 207–216.

d'Ovidio F, De Monte S, Della Penna A, Cotté C, Guinet C (2013) Ecological implications of eddy retention in the open ocean: a Lagrangian approach. Journal of Physics A: Mathematical and Theoretical 46: 254023.

d'Ovidio F, Isern-Fontanet J., López C, Hernández-García E, García-Ladona E (2009) Comparison between Eulerian diagnostics and finite-size Lyapunov exponents computed from altimetry in the Algerian basin. Deep Sea Research Part I: Oceanographic Research Papers 56: 15–31.

Dragon A-C, Monestiez P, Bar-Hen A, Guinet C (2010) Linking foraging behaviour to physical oceanographic structures: southern elephant seals and mesoscale eddies east of Kerguelen Islands. Progress in Oceanography 87: 61–71.

Druon J-N, Fromentin J-M, Aulanier F, Heikkonen J (2011) Potential feeding and spawning habitats of Atlantic bluefin tuna in the Mediterranean Sea. Marine Ecology Progress Series 439: 223–240.

Dunn DC, Maxwell SM, Boustany AM, Halpin PN (2016) Dynamic ocean management increases the efficiency and efficacy of fisheries management. Proceedings of the National Academy of Sciences 113(3): 668–673.

Elith J, Leathwick JR (2009) Species distribution models: ecological explanation and prediction across space and time. Annual Review of Ecology, Evolution, and Systematics 40: 677.

Embling CB, Illian J, Armstrong E, van der Kooij J, Sharples J, Camphuysen KC, Scott BE (2012) Investigating fine-scale spatio-temporal predator–prey patterns in dynamic marine ecosystems: a functional data analysis approach. Journal of Applied Ecology 49: 481–492.

Fedorov KN (1986) The Physical Nature and Structure of Oceanic Fronts. United States: Springer Verlag, New York, NY.

Fiechter J, Huff D, Martin B, Jackson D, Edwards C, Rose K, Curchitser E, Hedstrom K, Lindley S, Wells B (2015) Environmental conditions impacting juvenile Chinook salmon growth off central California: An ecosystem model analysis. Geophysical Research Letters 42: 2910–2917.

Forman RT, Godron M (1981) Patches and structural components for a landscape ecology. BioScience 31: 733–740.

Forney KA, Becker EA, Foley DG, Barlow J, Oleson EM (2015) Habitat-based models of cetacean density and distribution in the central North Pacific. Endangered Species Research 27: 1–20.

Franks PJ (1992) Phytoplankton blooms at fronts: patterns, scales, and physical forcing mechanisms. Reviews in Aquatic Sciences 6: 121–137.

Franks PJ, Di Lorenzo E, Goebel NL, Chenillat F, Rivière P, Edward CA, Miller AJ (2013) Modelling physical-biological responses to climate change in the California Current System. Oceanography 26: 26–33.

Godley B, Blumenthal J, Broderick A, Coyne M, Godfrey M, Hawkes L, Witt M (2008) Satellite tracking of sea turtles: Where have we been and where do we go next. Endangered Species Research 4: 3–22.

Godø OR, Samuelsen A, Macaulay GJ, Patel R, Hjøllo SS, Horne J, Kaartvedt S, Johannessen JA (2012) Mesoscale eddies are oases for higher trophic marine life. PLoS One 7(1): e30161.

Grantham BA, Chan F, Nielsen KJ, Fox DS, Barth JA, Huyer A, Lubchenco J, Menge BA (2004) Upwelling-driven nearshore hypoxia signals ecosystem and oceanographic changes in the northeast Pacific. Nature 429: 749–754.

Greer AT, Cowen RK, Guigand CM, McManus MA, Sevadjian JC, Timmerman AH (2013) Relationships between phytoplankton thin layers and the fine-scale vertical distributions of two trophic levels of zooplankton. Journal of Plankton Research 35: 939–956.

Guilford T, Meade J, Willis J, Phillips RA, Boyle D, Roberts S, Collett M, Freeman R, Perrins C (2009) Migration and stopover in a small pelagic seabird, the Manx shearwater Puffinus puffinus: insights from machine learning. Proceedings of the Royal Society of London B: Biological Sciences 276: 1215–1223.

Guinet C, Dubroca L, Lea MA, Goldsworthy S, Cherel Y, Duhamel G, Bonadonna F, Donnay J-P (2001) Spatial distribution of foraging in female Antarctic fur seals Arctocephalus gazella in relation to oceanographic variables: a scale-dependent approach using geographic information systems. Marine Ecology Progress Series 219: 251–264.

Hardy AC (1936) Observations on the uneven distribution of oceanic plankton. Discovery Rept. 11, 511–538.

Haury L, Fey C, Newland C, Genin A (2000) Zooplankton distribution around four eastern North Pacific seamounts. Progress in Oceanography 45(1): 69–105.

Hawkes LA, Broderick AC, Coyne MS, Godfrey MH, Godley BJ (2007) Only some like it hot-quantifying the environmental niche of the loggerhead sea turtle. Diversity and Distributions 13: 447–457.

Hays GC, Ferreira LC, Sequeira AM, Meekan MG, Duarte CM, Bailey H, Bailleul F, Bowen WD, Caley MJ, Costa DP (2016) Key questions in marine megafauna movement ecology. Trends in Ecology & Evolution 31(6): 463–475.

Hazen EL, Friedlaender AS, Goldbogen JA (2015) Blue whales (*Balaenoptera musculus*) optimize foraging efficiency by balancing oxygen use and energy gain as a function of prey density. Science Advances 1: e1500469.

Hazen EL, Maxwell SM, Bailey H, Bograd SJ, Hamann M, Gaspar P, Godley BJ, Shillinger GL (2012) Ontogeny in marine tagging and tracking science: technologies and data gaps. Marine Ecology Progress Series 457: 221–240.

Hazen EL, Nowacek DP, St Laurent L, Halpin PN, Moretti DJ (2011) The relationship among oceanography, prey fields, and beaked whale foraging habitat in the Tongue of the Ocean. PloS One 6: e19269.

Hazen EL, Suryan RM, Santora JA, Bograd SJ, Watanuki Y, Wilson RP (2013) Scales and mechanisms of marine hotspot formation. Marine Ecology Progress Series 487: 177–183.

He KS, Bradley BA, Cord AF, Rocchini D, Tuanmu MN, Schmidtlein S, Turner W, Wegmann M, Pettorelli N (2015) Will remote sensing shape the next generation of species distribution models? Remote Sensing in Ecology and Conservation 1: 4–18.

Hidalgo M, Reglero P, Alvarez-Berastegui D, Torres AP, Alvarez I, Rodriguez JM, Carbonell A, Balbin R, Alemany F (2015) Hidden persistence of salinity and productivity gradients shaping pelagic diversity in highly dynamic marine ecosystems. Marine Environmental Research 104: 47–50.

Hinchey EK, Nicholson MC, Zajac RN, Irlandi EA (2008) Preface: marine and coastal applications in landscape ecology. Landscape Ecology 23: 1–5.

Hobday AJ, Hartog JR (2014) Derived ocean features for dynamic ocean management. Oceanography 27: 134–145.

Hobday AJ, Maxwell SM, Forgie J, McDonald J (2014) Dynamic ocean management: integrating scientific and technological capacity with law, policy, and management. Stanford Environmental Law Journal 33(2): 125–165.

Hsu AC, Boustany AM, Roberts JJ, Chang JH, Halpin PN (2015) Tuna and swordfish catch in the US northwest Atlantic longline fishery in relation to mesoscale eddies. Fisheries Oceanography 24: 508–520.

Hussey NE, Kessel ST, Aarestrup K, Cooke SJ, Cowley PD, Fisk AT, Harcourt RG, Holland KN, Iverson SJ, Kocik JF (2015) Aquatic animal telemetry: A panoramic window into the underwater world. Science 348: 1255642.

Hyrenbach KD, Forney KA, Dayton PK (2000) Marine protected areas and ocean basin management. Aquatic conservation: Marine and Freshwater Ecosystems 10(6): 437–458.

Hyrenbach K, Veit RR, Weimerskirch H, Hunt GL (2006) Seabird associations with mesoscale eddies: the subtropical Indian Ocean. Marine Ecology Progress Series 324: 271–279.

Irigoien X, Huisman J, Harris RP (2004) Global biodiversity patterns of marine phytoplankton and zooplankton. Nature 429: 863–867.

Irwin AJ, Oliver MJ (2009) Are ocean deserts getting larger? Geophysical Research Letters 36: L18609.

Jelinski DE (2014) On a landscape ecology of a harlequin environment: the marine landscape. Landscape Ecology 30: 1–6.

Jouma'a J, Le Bras Y, Richard G, Vacquié-Garcia J, Picard B, El Ksabi N, Guinet C (2016) Adjustment of diving behaviour with prey encounters and body condition in a deep diving predator: the southern elephant seal. Functional Ecology 30: 636–648.

Kalvelage T, Lavik G, Lam P, Contreras S, Arteaga L, Löscher CR, Oschlies A, Paulmier A, Stramma L, Kuypers MM (2013) Nitrogen cycling driven by organic matter export in the South Pacific oxygen minimum zone. Nature Geoscience 6: 228–234.

Kappes MA, Shaffer SA, Tremblay Y, Foley DG, Palacios DM, Robinson PW, Bograd SJ, Costa DP (2010) Hawaiian albatrosses track interannual variability of marine habitats in the North Pacific. Progress in Oceanography 86: 246–260.

Kavanaugh MT, Hales B, Saraceno M, Spitz YH, White AE, Letelier RM (2014) Hierarchical and dynamic seascapes: A quantitative framework for scaling pelagic biogeochemistry and ecology. Progress in Oceanography 120: 291–304.

Kavanaugh MT, Oliver MJ, Chavez FP, Letelier RM, Muller-Karger FE, Doney SC (2016) Seascapes as a new vernacular for pelagic ocean monitoring, management and conservation. ICES Journal of Marine Science 73(7): 1839–1850.

Kobara S, Heyman WD, Pittman SJ, Nemeth RS (2013) Biogeography of transient reef-fish spawning aggregations in the Caribbean: a synthesis for future research and management. Oceanography and Marine Biology: An Annual Review 51: 281–326.

Lawson GL, Hückstädt LA, Lavery AC, Jaffré FM, Wiebe PH, Fincke JR, Crocker DE, Costa DP (2015) Development of an animal-borne 'sonar tag' for quantifying prey availability: test deployments on northern elephant seals. Animal Biotelemetry 3: 1.

Le Fevre J (1986) Aspects of the biology of frontal systems. Advances in Marine Biology 23: 163–299.

Letessier T, Pond DW, McGill RA, Reid WD, Brierley AS (2012) Trophic interaction of invertebrate zooplankton on either side of the Charlie Gibbs Fracture Zone/Subpolar Front of the Mid-Atlantic Ridge. Journal of Marine Systems 94: 174–184.

Levin, PS, Fogarty, MJ, Murawski, SA, & Fluharty, D (2009). Integrated ecosystem assessments: developing the scientific basis for ecosystem-based management of the ocean. PLoS Biology 7(1): e1000014.

Lewison R, Hobday AJ, Maxwell S, Hazen E, Hartog JR, Dunn DC, Briscoe D, Fossette S, O'Keefe CE, Barnes M, Abecassis M, Bograd S, Bethoney ND, Bailey H, Wiley D, Andrews S, Hazen L, Crowder LB (2015) Dynamic ocean management: identifying the critical ingredients of dynamic approaches to ocean resource management. BioScience 65: 486–498.

Lindenmayer D, Pierson J, Barton P, Beger M, Branquinho C, Calhoun A, Caro T, Greig H, Gross J, Heino J (2015) A new framework for selecting environmental surrogates. Science of the Total Environment 538: 1029–1038.

Longhurst A, Sathyendranath S, Platt T, Caverhill C (1995) An estimate of global primary production in the ocean from satellite radiometer data. Journal of Plankton Research 17: 1245–1271.

Louzao M, Pinaud D, Péron C, Delord K, Wiegand T, Weimerskirch H (2011) Conserving pelagic habitats: seascape modelling of an oceanic top predator. Journal of Applied Ecology 48: 121–132.

Lowther AD, Lydersen C, Biuw M, De Bruyn P, Hofmeyr GJ, Kovacs KM (2014) Post-breeding at-sea movements of three central-place foragers in relation to submesoscale fronts in the Southern Ocean around Bouvetøya. Antarctic Science 26: 533–544.

Macías D, Stips A, Garcia-Gorriz E (2014) The relevance of deep chlorophyll maximum in the open Mediterranean Sea evaluated through 3D hydrodynamic-biogeochemical coupled simulations. Ecological Modelling 281: 26–37.

Mahadevan A, D'Asaro E, Lee C, Perry MJ (2012) Eddy-driven stratification initiates North Atlantic spring phytoplankton blooms. Science 337: 54–58.

Manderson, JP (2016) Seascapes are not landscapes: an analysis performed using Bernhard Riemann's rules. ICES Journal of Marine Science 73(7): 1831–1838.

Mannocci L, Catalogna M, Dorémus G, Laran S, Lehodey P, Massart W, Monestiez P, Van Canneyt O, Watremez P, Ridoux V (2014) Predicting cetacean and seabird habitats across a productivity gradient in the South Pacific gyre. Progress in Oceanography 120: 383–398.

Marra J, Houghton R, Garside C (1990) Phytoplankton growth at the shelf-break front in the Middle Atlantic Bight. Journal of Marine Research 48: 851–868.

Martin AP (2003) Phytoplankton patchiness: the role of lateral stirring and mixing. Progress in Oceanography 57(2): 125–174.

Martin A (2005) The kaleidoscope ocean. Philosophical Transactions A Mathematics Physics Engineering Science 363: 2873–2890.

Matthiopoulos J (2003) The use of space by animals as a function of accessibility and preference. Ecological Modelling 159: 239–268.

Maxwell SM, Hazen EL, Lewison RL, Dunn DC, Bailey H, Bograd SJ, Briscoe DK, Fossette S, Hobday AJ, Bennett M (2015) Dynamic ocean management: Defining and conceptualizing real-time management of the ocean. Marine Policy 58: 42–50.

McGarigal K, Cushman S, Ene E (2012) FRAGSTATS v4: spatial pattern analysis program for categorical and continuous maps. University of Massachusetts, Amherst.

McGarigal K, Tagil S, Cushman SA (2009) Surface metrics: an alternative to patch metrics for the quantification of landscape structure. Landscape Ecology 24: 433–450.

Miller P (2009) Composite front maps for improved visibility of dynamic sea-surface features on cloudy SeaWiFS and AVHRR data. Journal of Marine Systems 78: 327–336.

Miller PI, Xu W, Carruthers M (2015) Seasonal shelf-sea front mapping using satellite ocean colour and temperature to support development of a marine protected area network. Deep Sea Research Part II: Topical Studies in Oceanography 119: 3–19.

Moore AM, Arango HG, Broquet G, Edwards C, Veneziani M, Powell B, Foley D, Doyle JD, Costa D, Robinson P (2011) The Regional Ocean Modelling System (ROMS) 4-dimensional variational data assimilation systems: Part II–Performance and application to the California Current System. Progress in Oceanography 91: 50–73.

Morato T, Hoyle SD, Allain V, Nicol SJ (2010) Seamounts are hotspots of pelagic biodiversity in the open ocean. Proceedings of the National Academy of Sciences 107: 9707–9711.

Morato T, Miller PI, Dunn DC, Nicol SJ, Bowcott J, Halpin PN (2015) A perspective on the importance of oceanic fronts in promoting aggregation of visitors to seamounts. Fish and Fisheries doi: 10.1111/faf.12126.

Muller-Karger FE, Kavanaugh MT, Montes E, Balch WM, Breitbart M, Chavez FP, Doney SC, Johns EM, Letelier RM, Lomas MW (2014) A framework for a marine biodiversity observing network within changing continental shelf seascapes. Oceanography 27: 18–23.

Nathan R, Getz WM, Revilla E, Holyoak M, Kadmon R, Saltz D, Smouse PE (2008) A movement ecology paradigm for unifying organismal movement research. Proceedings of the National Academy of Sciences 105: 19052–19059.

Nieto K, Demarcq H, McClatchie S (2012) Mesoscale frontal structures in the canary upwelling system: New front and filament detection algorithms applied to spatial and temporal patterns. Remote Sensing of Environment 123: 339–346.

Nishizawa B, Ochi D, Minami H, Yokawa K, Saitoh S-I, Watanuki Y (2015) Habitats of two albatross species during the non-breeding season in the North Pacific Transition Zone. Marine Biology 162: 743–752.

Nordstrom CA, Battaile BC, Cotte C, Trites AW (2013) Foraging habitats of lactating northern fur seals are structured by thermocline depths and submesoscale fronts in the eastern Bering Sea. Deep Sea Research Part II: Topical Studies in Oceanography 88: 78–96.

Nur N, Jahncke J, Herzog MP, Howar J, Hyrenbach KD, Zamon JE, Ainley DG, Wiens JA, Morgan KH, Ballance LT, Stralberg D (2011) Where the wild things are: predicting hotspots of seabird aggregations in the California Current System. Ecological Applications 21: 2241–2257.

Oliver MJ, Irwin AJ (2008) Objective global ocean biogeographic provinces. Geophysical Research Letters 35(15): L15601.

Orben RA, Paredes R, Roby DD, Irons DB, Shaffer SA (2015) Wintering North Pacific black-legged kittiwakes balance spatial flexibility and consistency. Movement Ecology 3: 1–14.

Oschlies A (2002) Can eddies make ocean deserts bloom? Global Biogeochemical Cycles 16(4): 53-1–53-11.

Paine RT, Levin SA (1981) Intertidal landscapes: disturbance and the dynamics of pattern. Ecological Monographs 51: 145–178.

Palumbi SR (2004) Marine reserves and ocean neighbourhoods: the spatial scale of marine populations and their management. Annual Review Environmental Resources 29: 31–68.

Pelletier L, Kato A, Chiaradia A, Ropert-Coudert Y (2012) Can thermoclines be a cue to prey distribution for marine top predators? A case study with little penguins. PLoS One 7(4): e31768.

Piatt JF, Wetzel J, Bell K, DeGange AR, Balogh GR, Drew GS, Geernaert T, Ladd C, Byrd GV (2006) Predictable hotspots and foraging habitat of the endangered short-tailed albatross (*Phoebastria albatrus*) in the North Pacific: Implications for conservation. Deep Sea Research Part II: Topical Studies in Oceanography 53: 387–398.

Pike GC (1962) Migration and feeding of the gray whale (*Eschrichtius gibbosus*). Journal of the Fisheries Board of Canada 19: 815–838.

Pirotta E, Thompson PM, Miller PI, Brookes KL, Cheney B, Barton TR, Graham IM, Lusseau D, Costa D (2014) Scale-dependent foraging ecology of a marine top predator modelled using passive acoustic data. Functional Ecology 28: 206–217.

Pittman SJ, Kneib RT, Simenstad CA (2011) Practicing coastal seascape ecology. Marine Ecology Progress Series 427: 187–190.

Pittman S, McAlpine C (2003) Movements of marine fish and decapod crustaceans: process, theory and application. Advances in Marine Biology 44: 205–294.

Platt T, Sathyendranath S (1999) Spatial structure of pelagic ecosystem processes in the global ocean. Ecosystems 2: 384–394.

Poloczanska ES, Brown CJ, Sydeman WJ, Kiessling W, Schoeman DS, Moore PJ, Brander K, Bruno JF, Buckley LB, Burrows MT (2013) Global imprint of climate change on marine life. Nature Climate Change 3: 919–925.

Polovina JJ, Howell E, Kobayashi DR, Seki MP (2001) The transition zone chlorophyll front, a dynamic global feature defining migration and forage habitat for marine resources. Progress in Oceanography 49: 469–483.

Polovina JJ, Howell EA, Kobayashi DR, Seki MP (2015) The Transition Zone Chlorophyll Front updated: Advances from a decade of research. Progress in Oceanography. Online doi: 10.1016/j.pocean.20105.01.006.

Powell JR, Ohman MD (2015) Covariability of zooplankton gradients with glider-detected density fronts in the Southern California Current System. Deep Sea Research Part II: Topical Studies in Oceanography 112: 79–90.

Powell TM, Okubo A (1994) Turbulence, diffusion and patchiness in the sea. Philosophical Transactions of the Royal Society of London B: Biological Sciences 343(1303): 11–18.

Queiroz N, Humphries NE, Noble LR, Santos AM, Sims DW (2012) Spatial dynamics and expanded vertical niche of blue sharks in oceanographic fronts reveal habitat targets for conservation. PLoS One 7: e32374.

Raymond B, Lea MA, Patterson T, Andrews-Goff V, Sharples R, Charrassin JB, Cottin M, Emmerson L, Gales N, Gales R (2015) Important marine habitat off east Antarctica revealed by two decades of multi-species predator tracking. Ecography 38: 121–129.

Reid P, Colebrook J, Matthews J, Aiken J, Team CPR (2003) The Continuous Plankton Recorder: concepts and history, from plankton indicator to undulating recorders. Progress in Oceanography 58: 117–173.

Reygondeau G, Longhurst A, Martinez E, Beaugrand G, Antoine D, Maury O (2013) Dynamic biogeochemical provinces in the global ocean. Global Biogeochemical Cycles 27: 1046–1058.

Robinson PW, Costa DP, Crocker DE, Gallo-Reynoso JP, Champagne CD, Fowler MA, Goetsch C, Goetz KT, Hassrick JL, Hückstädt LA (2012) Foraging behaviour and success of a mesopelagic predator in the northeast Pacific Ocean: insights from a data-rich species, the northern elephant seal. PloS One 7: e36728.

Romeiser R, Graber HC (2015) Advanced remote sensing of internal waves by spaceborne along-track InSAR – a demonstration with TerraSAR-X. IEEE Transactions on Geoscience and Remote Sensing 53(12): 6735–6751.

Ruiz J, Macías D, Rincón MM, Pascual A, Catalán IA, Navarro G (2013) Recruiting at the edge: kinetic energy inhibits anchovy populations in the western Mediterranean. PloS One 8: e55523.

Rutz C, Hays GC (2009) New frontiers in biologging science. Biology Letters 5: 289–292.

Ryan JP, Chavez FP, Bellingham JG (2005) Physical-biological coupling in Monterey Bay, California: topographic influences on phytoplankton ecology. Marine Ecology Progress Series 287: 23–32.

Sabarros PS, Ménard F, Lévénez J-J, Tew-Kai E, Ternon J-F (2009) Mesoscale eddies influence distribution and aggregation patterns of micronekton in the Mozambique Channel. Marine Ecology Progress Series 395: 101–107.

Sathyendranath S, Longhurst A, Caverhill CM, Platt T (1995) Regionally and seasonally differentiated primary production in the North Atlantic. Deep Sea Research Part I: Oceanographic Research Papers 42: 1773–1802.

Scales KL, Miller PI, Embling CB, Ingram SN, Pirotta E, Votier SC (2014a) Mesoscale fronts as foraging habitats: composite front mapping reveals oceanographic drivers of habitat use for a pelagic seabird. Journal of the Royal Society Interface 11: 20140679.

Scales KL, Miller PI, Hawkes LA, Ingram SN, Sims DW, Votier SC (2014b) On the Front Line: frontal zones as priority at-sea conservation areas for mobile marine vertebrates. Journal of Applied Ecology 51: 1575–1583.

Scales KL, Hazen EL, Maxwell SM, Dewar H, Kohin S, Jacox MG, Edwards CA, Briscoe DK, Crowder LB, Lewison RL, Bograd SJ. Fit to predict? Ecoinformatics for modeling the catchability of a pelagic fish in near real-time. Ecological Applications, In Press.

Scheffer A, Trathan PN, Edmonston JG, Bost C-A (2016) Combined influence of meso-scale circulation and bathymetry on the foraging behaviour of a diving predator, the king penguin (*Aptenodytes patagonicus*). Progress in Oceanography 141: 1–16.

Schick RS, Roberts JJ, Eckert SE, Halpin PN, Bailey H, Chai F, Shi L, Clark JS (2013) Pelagic movements of pacific leatherback turtles (*Dermochelys coriacea*) highlight the role of prey and ocean currents. Movement Ecology 1: 1–14.

Scott B, Webb A, Palmer M, Embling C, Sharples J (2013) Fine scale bio-physical oceanographic characteristics predict the foraging occurrence of contrasting seabird species; Gannet (*Morus bassanus*) and storm petrel (*Hydrobates pelagicus*). Progress in Oceanography 117: 118–129.

Shaffer SA, Tremblay Y, Weimerskirch H, Scott D, Thompson DR, Sagar PM, Moller H, Taylor GA, Foley DG, Block BA (2006) Migratory shearwaters integrate oceanic resources across the Pacific Ocean in an endless summer. Proceedings of the National Academy of Sciences 103: 12799–12802.

Sharples J, Scott BE, Inall ME (2013) From physics to fishing over a shelf sea bank. Progress in Oceanography 117: 1–8.

Shchepetkin AF, McWilliams JC (2005) The regional oceanic modelling system (ROMS): a split-explicit, free-surface, topography-following-coordinate oceanic model. Ocean Modelling 9: 347–404.

Shillinger GL, Palacios DM, Bailey H, Bograd SJ, Swithenbank AM, Gaspar P, Wallace BP, Spotila JR, Paladino FV, Piedra R (2008) Persistent leatherback turtle migrations present opportunities for conservation. PLoS Biology 6: e171.

Simpson JH, Sharples J (2012) Introduction to the Physical and Biological Oceanography of Shelf Seas. Cambridge University Press, New York, NY.

Sims DW, Quayle VA (1998) Selective foraging behaviour of basking sharks on zooplankton in a small-scale front. Nature 393: 460–464.

Sims D, Southall E (2002) Occurrence of ocean sunfish, *Mola mola* near fronts in the western English Channel. Journal of the Marine Biological Association of the UK 82: 927–928.

Sims DW, Southall EJ, Quayle VA, Fox AM (2000) Annual social behaviour of basking sharks associated with coastal front areas. Proceedings of the Royal Society of London B: Biological Sciences 267: 1897–1904.

Spalding MD, Agostini VN, Rice J, Grant SM (2012) Pelagic provinces of the world: a biogeographic classification of the world's surface pelagic waters. Ocean and Coastal Management 60: 19–30.

Spear LB, Ballance LT, Ainley DG (2001) Response of seabirds to thermal boundaries in the tropical Pacific: the thermocline versus the Equatorial Front. Marine Ecology Progress Series 219: 275–289.

Steele JH (1978) Spatial pattern in plankton communities (Vol. 3). Springer Science and Business Media.

Steele JH (1989) The ocean 'landscape'. Landscape Ecology 3: 185–192.

Stommel H (1963) Varieties of oceanographic experience. Science 139(3555): 572–576.

Stumpf RP, Culver ME, Tester PA, Tomlinson M, Kirkpatrick GJ, Pederson BA, Truby E, Ransibrahmanakul V, Soracco M (2003) Monitoring Karenia brevis blooms in the Gulf of Mexico using satellite ocean color imagery and other data. Harmful Algae 2(2): 147–160.

Suryan RM, Sato F, Balogh GR, David Hyrenbach K, Sievert PR, Ozaki K (2006) Foraging destinations and marine habitat use of short-tailed albatrosses: A multi-scale approach using first-passage time analysis. Deep Sea Research Part II: Topical Studies in Oceanography 53: 370–386.

Tew Kai E, Marsac F (2010) Influence of mesoscale eddies on spatial structuring of top predators' communities in the Mozambique Channel. Progress in Oceanography 86: 214–223.

Tew Kai E, Rossi V, Sudre J, Weimerskirch H, Lopez C, Hernandez-Garcia E, Marsac F, Garçon V (2009) Top marine predators track Lagrangian coherent structures. Proceedings of the National Academy of Sciences 106: 8245–8250.

Tittensor DP, Mora C, Jetz W, Lotze HK, Ricard D, Berghe EV, Worm B (2010) Global patterns and predictors of marine biodiversity across taxa. Nature 466: 1098–1101.

Tsuda A (1995) Fractal distribution of an oceanic copepod, *Neocalanus cristatus*, in the subarctic pacific. Journal of Oceanography 51(3): 261–266.

Tsuda A, Sugisaki H, Kimura S (2000) Mosaic horizontal distributions of three species of copepods in the subarctic Pacific during spring. Marine Biology 137(4): 683–689.

Turner MG, Gardner RH, O'Neill RV (2001) Landscape Ecology in Theory and Practice: Pattern and Process. Springer Science & Business Media, New York, NY.

Uuemaa E, Antrop M, Roosaare J, Marja R, Mander Ü (2009) Landscape metrics and indices: an overview of their use in landscape research. Living Reviews in Landscape Research 3: 1–28.

Wedding L, Friedlander A, Kittinger J, Watling L, Gaines S, Bennett M, Hardy S, Smith C (2013) From principles to practice: a spatial approach to systematic conservation planning in the deep sea. Proceedings of the Royal Society of London B: Biological Sciences 280: 20131684.

Wedding LM, Lepczyk CA, Pittman SJ, Friedlander AM, Jorgensen S (2011) Quantifying seascape structure: extending terrestrial spatial pattern metrics to the marine realm. Marine Ecology Progress Series 427: 219–232.

Weimerskirch H (2007) Are seabirds foraging for unpredictable resources? Deep Sea Research Part II: Topical Studies in Oceanography 54: 211–223.

Weng KC, Boustany AM, Pyle P, Anderson SD, Brown A, Block BA (2007) Migration and habitat of white sharks (*Carcharodon carcharias*) in the eastern Pacific Ocean. Marine Biology 152: 877–894.

Wiens JA, Milne BT (1989) Scaling of 'landscapes' in landscape ecology, or, landscape ecology from a beetle's perspective. Landscape Ecology 3: 87–96.

Wilson RP, Grémillet D, Syder J, Kierspel MA, Garthe S, Weimerskirch H, Schafer-Neth C, Scolaro JA, Bost C-A, Plotz J (2002) Remote-sensing systems and seabirds: their use, abuse and potential for measuring marine environmental variables. Marine Ecology Progress Series 228: 241–261.

Witt MJ, Broderick AC, Johns DJ, Martin C, Penrose R, Hoogmoed MS, Godley BJ (2007) Prey landscapes help identify potential foraging habitats for leatherback turtles in the NE Atlantic. Marine Ecology Progress Series 337: 231–244.

Woodworth PA, Schorr GS, Baird RW, Webster DL, McSweeney DJ, Hanson MB, Andrews RD, Polovina JJ (2012) Eddies as offshore foraging grounds for melon-headed whales (*Peponocephala electra*). Marine Mammal Science 28: 638–647.

Worm B, Sandow M, Oschlies A, Lotze HK, Myers RA (2005) Global patterns of predator diversity in the open oceans. Science 309: 1365–1369.

Wu J, Hobbs R (2002) Key issues and research priorities in landscape ecology: an idiosyncratic synthesis. Landscape Ecology 17: 355–365.

Yen PPW, Sydeman WJ, Bograd SJ, Hyrenbach KD (2006) Spring-time distributions of migratory marine birds in the southern California Current: Oceanic eddy associations and coastal habitat hotspots over 17 years. Deep Sea Research Part II: Topical Studies in Oceanography 53: 399–418.

Yen PP, Sydeman WJ, Hyrenbach KD (2004) Marine bird and cetacean associations with bathymetric habitats and shallow-water topographies: implications for trophic transfer and conservation. Journal of Marine Systems 50(1): 79–99.

4

Scale and Scaling in Seascape Ecology
David C. Schneider

4.1   Introduction

The builders of stone structures in the ancient world understood the problem of scale through bitter experience – a large temple or aqueduct cannot be built as a 1 to 1 replica of a smaller structure. The larger building collapses unless constructed with stronger materials and a different form. This fact, known to builders of fortified castles and vaulting cathedrals, was captured by Galileo in 1638 as the principle of similitude (Thompson 1915; Rosen 1989). Galileo illustrated the principle by drawing a series of bones, arranged from short to long, where the thickness increases more than length, easily seen as a change to a more robust shape at larger sizes. The principle is not intuitive. An example of intuitive scaling comes from a classic in English literature *Gulliver's Travels* by the satirist Jonathan Swift (Swift 1726). Gulliver reports the volume of food provided to him after capture by Brobdingnags, at ten times his height, as approximately one-tenth of their consumption of the same foods. The correct scaling according to Kleiber's Law (Kleiber 1932) is: E/day $\sim Mbody^{3/4}$ where E is daily energy required and *Mbody* is body mass. This becomes *Vfood*/day $\sim Vbody^{3/4}$ where food volume *Vfood* has the same energy density in Brobdingnag (*Vfood* proportional to E) and Gulliver has the same mass to volume ratio *Mbody* / *Vbody* as Brobdingnags (*Mbody* proportional to *Vbody*). Consumption relative to body mass (as a percentage) is proportional to *Vbody* and hence height as follows:

$$\% \,/\, \text{day} \sim Vbody^{-1/4} \sim (Length^3)^{-1/4} \sim Length^{-3/4} \tag{4.1}$$

At a height one-tenth of a Brobdingnag, Gulliver would require a far higher daily ration $(1/10)^{-3/4} = 5.6$ times that of a Brobdingnag, relative to body volume, assuming the same body shape and metabolism. In Lilliput, at ten times the height of a Lilliputian, Gulliver would need only a ration of $(10/1)^{-3/4} = 0.18$ times that of one of his captors.

The problem of scale in the environmental sciences (Levin 1992) is twofold. On the natural science side, physical and biogeochemical processes that prevail at any one space or time scale do not prevail at another. A physical example is the influence of the earth's rotation on fluid motion. The influence is vanishingly small at a local scale, yet at a broader scale, it determines the dynamics of coastal upwelling and ocean scale gyres. A biological example is the transition (at around 5 mm) from plankton to nekton. Plankton have too little mass and velocity to overcome the drag of viscosity, while

Seascape Ecology, First Edition. Edited by Simon J. Pittman.
© 2018 John Wiley & Sons Ltd. Published 2018 by John Wiley & Sons Ltd.

nekton have enough mass (and hence inertial motion) to overcome viscosity. On the societal side, human impacts occur at multiple space and time scales, from local to global. At the global scales of human impacts on the environment, scientifically rigorous modes of causal research found in the life and health sciences are often logistically limited or even impossible. In the ocean, causal experiments are compromised by uncontrollable variables both at the sea floor (Nowell & Jumars 1984) and in open water (Lalli 1990; Petersen & Kemp 2008). Adequately controlled experiments are limited by costs to small areas, are at best contentious at a regional scale, and are ethically fraught at the global scale of economic and power imbalances among nations. The consequences of scale-blind science in the public policy arena range from inadequate action (leaving Gulliver to starve in Brobdingnag) to overreaction – an unnecessarily large drain on the treasury to feed Gulliver in Lilliput. An example of scale-blind science in marine resource management is 1 to 1 scaling of catch to effort leading up to the collapse of the northern cod fishery in Newfoundland in 1992. Catch per unit effort (CPUE) was measured as boat number and size, a metric that missed decades long increases in catch efficiency due to changes in practice, time on the water, technological innovation and, most notably, increases in engine power (Neis *et al.* 1999).

4.1.1 The Development of the Concept of Scale in the Twentieth Century

In the early twentieth century, a physicist, Richard Tolman (1914), argued that 'the fundamental entities out of which the physical universe is constructed are of such a nature that from them a miniature universe could be constructed exactly similar in every respect to the present universe.' The scaling is Swiftian, a 1 to 1 similarity. In the same issue of the journal, another physicist, Edgar Buckingham (1914), argued against the introduction of the several postulates that Tolman used to obtain exactly similar Lilliputian universes. He then presents the key theorem in the dimensional analysis of similar systems, a method that remains in use today in physics and engineering. It is interesting to note that at about the same time, statisticians (William Gosset, Karl Pearson, Ronald Fisher) were developing (Fienberg 1992) the use of nondimensionalized ratios (notably likelihood ratios, F-ratios, the t-ratio, correlation coefficients). These, like the dimensional analysis of physically similar systems, remain in use over a century later. The salient distinction is that change in a nondimensionalized ratio with change in size or area is an analytic tool in engineering and physics; the same tool is rarely used in environmental biology (Schneider 2009), even though it is equally applicable.

Nondimensionalized ratios in statistics made their way, via agriculture, into the biologically oriented environmental sciences, beginning with Fisher's classic text (Fisher 1925). The method of nondimensionalized ratios developed in engineering and physics in the late nineteenth and early twentieth century is central to fluid mechanics, making its way into hydrology, meteorology and physical oceanography after World War II. Scaling concepts based on nondimensionalized ratios spread from physical to biological oceanography (Steele 1978; Platt 1981; Mackas *et al.* 1985; Schneider & Methven 1988) and from there to ecology in qualitative (Wiens 1989) and eventually quantitative form (Levin 1992; Schneider 1994).

The concept of scale, as it has developed in engineering, geophysics, hydrology, meteorology, and oceanography differs in several ways from that in cartography, geography, geostatistics and much of environmental biology and ecology. Terms differ. For example, in the earth and ocean sciences, 'large scale' refers to a large area; in cartography and geography large scale refers to a small area shown in detail. Another example is 'support,' defined in geography as the length, area, or volume from which measured data is taken. Support in this sense is usually implicit in hydrology, meteorology, oceanography, and ecology. In geostatistics, support often refers to weight of evidence, as measured by the logarithm of a likelihood ratio. The role of spatial scale differs. In the earth and ocean sciences spatial scale as it governs dynamics is central. Spatial variance in quantities such sea level height (Stommel 1963) or water velocity (Kolmogorov 1941) is interpreted relative to potential or kinetic energy at multiple space and time scales. In cartography and geography, correlation of quantities as it depends on arbitrary size of political units of aggregation is taken as a statistical issue (Gehlke & Biehl 1934) called MAUP – the modifiable unit area problem (Openshaw & Taylor 1979). In geostatistics (Matheron 1962; Ripley 1981; Cressie 1993) unit size as it conditions the representation of georeferenced data is a central topic. This primarily statistical treatment, as either a problem stemming from arbitrary units, or as a necessary tool in geostatistics, carries over into ecology (Legendre 1993). Emphasis on socio-economic impacts differs. In landscape ecology, in the present century, spatial scale as it bears on our capacity to quantify human impacts on ecosystems at multiple scales is a key topic (Wu & Hobbs 2007) and a focal issue (Wiens & Moss 2005). This extension to socio-economic components is one of the defining features of seascape ecology (Hinchey *et al.* 2008; Manderson 2016, Chapter 14).

4.1.2 Prevalence and Usage of 'Scale' in the Scientific Literature

The prevalence of scaling concepts differs in physical oceanography, biological oceanography, marine ecology, and terrestrial ecology, based on full text searches of journals with long publication runs in each of the four fields. An ecological journal (*Landscape Ecology*) with explicit focus on georeferenced ecological data was included for comparison, despite a shorter journal run. Prevalence of the term 'scale' increased in all five journals in the late twentieth century, but at different rates and with different histories (Figure 4.1). Prevalence (articles with 'scale' per hundred articles published) was already high in the first years of issues of the *Journal of Physical Oceanography* at 76 articles per 100 published between 1971 and 1976 (Figure 4.1). The increase was weakly exponential at 0.7%/year from 1971 through 2006. Prevalence of the word 'scale' in *Marine Geology*, an older journal in marine science, more than trebled from 1964 and 1970 (16/100) to the period from between 2010 and 2015 (58/100). In *Limnology and Oceanography*, prevalence more than doubled from 26/100 in the first years of publication (1956 to 1960) to 68/100 between 2010 and 2015. This journal publishes articles in physical, chemical, and biological oceanography, with a strong emphasis on the latter. In *Marine Ecology – Progress Series*, prevalence rose from 27/100 in the first years of publication (1981 to 1985) to 59/100 at an exponential rate of 3.7% / year from 1980 through 2006. The history of the use of the word 'scale' in oceanography journals before 1980 stands in contrast to the near absence of the term in the journal *Ecology* before 1985 (Figure 4.1). Prevalence rose from 2/100 between 1971 and 1985 to 21/100 between 2010 and 2015 (Figure 4.1) at an exponential rate of 8.6% / year from 1980 until 2006 with little change afterward. The word 'scale' was well established in the first issue (1987) of *Landscape Ecology*, a journal explicitly founded on recognition of the problem of scale in terrestrial ecology. Prevalence of the word 'scale' in this journal was 88/100 in 1987–1990, declining between 2001 and 2005, then rising between 2010 and 2015.

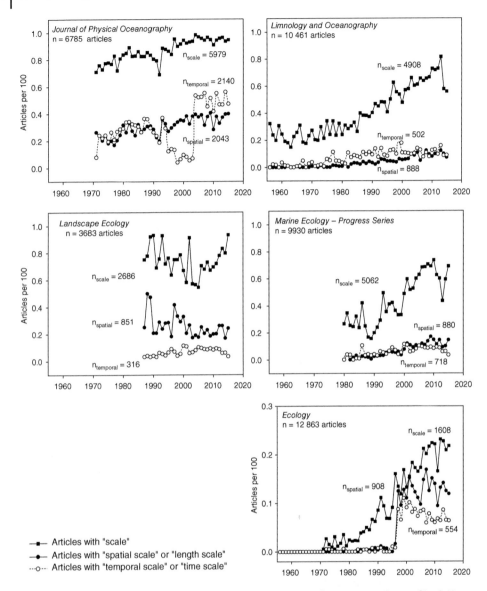

Figure 4.1 Frequency (n = articles / year) of use of 'scale' and use of two restricted uses of 'scale' in two oceanographic and three ecological journals. Frequency adjusted as *n* per 100 articles published in a year. Note threefold expansion of vertical axis for one journal (*Ecology*) relative to other journals.

The word 'scale' has somewhat different usages in journals shown in Figure 4.1. In *Marine Ecology – Progress Series*, the word often refers to measurement or to the scale of a graph; in *Landscape Ecology* 'scale' often refers to map scale and sometimes to organizational level (*i.e.*, individuals within species, species within populations, populations within ecosystems).

The word 'scale' typically refers to both spatial and temporal applications so a search on 'spatial scale' and on 'temporal scale' was conducted in all five journals.

Both terms appear infrequently in the *Journal of Physical Oceanography* (413/6785 articles for 'spatial scale' and 45/6785 articles for 'temporal scale') where they are replaced by 'length scale' (1880/6785) and 'time scale' (2020/6785). 'Length scale' occurs infrequently in the ecology journals (*e.g.,* 47/10400 in *Marine Ecology – Progress Series* and 14/3470 articles in *Landscape Ecology*). A more focused Boolean search was then conducted on ('spatial scale' or 'length scale') and on ('temporal scale' or 'time scale'). In the *Journal of Physical Oceanography* frequency of articles with 'spatial' or 'length' scale doubled from 20/100 to 40/100 over 45 years (Figure 4.1) while articles with 'time' or 'temporal' scale nearly tripled over the same period (Figure 4.1). In *Limnology and Oceanography* articles with 'spatial' or 'length' scale rose from less than 1/100 in the first two decades of the journal (1956–1976) to 10/100 in 2011–2015. Articles with 'spatial' or 'length' scale rose in a similar fashion to 10/100 in 2011–2015. In *Marine Ecology – Progress Series* articles with 'spatial' or 'length' scale rose from 0/100 to around 15/100 over 35 years (Figure 4.1); articles with 'time' or 'temporal' scale rose to 10/100 then fell over the same period (Figure 4.1). Similar patterns of rapid increase in the use of 'spatial scale' after 1985 were found in the *Journal of Experimental Marine Ecology and Biology* and the *Journal of the North American Benthological Society* (Ellis & Schneider 2008). In *Ecology* articles with 'spatial' or 'length' scale rose from 0/100 to nearly 20/100 over a 45-year period (Figure 4.1); articles with 'time' or 'temporal' scale also rose from 0/100 to 20/100 (Figure 4.1). In *Landscape Ecology* articles with 'spatial' or 'length' scale accounted for most of the articles with 'scale' In contrast articles with 'time' or 'temporal' scale rose from 0/100 to 10/100 than fell (Figure 4.1). The lower prevalence reflects the dominance of mapping (spatial) concepts in this journal, compared to quantitative treatment of rates and dynamics in hydrology and physical and biological oceanography.

Yet another difference, obvious from scanning articles in these journals, is the use of mathematical expression of physical or biological concepts. Not unexpectedly, mathematical expression is standard in the *Journal of Physical Oceanography*. Mathematical expression of the concept of scale was generally absent in the other journals. This long recognized lack of quantitative treatment of spatial scale has been attributed to the absence of any rationale for choice of scale (Meentemeyer 1989; Wiens 1989). Unfamiliarity with applicable methods from physics, oceanography and hydrology (Rodriguez-Iturbe & Rinaldo 1997) is also possible.

The broad conclusion from these comparisons is that the concept of scale increased in frequency in the environmental sciences in the late twentieth and early twenty-first centuries.

The conclusion is supported by similar searches in other journals, such as *Trends in Ecology and Evolution*. A second conclusion is that the concept of scale appeared recently in terrestrial ecology, compared to oceanography. From an historical perspective, the concept of scale in seascape ecology is a fusion of older concepts rooted in geophysical fluid dynamics with statistical concepts and socio-economic components from geography, geostatistics and landscape ecology.

4.1.3 Definition of Scale

The word 'scale' has many definitions, arising from two different roots. The Old Norse root in *skal* or 'bowl' gives rise to fish scales, the scales of justice and measurement via comparison to a standard. The Latin root in *scala*, or 'ladder', gives rise to musical scales, scaling a wall, and measurement by counting steps. Both roots differ from that of 'landscape' and by extension 'seascape'. The Dutch root *-schap* is cognate with *-schaft* (German) and *-ship* (English), giving rise to landscape as a noun (a painting from a particular viewpoint) and landscape as a verb (reshaping the land).

In the environmental sciences, 'scale' often refers to the Latin root. One common use is the phrase *scale of...* referring to spatial or temporal extent. Examples are the scale of a hurricane (distance across the weather system) and the time scale of El Niño recurrence (2 to 7 years). Technical definitions often entail the taking of ratios. Box 4.1 shows commonly used technical definitions.

Box 4.1 Technical definitions of scale in the environmental sciences

Cartographic scale is the ratio of the distance on a map to the distance on the ground. A metre-wide map of the world will have a scale of about 1 : 39 000 000.

Spatial or temporal scale of a measured variable refers to the resolution and extent. An example is a variable mapped at a resolution of $2\,m^2$ within an extent of 1 ha and so with a resulting scope of $(100\,m^2\,/2\,m^2) = 50$. Resolution is measured in the distance domain (distance between samples), or frequency domain (samples/distance as on a transect). Resolution is also expressed as the support, *i.e.*, the duration and spatial extent (length, area, or volume) of fully censused units. In landscape ecology, grain refers to the finest spatial resolution while extent refers to the area of study.

Scaling refers to the use of the principle of similitude to relate one variable to another, often expressed as a power law. An example is the scaling of species number to $Area^{\beta}$. Where β has a typical value of 0.3, species number increases by a factor of $2^{0.3} = 1.2$ with a doubling in area.

Multiscale analysis refers to change in a measure of variability with change in resolution, extent, measurement frequency, or separation (lag). Examples include spectral analysis, spatial autocorrelation and techniques based on semivariance.

4.2 Expressions of Scale

4.2.1 Graphical Expression of Scale

In 1963, a physical oceanographer, Henry Stommel, published a conceptual diagram that was to have a profound effect on all of the environmental sciences (Vance & Doel 2010). The diagram (Stommel 1963) depicted variation in sea level height at spatial scales from centimetres to that of the planet and at time scales from seconds to tens of millennia. The units of variability were not explicitly stated, but a footnote suggests that variability in Stommel's figures was envisioned as the standard deviation in sea level height $Var(h)^{1/2}$, a quantity having the same units as height h (cm). Vance & Doel (2010) traced the influence of the diagram as it spread from physical oceanography to biological oceanography (Haury *et al.* 1978; Steele 1978) and from there into the other environmental sciences, including landscape ecology (Steele 1989) and ecology (Levin 1992).

The first step toward practical application of Stommel's conceptual diagram was taken by John Steele, a biological oceanographer, who used Stommel's logarithmic spatial and temporal axes (Steele 1978) to construct two novel diagrams. The first diagram was conceptual, like those of Stommel (1963), with names of phenomena placed in axes showing the time (1–1000 days) and space (1–1000 km) scales of patchiness of phytoplankton, zooplankton, and fish. Steele's diagram went beyond that of Stommel by placing labels inside delimited circles showing the temporal and spatial extent of phenomena. Haury *et al.* (1978) took a different approach by reproducing Stommel's three-dimensional diagram with overlays identifying physical processes thought to influence plankton patchiness at multiple scales. The juxtaposition is a stretch, in that the connection between variation in sea level height and plankton patchiness is at best indirect, to the degree that fluid dynamics drive variance in both sea level height and plankton density. Five years later, Delcourt *et al.* (1983) published the first conceptual space-time (ST) diagram in terrestrial ecology, using labels in defined areas, as in Steele's diagram. Steele's conceptual ST diagram was accompanied by an instrumental ST diagram showing space and time scales covered by data from a single ship, by data from multiple ships working in a directed research program, and by a standard fish stock survey. Figure 4.2 shows a recent Stommel diagram (Godø *et al.* 2014) with

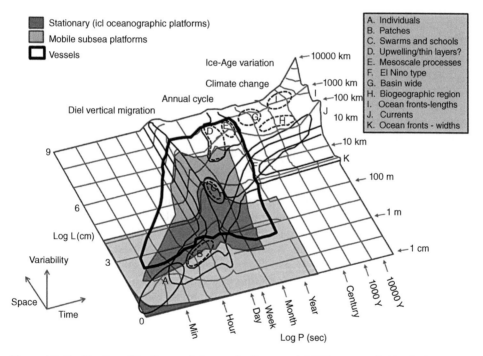

Figure 4.2 Modification of the Stommel diagram by Haury *et al.* (1978), as extended by Godø *et al.* (2014) to illustrate coverage potential, overlap and uniqueness in time and space of acoustic data from stationary and oceanographic platforms. Vessels include vessel and vessel operated tethered platforms. Stommel diagram shows increasing variance in sea surface height from small scale (lower left) to large scale (upper right). Note large variance at days to week due to tides and ice age variations. Gravity waves, tsunamis, geostrophic turbulence and meteorological effects (in Stommel 1963) replaced with processes relevant to plankton – diel vertical migration, annual cycle and climate change. Phenomena relative to plankton dynamics are shown as circled areas labelled by letters A – K. *Source*: Reprinted with permission from Godø *et al.* (2014).

overlays from Haury *et al.* (1978) and instrumental coverage as an overlay, rather than a separate diagram. Published ST diagrams, almost all conceptual, increased at an exponential rate of 13.3%/year from 1980 to 1998 (Schneider 2009). These diagrams lacked the dynamical linkages shown in Steele's 1978 diagram.

Stommel (1963) clearly intended that variability in any measured variable be considered relative to both space and time axes. Stommel mentioned variability in sea level, spectral density in water velocity, spectral density in water density (sigma-t), and spectral density in Reynolds stress (inertial relative to viscous forces in the water column). Spectral density is a smoothed estimate of variance in a quantity, per unit of distance or time, as a function respectively of spatial or temporal resolution. Spectral density summarizes, in different form, the same information as the autocorrelation function widely used in geostatistics and landscape ecology. Turning to spectral density in water velocity *SpD(v)* and Reynolds stresses *SpD(Re)*, these dynamic quantities are more directly relevant to biological dynamics than spectral density in sea-level height. The relevance is readily seen from consideration of units and dimensions. *SpD(v)* has units of $(velocity)^2 (cycle)^{-1}$. It is a measure of dynamics, because the square of water velocity v^2 is proportional to the kinetic energy of water parcels. The spectral density of v^2 expresses the kinetic energy of water circulation at spatial and temporal scales from global to that of water molecules.

Weber *et al.* (1986) took the first step toward quantification of the Stommel diagram by dropping the temporal axis, resulting in a two-dimensional cut through Stommel's three dimensional diagram. Weber *et al.* (1986) replaced spectral density in water velocity *SpD(v)* with spectral density in organism abundance (Platt & Denman 1975). Spectral density in phytoplankton abundance (as measured by chlorophyll fluorescence) is a measure of patchiness, which in turn is expected to scale in time and space in the same way as a passive physical tracer of the fluid, such as temperature and salinity. As expected, spectral density in phytoplankton abundance deviated little from that of a passive tracer (Weber *et al.* 1986). In contrast, spectral density of nekton abundance deviated substantially. Figure 4.3 reproduces Weber's diagram with overlays for marine fish and birds. The results are consistent with the expected transition in dynamics with increase in organism size from passively drifting (planktonic) organisms where density is structured by flow regimes, to actively swimming (nektonic) organisms capable of velocities greater than that of the surrounding fluid. A similar result holds at the seafloor (Figure 4.3), where sedentary organisms (scallops *Chlamys islandica*, sand dollars *Echinarachnius parma*, and urchins *Strongylocentrotus pallidus*) scale closely with substrate roughness as quantified by spectral analysis. *Buccinum* whelks, an active predator, do not scale in the same way as environmental gradients.

4.2.2 Graphical Expression of Scale in Research Planning

Stommel's diagram was developed in the context of planning research in the ocean, where the logistical costs of *in situ* data are substantial, at tens of thousands of dollars per day for ship time. The Stommel diagram did not appear in publication as a planning tool (Vance & Doel 2010) and this is almost certainly due to the formidable challenge of putting numbers in a Stommel diagram. Steele's (1978) instrumental diagram rectified this, by using the temporal and spatial extent of phenomena or research activities to circumscribe known values within a square (or ellipse) on the two axes. Yet another step

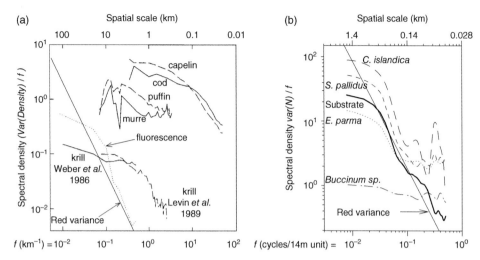

Figure 4.3 Spectral density of pelagic (a) and benthic (b) organisms compared to spectral density of water column fluorescence and seafloor substrate on a roughness scale of 1 to 9 (sand through gravel and cobble). Spectral density is the variance per unit of frequency measurement. Periods corresponding to measurement frequency are shown on the upper axis. Fluorescence is a measure of the concentration of passively drifting phytoplankton. *Source:* Panel (a) redrawn from Horne & Schneider (1997). Panel (b) from data in Schneider & Haedrich (1991).

was to display data support and the target of statistical inference in conceptual ST diagrams (Schneider *et al.* 1997; Schneider 2009). Figure 4.4a shows a conceptual diagram for the problem of detecting and monitoring the effects of release of a contaminant, chlordane, on benthic invertebrates in a coastal lagoon. Data support is the length, area, or volume over which a measurement is made. The definition can be extended to the duration of a measurement. Figure 4.4b shows the spatial and temporal support (cumulative area of core samples and cumulative time to obtain samples) for a statistical model of bivalve density in a 250 × 500 m (12.5 ha) area gridded into 200 cells, each 25 × 25 m. Randomly placed samples from each cell were used to estimate the density-scape of the bivalve *M. lilliana* (Legendre *et al.* 1997) in the 12.5 ha target of inference. Experimental treatments were placed at locations with high, intermediate, or low density. A regression model of the density-scape (Legendre *et al.* 1997) allows experimental results to be upscaled by integration to an estimate of density and dynamics for the 12.5 ha area.

A further step was to extend the calculation of a critical scale (where two rates are equal) to identifying those regions of ST diagrams where one rate prevails over another. The classical example of a critical scale is patch size of red tide blooms. The dynamics are modelled as the antagonistic effects of growth, which create patches, and eddy motions, which tend to mix and destroy patchiness. At the critical patch size these are equal (Kierstead & Slobodkin 1953; Platt 1981). Horne & Schneider (1994) extended the critical scale calculation to cross-scale depiction at multiple space and time scales, using the graphical format developed by Steele (1978). In these graphs, a contour line marks the space and time scales at which two rates are of the same order of magnitude. At larger scales, one rate prevails, while at smaller scales the other rate prevails. Figure 4.4c shows the spatial and temporal scales at which mortality prevails

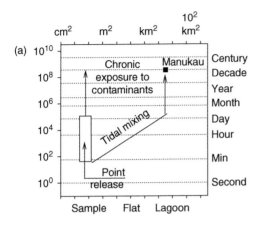

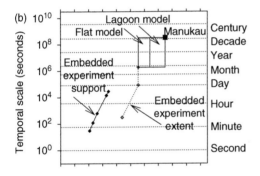

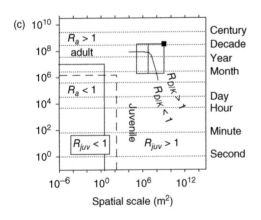

$$R_{juv} = b/F \qquad R_a = T_g/F \qquad R_{D/K} = \frac{\text{mortality rate}}{\text{movement rate}}$$

Figure 4.4 Space-time diagrams for the problem of monitoring the effect of chronic release of contaminants in Manukau harbour, New Zealand. (a) Spatial and temporal scale of release and subsequent tidal mixing to the scale of the lagoon; (b) Support and extent of embedded experiment (Legendre *et al.* 1997) compared to computational model of bedload transport at the scale of a single flat and the scale of the lagoon; (c) Dimensionless ratios that separate space and time scales where demographic rates (recruitment and death) prevail over kinematic rates (locomotion and passive drift). See text for definition of R_{juv}, R_a and $R_{D/K}$. *Source:* Redrawn from Schneider *et al.* (1997).

over kinematics (movement due to passive and active transport) of the intertidal bivalve *Macomona liliana*. Figure 4.4c also shows the spatial and temporal scales at which active movements prevail over passive transport due to bedload dynamics on intertidal flats inhabited by *M. liliana*. For details of the calculation, see Schneider *et al.* (1997).

4.2.3 Formal Expression of Scale: Scope, Similarity and Power Laws

Formal expression of the concept of scale begins with its components: resolution (or grain or support) and extent. Taking the ratio of extent to resolution yields a scope. The scope of a meter stick is $1\,m/1\,mm = 10^3$. Scope can also be taken relative to support: the spatial scope of the *M. liliana* experiment listed above, relative to support (a 13 cm diameter core), was $12.5\,ha/(\pi\,(13\,cm/2)^2) = 9.4 \times 10^4$. In Figure 4.4b, support is shown as a line connecting a single core to the cumulative number of cores taken. The line runs at an angle set by the duration of the time to take a sample. The spatial scope of the experiment relative to the area of experimentally altered 50 cm × 50 cm plots was less at $12.5\,ha / (50\,cm)^2 = 5{\times}10^3$.

The principle of similitude equates the scope of one quantity to another via an exponent (Schneider 2001a).

$$\frac{Q}{Q_{ref}} = \left(\frac{X}{X_{ref}}\right)^{\beta} \tag{4.2a}$$

With some algebraic rearrangement, the result becomes a power law scaling function:

$$Q = (Q_{ref}X_{ref}^{-\beta})\ X^{\beta} \tag{4.2b}$$

An example is the scaling of species number S to area A:

$$\frac{S}{S_{ref}} = \left(\frac{A}{A_{ref}}\right)^{\beta} \tag{4.3a}$$

where species number S in an area, A scales with species number S_{ref} in a reference area A_{ref} according to the exponent β. When the exponent is estimated by regression, the reference area is the unit of measurement (*e.g.*, 1 hectare) and S_{ref} is an estimate of the number of species in a single reference unit. With rearrangement, Equation 4.3a becomes the species area curve, expressed in traditional notation as

$$S = c \cdot A^z \tag{4.3b}$$

where $c = (S_{ref}A_{ref}^{-z})$. Another example is the scaling of coastline length CL to the step length SL used to measure a coastline. The number of steps n decreases with increasing step length SL, according to the fractal dimension D_f.

$$\frac{n}{n_{grain}} = \left(\frac{SL}{SL_{grain}}\right)^{-D_f} \tag{4.4}$$

Coastline length CL is the product of step number n and SL:

$$\frac{CL}{CL_{grain}} = \frac{n}{n_{grain}}\left(\frac{SL}{SL_{grain}}\right) = \left(\frac{SL}{SL_{grain}}\right)^{1-D_f} \tag{4.5}$$

For simple coastlines the fractal dimension (Mandelbrot 1977) is on the order of $D_f = 1.3$ and the exponent becomes $1 - D_f = -0.3$. For coastlines with islands and complicated topography, the exponent decreases as fractal dimension rises toward values of 1.7 or more. The calculated length decreases with loss of resolution at coarser scales; the rate of decrease depends on the degree of convolution of the coastline. In Equation 4.5 the fractal dimension is interpreted as a line more convoluted than a straight line (dimension greater than 1), but not so convoluted as to fill the whole plane (dimension approaching 2). The scaling relation in Equation 4.4 applies to any counting procedure, such as multiscale characterization of habitat area by counting the number of grid boxes with a particular habitat.

Many geographically distributed variables exhibit fractal properties for which the scaling relation is:

$$\gamma(h) = c \cdot h^{\omega} \tag{4.6a}$$

where h is the separation between two points in space and $\gamma(h)$ is the average squared difference of a measured variable between those points. The coefficient c is a constant and ω is the scaling exponent that ranges upwards from zero (*i.e.*, no change in spatial gradient with increasing lag). The function $\gamma(h)$ is the variogram and has units of the square of the measured variable. The square root of the variogram (semivariance) divided by the lag is thus the average gradient of that variable at lag h:

$$\gamma(h)^{1/2} h^{-1} = c^{1/2} h^{(\omega/2 - 1)} \tag{4.6b}$$

For closely spaced points in space, spatial autocorrelation decreases and the spatial gradient increases with increasing separation. The semivariance is more than a statistical measure, it is a physically interpretable quantity – it is the change in the average gradient with increasing separation. Landscape ecologists use a variety of models to characterize $\gamma(h)$ in relation to h, including nugget, spherical, exponential, power law, and Gaussian. For the power law model, the scaling relation of the average gradient to separation h is

$$\left(\frac{\gamma}{\gamma_{grain}}\right)^{(1/2)} h^{-1} = \left(\frac{h}{n_{grain}}\right)^{\omega/2 - 1} \tag{4.6c}$$

The scaling exponent ω summarizes the change in gradient of a measured quantity, with change in distance over which the gradient is calculated.

Scaling relations emerge from cumulative distributions. The commonest application in ecology is the collector's curve describing increase in species number with increase in some measure of effort, such as number of individuals collected or number of surveys (Schneider 2001b). At least a dozen scaling functions have been developed for collector's curves; these are reviewed by Thompson *et al.* (2003). Another application is cumulative catch relative to fishing effort, usually resulting in a flux, defined as number of fish intercepted per unit area per unit time. For cumulative data ΣX, relative to some measure of cumulative effort ΣE, the scaling relation is

$$\frac{\sum_1^n X}{X_{grain}} = \left(\frac{\sum_1^n E}{E}\right)^{\beta} \tag{4.7}$$

where, E is defined within a fully censused unit, such as a plot (area), a core sample (volume), or a period of observation. In geostatistics, support is sometimes defined as the average over samples within a larger unit of space, such as a block. When two variables of interest do not scale with each other in some simple fashion, a scaling relation will often emerge when each is scaled to cumulative form, as in Equation 4.7.

4.2.4 Scaling Manoeuvres

Equations 4.3–4.7 illustrate four different scaling manoeuvres (Figure 4.5). In Equation 4.3, scaling is with reference to isolated units that differ in size. Scaling by rating the quantity of interest to unit size has a long history in ecology going back to the species-area curves of Arrhenius (1921). Mesocosm studies use rating with respect to both surface area and volume to investigate ecological dynamics in both field (Petersen & Kemp 2008) and laboratory (Sanford 1997; Petersen & Kemp 2008) settings. In Equation 4.4, scaling is by coarse graining (Figure 4.5). A stretch of coast is measured repeatedly using segment lengths that range from short (high resolution) to long (low resolution and coarse scale). Scaling by coarse graining has a long history in agricultural science (Mercer & Hall 1911) and plant ecology (Greig-Smith 1952). Coarse-graining requires contiguous units. It has long been used to discover the scale at which populations are associated with key habitat features. Marine examples include zooplankton in relation to phytoplankton (Mackas & Boyd 1979), seabirds in relation to pelagic fish prey (Schneider & Piatt 1986), and benthic epifauna in relation to substrate (Schneider *et al.* 1987). More recently, coarse graining has been used to investigate how models of habitat selection respond to changes in the resolution scale of input variables (Pittman & Brown 2011; Bartolino *et al.* 2012; Alvarez-Berastegui *et al.* 2014; Scales *et al.* 2017).

Lagging (Figure 4.5) employs measurements separated in space. In Equation 4.6a scaling is relative to the distance between points (lag). Scaling by lagging (Figure 4.5) differs from coarse graining in how averaging occurs. The former takes averages across pairs of measurements having the same spatial separation; the latter takes averages within spatial units that differ in size. The spatial information summarized by the semivariance (in effect the average gradient as a function of spatial separation) is used to interpolate between measured points. The semivariance was developed in the middle of the twentieth century (Krige 1951) to estimate gold reserves from sparse data where spatially continuous data were not possible. Matheron (1962) developed the mathematical basis for Krige's method of interpolation.

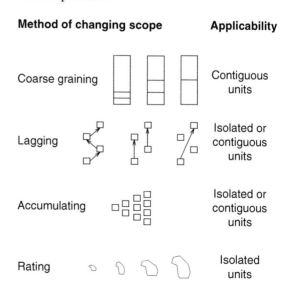

Figure 4.5 Comparison of scaling manoeuvres and applicability. Accumulation, coarse graining and lagging occur by iterative calculation on data, rating occurs by noniterative measurement of units of different size.

Method of changing scope		Applicability
Coarse graining		Contiguous units
Lagging		Isolated or contiguous units
Accumulating		Isolated or contiguous units
Rating		Isolated units

In Equation 4.7, scaling is by accumulation of units of effort (Figure 4.5). Figure 4.6a shows the emergence of a scaling relation of cumulative species to cumulative effort over time, where there was a complex and noisy relation of species number with time. The number of polychaete species known to occur in the Bodega Marine Reserve in California increased over time, based on 48 reports in the Cadet Hand Library at the Bodega

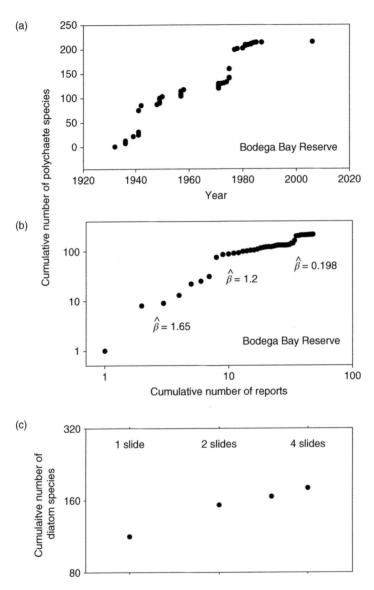

Figure 4.6 Cumulative number of species with respect to time, effort, and number of samples. (a) Polychaete species on an arithmetic scale, from 1932 to 2006 from the Bodega Marine Reserve in a fixed area, Shell Beach to Doran Beach. (b) Polychaete species number with respect to effort (number of reports) on logarithmic scale. (c) Diatom species number from Patrick (1968) on a logarithmic scale. Polychaete data from Schneider *et al* (2007). Figure 6c redrawn from Schneider (2001b).

Marine Laboratory, from 1932 to 2006 (Finity 2006; Schneider *et al.* 2007). The number of reports varied from year to year, ranging from none in many years to six in 1971. Most of the reports were student projects. Two reports (Pettibone 1941; Pacific Marine Station Staff 1977) increased the number of species by factors of 75/31 and 199/160 respectively. When cumulative species number is plotted against cumulative number of reports on a logarithmic scale (Figure 4.6b), three scaling relations emerge. The decline in the scaling exponent from $\beta = 1.656$ in 1932–1941 to $\beta = 0.198$ in 1978–2006 is consistent with the well-known tendency of a collector's curve to rise ever more slowly with time, as the additions to the species list shift from common residents to rare residents and vagrants. The likelihood ratio for the regression model is $(1 - r^2)^{-ntot/2} = (1 - 0.9146)^{-12/2} = 2.77 \times 10^6$ – the weight of evidence is against the no relation model compared to an increase in species number with effort (*i.e.*, number of reports). In a decision theoretic, instead of weight of evidence context, the standard error on the 1978–2006 estimate of β is 0.019, which yields a lower 95% confidence limit LCL of $0.198 - 2.093*0.019 = 0.159$, a value for which the decision at the 95% level is to reject $\beta = 0$, that of no relation between cumulative species and cumulative effort. A cursory glance at the polychaete species list for San Francisco Bay, with an entrance 80 km south of Bodega Bay, points at a very different scaling. The San Francisco Bay list since 1978 is heavily impacted by exotic polychaete species generally absent from Bodega Bay. The continued introduction of exotic species would be expected to maintain a higher scaling exponent than in Bodega Bay since 1978.

An important conclusion from Figures 4.6a and 4.6b is that power law scaling relations apply only within the stated extent (in this case, within three different time periods). The conclusion follows from the well-known limits on the use of regression. The limited extent over which power law scaling applies was reported by Avnir *et al.* (1998) and is consistent with the finite size effect of scaling properties in physical and biological systems.

Figure 4.6c shows the scaling relation of cumulative species to cumulative effort in a laboratory setting, where diatom number from a single location increases with the number of slides (and hence area) examined. The exponent for the scaling relation in Figure 4.6c is estimated at $\beta = 2.917$, for which the likelihood ratio is $(1 - 0.9731)^{-4/2} = 1400$. The weight of evidence is strongly in favour of a coefficient of 2.9 relative to a coefficient of zero. The conclusion from this analysis is that far more than four slides (samples) are needed to characterize the number of diatom species from similar ecological conditions. The weight of evidence analysis puts a number on what is evident from the graph. In this case the weight of evidence, as a likelihood ratio, suffices; a decision against a fixed tolerance for Type I error adds little.

4.2.5 Ratio of Rates in Research Planning

Taking the ratio of two rates expressed in the same units results in a dimensionless number free of the influence of change in unit size. Curves of these ratios in space and time (Schneider *et al.* 2007) are useful in evaluating research hypotheses and planning scale relevant research programs. They allow identifying the space and time domains where one rate prevails over another. Dimensionless ratios of rates are routinely used in physical oceanography. An example is the Reynolds number, the ratio of inertial forces to viscous forces in a fluid. In biological oceanography several dimensionless numbers

(Schneider *et al.* 1992, 1997) are needed to scale demographic (recruitment and mortality) to kinematic (active and passive movement) rates. Figure 4.4c shows curves that separate those space and time scales where demographic rates prevail and where kinematic rates prevail in driving changes in organism density. According to Stigler's Law of Eponymy (Stigler 1980), neither the r/F (demographic to kinematic) ratio nor the F_{loc}/F_{fluid} (active to passive kinematic) ratio will be named according to their first (1992) or subsequent (1997) appearances in the published literature.

Dimensionless ratios of rates at a single scale, or as curves in Figure 4.4c, are formed relative to the research context and goals. An example of hypothesis evaluation and scale relevant research planning using a dimensionless ratio comes from the trophodynamics of large seabird colonies, an impressive but rare component of coastal seascapes. Some colonies reach millions in numbers during the breeding season, when adults must draw prey from the ocean for both themselves and their chicks. Ashmole (1963) hypothesized that tropical seabird populations are limited at colonies by scramble competition for prey during the breeding season in the relatively unproductive waters around tropical colonies. Initial support for the hypotheses came, somewhat surprisingly, from colonies in more productive shelf ecosystems at mid (Wiens & Scott 1975) and high latitudes (Furness & Barrett 1985). These authors took the ratio of prey demand to prey standing stock or to prey productivity as estimated at several trophic levels above primary production. The ratios pointed at substantial draws on prey. However, the prey taken by seabirds come largely from the water column, which moves past colonies at rates governed by daily tides and longer term advection. Bourne (1983) argued that advective resupply makes scramble competition via prey depletion unlikely. If we take a Bourne ratio (seabird food demand / advective resupply), we find values well below unity (Schneider *et al.* 1992). The Bourne ratio points at prey depletion and hence scramble competition as inapplicable for seabirds feeding in the water column. Conversely, prey depletion is applicable to immobile prey, such as crevice living benthic fish. Birt *et al.* (1987) reported depletion of prey around a colony of cormorants feeding on sedentary bottom living fish. To date this is the only direct evidence that seabirds have any impact on prey density or supply.

This seabird example illustrates several characteristics of the use of dimensionless ratios in hypothesis evaluation and scale-relevant research planning. First, components of a ratio are no more than an order of magnitude and thus the ratio is no better than one or even two orders of magnitude. A second characteristic of the use of dimensionless ratios is that they require knowledge of the system and its dynamics (Barenblatt 1996). In the case of seabird colonies, knowledge of the source of seabird prey (advectively supplied pelagic prey versus sedentary benthic prey) was key to forming a biologically based Bourne ratio. A third characteristic is that the scale at which one rate prevails over another is not necessarily intuitive. Mean rates of advection by major ocean currents are on the order of tens of centimetres per second, making advective rate less than obvious at the appropriate time scale for food demand by a breeding colony, which is on the order of months for seabirds. Advective rates from the literature need to be converted to % / month, so advective resupply can be scaled as a dimensionless number relative to food demand by a colony.

Dimensional analysis yields ratios of rates that separate space and time domains where one rate prevails over another. The classic example in oceanography is the formation of red tide blooms (Kierstead & Slobodkin 1953). These authors defined the 'characteristic scale' $d = (k/r)^{1/2}$ at which patch size d is large enough to balance the rate of dinoflagellate proliferation r, tending to create a local bloom, against the diffusion rate k by eddy mixing, tending to suppress a bloom by dispersal and dilution. The ratio via dimensional analysis (Platt 1981) is $\Pi = d^{-1} (k/r)^{1/2}$. At typical rates of eddy diffusivity and growth rate in the ocean, patch scale d is on the order of 1–10 km. Where growth rate is high, such as nutrient input from coastal runoff, patches are capable of maintaining themselves at high concentrations against eddy diffusivity. Where eddy diffusivity is high, such as strong wind conditions at the sea surface, patches are mixed and rapidly dispersed.

The term 'characteristic scale.' has an unfortunate history in evolving to a new meaning, that of the search for the 'right' (= characteristic) scale at which to plan research. In the journal *Landscape Ecology*, the term 'characteristic scale' occurs in approximately 20 articles from 1987 to 2015. It usually occurs in the context of choosing the best or correct space and time scales for research activity. In the *Journal of Physical Oceanography*, the term occurs approximately 70 times from 1971 to 2015. It usually occurs with an order of magnitude calculation that qualifies the discussion of a parameter or phenomenon in light of the relative importance of rates, as in Figure 4.4c. Quantitative use of the term 'characteristic scale,' via dimensional analysis, is of utility in the planning of ecological research and in the reporting of research results as in Figure 4.4c.

The ratio of rates in Figure 4.4c was obtained (Schneider *et al.* 1992) from a conservation equation for population numbers expressing demographic components (birth and death) and kinematic components (active and passive movements). Table 4.1 shows conservation equations for population numbers in a fluid environment. The local rate of change in population number N is due to the joint action of the realized rate of demographic change r and of movements, which are described as a flux (Schneider *et al.* 1997) that has the same units (% / unit time) as the demographic rate of change.

$$\dot{N}_o = r + F \tag{4.8}$$

In Equation 4.8, all three terms have units of per capita rate of change $N^{-1} dN/dt$, where dN/dt is the instantaneous rate of change in with change in time t.

The realized rate of demographic change has two components, birth rate B and death rate D.

$$r = B - D \tag{4.9}$$

Table 4.1 Population dynamics of plankton [$\dot{N}_{Plankton}$] and nekton [$\dot{N}_{Nekton}$] as a function of demographic rates (r = recruitment [$\dot{B}$] – mortality [$\dot{M}$]) and as a function of kinematic rates due to locomotory behaviour ($F_{loc} = \nabla_h \cdot u_{Nekton}$) and to passive motion with the ocean circulation ($F_{fluid} = \nabla_h \cdot u_{fluid}$). An additional kinematic term ($F_{atmos} = \nabla_h \cdot u_{atmos}$) is required for marine birds and fishing vessels, whose motions interact with atmospheric dynamics (Schneider *et al.* 1992). ∇_h is the horizontal gradient operator, u = organism velocity in directions x and y, with respect to fixed Cartesian grid on the surface of the earth. $u = (\delta x/\delta t \; \delta y/\delta t)$.

Population	Distribution		Demography		Behaviour	Passive
Plankton	[$\dot{N}_{Plankton}$]	=	[$\dot{B}_{Plankton}$]	$-[\dot{M}_{Plankton}]$		$-\nabla_h \cdot u_{Nekton}$
Nekton	[$\dot{N}_{Nekton}$]	=	[$\dot{B}_{Nekton}$]	$-[\dot{M}_{Nekton}]$	$-\nabla_h \cdot u_{Nekton}$	$-\nabla_h \cdot u_{Nekton}$

The kinematic term F can be partitioned into a passive component due to drift with currents, plus an active component due to locomotion.

$$F = F_{loc} + F_{fluid} \tag{4.10}$$

The local rate of change then expands to

$$\dot{N}_o = B - D + F_{loc} + F_{fluid} \tag{4.11}$$

The dynamics of plankton are captured by two dimensionless numbers – the demographic to kinematic number r/F and the population balance number B/D. The dynamics of nekton are captured by three dimensionless numbers the r/F number, the B/D number and the ratio of locomotory to passive motions, the F_{loc}/F_{fluid} number (Schneider *et al.* 1992). An example of the application of these dimensionless numbers is the dynamics of the benthic bivalve, *Macomona liliana*, during the Manukau study in New Zealand. In this example, the bedload transport due to tidal currents (F_{fluid}) was greater for recently settled juveniles than for larger adults. Consequently, it was of interest to estimate a modified r/F number, the ratio of benthic recruitment rate B to bedload transport ($R_{juv} = B/F_{fluid}$) for juveniles, and the ratio of demographic change r to bedload transport for adults ($R_a = r/F_{fluid}$). In the absence of an estimate of r in *M. liliana*, the inverse of generation time T_g^{-1} was used. Calculations in Figure 4.4c show r/F numbers less than 1 at the time and space scale of experimental units, where kinematics prevail over demographic rates. At the time and space scales of a geophysical model of sediment dynamics for the entire lagoon the r/F number exceeds 1 and demographic rates drive the dynamics of this species.

4.3 Spatial and Temporal Scaling in Estimating Uncertainty

The patterns of spatial variance in Figure 4.3 have implications for estimating uncertainty of an estimate of organism density or abundance. The pattern common to both benthic and pelagic seascapes was that spatial variance in both passively drifting organisms and sedentary organisms at the seafloor followed the rate of change in variance in physical gradients. In contrast, spatial variance in mobile organisms did not follow the rate of change in physical gradients in the environment. The rate of change in spatial variance with change in scale (measurement frequency) was gradual in mobile organisms, compared to sedentary organisms at the seafloor, or to organisms too small to escape viscous forces that govern density and abundance in the water column. Where the rate of change in variance with change in spatial scale is small (mobile organisms) variance at one scale provides a defensible estimate of variance at scales a few orders of magnitude above or below the scale of the estimate. Where the rate of change in variance with change in spatial scale is large (passively drifting and sedentary organisms) the variance estimated at one scale rapidly becomes inapplicable at another. Consequently, an estimate of the standard deviation at one spatial scale is a poor choice to estimate the sampling distribution and hence uncertainty on an estimate at another scale. While change in variance with change in spatial scale does not apply to the design of experiments and surveys where a prior estimate of the standard deviation exists at the same spatial resolution and extent, it does apply where a standard deviation is only available at another spatial scale. Where change in variance with spatial scale is to be expected and estimates of variance are not available prior to implementing a survey or experiment, the spectral density *Var(N)/f* can be used to estimate variance at one scale, from an estimate at another scale (Schneider 2009).

4.4 Spatial and Temporal Scaling in the Pelagic and Benthic Realms

Seascape ecology is founded on maps that can now, in the twenty-first century, be generated from spatially continuous data acquired from a wide range of sensors deployed both above and in the sea (Lecours *et al.* 2015; see also Chapter 2 in this book). Oceanography is necessarily founded on computational models of flow regimes, nutrient fluxes, and population dynamics. These models have long been informed by remotely sensed data of the sea surface and seafloor. Seascape ecology stands at the intersection of computational models that capture fleeting features in the pelagic realm and seafloor maps that capture more permanent features of habitat and organism distribution.

The challenge in both the pelagic and benthic realms is to scale from *in situ* measurements (*e.g.*, vertical casts, rate measurements at the seafloor, organism concentration in the water column) to the larger area of seafloor maps and water column models. Sea surface features are fugacious compared to mapped features of the seafloor or the land. The degree of fugacity is readily judged from a series of vertical casts at closely spaced intervals along a transect, then repeating them by backtracking along the transect. The scaling challenge is the degree to which a series of casts at points along the transect represents water masses along a transect. For marine birds, the scaling challenge is different. The transect is fully censused, but the repeatability by backtracking is almost nil. The degree of correlation between first pass and backtrack along the same transect deteriorates rapidly to zero after a distance of only a kilometre.

In contrast, benthic cores give repeatable measures of organism density at time scales of weeks or even months. Similarly, maps of the physical features of the seafloor are usable over longer terms, depending on the variable. Depth, gradient, and rugosity obtained by sonar backscatter from the seafloor are not expected to change over days to years or even decades, again depending on the variable. Temporally stable maps of the seafloor from remotely sensed (backscatter) data are readily trained to *in situ* point measurements of the physical properties of the sediment. Variables measured *in situ* are readily pegged to remotely sensed maps of the seafloor, over large areas with substantial support (area measured relative to area of interest). *In situ* measurements include density of organisms, species diversity, chemical fluxes into and out of the sediment, particle fluxes to the seafloor, and benthic infaunal density. A rapidly growing catalogue of statistical techniques exists to relate *in situ* measurements to acoustically mapped variables (Lecours *et al.* 2015). There is much to be said for using biologically informed models to relate *in situ* variables to seafloor maps. Once a mechanistically plausible model is used to peg *in situ* data to a map, that model can be used via integration to calculate as a first approximation, the dynamics of *in situ* variables, whether these be a standing stock (*e.g.,* number of species) or a rate.

4.5 Looking to the Future: Scaling Concepts and Practice in Seascape Ecology

Scaling concepts and practice in engineering and oceanography have already transformed ecology, via the journey of the Stommel diagram into ecology and the biological environmental sciences (Vance & Doel 2010). Concepts from oceanography can further transform research practice in seascape ecology, where dynamics in the water column differ fundamentally from those on land (Steele 1989; Kavanaugh *et al.* 2016). Here is a list of candidates for transformation.

4.5.1 From Useful Fictions to Calculation

Figure 4.2 shows three superimposed fictions, all useful. The first is the Stommel diagram itself, showing lumps where physical processes were thought to have relatively large effects on variance in sea surface height. The second is the biological overlay, showing regions where biological processes were thought to have relatively large effects on horizontal variance in concentration of plankton. The third is the overlay showing potential coverage by acoustic measurement from three different types of platform. The third overlay is a fiction relative to the coverage obtained by calculation as in Figure 4.4. Spatiotemporal support by calculation (Figure 4.4b) typically presents as a diagonal line, where the angle of the diagonal depends on the velocity of the sampling program and on lags between samples (often logistically driven). Inference from sample support to the target of inference appear in an instrumental ST diagram as rectangular areas bounded by the spatial and temporal constraints of a sampling program, by constraints of a technique such as mesocosm experiments (Sanford 1997; Petersen & Kemp 2008), or by computational limits on a dynamic model of ecosystem processes. Instrumental ST diagrams from Steele (1978) to recent (Godø *et al.* 2014) tend to show targets of inference, not support. The magnitude of scale-up becomes evident when targets of inference are displayed along with a calculation of the spatial and temporal support as in Figure 4.4.

4.5.2 From Comparative to Confirmatory Modes of Investigation

The style of inquiry in the earth and ocean sciences, notably geophysics, physical and biological oceanography, and meteorology differs from that in ecology. The former uses quantitative expression of dynamics as in Equations 4.8 to 4.10 above; ecology relies more on comparative studies, including experimental studies organized around one or more treatments relative to controls. The former rely on challenging quantitative predictions with data, while ecology remains firmly bound to statistical hypothesis testing. Is the style of inquiry found in the earth and ocean sciences, where predictions are challenged with data, possible in ecology? Here are two examples relevant to seascape ecology.

Coastal upwelling governs the rate of nutrient supply into the euphotic zone. The width of the upwelling zone is thus a determinant of primary production in coastal ecosystems. The lateral scale of the upwelling zone also influences the movement and aggregation of fish that respond to thermal gradients (Schneider & Methven 1988). The width of the upwelling zone can be predicted from the balance between wind stress (tending to spin cold nutrient-rich water upward in conjunction with Coriolis forces) and buoyancy (tending to suppress upwelling). Five parameters are needed to calculate the scale of upwelling as estimated by the internal radius of deformation R2 (Csanady 1982; Equation 3.96). For the Avalon channel east of Newfoundland in summer, R2 has a value of 4.7 km. The calculated value correctly predicts observed values, which are on the order of 5 km (Schneider 1989).

A second example of what might be called predictive seascape science comes from habitat selection by juvenile cod in the coastal zone. Thistle (2006) and Thistle *et al.* (2010) found that maximum densities of juvenile cod occur at intermediate eelgrass complexity, as measured by the fractal scaling exponent expressing the ratio between perimeter and area. One explanation, from experimental work by Gorman *et al.* (2009), is that juvenile cod have higher foraging success at intermediate complexity. In eelgrass meadows (low complexity) cod face high risk of predation by hidden predators, which results in reduced activity and reduced foraging success. In highly fragmented eelgrass (high complexity) cod face high mortality risk from paucity of shelter. The concept of higher foraging success in eelgrass of intermediate complexity is captured by three equations. The first is the parabolic relation (Thistle *et al.* 2010) of cod density D to habitat complexity β_{PA}:

$$D = e^{\eta 1} \quad \eta 1 = \alpha_o + \alpha_1 \ \beta_{PA} + \alpha_2 \ \beta_{PA}^{2} \tag{4.12}$$

The next is density taken as a function of foraging success.

$$D = f(FS) \tag{4.13}$$

The third is a parabolic relation of foraging success habitat complexity, from Equations 4.12 and 4.13:

$$FS = e^{\eta 2} \quad \eta 2 = \gamma_o + \gamma_1 \ \beta_{PA} + \gamma_2 \ \beta_{PA}^{2} \tag{4.14}$$

Thistle (2006) tested the predicted parabolic relation by choosing sites that ranged from very low to very high values of β_{PA}, then measured cod density D, habitat complexity β_{PA} and cod foraging success FS, the latter from stomach contents as a percentage by weight of fish weight.

The observed relation of cod density to habitat complexity was

$$D = e^{\eta 1} \quad \text{where} \quad \eta 1 = 1.66 + 13.95\beta_{PA} - 13.78\beta_{PA}^{2} \tag{4.15}$$

The observed relation of cod density to foraging success was

$$D = e^{\eta 3} \quad \text{where} \quad \eta 3 = 1.789 \quad \ln(FS + 5.910) \tag{4.16}$$

From (4.15) and (4.16) the predicted relation of foraging success to habitat complexity is

$$FS = -5.91 + e^{\eta 4} \quad \text{where} \quad \eta 4 = 0.928 + 7.97\beta_{PA} - 7.703\beta_{PA}^{2} \tag{4.17}$$

The observed relation was parabolic

$$FS = -5.91 + e^{\eta 5} \quad \text{where} \quad \eta 5 = -0.3065 + 6.686\beta_{PA} - 6.164\beta_{PA}^{2} \tag{4.18}$$

The observed coefficients (4.18) were similar in magnitude to the predicted coefficients (4.17) for β_{PA}, resulting in closely parallel curves when plotted.

4.5.3 From Hypothesis Testing to Likelihood

Decision-theoretic hypothesis testing against a fixed tolerance for Type I error (the P-value) is standard practice in ecology, rare in physical oceanography (Table 4.2) and in the related fields of meteorology and hydrology. Criticisms of the use of null hypothesis testing are endemic across disciplines where the practice is standard, to the point that the American Statistical Association recently issued a statement concerning statistical significance and P-values (Wasserstein & Lazar 2016). The last of six statements is of particular note, relative to the contrast in practice in the use of hypothesis testing in ecology compared to oceanography. Here is the sixth statement:

By itself, a P-value does not provide a good measure of evidence regarding a model or hypothesis.

The statement is incontrovertible, given the technical definition of a P-value. The statement begs the question: What standard should we use in research where we need a measure of the strength of evidence? What standard is reasonable, as to weight of evidence? An answer, curiously missing from the ASA statement, lies in a well-reasoned book written two decades ago. Royall (1997) argued that P-values are appropriate where a decision between a false positive (Type I error) and false negative (Type II error) is required, while likelihood ratios are appropriate measures of weight of evidence. The impediments to the adoption of likelihood ratios as a measure of weight of evidence are several (Table 4.2). First, the concept is unfamiliar to natural scientists. Likelihood ratios are not part of the standard statistical curriculum for natural scientists. In addition, likelihood ratios by themselves are of limited utility. Likelihood rises as more parameters are added and more data adduced to a model. The problem is addressed by likelihood ratios that penalize for more parameters, as does the Information Criterion (AIC) of Akaike (1974). It is further addressed by ratios that penalize for more data, as does the so-called 'Bayesian' information criterion (BIC). A third impediment is that standard output from widely used software reports F-ratios and P-values. Happily, likelihood and penalized likelihood ratios are readily calculated from standard statistical outputs in an ANOVA table, including t and F statistics (Azzalini 1996) and the

Table 4.2 Prevalence of terms in four representative journals: *Journal of Physical Oceanography (JPO), Marine Ecology – Progress Series (MEPS), Ecology (ECOL)* and *Landscape Ecology* (LECOL).

Journal	JPO	MEPS	ECOL	LECOL
Number of abstracts (Nabs)	6785	9930	12863	3673
Prevalence of terms (% of Nabs)				
mosaic (%)	0.3	0.0	4.3	25.2
fragmentation (%)	0.4	0.0	9.1	34.6
patch or patchiness (%)	8.8	17.0	18.5	40.6
spectrum (%)	32.7	4.4	5.2	7.1
P-value (%)	0.6	4.5	43.3	6.0
GIS (%)	0.2	4.1	2.6	32.1

explained variance r^2 as shown above. The issue of mode of inference (decision? weight of evidence?) and acceptable standards can only be thrashed out once natural scientists become acquainted with and begin using penalized likelihoods as a measure of weight of evidence.

4.5.4   From Scaling on a Mosaic to Scaling on the Continuum

The patch-mosaic model of landscape structure (Forman & Godron 1986) guides experimental design, analysis and management in landscape ecology. It is the foundation of a central tenet in landscape ecology that pattern varies with scale (Cushman *et al.* 2010). This assessment is borne out by prevalence of categorical terms (mosaic, patch, fragmentation) in abstracts in *Landscape Ecology*, compared to *Ecology*, *Marine Ecology – Progress Series* and *Journal of Physical Oceanography* (Table 4.2). The prevalence of a term for analysis on a continuum (spectrum) in the latter journal is notable, compared to the other three. Mayor *et al.* (2007) showed that habitat selection could be quantified on a continuum for a highly migratory species (woodland caribou *Rangifer tarandus caribou*) using two key components of winter habitat, *Cladina* lichens and snow depth. Mayor *et al.* (2009) extended this Eulerian approach (locations through which the animal moves) to a LaGrangian approach (series of locations encountered by the animal, as with a geolocation tag). Cushman *et al.* (2010) proposed a similarly continuous characterization of habitat selection, listing the problems generated by categorical treatment of habitat selection. Zhang *et al.* (2014), in a brief section of a text in geostatistics devoted to analysis on the continuum, list the technical deficiencies that arise from categorical characterization of the landscape. Detailed description of ecological analysis on a continuum can be found in a text emphasizing biological dynamics over statistical modelling (Schneider 1994, 2009). More recently, from two different starting points, Kavanaugh *et al.* (2016) and Manderson (2016) arrived at the conclusion that seascape ecology must shed the concept of habitat selection on a patch mosaic in favour of habitat selection on gradients in the water column, where boundaries shift in location and strength and large organisms rapidly coalesce and disperse through passive and active motions.

4.6   From *Ceteris Paribus* to Dimensional Thinking

It is natural to assume, as with the satirist Jonathan Swift, that what we see and observe at our human scale of days to lifetimes and of daily ambits will hold in proportion at some broader scale, all other things being equal (*ceteris paribus*). In ecology, the first response to the growing recognition that *ceteris paribus* does not apply across spatial and temporal scales was comparative investigation. An example is comparing the relation of two variables at several different resolutions, or at several different extents. The logical extension of the comparative approach is to put survey and experimental results on a continuum (Mayor *et al.* 2007, 2009). Yet another extension, beyond systematically altering resolution, extent, unit size, or spatial and temporal separation (Figure 4.5), is to use dimensional analysis to make cross-scale calculations. An example is the use of dimensional analysis to calculate the expected change in an experimentally manipulated variable, going from small to large mesocosms (Sanford 1997; Petersen & Kemp 2008).

Intuition is an unreliable guide to cross-scale research. The computational alternative, dimensional analysis, is neither a panacea nor easy. It requires a sound grasp of biological and physical processes (Barenblatt 1996). Dimensional thinking (Platt 1981), which relates one variable to another according to the principle of similitude (Equation 4.2a), is demonstrably better than intuitive notions of scale in the construction of medieval buildings (Galileo), in calculating caloric intakes (Kleiber), in survey design, in experimental design, in developing reliable models of habitat selection by mobile marine organisms (Louzao *et al.* 2011; Pittman & Brown 2011; Alvarez-Berastegui *et al.* 2014), and in research planning in the environmental sciences.

4.7 Acknowledgements

I thank Yolanda Wiersma for a productive discussion of landscape ecology and for suggesting an emphasis on research planning. I thank Diego Alvarez-Berastegui and Simon Pittman for comments on the manuscript. A sustained program of scale-focused research was made possible by a continuing sequence of grants over three decades by the Natural Sciences and Engineering Research Council (NSERC) under its individual grant program.

References

Akaike H (1974) A new look at the statistical model identification. IEEE Transactions on Automatic Control 19: 716–723.

Alvarez-Berastegui D, Ciannelli L, Aparicio-Gonzalez A, Reglero P, Hidalgo M, López-Jurado JL, Tintoré J. Alemany F (2014) Spatial scale, means and gradients of hydrographic variables define pelagic seascapes of bluefin and bullet tuna spawning distribution. PloS One 9(10): e109338.

Arrhenius O (1921) Species and area. Journal of Ecology 9: 95–99.

Ashmole NP (1963) The regulation of numbers of tropical oceanic birds. Ibis 103: 458–473.

Avnir D, Biham O, Lidar D, Malca O (1998) Is the geometry of nature fractal? Science 279: 39–40.

Azzalini A (1996) Statistical Inference Based on the Likelihood. Chapman & Hall / CRC, Boca Raton, FL.

Barenblatt GI (1996) Scaling, Self-Similarity, and Intermediate Asymptotics. Cambridge University Press, Cambridge.

Bartolino V, Ciannelli, Spencer P, Wilderbuer TK, Chan KS (2012) Scale-dependent detection of the effects of harvesting a marine fish population. *Marine Ecology – Progress Series* 444: 251–261.

Birt VL, Birt TP, Goulet D, Cairns DK, Montevecchi WA (1987) Ashmole's halo: Direct evidence for prey depletion by a seabird. *Marine Ecology – Progress Series* 40: 205–208.

Bourne WRP (1983) Birds, fish and offal in the North Sea. Marine Pollution Bulletin 14: 294–296.

Buckingham E (1914) On physically similar systems; illustrations of the use of dimensional equations. Physical Review 4(4): 345–376.

Cressie NAC (1993). Statistics for Spatial Data. John Wiley & Sons, Inc., New York, NY.

Csanady G (1982) Circulation in the Coastal Ocean. Reidel, Dordrecht.

Cushman SA, Gutzweiler K, Evans JS, McGarigal K (2010) The gradient paradigm: a conceptual and analytical framework for landscape ecology.In Cushman SA & Hetman F (eds) Informatics, and wildlife conservation. Springer, Japan, pp. 83–108.

Delcourt HR, Delcourt PA, Webb T (1983) Dynamic plant ecology: The spectrum of vegetational change in space and time. Quaternary Science Reviews 1: 153–175.

Ellis J, Schneider DC (2008) Spatial and temporal scaling in benthic ecology. Journal of Experimental Marine Biology and Ecology 366: 92–98.

Fienberg SE (1992) A brief history of statistics in three and one-half chapters: A review essay. Statistical Science 7: 208–225.

Finity L (2006) Polychaete Biodiversity over Time: A Compilation of Species Reported from Bodega Harbor and Adjacent Areas, http://www.bml.ucdavis.edu/bmr/BiodiversityOverTimeFinal6Jan.pdf (accessed 25 May 2017).

Fisher RA (1925) Statistical Methods for Research Workers. Oliver & Boyd, Edinburgh (13th edition 1958).

Forman RTT, Godron M (1986) Landscape Ecology. John Wiley & Sons, Inc., New York, NY.

Furness RW, Barrett RT (1985) The food requirements and ecological relationships of a seabird community in North Norway. Ornis Scandinavica 16: 305–313.

Gehlke CE, Biehl K (1934) Certain effects of grouping upon the size of the correlation coefficient in census tract material. Journal of the American Statistical Association 29: 169–170.

Godø OR, Handegard NO, Browman HI, Macaulay GJ, Kaartvedt S, Giske J, Ona E, Huse G, Johnsen E (2014) Marine ecosystem acoustics (MEA): quantifying processes in the sea at the spatio-temporal scales on which they occur. ICES Journal Marine Science 71: 2357–2369.

Gorman AM, Gregory RS, Schneider DC (2009) Eelgrass patch size and proximity to the patch edge affect predation risk of recently settled age 0 cod (*Gadus*). Journal of Experimental Marine Biology and Ecology 371: 1–9.

Greig-Smith P (1952) The use of random and contiguous quadrats in the study of the structure of plant communities. Annals of Botany 16: 293–316.

Haury LR, McGowan JA, Wiebe PH (1978) Patterns and processes in the time-space scales of plankton distributions. In Steele JH (ed.) Spatial pattern in plankton communities. Plenum Press, New York, NY, pp. 277–327.

Hinchey EK, Nicholson MC, Zajac RN, Irlandi EA (2008) Preface: marine and coastal applications in landscape ecology. Landscape Ecology 23: 1–5.

Horne JK, Schneider DC (1994) Analysis of scale dependent processes with dimensionless ratios. Oikos 70: 201–211.

Horne JK, Schneider DC (1997) Spatial variance of mobile aquatic organisms: capelin and cod in Newfoundland coastal waters. Philosophical Transactions of the Royal Society of London B: Biological Sciences 352(1353): 633–642.

Kavanaugh MT, Oliver MJ, Chavez FP, Letelier RM, Muller-Karget FE, Doney SC (2016) Seascapes as a new vernacular for ocean monitoring, management and conservation. ICES Journal of Marine Science 73: 1839–1850.

Kierstead H, Slobodkin LB (1953) The size of water masses containing plankton blooms. Journal of Marine Research 12: 141–147.

Kleiber M (1932) Body size and metabolism. Hilgardia 6: 315–351.

Kolmogorov AN (1941) The local structure of turbulence in incompressible viscous fluids at very large Reynolds numbers. Doklady Akademii Nauk SSSR 30: 299-303. Reprinted in Proceedings of the Royal Society London A 434, 9–13 (1991).

Krige DE (1951) A statistical approach to some basic mine valuation problems on the Witwatersrand. Journal of the Chemical Metal and Mining Society of South Africa 52: 119–139.

Lalli CM (ed.) (1990) Enclosed experimental marine ecosystems: A review and recommendations. Coastal and Estuarine Studies 37. Springer-Verlag, New York, NY.

Lecours V, Devillers R, Schneider DC, Lucieer VL, Brown CJ, Edinger EN (2015) Spatial scale and geographic context in benthic habitat mapping: review and future directions. *Marine Ecology – Progress Series* 535: 259–284.

Legendre P (1993) Spatial autocorrelation: Trouble or new paradigm? Ecology 74: 1659–1673.

Legendre P, Thrush SF, Cummings VJ, Dayton PK, Grant J, Hewitt JE, Hines AH, McArdle BH, Pridmore RD, Schneider DC, Turner SJ (1997) Spatial structure of bivalves in a sandflat: Scale and generating processes. Journal of Experimental Marine Biology and Ecology 216: 99–128.

Levin SA (1992) The problem of pattern and scale in ecology. Ecology 73: 1943–1967.

Levin SA, Morin A, Powell TM (1989) Patterns and processes in the distribution and dynamics of Antarctic krill. SC-CAMLR-VII/BG/20: 281–296.

Louzao M, Pinaud D, Peron C, Delord K, Wiegand T, Weimerskirch H. (2011). Conserving pelagic habitats: seascape modelling of an oceanic top predator. Journal of Applied Ecology 48: 121–132.

Mackas DL, Boyd CM (1979) Spectral analysis of zooplankton spatial heterogeneity. Science 204: 62–64.

Mackas DL, Denman KL, Abbott MR (1985) Plankton patchiness: Biology in the physical vernacular. Bulletin of Marine Science 37: 652–674.

Mandelbrot BB (1977) Fractals: Form, Chance, and Dimension. Freeman, San Francisco, CA.

Manderson J (2016) An essay exploring differences between seascapes and landscapes using Berhard Riemann's rules for analysis. ICES Journal Marine Science 73: 1831–1838.

Matheron G (1962) Traité de géostatistique appliquée. Editions Technip, Paris.

Mayor SJ, Schaefer JA, Schneider DC, Mahoney SP (2007) Spectrum of selection: new approaches to detecting the scale-dependent response to habitat. Ecology 88: 1634–1640.

Mayor SJ, Schaefer JA, Schneider DC, Mahoney SP (2009) The spatial structure of habitat selection: a caribou's-eye-view. Acta Oecologia 35: 253–260.

Meentemeyer V (1989) Geographical perspectives of space, time, and scale. Landscape Ecology 3: 163–173.

Mercer WB, Hall AD (1911) The experimental error of field trials. Journal of Agricultural Science 4: 107–132.

Neis B, Schneider DC, Felt L, Haedrich RL, Fischer J, Hutchings JA (1999) Stock assessment: What can be learned from interviewing resource users? Canadian Journal of Fisheries and Aquatic Sciences 56: 1949–1963.

Nowell ARM, Jumars PA (1984) Flow environments of aquatic benthos. Annual Review of Ecology and Systematics 15: 303–328.

Openshaw S, Taylor PJ (1979) A million or so correlation coefficients: three experiments on the modifiable areal unit problem. In Wrigley N (ed.) Statistical Applications in the Spatial Sciences. Pion, London, pp. 127–144.

Pacific Marine Station Staff (1977) A biological and chemical monitoring study of Bodega Harbor, California. Research Report Number 14, University of the Pacific, presented to the Sonoma County Board of Supervisors.

Patrick R (1968) The structure of diatom communities in similar ecological conditions. The American Naturalist 102: 173–183.

Petersen JE, Kemp WM (2008) Mesocosms: Enclosed experimental ecosystems in ocean science. In Steele JH, Thorpe SA, Turekian KK (eds) Encyclopedia of Ocean Sciences. Academic Press, New York, NY, pp. 732–747.

Pettibone M (1941) Polychaetes collected during intersession, 1941. Student report. Bodega Marine Laboratory, Bodega Bay, CA.

Pittman SJ, Brown KA (2011) Multi-scale approach for predicting fish species distributions across coral reef seascapes. PloS One 6(5): e20583.

Platt TR (1981) Thinking in terms of scale: Introduction to dimensional analysis. In Platt TR, Mann KH, Ulanowicz RE (eds) Mathematical Models in Biological Oceanography. UNESCO Press, Paris, pp. 112–121.

Platt TR, Denman KL (1975) Spectral analysis in ecology. Annual Review of Ecology and Systematics 6: 189–210.

Ripley BD (1981) Spatial Statistics. John Wiley & Sons, Inc., New York, NY.

Rodriguez-Iturbe I, Rinaldo A (1997) Fractal River Basins. Chance and Self-Organization. Cambridge University Press, Cambridge.

Rosen R (1989) Similitude, similarity, and scaling. Landscape Ecology 3: 207–216.

Royall RM (1997) Statistical evidence: a likelihood paradigm. Chapman & Hall / CRC, Boca Raton, FL.

Sanford LP (1997) Turbulent mixing in experimental ecosystem studies. *Marine Ecology – Progress Series* 161: 265–293.

Scales KL, Hazen EL, Jacox MG, Edwards CA, Boustany AM, Oliver MJ, Bograd SJ (2017). Scale of inference: On the sensitivity of habitat models for wide-ranging marine predators to the resolution of environmental data. Ecography 40(1): 210–220.

Schneider DC (1989) Identifying the spatial scale of density dependent interaction of predators with schooling fish in the southern Labrador Current. Journal of Fish Biology 35: 109–115.

Schneider DC (1994) Quantitative Ecology: Spatial and Temporal Scaling. Academic Press, San Diego, CA.

Schneider DC (2001a) The rise of the concept of scale in ecology. Bioscience 51: 545–553.

Schneider DC (2001b) Scale. Concept and effects. In Levin SA (ed.) Encyclopedia of Biodiversity, Vol. 5. Academic Press, San Diego, CA, pp. 245–254.

Schneider DC (2009) Quantitative Ecology, Measurements, Models and Scaling. 2nd edition. Academic Press, San Diego, CA.

Schneider DC, Duffy DC, MacCall AD, Anderson DW (1992) Seabird-fisheries interactions: Evaluation with dimensionless ratios. In McCullough DR, Barrett RH (eds) Wildlife 2001: Populations. Elsevier, London, pp. 602–615.

Schneider DC, Finity L, Schneider R (2007) Biodiversity benchmarks in space and time: Polychaetes at a single location in central California. Unpublished report prepared for the Bodega Marine Reserve http://www.bml.ucdavis.edu/bmr/PolyReport.pdf (accessed 25 May 2017).

Schneider DC, Gagnon JM, Gilkinson KD (1987) Patchiness of epibenthic megafauna on the outer Grand Banks of Newfoundland. *Marine Ecology – Progress Series* 39: 1–13.

Schneider DC, Haedrich RL (1991) Post-mortem erosion of fine-scale spatial structure of epibenthic megafauna on the outer Grand Bank of Newfoundland. Continental Shelf Research 11: 1223–1236.

Schneider DC, Methven DA (1988) Response of capelin to wind-induced thermal events in the southern Labrador Current. Journal of Marine Resources 46: 105–118.

Schneider DC, Piatt JF (1986) Scale-dependent correlation of seabirds with schooling fish in a coastal ecosystem. *Marine Ecology – Progress Series* 32: 237–246.

Schneider DC, Walters R, Thrush SF, Dayton PK (1997) Scale-up of ecological experiments: Density variation in the mobile bivalve *Macomona liliana*. Journal of Experimental Marine Biology and Ecology 216: 129–152.

Steele JH (1978) Some comments on plankton patches. In Steele JH (ed.) Spatial pattern in plankton communities. Plenum, New York, NY, pp. 1–20.

Steele JH (1989) The ocean 'landscape'. Landscape Ecology 3: 185–192.

Stigler SM (1980) Stigler's law of eponymy. In Science and social structure: a festschrift for Robert K. Merton. Transactions of the New York Academy of Science 39: 147–158.

Stommel H (1963) Varieties of oceanographic experience. Science 139: 572–576.

Swift J (1726) Travels into Several Remote Nations of the World. In Four Parts. By Lemuel Gulliver, First a Surgeon, and then a Captain of Several Ships. Benjamin Motte, London.

Thistle ME (2006) Distribution and Risk-Sensitive Foraging of Juvenile Gadids in Relation to Fractal Complexity of Eelgrass Habitat. MSc thesis, Memorial University, St John's, Canada.

Thistle ME, Schneider DC, Gregory RS, Wells NJ (2010) Fractal measures of habitat structure: Maximum densities of juvenile cod occur at intermediate eelgrass complexity. *Marine Ecology – Progress Series* 405: 39–56.

Thompson, DAW (1915) Galileo and the principle of similitude. Nature 95: 426–427.

Thompson GG, Withers PC, Pianka ER, Thompson SA (2003) Assessing biodiversity with species accumulation curves; inventories of small reptiles by pit-trapping in Western Australia. Australian Ecology 28: 361–383.

Tolman RC (1914) The principle of similitude. Physical Review 3: 244–255.

Vance TC, Doel RE (2010) Graphical methods and cold war scientific practice: The Stommel diagram's intriguing journey from the physical to the biological environmental sciences. Historical Studies in the Natural Sciences 40: 1–47.

Wasserstein RL, Lazar NA (2016) The ASA's statement on P-values: context, process, and purpose. The American Statistician 70(2): 129–133.

Weber LH, El-Sayed SZ, Hampton I (1986) The variance spectra of phytoplankton, krill and water temperature in the Antarctic Ocean south of Africa. Deep-Sea Research 33: 1327–1343.

Wiens JA (1989) Spatial scaling in ecology. Functional Ecology 3: 385–397.

Wiens JA, Moss MR (2005) Issues and Perspectives in Landscape Ecology. Cambridge University Press, Cambridge.

Wiens JA, Scott JM (1975) Model estimation of energy flow in Oregon coastal seabird populations. Condor 71: 439–452.

Wu J, Hobbs RJ (2007) Key Topics in Landscape Ecology. Cambridge University Press, Cambridge.

Zhang J, Atkinson PM, Goodchild MF (2014) Scale in Spatial Information and Analysis. CRC Press, Boca Raton.

PART II

Linking Seascape Patterns and Ecological Processes

5

Ecological Consequences of Seagrass and Salt-Marsh Seascape Patterning on Marine Fauna

Christoffer Boström, Simon J. Pittman and Charles Simenstad

5.1   Introduction

Perhaps you have swum over the lush vegetation of a seagrass meadow or observed the patchwork mosaics of marine vegetation and sand from the window of an airplane flying along the coast. You may have been fortunate enough to have trudged across a salt marsh and remember the diverse plant communities arranged in patches, zones and gradients along the shore. From the air, the patterning of salt marshes is even more intricate with complex branching (dendritic) networks of tidal channels meandering across the seascape. It is clear that seagrass meadows and salt marshes seldom form continuous vegetated areas. Instead, they are structurally heterogeneous, typically occurring as spatial mosaics of different types of plants and sediments. When viewed as interacting mosaics of habitat forming a seascape, these patterns elicit new and important questions for ecologists. How are the great variety of dynamic processes, such as the tidal movements of water, storm events and activities of organisms responsible for these patterns? What are the consequences of these spatial patterns on animal distributions and biological processes? How can we measure and analyse spatial patterns and ecological processes to gain a better understanding of how coastal ecosystems work? How will global change influence the distribution and function of these habitats?

Our primary objective in this chapter is to explain, using a seascape ecology perspective, how the self-organizing properties of seagrass and salt marsh seascapes shape the communities and interactions of organisms that occupy them. Firstly, we present information on the global distribution and functions of seagrasses and salt marshes, and highlight their present and future threats and stressors. Secondly, we present intrinsic and extrinsic processes maintaining seagrass and salt marsh seascapes. Thirdly, we describe a range of distinctive structural features that have been mapped and measured using a landscape ecology approach and discuss, with examples, the ecological consequences of structural features such as patch size and edges on invertebrate and fish distribution patterns. We include examples of how vegetation patterns influence processes including carbon and nutrients flows; settlement of organisms; animal movement; predation and growth. Fourthly, we demonstrate and discuss the importance of physical and biological connectivity between seagrass meadows and salt marshes. We conclude the chapter by highlighting some key research advances in seagrass and salt marsh ecology and identify

Seascape Ecology, First Edition. Edited by Simon J. Pittman.
© 2018 John Wiley & Sons Ltd. Published 2018 by John Wiley & Sons Ltd.

research priorities that need to be addressed in order to improve our understanding of the structure and function of coastal vegetated seascapes.

5.1.1   Seagrasses and Salt Marshes: Global Distributions and Ecosystem Functions

Seagrasses and salt marshes form ecologically and economically important habitats of shallow subtidal and intertidal coastal ecosystems (Figure 5.1). Mapping studies have estimated the global cover of seagrasses at between 177 000 and 600 000 km^2 and salt marshes between 22 000 and 400 000 km^2 (Chmura *et al.* 2003; Green and Short 2003; Duarte *et al.* 2005a, b; McLeod *et al.* 2011). Seagrasses are marine flowering plants of 72 species distributed over 13 genera and five families (Unsworth & Cullen-Unsworth 2014). Most species are found in shallow water down to 10 m depth, but extensive deep water (45 m) meadows have also been recorded (Procaccini *et al.* 2003). With records from >120 countries, seagrasses occupy both brackish and marine coastal waters from the tropics to the shores of Greenland and Iceland (Green *et al.* 2003). The Philippines, Papua New Guinea and Indonesia forms the centre of seagrass biodiversity (14 species), while the Atlantic coasts of the northern hemisphere are dominated by *Zostera* (Figure 5.2) and *Ruppia* species (Green *et al.* 2003). For further biogeographic information on seagrasses, see Den Hartog & Kuo (2006) and Short *et al.* (2007).

Salt marshes form along low-energy shorelines in temperate and high latitudes influenced largely by sea level change and tidal range, availability of sediments, wind and wave energy (Scott *et al.* 2014). Salt marshes do occur in some tropical settings, although mangroves tend to dominate intertidal areas. Globally, salt marsh plant richness is high with >500 known species (Silliman 1994). Salt marshes are distinctive in different tidal regimes: across broad coastal plain estuaries and wave-built features along microtidal coasts, behind barrier islands in mesotidal settings and within diverse types of funnel-shaped estuaries on macrotidal coasts (Eisma *et al.* 1998). Seagrasses and salt marshes provide multiple ecosystem services including provisioning of food and shelter for diverse food webs, coastal sediment stabilization, particle trapping and nutrient cycling (Weslawski *et al.* 2004; Cullen-Unsworth *et al.* 2014). Compared to terrestrial ecosystems, seagrasses, salt marshes and mangroves also play a disproportionally large role in carbon dioxide sequestration (Nellemann *et al.* 2009; McLeod *et al.* 2011). In particular, carbon burial rates in salt marshes is much higher than in seagrass meadows and sediments in vegetated coastal ecosystems may approach 1000 g C m^{-2} year^{-1} compared to 10 g C m^{-2} year^{-1} for tropical, temperate and boreal forests (McLeod *et al.* 2011).

5.2   Structural Processes and Change in Coastal Seascapes

Increasing anthropogenic pressures have resulted in seagrass meadows and salt marshes being lost at much faster rates than tropical rain forests (Achard *et al.* 2002). The global loss of seagrasses is estimated to be 29% or >50 000 km^2 since 1879, with an annual loss rate of 7% since 1990 (Waycott *et al.* 2009). The corresponding estimate for changes in salt marshes is approximately a 30% global loss (Bridgeham *et al.* 2006; Intergovernmental Panel on Climate Change 2007; McLeod *et al.* 2011). For a summary of drivers

(a) Saltmarsh

(b) Seagrass

(c)

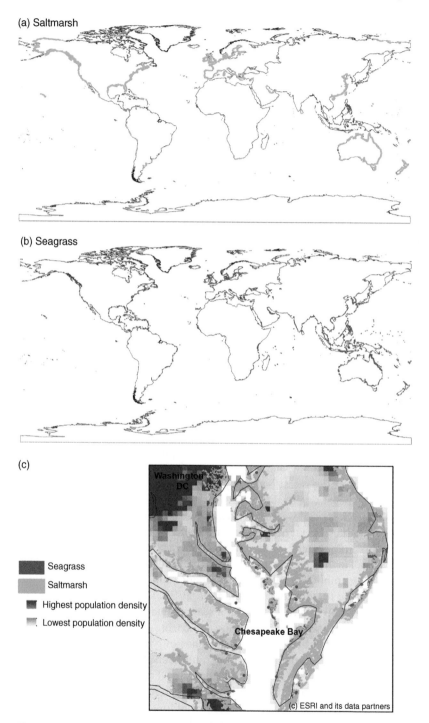

Figure 5.1 Global distribution of (a) seagrasses and (b) saltmarshes; and (c) a close up of Chesapeake Bay to highlight the juxtaposition of seagrasses and saltmarshes in close proximity to the densely populated east coast of the United States. *Source*: Adapted from habitat data provided by UNEP World Conservation Monitoring Centre and human population data provided by ESRI.

Figure 5.2 Seagrass seascape dominated by eelgrass (*Zostera marina*) in SW Finland, northern Baltic Sea. *Source*: Photo: Christoffer Boström.

responsible for these alarming declines, see McLeod *et al.* (2011) and Boström *et al.* (2011). Although there are ample projections of quantitative changes in seagrass meadows and salt marshes with accelerated climate change (McKee *et al.* 2012; Osland *et al.* 2016), evaluation of qualitative changes in their spatially explicit geometry are rare (Moffett & Gorelick 2016). Consequently, quantifying and understanding the ecosystem consequences of spatial patterning is fast becoming a priority for applied coastal ecology, planners and engineers because global loss has already resulted in significant declines in economic benefits from coastal wetlands (Costanza *et al.* 2014). For seagrasses, negative changes include reduced percentage cover and habitat fragmentation, *i.e.* splitting of continuous habitats into (smaller) fragments causing increased isolation of patches and smaller patch sizes. Furthermore, spatial changes typically induce nonlinear positive feedbacks, which further amplify the effects of disturbance and complicate the detection of the ecological response to change (Maxwell *et al.* 2016). Construction and dredging activities may, in addition to direct removal of seagrass plants, cause reduced water clarity and increased sedimentation. Mapping and comparing spatial pattern metrics of tidally open salt marshes to tide-restricted salt marshes in New Jersey, United States, Artigas & Yang (2004) found that the marshes' area, number of patches and total edge was higher in tidally open sites than in tide-restricted sites. From an areal perspective, SLR will primarily impact salt marsh by reducing patch size (Craft *et al.* 2009). For the southeastern United States, Hunter *et al.* (2015) assessed forecasted habitat change up to year 2100 using three structural habitat metrics: total area, patch size and habitat permanence. Salt marsh and species that depended on marshes are expected to experience greatest change across these three pattern metrics.

5.2.1 Processes Creating and Maintaining Seagrass Seascapes

Seagrass seascapes exhibit a wide range of patches of different sizes and shapes at different spatial scales (cm–km) especially in physically exposed locations where small and narrow patches with low core area, or doughnut-shaped circular patches (Figure 5.2) with large, central openings or 'halos' exist (Fonseca & Bell 1998; Fonseca *et al.* 2002; Frederiksen *et al.* 2004). Both immediate and long-term responses of seagrass seascape structure to physical disturbance has been extensively studied. Compared to sheltered sites, seagrass seascapes exposed to currents, migrating sand dunes or wave surge tend to be less aggregated (larger sand gaps) with more elongated patch shapes, accumulate less detrital material and show lower leaf nitrogen concentrations (Fonseca & Bell 1998; Bell *et al.* 1999; Hovel *et al.* 2002; Frederiksen *et al.* 2004; Ricart *et al.* 2015). Fragmentation and extreme sedimentation events can cause an often irreversible bare state with altered sediment and turbidity conditions (Carr *et al.* 2015) and limited restoration success (Paling *et al.* 2009). There is also some evidence of faunal interactions influencing the structure and dynamics of the seagrass-sand mosaic. For example, rays, crabs, birds, urchins and polychaetes may through their burrowing activity, or direct consumption, contribute to gap formation and hamper seed recruitment (Valentine & Heck 1999; Fishman & Orth 2001; Valdemarsen *et al.* 2011).

Intrinsic mechanisms for seagrass patch expansion and thus landscape development, include two strategies; clonal growth through ramets and / or infilling of unvegetated gaps through seedling recruitment. Patch expansion is typically fastest at edges through runners occupying the unvegetated substrate and slower towards the patch interior (Marba & Duarte 1998). Small patches show greater mortality (Olesen & Sand Jensen 1994) and there is limited, but increasing evidence that patch expansion is a nonlinear and self accelerative process where colonization rates increase with increasing patch size (Duarte *et al.* 2006). The factors determining the outcome of this process are rhizome elongation rate, branching frequency and branching angle (Marba & Duarte 1998; Sintes *et al.* 2005). Linking local scale ramet interactions with larger scale patch and seascape patterns and physical setting should provide new insight into the mechanisms behind seagrass landscape dynamics and inform seagrass restoration, but this area of research has received limited attention (Kendrick *et al.* 2005).

5.2.2 Processes Creating and Maintaining Salt-Marsh Seascapes

At regional scales, salt marshes are shaped by the interaction of the coastal geomorphology, wave climate, sediment supply and tidal range, predicated by antecedent processes that effected marsh formation and evolution, *e.g.*, primarily sea level rise (Figure 5.3). Vertical zonation of salt marsh vegetation is determined largely by the tidal regime, which regulates flooding frequency and duration, substrate saturation and sediment structure relative to the physiological tolerances of the different halophytes (Figure 5.4). In general, the greater the tidal range, the more complex the zonation. Given appropriate conditions for salt marsh transgression (landward migration) with sea level rise, low wind wave / current erosion and sufficient sediments, the determining factors will be the distribution and shape of accommodation space for horizontal landward movement. In low topography coasts, marshes have formed and persisted as broad contiguous patches, especially so in large river deltas. Conversely, along topographically complex coasts with variable sediment supply and erosion regimes, salt

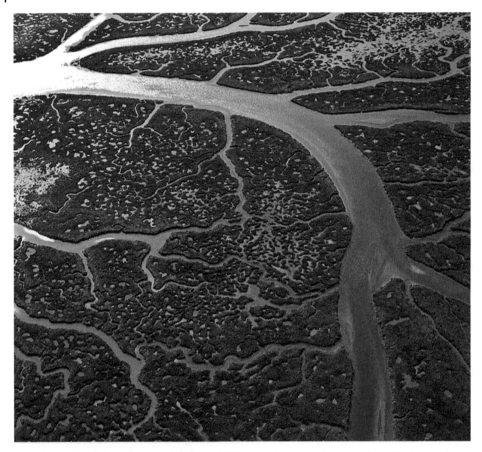

Figure 5.3 Salt marsh seascape in Bahia de Cadiz Natural Park, Huelva, Andalucía, Spain. *Source*: Photo: Juan Carlos Munoz 2008.

marshes exhibit spatially complex distributions and varying patch sizes dependent on the availability of suitable space in fjords, estuaries and bays, barrier beaches lagoons and varieties of estuaries (Scott *et al.* 2014).

While vegetation patches and tidal channel features are relatively stable in mature salt marshes, some marsh settings have unique local features that are ephemeral over space and time. Some processes that affect spatial patterning are more stochastic and dynamic. Various natural disturbance processes are responsible for patch formation and change in salt marshes. For example, tidally deposited plant debris or sediment that smothers underlying plants, and produces open bare patches (Bertness & Ellison 1987; Bertness & Shumway 1993). Such bare patches, which often take the shape of ponds or 'pannes' (Redfield 1972; Pethick 1974; see also Figure 5.3), may be colonized by pioneer plants, but their persistence is usually short term (2–3 years) before zonal dominants succeed within 3–4 years. In some temperate regions, fluvial or tidal flooding often recruits floating dead trees into salt marshes where smothering or 'ploughing' of the marsh vegetation and surface sediments creates complex disturbance patches that produce diverse successional patterns (Hartman 1988; Bertness & Shumway 1993), as

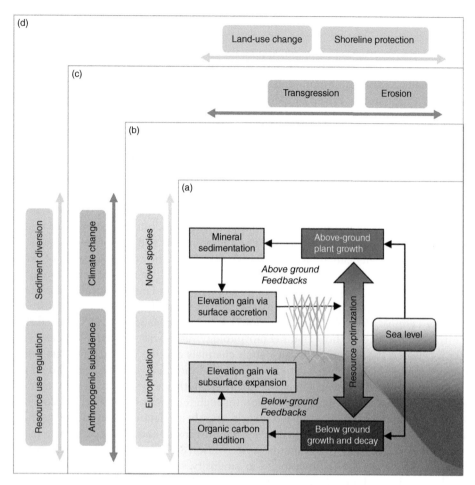

Figure 5.4 Natural horizontal and vertical processes and feedback mechanisms in a salt marsh. Effects of sea-level rise may increase horizontal erosion in shallow intertidal environments or increase vertical accretion in submerged salt marsh seascapes by increased sediment delivery and deposition (a). Examples of local (b), broad-scale (c) and socio-economic (d) drivers that influence above and below-ground accretion and erosion processes are shown in green, blue and yellow boxes, respectively. *Source*: Figure modified from Kirwan & Megonigal (2013).

well as enhancing vegetation zonation (Hood 2007a). Major storm events can also cause lateral erosion of marsh edges, dissection by more or wider tidal channels (Pethick 1992), or overwash deposits may cover the marsh surface, especially along barrier beaches (Donnelly *et al.* 2004).

Salt marshes are also structured by intrinsic processes and biotic interactions within vegetation patches. The distribution of plant associations, tidal channels and unvegetated sediments in salt marshes are organized in part by the feedback between the tidal range, intertidal gradient, disturbance and patch size (see Figure 5.4). Interactions among plants within and among marsh patches include both successional progression of plant dominance resulting from either scouring of vegetation or changes in substrate elevation, usually in response to natural disturbance such as channel erosion or large floating wood, as well as shifts in assemblages due to both indigenous and nonindigenous invasive species. Perennial pepperweed (*Lepidium latifolium*) and several species of cordgrass (*Spartina* spp.) have been found to change marsh microtopography and soil salinity and chemistry (Tiner 2013). Some shifts in marsh vegetation can be even just as dramatic among endemic species, such as the invasion of salt marshes by mangroves apparently driven by global climate change (Saintilan *et al.* 2014).

5.2.2.1 Tidal Channel Networks in Salt Marshes

The network of tidal channels that flush water and its constituents across the marsh seascape strongly influences patch size, shape, patch dynamics and therefore ecological function. Eisma *et al.* (1998) classify three (straight, sinuous, meandering) channel types and six or more (parallel, dendritic, elongated, distributary, braided, interconnecting) types of channel systems. The complex factors underlying the different types of channels and channel systems are still somewhat poorly known, but the interplay of tides, waves, tidal prism, slope, type of sediment and its transport and consolidation, vegetation and faunal activity can all be involved (Eisma 1993). The erosive capacity of tidal flow, as a function of the tidal regime, marsh area and marsh area / tidal prism (volume of water stored in a drainage area between mean high tide and mean low tide), is related to channel plan form, complexity and cross section by hydraulic geometry relationships (Myrick & Leopold 1963; Williams *et al.* 2002). Leopold & Maddock (1953) coined the term hydraulic geometry (how channel morphology varies with discharge) to describe the link between the structure of salt marsh channels and the dynamic properties of hydrology. Since then, ecological studies of channel morphology have been shown to scale strongly with a number of ecological functions, both in terms of direct nekton densities but also a number of indirect function through food web support (Hood 2007b). The study of the morphology of tidal networks and their relation to salt marsh vegetation is an active area of research with modern remote sensing and spatial pattern metrics being applied to characterize stream networks (Mason & Scott 2004). Examples of spatial pattern metrics include bifurcation ratio, sinuosity, stream order, perimeter-to-area ratio, shape index and fractal dimensions (Zeff 1999).

5.3 Ecological Consequences of Seascape Structure

5.3.1 Seagrass Patch-size Effects on Epifauna and Fish

Ecological theory suggests that patch size should have a profound influence on patterns of species distribution, assemblage diversity and biological processes (Williams 1964; Connor & McCoy 1979; Freemark & Merriam 1986). In general, terrestrial landscape ecology has found strong patch size effects for edge-specific and interior-specific species and negligible effects for generalist species using both interior and edge as habitat (Bender *et al.* 1998). Consequently, patch size became one of the first pattern metrics applied through a landscape ecology approach to shallow water vegetated seascapes. Influenced by the theory of island biogeography (MacArthur & Wilson 1967), studies in seagrass beds first sought to understand the effect of patch size on marine fauna to address the SLOSS (single large or several small areas) debate and guide marine protected area

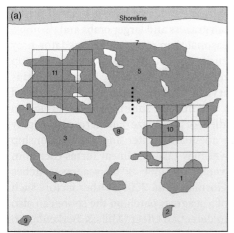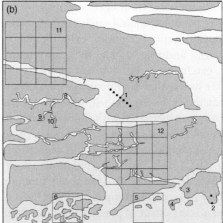

Figure 5.5 Vertical view of a seagrass (a) and a saltmarsh (b) seascape illustrating commonly investigated metrics and comparisons influencing distribution and abundance of marine fauna. (i) 1 versus 2: patch size effects, 3 versus 4: patch shape effects (differing edge : area ratios), 5 versus 6 and 7: interior versus edge effects (dots at 6 illustrate a transect sampling across the edge), 8 versus 9: patch isolation; and 10 versus 11: grid to measure percentage cover. (ii) 1: transect for sampling across channel edge, 2: edge : interior comparison in marsh patch, 3 versus 4: patch size effect, 5 versus 6: patch density (percentage cover), 7 versus 8, 9 and 10: effect of larger, higher order channel / corridor versus lower smaller order channels; and 11 versus 12: effects of amount of edge per unit area.

planning (McNeill & Fairweather 1993). The question has been addressed by either sampling natural patches of different sizes (Figure 5.5), or utilizing different sized artificial seagrass units (ASUs). In general, many marine fauna seem not to respond strongly to differences in seagrass patch size. In fact, about three-quarters of the available studies to date show no significant influence of patch size on fishes and invertebrates (Bell *et al.* 2001; Connolly & Hindell 2006; Boström *et al.* 2006, 2011). In the few cases where a significant positive patch size effect was found (*i.e.*, increasing abundance or richness with increasing patch size), the relationship appears to be more common among invertebrates than among fishes (Boström *et al.* 2011).

To study seagrass patch size effects in seascapes is challenging and there are numerous examples of inconsistent species-patch size relationships. Typically, effects are species, or life stage specific, or dependent on patch quality variables, including but not limited to, shoot density, amount of drift algae or sediment properties. Reusch (1998) found more mussel recruits (*Musculista senhousia*) in medium sized patches than in small and large patches. For seagrass epifauna in a South African estuary, Källén *et al.* (2012) found that epifaunal richness and gastropod abundance was significantly greater in large beds compared to small ones, but also that increasing patch size increased the difference in faunal communities sampled at patch edges versus interiors. Patch size effects on growth have been studied for bivalves. For filter-feeding bivalves dependent on water flow rates for adequate particle / food supply, less vegetation and smaller patches may be more beneficial (Carroll & Peterson 2013). Thus, growth rates of filter feeding bivalves is maximized outside vegetation and decrease with increasing patch size (Reusch & Williams 1999; Carroll & Peterson 2013), but see Irlandi (1996) for conflicting results.

Studies of predation effects in relation to seagrass patch size have generated mixed results. Mobile epifauna, such as small crustacean grazers and larger crabs and shrimps, exhibit substantial temporal variation due to seasonality in recruitment, predator-prey dynamics and migrations. Thus, amphipod density may show a positive relationship with patch size in fall, but not in spring (Bell *et al.* 2001). Similarly, Hovel & Lipcius (2001) noted that fragmentation and crab survival varied temporally and recorded increasing crab survival with decreasing patch size in early summer, while survival in late summer increased with shoot density regardless of patch size. Furthermore, a highly fragmented seascape may not always provide adequate shelter and the relationship between predation intensity and vegetation cover may take nonlinear forms (Boström *et al.* 2011). For example, predation risk of juvenile cod in 1–80 m^2 seagrass patches was best described with a parabolic function (Gorman *et al.* 2009). Other factors such as the within-patch composition and location of a seagrass patch on the shore can also influence the effect of patch size or the ability to detect an effect (Mills & Berkenbusch 2009). Some evidence suggests that larger, more continuous seagrass patches provide greater long-term stability and harbour more stable nekton communities across time than smaller patches (Hensgen *et al.* 2014). These patterns may be depth related, as deeper patches have been show to support higher density and diversity of fish (Jackson *et al.* 2006) and influence edge effects on fish (Smith *et al.* 2008, 2010).

Instead of measuring patch-centred metrics such as patch size and isolation, Fahrig (2013) argued for ecologists to measure the amount of habitat in the surrounding seascape. Such measurements, *i.e.* the percentage of the seascape, or areal cover of seagrasses, can be quantified from benthic habitat maps at a range of spatial scales. Recognizing the importance of seascape context, several studies have found that the amount of seagrass in the surrounding seascape influences the abundance of fish species, crustaceans and assemblage diversity sampled within individual patches (Pittman *et al.* 2004, 2007; Grober-Dunsmore *et al.* 2007; Yeager *et al.* 2011, 2012).

5.3.2 Patch Edges: Conceptual Framework and Application

In landscape ecology, habitat patches may be divided into distinct core or edge areas (Forman 1995). The core or patch interior is often defined as the area located at a certain distance from the patch border and the edge is the transition zone between two different, adjoining habitat types, which in seagrass seascapes is typically between bare sand and vegetation. In theory the outcome of edge responses of a patch-interior or patch-exterior species can be positive (increase), neutral (no effect) or negative (decrease). Thus, the shape of the response curve crossing two habitats varies depending on the resource distribution between the habitats compared (Ries & Sisk 2004). However, only one example currently exists in marine systems where the model has been tested using artificial seagrass patches (Macreadie *et al.* 2010a and b). The characterization of edge versus core habitat is problematic as the distance between the core area and the edge area is dependent on the species or process of interest, as well as the distance to neighbouring patches (Figure 5.6). The underlying assumption of edge effects is the physical or abiotic differences between the edge and the interior. Regardless of the way boundary function is classified, a focus on boundary function in addition to physical boundary properties will help to understand structure-function relationships and could be applied within a spatial context to better understand and categorize fluxes of larvae and animal movements across the seagrass / sand boundary.

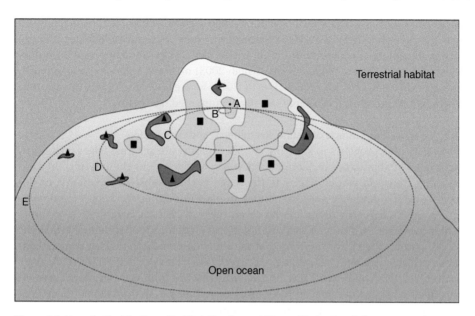

Figure 5.6 Hypothetical Scaling of habitat. Species mobility and behaviour influence organism responses to seascape connectivity and configuration. Habitat perception is species and process specific and a fragmented habitat may appear fragmented for one species but continuous for another species. A = a sessile invertebrate, *e.g.* a bivalve, perceives a small patch as a continuum, B = semimobile invertebrate, *e.g.* a shrimp, with a small perceptual range utilizing the edge and the interior of a small patch, C = a fish species with a limited home range can easily cross bare sand and perceives its surroundings as almost continuous vegetation, D = a fish with medium sized home range may effectively utilize habitat edges and perceive its foraging area as either fragmented or continuous, E = a large, highly mobile species may perceive the entire bay as an unvegetated area. Dark green elongated patches with high P : A ratios are suitable only for edge specialists (black triangles), while the rest of the patches (light green) have lower P : A ratios able to support species dependent on core areas (black squares).

The study of edge effects has been advanced through the integration of landscape ecology concepts with conventional marine ecology field experiments and visual surveys (Connolly & Hindell 2006; Boström *et al.* 2011). Landscape ecology studies have often used the perimeter-to-area ratio (P : A) as a metric to quantify patch shape and as an indicator of fragmentation (Calabrese & Fagan 2004). It follows that two patches of the same area (*e.g.*, 9 m^2), but with differing shape, *e.g.* quadratic 3 × 3 m versus elongated 0.5 × 18 m, have widely differing P : A ratios (*i.e.*, 4.1 and 1.33), respectively. Experimental designs using artificial seagrass patches arranged in different spatial configurations suggest that the spatial distribution of seagrasses can influence edge effects in predatory fishes because of greater abundance of prey at edges. For instance, Macreadie *et al.* (2010b), found higher abundance of pipefish at the edges (0–0.5 m) of continuous (9 m^2) seagrass beds than patchy seagrass beds (4 × 1 m^2 patches), an edge effect that was driven by higher prey abundance at the edges of continuous patches. In addition, it is important to recognize that although edges have most commonly been defined and measured at relatively fine spatial scales and often for single patches, the amount of edge, edge density and even the contrast between adjacent patches (contrast-weighted edge density) can be quantified across entire seascapes scaled by organism movements (Hitt *et al.*

2011; Kendall *et al.* 2011). Distinct vegetation edges are a prominent feature in many salt marsh seascapes (Figures 5.3 and 5.5). Some of the more well-known linkages between spatial patterning and ecological function within salt marsh seascapes have been documented for tidal channel edge density (Kneib 2003) or the marsh edge interface with open water (Minello *et al.* 2008; Nevins *et al.* 2014) where the highest densities and taxa richness of fish and crustaceans are typically documented (see below).

5.3.2.1 Seagrass Edge Effects on Faunal Recruitment and Distribution

The majority of approaches to studying animal responses to seagrass edges include: (i) comparisons of abundance or biomass at the patch edge versus the interior; (ii) sampling along transects perpendicular to the seagrass / bare substrate boundary; or (iii) comparisons of abundance and diversity of animals in fragmented versus continuous seagrass (Figure 5.5). The literature, however, suggests that the magnitude and direction of the interaction among edges and functional attributes is often influenced by the choice of pattern metric, or the timing of the study (seasonality), making general conclusions difficult (Irlandi 1997; Bologna & Heck 2000; Hovel 2003; Smith *et al.* 2010; Carroll *et al.* 2012; Rossi *et al.* 2013). The majority of work conducted show inconsistent or no effects of seagrass edges on the density or richness of marine fauna. In cases of significant edge effects, however, positive responses appear more common than negative responses, especially for fishes and particularly for larger more generalist predatory fishes foraging at edges (Connolly & Hindell 2006; Boström *et al.* 2006, 2011; Smith *et al.* 2011, 2012).

There is mounting evidence that seagrass edges influence the processes that regulate settlement and recruitment of benthic invertebrates. Seagrasses interact with water flow at multiple scales from single shoots to patches and entire seascapes (Bell *et al.* 1999). It is well known that dense seagrass canopies reduce water flow and trap sediment and organic particles (Gambi *et al.* 1990). Consequently, above-ground structure also modifies passive bedload transport, settlement and recruitment of larval stages of many benthic species (Butman 1987; Boström & Bonsdorff 2000; Boström *et al.* 2010). Compared to the usually species poor unvegetated sediments (*e.g.*, Boström & Bonsdorff 1997), this depositional process may build up a pool of organic carbon (Greiner *et al.* 2013) which, in turn, fuels diverse infaunal and epifaunal food webs (Boström & Bonsdorff 1997). The seagrass patch edge disrupts bedload transport and thus functions as a depositional environment for particles and larvae. Similar to butterfly interactions when encountering terrestrial grasses (Haddad & Baum 1999), seagrass will likely function as a 'drift fence', intercepting dispersing individuals from the water column. The depletion of larval density into a seagrass bed was first noticed by Orth (1992), who coined the term 'settlement shadow'. Indeed, several studies from different regions show that settlement of invertebrate larvae is highest along patch edges compared to the interior (Bologna & Heck 2000; Boström *et al.* 2010).

Seagrass patch edges are regions of intensified predator-prey interactions with higher mortality of macrofauna and fish along seagrass edges compared to interiors (Irlandi *et al.* 1995, 1999; Bologna & Heck 1999; Gorman *et al.* 2009; Smith *et al.* 2011). For example, due to predation causing postsettlement mortality an initially positive edge effect on bay scallop settlers (*Argopecten irradians*) may be transient and later manifest as a neutral effect, *i.e.*, no difference between edge and interior on the actual recruitment patterns (Carroll *et al.* 2012). Seagrass patch edges may also influence the distribution of actively moving adult species such as mesograzers. Some epifaunal crustaceans, such as gammaridean amphipods and tanaids, show higher abundances at the edges compared to interior parts (Tanner 2005). Infaunal responses to seagrass edges are still under-studied, but polychaetes have in several studies shown elevated densities along edges (Bowden *et al.* 2001; Bologna & Heck 2002; Tanner 2005). Smith *et al.* (2011) sampled fish communities at multiple patchy seagrass seascapes in Australia and found a strong and consistent pattern of predatory fish using sand edges adjacent to seagrass more than the seagrass interior. Experimental manipulations of food availability at edges and interiors of artificial seagrass patches revealed that greater food availability at edges explained the positive edge effect observed for a pipefish (Macreadie *et al.* 2010a).

5.3.3 Effects of Salt-Marsh Patch Size, Edges and Connectivity on Faunal Patterns and Processes

Tidal salt marshes exhibit complex spatial heterogeneity in geomorphological and vegetation patterns at a range of spatial scales that influence wetland functions (Dunning *et al.* 1992). Salt marshes may occur as multiple isolated patches (*e.g.*, in volcanic or tectonic landscapes), highly channelized interconnected patches (*e.g.*, deltas), or long narrow patches (*e.g.*, lagoons behind barrier sand bars and cuspate beaches). Salt marsh patch size and shape influence the distribution of marine animals and ecological processes (Green *et al.* 2012; Meyer & Posey 2014). The size, amount of edge, connectivity and spatial arrangement of salt marsh patches have been documented to affect the distribution of fish and crustaceans in Australia (Olds *et al.* 2012; Gibbes *et al.* 2014), Gulf of Mexico (Roth *et al.* 2008), and various estuarine seascapes of the southeastern coast of North America (Meyer and Posey 2014; Baillie *et al.* 2015). In addition, patch area effects have been documented for diverse avifauna and even insects; for instance, Raupp & Denno (1979) found that the densities of sap-feeding homopteran and hemipteran insects varied positively with marsh (*Spartina patens*) patch size, although taxa diversity did not. The total size of island marshes can also influence marine faunal distributions. Densities of marsh dependent species, including mummichog (*F. heteroclitus*), were significantly higher within large (3000–10 000 m^2) island marshes compared to small (<1000 m^2) island marshes (Meyer & Posey 2014). Other salt marsh patch attributes that have been related to nekton density or taxa richness have included shape, area and perimeter : area ratio of tidal channels (Green *et al.* 2012).

Edges of marsh patches constitute important ecotones with channels and unvegetated flats. Fish and crustacean species richness and abundance sampled at the marsh edges are typically greater than those sampled in the patch interior (Peterson & Turner 1994; Kneib 2000; Minello *et al.* 2003). Kneib (2003) reported a strong sigmoid relationship between nekton production and the amount of intertidal marsh channel edge within a 200 m radius of a site for both resident ($r^2 = 0.82$) and migrant ($r^2 = 0.68$) marine faunal species. Minello (1999) and Whaley & Minello (2002) sampled highly fragmented salt marsh in Texas and observed a high abundance of predatory fish and crustaceans at marsh edges due to high availability of infaunal prey. Spatial simulation models have also been used to examine salt marsh edge effects on fish and fisheries species abundance. Browder *et al.* (1989) modelled increases in the amount of edge (water-marsh interface) and the effect on brown shrimp (*Farfantepenaeus aztecus*) productivity in Louisiana salt marshes, which predicted a statistically significant positive linear relationship between shrimp catch and total edge length. Haas *et al.* (2004) used a spatially explicit, individual-based simulation model (IBM) to investigate the role of marsh edge habitat on brown shrimp movement, mortality and growth. The model predicted that seascapes with more edge produced higher shrimp survival rates (see also Chapter 8 in this book). Lowe & Peterson (2014) sampled salt marsh seascape structure and nekton and macroinvertebrate fauna across a gradient of human impact showing that both faunal assemblages were most strongly correlated with landscape features, especially total edge, the proportion of salt marsh in the seascape and tidal channel connectivity (proximity of channel mouths). Marsh plant structure and species composition may also have strong effects on the associated community. Some of the more prominent examples of invasive species changes involve almost complete conversion of naturally diverse salt marsh assemblages to monocultures of common reed (*Phragmites australis*), which has been found to negatively affect larval and juvenile fish that would normally benefit from the natural marsh structure (Able & Hagan 2000).

Connectivity among different marsh features is important to organisms that must move with the changes of the tide, whereby tidal channel networks provide the corridors for access to and from the marsh surface and connected pannes that provide refugia at lower tides (McIvor & Odum 1988; Webb & Kneib 2002). Because the form and size of tidal channel systems in salt marshes dictate the extent of that connectivity among the different elevation and vegetation zones, especially in macrotidal regimes, ecological interactions among organisms and the different zonal communities vary by tidal pulses (Rozas 1995). Another spatial variable found to be ecologically important in determining patterns of faunal community composition is geographical location of features relative to tidal connectivity. Davis *et al.* (2014) surveyed marine faunal communities in salt marsh ponds and reported that a distinct faunal group including salt marsh residents and juvenile marine-spawned taxa occurred in greater abundances in more isolated, higher elevation pools, which connect to the estuary channel or other pools only on larger spring high tides. In addition to within marsh connectivity, many salt marshes are critically linked to other intertidal or shallow subtidal habitats, particularly seagrass meadows and mangroves.

Although tides and daylight often dominate patterns of fish movement with water movement along channel systems in salt marsh seascapes, fish have been shown to display both strong site fidelity and intramarsh movement (Able *et al.* 2012). For instance, Dresser & Kneib (2007) determined that red drum (*Sciaenops ocellatus*) implanted with ultrasonic transmitters and released in a coastal estuary in Georgia, United States, illustrated both site fidelity and extensive movement, including 'site skipping' through the marshes instead of following channels. Connectivity to other habitat types can also affect the scale of predation in salt marshes, as found for survivorship of clams preyed upon by blue crabs (*Callinectes sapidus*) that can utilize seagrasses as a movement corridor between oyster reefs and salt marsh vegetation (Micheli & Peterson 1999). Rountree & Able (2007) surmise fish movement among 'ecological habitat' gradients of salt marshes to reflect shifts (tidal, diel, seasonal) to maintain optimum conditions of predation risk, food availability and environmental conditions.

5.3.4   Faunal Linkages between Salt Marshes and Seagrass Meadows

For many salt marsh organisms, the connectivity between the marsh pools, intertidal and subtidal creeks suggest that all of these subhabitats are interlinked components of the salt marsh seascape (Kneib 2000). However, marshes and seagrasses do not function in isolation, instead they are often functionally connected to a broader mosaic of habitat types (Kneib 2000; Heck *et al.* 2008; see also Figure 5.7). Although seagrass meadows and salt marshes commonly co-occur in many estuaries (see Chesapeake Bay example in Figure 5.1), our understanding of how these habitats are connected is still not fully understood. However, there are several mechanisms by which these habitat types are interlinked, including the exchange of nutrients, dissolved and particulate forms of carbon, flows of animal larvae, as well as by active movements or predator-prey interactions between estuarine organisms (Kneib 1997; Bouillon & Connolly 2009). Consequently, seascapes are increasingly being perceived and managed as spatial mosaics of interconnected patch types instead of isolated patches (Pittman *et al.* 2011; Weinstein & Litvin 2016). This is particularly evident for fisheries management and restoration where seagrass and salt marsh ecosystems are often perceived as a part of a broader seascape nursery for marine fauna (Kneib 2000; Heck *et al.* 2008; Nagelkerken *et al.* 2013). For example, Able *et al.* (2007) investigated marsh dependency by sampling the distribution of 14 fish and crab species across multiple habitat types over seven years in a single

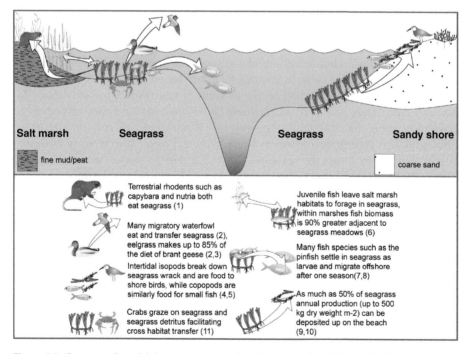

Figure 5.7 Conceptual model showing routes and mechanisms of trophic transfers between marine and terrestrial habitats in a temperate ecosystem. References: 1. Kantrud (1991), 2. Ganter (2000), 3. Cottam *et al.* (1944), 4. Lenanton *et al.* (1982), 5. Lenenton & Caputi (1989), 6. Irlandi & Crawford (1997), 7. Hanson (1969), 8. Darcey (1985), 9. Mateo *et al.* (2003), 10. Mateo *et al.* (2006), 11. Gunter (1967). *Source*: Figure adapted from Heck *et al.* (2008) with permission from Springer.

estuary revealing that there are few marsh-dependent species because almost all species simultaneously use a variety of other habitat types. In Australia, combining spatial pattern analyses of coastal seascapes with fisheries catch data showed that at a broad scale (>5000 kilometres of coastline) connectivity of tidal wetlands (mangroves, saltmarsh and channels) was important for fisheries productivity (Meynecke *et al.* 2008).

Few direct comparisons of the distribution of nekton in salt marshes and adjacent seagrass meadows are available, but Rozas & Minello (1998) sampled both habitats simultaneously and noted that seagrass and salt marsh habitats are utilized by the same species, but their dynamics are species specific and driven by habitat complexity, seasonality and water level fluctuations. For example, in both fall and spring, blue crabs and grass shrimps were more abundant in salt marsh than in seagrass, while brown shrimp showed higher abundance in seagrass in spring. They also recorded higher salt marsh vegetation density in fall than in spring, which might explain the habitat use patterns. The energy of marsh-derived secondary production is transported through passive flow, active movements or trophic interactions to adjacent seagrass habitats and deeper areas (Kneib 1997). Apparently, this two-way interaction is driven by tradeoffs among food, shelter, access to mates and water level fluctuations. In a novel field study designed to examine marsh-seagrass connectivity for fish, Irlandi & Crawford (1997) found that pinfish (*Lagodon rhomboides*) were more than twice as abundant and experienced greater growth rate in salt marsh channels adjacent to subtidal seagrass meadows compared to marshes adjacent to subtidal unvegetated bottom. They hypothesized that access to abundant food supply throughout the tidal cycle explained why pinfish grew better in salt marshes adjacent to subtidal seagrass compared to salt marshes bordering subtidal unvegetated areas.

Understanding the role of fauna in the transfer of trophic energy across tidal wetlands and seagrasses requires information on the patterns of faunal space use over time. To incorporate faunal movement dynamics, Kneib (2000) applied a landscape ecology perspective to examine the interactions between faunal processes and spatial patterns in wetland structure, including constraints to mobility and feeding, to identify hotspots of trophic interaction across the seascape. Figure 5.8 shows a map of potential places and times during the tidal cycle when trophic transfers involving estuarine organisms are likely to occur. Much of the trophic acquisition of marsh productivity will take place with the flooding tide when higher elevation parts of the seascape become accessible to organisms (Figure 5.8a). On the ebbing tide (Figure 5.8b), the important interaction zones shift to lower portions of the seascape. Marsh edge habitats, especially near intertidal-subtidal corridors, *i.e.* rivulets and creek mouths, and deeper water vegetation such as seagrasses are likely to experience high predator-prey interactions. For example, using acoustic telemetry to map individual striped bass movements in estuaries, Kennedy *et al.* (2016) identified interaction hotspots for migrating fish that were linked to physical discontinuities created by water depth and coastal geomorphology (amount of sandbar, confluence diversity, channel proximity, median drop-off size). The highest concentrations of fish were found in locations where multiple discontinuities occurred.

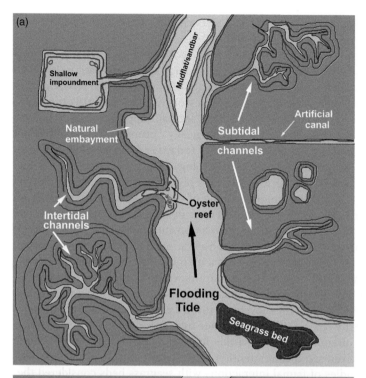

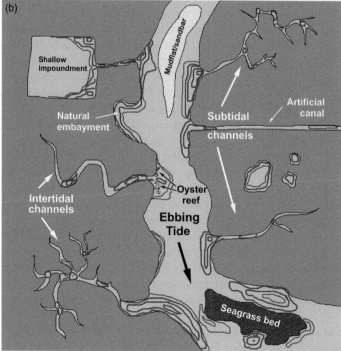

Figure 5.8 Shifting interaction zones where marsh nekton may be involved in transfers of intertidal production via predator-prey relationships during (i) high and flooding tide stages and (ii) ebbing and low tide stages. Red lines are the hypothetical 'hot spots' of production transfers across the seascape at each tidal stage. *Source*: Figure adapted from Kneib (2000).

5.4   Challenges and Opportunities in Seascape Ecology

Although building on the same principles, seascape ecology has not yet reached the same impact as landscape ecology has in terrestrial biology and conservation (Pittman *et al.* 2011). Given the vast spatial scales and character of environmental stressors and the complex interplay between present and future stressors (Bopp *et al.* 2013), the problem is indeed a seascape issue and we propose that a seascape approach is essential for understanding impacts and for the design of ecologically optimal habitat restoration actions. The grand challenge facing seascape ecology today is raising awareness of useful approaches and future possibilities of spatial ecology to manage changing marine plant communities in a spatially explicit context. As highlighted in this chapter, there is still a huge knowledge gap in our understanding of the ecological consequences of changes to seascape patterning and therein also exist many exciting opportunities to ask new questions and gain fresh insights on the relationships between seascape structure, function and change. To achieve effective conservation and address sustainability, seascape ecology must now rapidly develop from a marginal discipline in marine ecology to a central research field offering an attractive framework and toolbox for ecologists and managers to explore, resolve and communicate complex issues in rapidly changing coastal environments. To facilitate the movement toward a seascape ecology approach, we here offer some priority research themes to guide future studies.

Edges and structural discontinuities are visible features in shallow-water vegetated seascapes and have received significant attention in ecological studies of salt marshes and seagrasses. Early studies of faunal interactions with seagrass beds considered the shape and orientation of a patch relative to current flow, as well as the location of the patch relative to larval sources, as potentially important influences on recruitment patterns (Bell *et al.* 1988). This area of inquiry is still rarely studied and would benefit from field sampling and experiments focusing on the mechanisms and spatial and temporal variability of edge effects and the influence of patch shape, amount and seascape context on larval distribution and settlement behaviour. It is clear that additional variables such as predator abundance, food supply and physical complexity, will influence the long-term outcome of an initial edge effect on settlement. Given that edge morphology and position is sufficiently stable, more studies on the long-term influence of habitat edges on settlement and recruitment are needed (Carroll *et al.* 2012). Despite the persuasive evidence that density and composition of faunal assemblages are often enhanced by a variety of regional and local seascape characteristics, studies that document how salt marshes function to provide recruits to adult populations (Beck *et al.* 2001) are still rare. More specifically, the complex, dynamic mechanisms that promote advantageous growth and survival are still poorly documented (Sheaves *et al.* 2015). In general, several implications emerge: (i) all species are predicted to show positive, neutral and negative edge responses, depending on the edge type encountered, (ii) the habitat quality on either side of the edge must be considered to understand the response to edges and (iii) temporal differences in resource use can alter an organism's response to edge.

The majority of seagrass and salt marsh stress experiments are still short-term laboratory or mesocosm tests with limited value for predicting long-term changes across the seascape. New innovative approaches are needed as interactions between abiotic stressors (see above) jointly influence the structure of the coastal seascapes and their associated animal communities (Morzaria-Luna *et al.* 2014). To investigate and separate the effects of seagrass habitat fragmentation and total area on recruitment, diversity and mobility of fish at spatial scales that are operationally meaningful for management remains a challenging field. Innovative broad scale natural field observations and experiments using seascapes of differing configuration (*e.g.*, Yeager *et al.* 2016; Williams *et al.* 2016) may provide new insight into the role of seascape structure for coastal fisheries and food webs. More studies on tracking the movement patterns of fauna across the seascape are needed and, more generally, scaling in seascape ecology, particularly in marine animal ecology, requires a greater understanding of the movements of organisms.

Combining global change scenario models with spatial predictive modelling for specific species may predict local, regional and biogeographical range shifts in seagrasses and salt marsh vegetation (Valle *et al.* 2014). Such modelling efforts complemented with field experiments testing the physiological plasticity of coastal vegetation may further provide valuable information about plant adaption capacity and critical tolerance limits (Reusch *et al.* 2005; Gustafsson & Boström 2014; Cherry *et al.* 2009; Salo *et al.* 2015). Future modelling in marshes should combine ecological models of the upper marsh-forest ecotone with geomorphic models of the retreat dynamics of the seaward edge (Kirwan & Megonigal 2013). For salt marshes, these research approaches should be combined with monitoring of: (i) landscape level changes in habitat cover and species composition; (ii) accretion dynamics (sedimentation and organic matter accumulation); (iii) abundances of shellfish and calcifying species; (iv) trends in commercial fishery species (reproductive females, distribution of fish larvae); and (v) long-term trends in wetland dependent crustaceans and fishes (Morzaria-Luna 2014).

As return of ecosystem services is an important goal of restoration science, several restoration actions can provide valuable, often broad (100's meters to kilometres) scale experiments (Katwijk *et al.* 2016), that could, to a much larger extent, be used to address many of the uncertainties about the functional role of seascape features. For example, faunal responses to planted seagrass plots of differing degree of fragmentation can be followed over several years and effects can be compared between sites (Lefcheck *et al.* 2016; McSkimming *et al.* 2016).

A meta community approach provides a framework for a better understanding of connectivity processes across seascapes (Leibold *et al.* 2004). Here, communities inhabiting distinct patches embedded in a hostile matrix are connected through dispersal processes and the role of geographic distance *per se*, flow connectivity and species traits can be assessed (De Bie *et al.* 2012; Moritz *et al.* 2013). The initial four separate models *i.e.* neutrality, patch dynamics, species sorting and mass effect, see Leibold *et al.* (2004), are now considered oversimplified and an integrated approach focusing on quantifying the relative roles of dispersal, species interactions and environmental gradients (filtering) is now used to explain variance in marine community structure over large geographical scales (Bell 2006; Moritz *et al.* 2013). Presently theoretical metacommunity model predictions are still largely untested, but hold promise for understanding connectivity in salt marsh tidal pools (Davis *et al.* 2014), seagrass communities (Melià *et al.* 2016; Yamada *et al.* 2014), coral reefs (MacNeill *et al.* 2009) can support marine reserve design (Guichard *et al.* 2004; Melià *et al.* 2016).

References

Able KW, Balletto JH, Hagan SM, Jivoff PR, Strait K (2007) Linkages between salt marshes and other nekton habitats in Delaware Bay, United States. Reviews in Fisheries Science 15(1–2): 1–61.

Able KW, Hagan, SM (2000) Effects of common reed (*Phragmites australis*) invasion on marsh surface macrofauna: response of fishes and decapod crustaceans. Estuaries 23(5): 633–646.

Able KW, Vivian DN, Petruzzelli G, Hagan SM (2012) Connectivity among salt marsh subhabiatats: residency and movements of the mummichog (*Fundulus heteroclitus*). Estuaries and Coasts 35: 743–753.

Achard F, Eva DH, Stibig HJ, Mayaux P, Gallego J, Richards T, Malingreau J-P (2002) Determination of deforestation rates of the world's humid tropical forests. Science 297: 999–1002.

Artigas FJ, Yang J (2004) Hyperspectral remote sensing of habitat heterogeneity between tide-restricted and tide-open areas in New Jersey Meadowlands. Urban Habitat 2(1): 1–18.

Baillie CJ, Fear JM, Fodrie FJ (2015) Ecotone effects on seagrass and salt marsh habitat use by juvenile nekton in a temperate estuary. Estuaries and Coasts 38: 1414–1430.

Beck MW, Heck Jr. KL, Able KW, Childers DL, Eggleston DB, Gillanders BM, Halpern B, Hays CG, Hoshino K, Minello TJ, Orth RJ, Sheridan PF, Weinstein MP (2001) The identification, conservation, and management of estuarine and marine nurseries for fish and invertebrates. Bioscience 51: 633–641.

Bell SS (2006) Seagrasses and the metapopulation concept: Developing a regional approach to the study of extinction, colonization, and dispersal. In Kritzer JP, Sale PF (eds) Marine Metapopulations. Elsevier Academic Press, Amsterdam, pp. 387–407.

Bell JD, Steffe AS, Westoby M (1988) Location of seagrass beds in estuaries: effects on associated fish and decapods. Journal of Experimental Marine Biology and Ecology 122(2): 127–146.

Bell SS, Brooks RA, Robbins BD, Fonseca MS, Hall MO (2001) Faunal response to fragmentation in seagrass habitats: implications for seagrass conservation. Biological Conservation 100(1): 115–123.

Bell SS, Robbins BD Jensen SL (1999) Gap dynamics in a seagrass landscape. Ecosystems 2(6): 493–504.

Bender DJ, Contreras TA, Fahrig L (1998) Habitat loss and population decline: a meta-analysis of the patch size effect. Ecology 79(2): 517–533.

Bertness MD, Ellison AM (1987) Determinants of pattern in a New England salt marsh plant community. Ecological Monographs 57: 129–147.

Bertness MD, Shumway SW (1993) Competitiion and facilitation in marsh plants. The American Naturalist 142(4): 718–724.

Bologna PAX, Heck KL Jr. (1999) Differential predation and growth rates of bay scallops within a seagrass habitat. Journal of Experimental Marine Biology and Ecology 239: 299–314.

Bologna PAX, Heck KL Jr. (2000) Impacts of seagrass habitat architecture on bivalve settlement. Estuaries 23: 449–457.

Bologna PAX, Heck KL Jr. (2002) Impact of habitat edges on density and secondary production of seagrass-associated fauna. Estuaries 25: 1033–1044.

Bopp L, Resplandy L, Orr JC, Doney SC, Dunne JP, Gehlen P, Halloran C, Heinze T, Ilyina R, Séférian , Tjiputra J, Vichi M (2013) Multiple stressors of ocean ecosystems in the 21st century: projections with CMIP5 models. Biogeoscience 10: 6225–6245.

Boström C, Bonsdorff E (1997) Community structure and spatial variation of benthic invertebrates associated with *Zostera marina* (L.) beds in the northern Baltic Sea. Journal of Sea Research 37: 153–166.

Boström C, Bonsdorff E (2000) Zoobenthic community establishment and habitat complexity – the importance of seagrass shoot density, morphology and physical disturbance for faunal recruitment. Marine Ecology Progress Series 205: 123–138.

Boström C, Jackson EL, Simenstad CA (2006) Seagrass landscapes and their effects on associated fauna: a review. Estuarine Coastal Shelf Science 68: 383–403.

Boström C, Pittman S, Kneib R, Simenstad C (2011) Seascape ecology of coastal biogenic habitats: advances, gaps and challenges. Marine Ecology Progress Series 427: 191–217.

Boström C, Törnroos A, Bonsdorff E (2010) Invertebrate dispersal and habitat heterogeneity: expression of biological traits in a seagrass landscape. Journal of Experimental Marine Biology and Ecology 390: 106–117.

Bouillon S, Connolly RM. Carbon exchange among tropical coastal ecosystems. In Nagelkerken I. (ed.) Ecological Connectivity among Tropical Coastal Ecosystems. Springer, Dordrecht, pp. 45–70.

Bowden, DA, Rowden AA, Attrill MJ (2001) Effects of patch size and in-patch location on the infaunal macroinvertebrate assemblages of *Zostera marina* seagrass beds. Journal of Experimental Marine Biology and Ecology 259: 133–154.

Bridgeham SD, Megonigal JP, Keller JK, Bliss NB, Trettin C (2006) The carbon balance of North American wetlands. Wetlands 26: 889–916.

Browder JA, May LN Jr., Rosenthal A, Gosselink JG, Baumann RH (1989) Modelling future trends in wetland loss and brown shrimp production in Louisiana using thematic mapper imagery. Remote Sensing Environment 28: 45–59.

Butman CA (1987) Larval settlement of soft-sediment invertebrates: The spatial scales of pattern explained by active habitat selection and the emerging role of hydrodynamical processes. Oceanography and Marine Biology Review 25: 113–165.

Calabrese JM, Fagan WF (2004) A comparison-shopper's guide to connectivity metrics. Frontiers in Ecology and the Environment 2(10): 529–536.

Carr JA, D'Odorico McGlathery KJ, Wiberg PL (2015) Spatially explicit feedbacks between seagrass meadow structure, sediment and light: Habitat suitability for seagrass growth. Advances in Water Resources 93: 315–325.

Carroll JM, Furman BT, Tettelbach ST, Peterson BJ (2012) Balancing the edge effects budget: bay scallop settlement and loss along a seagrass edge. Ecology 93(7): 1637–1647.

Carroll JM, Peterson BJ (2013) Ecological trade-offs in seascape ecology: bay scallop survival and growth across a seagrass seascape. Landscape Ecology 28: 1401–1413.

Carson M, Kohl A, Stammer D, Slangen ABA, Katsman CA, van de Wal RSW, Church JA, White N (2016) Coastal sea level changes, observed and projected during the twentieth and twenty-first century. Climate Change 134: 269–281.

Cherry JA, McKee KL, Grace JB (2009) Elevated CO2 enhances biological contributions to elevation change in coastal wetlands by offsetting stressors associated with sea-level rise. Journal of Ecology 97: 67–77.

Chmura GL, Anisfeld SC, Cahoon DR, Lynch JC (2003) Global carbon sequestration in tidal, saline wetland soils. Global Biogeochemical Cycles 17(4): 1125.

Connolly RM, Hindell JS (2006) Review of nekton patterns and ecological processes in seagrass landscapes. Estuarine Coastal Shelf Science 68: 433–444.

Connor EF, McCoy ED (1979) The statistics and biology of the species-area relationship. American Naturalist 113(6): 791–833.

Costanza R, de Groot R, Sutton P, van der Ploeg S, Anderson SJ, Kubiszewski I, Farber S, Turner RK (2014) Changes in the global value of ecosystem services. Global Environmental Change 26: 152–158.

Cottam C, Lynch JJ, Nelson AL (1944) Food habits and management of American sea brant. Journal of Wildlife Management 8: 36–56.

Craft C, Clough J, Ehman J, Joye S, Park R, Pennings S, Guo H, Machmuller M (2009) Forecasting the effects of accelerated sea-level rise on tidal marsh ecosystem services. Frontiers in Ecology and Environment 7: 73–78.

Cullen-Unsworth LC, Mtwana Nordlund L, Paddock J, Baker S, McKenzie LJ, Unsworth RKF (2014) Seagrass meadows globally as a coupled social-ecological system: Implications for human wellbeing. Marine Pollution Bulletin 83: 387–397.

Darcey GH (1985) Synopsis of Biological Data on the Pinfish, *Lagodon rhomboides* (Pisces: Sparidae). National Oceanic and Atmospheric Administration Technical Report NMFS 23. US Department of Commerce, Washington, DC.

Davis B, Baker R, Sheaves M (2014) Seascape and metacommunity processes regulate fish assemblage structure in coastal wetlands. Marine Ecology Progress Series 500: 187–202.

De Bie T, De Meester L, Brendonck L, Martens K, Goddeeris B, Ercken D, Hampel H, Denys L, Vanhecke L, Van der Gucht K, Van Wichelen J, Vyverman W, Declerck SAJ (2012) Body size and dispersal mode as key traits determining metacommunity structure of aquatic organisms. Ecology Letters 15: 740–747.

Den Hartog C, Kuo, J (2006) Taxonomy and biogeography in seagrasses. Larkum EWD, Orth RJ, Duarte CM (eds) Seagrasses: Biology, Ecology and Conservation. Springer, Dordrecht, pp. 1–23.

Donnelly JP, Butler J, Roll S, Wengren M, Webb III, T (2004) A backbarrier overwash record of intense storms from Brigantine, New Jersey. Marine Geology 210: 107–121.

Dresser BK, Kneib RT (2007) Site fidelity and movement patterns of wild subadult red drum, *Sciaenops ocellatus* (Linnaeus), within a salt marsh-dominated estuarine landscape. Fisheries Management and Ecology 14: 183–190.

Duarte CM, Borum J, Short FT, and Walker DI (2005b) Seagrass ecosystems: their global status and prospects. In Polunin NVC (ed.). Aquatic ecosystems: trends and global prospects. Cambridge University Press, Cambridge.

Duarte CM, Fourqurean JW, Krause-Jensen D, Olesen B (2006) Dynamics of Seagrass Stability and Change. In Larkum EWD, Orth RJ, Duarte CM (eds) Seagrasses: Biology, ecology and conservation. Springer, Dordrecht, pp. 1–23.

Duarte CM, Middelburg J, Caraco N (2005a) Major role of marine vegetation on the oceanic carbon cycle. Biogeosciences 2(1): 1–8.

Dunning JB, Danielson BJ, Pulliam RH (1992) Ecological processes that affect populations in complex landscapes. Oikos 65(1): 169–175.

Eisma D (1993) Suspended matter in the aquatic environment. Springer Verlag, Berlin.

Eisma D, de Boer PL, Cadée GC, Dijkema K, Ridderinkhof H, Philippart C (1998) Intertidal Deposits: River Mouths, Tidal Flats, and Coastal Lagoons. CRC Press, Boca Raton, FL.

Fahrig L (2013) Rethinking patch size and isolation effects: the habitat amount hypothesis. Journal of Biogeography 40: 1649–1663.

Fishman JR, Orth RJ (1996) Effects of predation on *Zostera marina* L. seed abundance. Journal of Experimental Marine Biology and Ecology 198: 11–26.

Fonseca MS, Bell SS (1998) Influence of physical setting on seagrass landscapes near Beaufort, North Carolina, USA. Marine Ecology Progress Series 171: 109–121.

Fonseca MS, Whitfield PE, Kelly NM, Bell SS (2002) Modelling seagrass landscape pattern and associated ecological attributes. Ecological Applications 12(1): 218–237.

Forman RTT (1995) Land Mosaics. The Ecology of Landscapes and Regions. Cambridge University Press, Cambridge.

Frederiksen M, Krause-Jensen D, Holmer M, Sund Laursen J (2004) Spatial and temporal variation in eelgrass (*Zostera marina*) landscapes: influence of physical setting. Aquatic Botany 78: 147–165.

Freemark KE, Merriam HG (1986) Importance of area and habitat heterogeneity to bird assemblages in temperate forest fragments. Biological Conservation 36(2): 115–141.

Gambi MC, Nowell ARM, Jumars PA (1990) Flume observations on flow dynamics in *Zostera marina* (eelgrass) beds. Marine Ecology Progress Series 61: 159–169.

Ganter B (2000) Seagrass (*Zostera* spp.) as food for brent geese (*Branta bernicla*): an overview. Helgoland Marine Research 54: 63–70.

Gibbes B, Grinham A, Neil D, Olds A, Maxwell P, Connolly R, Weber T, Udy N, Udy J (2014) Moreton Bay and its estuaries: a sub-tropical system under pressure from rapid population growth. In E. Wolanski (ed.), Estuaries of Australia in 2050 and Beyond, Estuaries of the World, Springer Science+Business Media, Dordrecht.

Gorman AM, Gregory RS, Schneider DC (2009) Eelgrass patch size and proximity to the patch edge affect predation risk of recently settled age 0 cod (*Gadus*). Journal of Experimental Marine Biology and Ecology 371: 1–9.

Green BC, Smith DJ, Underwood GJ (2012) Habitat complexity and spatial complexity differentially affect mangrove and salt marsh fish assemblages. Marine Ecology Progress Series 466: 177–192.

Green EP, Short FT (2003) World Atlas of Seagrasses. California University Press, Berkeley, CA.

Greiner JT, McGlathery KJ, Gunnell J, McKee BA (2013) Seagrass restoration enhances 'Blue Carbon' sequestration in coastal waters. PLoS One 8(8): e72469.

Grober-Dunsmore R, Frazer TK, Lindberg WJ, Beets J (2007) Reef fish and habitat relationships in a Caribbean seascape: the importance of reef context. Coral Reefs 26: 201–216.

Guichard F, Levin SA, Hastings A, Siegel D (2004) Toward a dynamic metacommunity approach to marine reserve theory. BioScience 54: 1003–1011.

Gunter G (1967) Some relationships of estuaries to the fisheries of the Gulf of Mexico. In Lauff GH (ed.). Estuaries. American Association for the Advancement of Science, Washington, DC, pp. 621–638.

Gustafsson C, Boström C (2014) Algal mats reduce eelgrass (*Zostera marina* L.) growth in mixed and monospecific meadows. Journal of Experimental Marine Biology and Ecology 461: 85–92.

Haas HL, Rose KA, Fry B, Minello TJ, Rozas LP (2004) Brown shrimp on the edge: linking habitat to survival using an individual-based simulation model. Ecological Applications 14(4): 1232–1247.

Haddad NM, Baum K (1999) An experimental test of corridor effects on butterfly densities. Ecological Applications 9: 623–633.

Hanson DJ (1969) Food, growth, reproduction and abundance of pinfish, *Lagodon rhomboides*, and Atlantic croaker, *Micropogon undulatus*, near Pensacola, Florida 1963–1965. Fisheries Bulletin 68: 135–156.

Hartman JM (1988) Recolonization of small disturbance patches in a New England salt marsh. American Journal of Botany 75(11): 1625–1631.

Heck Jr, KL, Carruthers TJ, Duarte CM, Hughes AR, Kendrick G, Orth RJ, & Williams SW (2008) Trophic transfers from seagrass meadows subsidize diverse marine and terrestrial consumers. Ecosystems 11(7): 1198–1210.

Hensgen GM, Holt J, Holt SA, Williams JA, Stunz GW (2014) Landscape pattern influences nekton diversity and abundance in seagrass meadows. Marine Ecology Progress Series 507: 139–152.

Hitt S, Pittman SJ, Nemeth RS (2011) Diel movements of fishes linked to benthic seascape structure in a Caribbean coral reef ecosystem. Marine Ecology Progress Series 427: 275–291.

Hood WG (2007a) Perspectives in Estuarine and Coastal Science: Landscape allometry and prediction in estuarine ecology: linking landform scaling to ecological patterns and processes. Estuaries and Coasts 30(5): 895–900.

Hood WG (2007b) Large woody debris influences vegetation zonation in an oliohaline tidal marsh. Estuaries and Coasts 30(3): 441–450.

Hovel KA (2003) Habitat fragmentation in marine landscapes: relative effects of habitat cover and configuration on juvenile crab survival in California and North Carolina seagrass beds. Biological Conservation 110: 401–412.

Hovel KA, Fonseca MS, Myer DL, Kenworthy WJ, Whitfield PE (2002) Effects of seagrass landscape structure, structural complexity and hydrodynamic regime on macrofaunal densities in North Carolina seagrass beds. Marine Ecology Progress Series 243: 11–24.

Hovel KA, Lipcius RN (2001) Habitat fragmentation in a seagrass landscape: patch size and complexity control blue crab survival. Ecology 82: 1814–1829.

Hunter EA, Nibbelink NP, Alexander CR, Barrett K, Mengak LF, Guy RK, Moore CT, Cooper RJ (2015) Coastal Vertebrate Exposure to Predicted Habitat Changes Due to Sea Level Rise. Environmental Management 56: 1528–1537.

Intergovernmental Panel on Climate Change (2007) Climate Change 2007: The Physical Science Basis: Contribution of Working Group I to the Fourth Assessment Report of the Intergovernmental Panel on Climate Change. S Solomon, D Qin, M Manning, Z Chen, M Marquis, KB Averyt, M Tignor and H . Miller (eds). Cambridge University Press, Cambridge.

Irlandi EA (1996) The effects of seagrass patch size and energy regime on growth of a suspension-feeding bivalve. Journal of Marine Research 54: 161–185.

Irlandi EA (1997) Seagrass patch size and survivorship of an infaunal bivalve. Oikos 78: 511–518.

Irlandi EA, Ambrose WG, Orlando, BA (1995) Landscape ecology and the marine environment how spatial configuration of seagrass habitat influences growth and survival of the Bay Scallop. Oikos 72: 307–313.

Irlandi EA, Crawford MK (1997) Habitat linkages: the effect of intertidal salt marshes and adjacent subtidal habitats on abundance, movement, and growth of an estuarine fish. Oceologia 110: 222–230.

Irlandi EA, Orlando BA, Ambrose WG (1999) Influence of seagrass habitat patch size on growth and survival of juvenile bay scallops, Argopecten irradians concentricus (Say) Journal of Experimental Marine Biology and Ecology 235: 21–43.

Jackson EL, Attrill, MJ, Jones, MB (2006) Habitat characteristics and spatial arrangement affecting the diversity of fish and decapod assemblages of seagrass (*Zostera marina*) beds around the coast of Jersey (English Channel). Estuarine Coastal and Shelf Science 68: 421–432.

Källén J, Muller H, Franken ML, Crisp A, Stroh C, Pillay D, Lawrence C (2012) Seagrass-epifauna relationships in a temperate South African estuary: Interplay between patch-size, within-patch location and algal fouling. Estuarine, Coastal and Shelf Science 113: 213–220.

Kantrud HA (1991) Widgeon Grass (*Ruppia maritima*): A Literature Review. Fish and wildlife research report no. 10. US Fish and Wildlife Service, Washington, DC.

Katwijk van MM, Thorhaug A, Marba N, Orth RJ, Duarte CM, Kendrick GA, Althuizen IHJ, Balestri E, Bernard G, Cambridge ML, Cunha A, Durance C, Giesen W, Han Q, Hosokawa S, Kiswara W, Komatsu T, Lardicci C, Lee K-S, Meinesz A, Nakaoka M, O'Brien MR, Paling EI, Pickerell C, Ransijn AMA, Verduin JJ (2016) Global analysis of seagrass restoration: the importance of large-scale planting. Journal of Applied Ecology 53: 567–578.

Kendall MS, Miller TJ, Pittman SJ (2011) Patterns of scale-dependency and the influence of map resolution on the seascape ecology of reef fish. Marine Ecology Progress Series 427: 259–274.

Kendrick GA, Duarte CM, Marbà N (2005) Clonality in seagrasses, emergent properties and seagrass landscapes. Marine Ecology Progress Series 290: 291–296.

Kennedy CG, Mather ME, Smith JM, Finn JT, Deegan LA (2016) Discontinuities concentrate mobile predators: quantifying organism–environment interactions at a seascape scale. Ecosphere 7(2): e01226.

Kirwan ML, Megonigal JP (2013) Tidal wetland stability in the face of human impacts and sea-level rise. Nature 504: 53–60.

Kneib RT (1997) The role of tidal marshes in the ecology of estuarine nekton. Oceanography and Marine Biology: An Annual Review 35: 163–220.

Kneib RT (2000) Salt marsh ecoscapes and production transfers by estuarine nekton in the southeastern US In Weinstein MP and Kreeger DA (eds) Concepts and Controversies in Tidal Marsh Ecology. Kluwer Academic Publishers, Dordrecht, pp. 267–291.

Kneib RT (2003) Bioenergetic and landscape considerations for scaling expectations of nekton production from intertidal marshes. Marine Ecology Progress Series 264: 279–296.

Lefcheck JS, Marion SR, Lombana AV, Orth RJ (2016) Faunal Communities Are Invariant to Fragmentation in Experimental Seagrass Landscapes. PLoS One 11(5): e0156550.

Leibold MA, Holyoak M, Mouquet N, Amarasekare P, Chase JM, Hoopes MF, Holt, RD, Shurin JB, Law R, Tilman D, Loreau M, Gonzalez A (2004) The metacommunity concept: a framework for a multi-scale community ecology. Ecology Letters 7: 601–613.

Lenanton RCJ, Caputi N (1989) The roles of food supply and shelter in the relationship between fishes, in particular *Cnidoglanis macrocephalus* (Valenciennes), and detached macrophytes in the surf zone of sandy beaches. Journal of Experimental Marine Biology and Ecology 128: 165–176.

Lenanton RCJ, Roberson AI, Hansen JA (1982) Nearshore accumulations of detached macrophytes as nursery areas for fish. Marine Ecology Progress Series 9: 51–57.

Leopold LB, Maddock Jr T (1953) The Hydraulic Geometry of Stream Channels and some Physiographic Implications. US Geological Survey Professional Paper. United States Government Printing Office, Washington, DC.

Lowe MR, Peterson MS (2014) Effects of coastal urbanization on salt-marsh faunal assemblages in the northern Gulf of Mexico. Marine and Coastal Fisheries 6(1): 89–107.

MacArthur RH, Wilson EO (1967) The Theory of Island Biogeography. Princeton University Press, Princeton, NJ.

MacNeill MA, Graham NAJ, Polunin NVC, Kulbicki M, Galzin R, Harmelin-Vivien M, Rushton SP (2009) Hierarchical drivers of reef-fish metacommunity structure. Ecology 90: 252–264.

Macreadie PI, Connolly RM, Jenkins GP, Hindell JS, Keough MJ (2010a) Edge patterns in aquatic invertebrate explained by predictive models. Marine and Freshwater Research 61: 214–218.

Macreadie PI, Hindell JS, Keough MJ, Jenkins GP, Connolly RM (2010b) Resource distribution influences positive edge effects in a seagrass fish. Ecology 91(7): 2013–2021.

Marbá N, Duarte CM (1998) Rhizome elongation and seagrass clonal growth. Marine Ecology Progress Series 174: 269–280.

Mason DC, Scott TR (2004) Remote sensing of tidal networks and their relation to vegetation. In Fagherazzi S, Marani M, Blum LK (eds) The Ecogeomorphology of Tidal Marshes. American Geophysical Union, Washington, DC, pp. 27–46.

Mateo MA, Cebrian J, Dunton K, Mutchler T (2006) Carbon flux in seagrass ecosystems. In Larkum AWD, Orth RJ, Duarte C (eds). Seagrasses: Biology, Ecology and Conservation. Springer, Dordrecht, pp. 159–192.

Mateo MA, Sanchez-Lizaso JL, Romero J (2003) *Posidonia oceanica* 'Banquettes': a preliminary assessment of the relevance for meadow carbon and nutrient budgets. Estuarine Coastal and Shelf Science 56: 85–90.

Maxwell PS, Eklöf JS, van Katwijk MM, O'Brien KR, de la Torre-Castro M, Boström C, Bouma TJ, Krause-Jensen D, Unsworth RKF, van Tussenbroek BI, van der Heide T (2016) The fundamental role of ecological feedback mechanisms for the adaptive management of seagrass ecosystems – a review. Biological Reviews doi: 10.1111/brv.12294.

McIvor DD, Odum WE (1988) Food, predation risk, and microhabitat selection in a marsh fish assemblage. Ecology 69: 1341–1351.

McKee K, Cherry JA (2009) Hurricane Katrina sediment slowed elevation loss in subsiding brackish marshes of the Mississippi River Delta. Wetlands 29: 2–15.

McKee K, Rogers K, Saintilan N (2012) Response of salt marsh and mangrove wetlands to changes in atmospheric CO_2, climate, and sea level. In BA Middleton (ed.), Global Change and the Function and Distribution of Wetlands: Global Change Ecology and Wetlands. Springer, Dordrecht, pp. 63–96.

McLeod E, Chmura GL, Bouillon S, Salm R, Bjork M, Duarte CM, Lovelock CE, Schlesinger WH, Silliman BR (2011) A blueprint for blue carbon: toward an improved understanding of the role of vegetated coastal habitats in sequestering CO2. Frontiers in Ecology and Environment 9: 552–560.

McNeill SE, Fairweather PG (1993) Single large or several small marine reserves? An experimental approach with seagrass fauna. Journal of Biogeography 20: 429–440.

McSkimming C, Connell SD, Russell BD, Tanner JE (2016) Habitat restoration: Early signs and extent of faunal recovery relative to seagrass recovery. Estuarine Coastal Shelf Science 171: 51–57.

Melià P, Schiavina M, Rossetto M, Gatto M, Fraschetti S, Casagrandi R (2016) Looking for hotspots of marine metacommunity connectivity: a methodological framework. Scientific Reports. 6: 23705.

Meyer DL, Posey MH (2014) Influence of salt marsh size and landscape setting on salt marsh nekton populations. Estuaries and Coasts 37: 548–560.

Meynecke JO, Lee SY, Duke NC (2008) Linking spatial metrics and fish catch reveals the importance of coastal wetland connectivity to inshore fisheries in Queensland, Australia. Biological Conservation 141: 981–996.

Micheli F, Peterson CH (1999) Estuarine vegetated habitats as corridors for predatory movements. Conservation Biology 13: 869–881.

Mills VS, Berkenbusch K (2009) Seagrass (*Zostera muelleri*) patch size and spatial location influence infaunal macroinvertebrate assemblages. Estuarine, Coastal and Shelf Science 81(1): 123–129.

Minello TJ (1999) Nekton densities in shallow estuarine habitats of Texas and Louisiana and the identification of essential fish habitat. In Benaka LR (ed.) Fish habitat: essential fish habitat, and rehabilitation. American Fisheries Society Symposium 22: 438–454.

Minello TJ, Able KW, Weinstein MP, Hays CG (2003) Salt marshes as nurseries for nekton: testing hypotheses on density, growth and survival through meta-analysis. Marine Ecology Progress Series 246: 39–59.

Minello TJ, Matthews GA, Caldwell PA, Rozas LP (2008) Population and production estimates for decapods crustaceans in wetlands of Galveston Bay, Texas. Transactions of the American Fisheries Society 137: 129–146.

Moffett KB, Gorelick SM (2016) Alternative stable states of tidal marsh vegetation patterns and channel complexity. Ecohydrology (online). doi: 10.1002/eco.1755.

Moritz C, Meynard CN, Devictor V, Guizien K, Labrune C, Guarini J-M, Moucet N (2013) Disentangling the role of connectivity, environmental filtering, and spatial structure on metacommunity dynamics. Oikos 122: 1401–1410.

Morzaria-Luna N, Turk-Boyer P, Rosemartin A, Camacho-Ibar VF (2014) Vulnerability to climate change of hypersaline salt marshes in the Northern Gulf of California. Ocean and Coastal Management 93: 37–50.

Myrick RM, Leopold LB (1963) Hydraulic geometry of a small tidal estuary. USGS Professional Paper 411-B. US Government Printing Office, Washington, DC.

Nagelkerken I, Sheaves M, Baker R, Connolly RM (2013) The seascape nursery: a novel spatial approach to identify and manage nurseries for coastal marine fauna. Fish and Fisheries 16: 362–371.

Nellemann C, Corcoran E, Duarte CM, Valdés L, De Young C, Fonseca L, Grimsditch G. (eds) (2009) Blue Carbon. A Rapid Response Assessment. United Nations Environment Programme, GRID-Arendal, Arendal.

Nevins JA, Pollack JB, Stunz GW (2014) Characterizing nekton use of the largest unfished oyster reef in the United States compared with adjacent estuarine habitats. Journal of Shellfish Research 33(1): 227–238.

Olds AD, Connolly RM, Pitt KA, Maxwell PS (2012) Primacy of seascape connectivity effects in structuring reef fish assemblages. Marine Ecology Progress Series 462: 191–203.

Olesen B, Sand Jensen K (1994) Patch dynamics of eelgrass *Zostera marina*. Marine Ecology Progress Series 106: 147–156.

Orth RJ (1992) A perspective on plant-animal interactions in seagrasses: physical and biological determinants influencing plant and animal abundance. In John DM, Hawkins SJ, Price JH (eds) Plant-Animal Interactions in the Marine Benthos. Clarendon Press, Oxford.

Osland MJ, Enwright NM, Day RH, Gabler CA, Stagg CL, Grace JB (2016) Beyond just sea-level rise: considering macroclimatic drivers within coastal wetland vulnerability assessments to climate change. Global Change Biology 22: 1–11.

Paling EI, Fonseca M, van Katwijk MM, van Keulen M (2009) Seagrass restoration. In Perillo G, Wolanski E, Cahoon D, Brinson M (eds) Coastal Wetlands: An Integrated Ecosystem Approach. Elsevier, Amsterdam, pp. 687–713.

Peterson GW, Turner RE (1994) The value of salt marsh edge vs interior as a habitat for fish and decapod crustaceans in a Louisiana tidal marsh. Estuaries 17(1): 235–262.

Pethick JS (1974) The distribution of salt pannes on tidal salt marshes. Journal of Biogeography 1: 57–62.

Pethick JS (1992) Saltmarsh geomorphology. In Allen JRL, Pye K (eds), Saltmarshes: Morphodynamics, Conservation and Engineering Significance. Cambridge University Press, Cambridge, pp. 41–62.

Pittman SJ, Hile SD, Caldow C, Monaco ME (2007) Using seascape types to explain the spatial patterns of fish using mangroves in Puerto Rico. Marine Ecology Progress Series 348: 273–284.

Pittman SJ, Kneib RT, Simenstad CA (2011) Practicing coastal seascape ecology. Marine Ecology Progress Series 427: 187–190.

Pittman SJ, McAlpine CA, Pittman KM (2004) Linking fish and prawns to their environment: A hierarchical landscape approach. Marine Ecology Progress Series 283: 233–254.

Procaccini G, Buia MC, Gambi MC, Perez M, Pergent G, Pergent-Martini C, Romero J (2003) The seagrasses of the western Mediterranean. In Green EP, Short FT (eds). World Atlas of Seagrasses, UNEP World Conservation Monitoring Centre. University of California Press, Berkeley, CA, pp. 48–58.

Raupp MJ, Denno RF (1979) The influence of patch size on a guild of sap-feeding insects that inhabit the salt marsh grass *Spartina patens*. Environmental Entomology 8: 412–417.

Redfield AC (1972) Development of a New England salt marsh. Ecological Monographs 42(2): 201–237.

Reusch TBH (1998) Differing effects of eelgrass *Zostera marina* on recruitment and growth of associated blue mussels *Mytilus edulis*. Marine Ecology Progress Series 167: 149–153.

Reusch TBH, Ehlers A, Hämmerli A, Worm B (2005) Ecosystem recovery after climatic extremes enhanced by genotypic diversity. Proceedings of the National Academy of Science USA 102: 2826–2831.

Reusch, TBH, Williams SL (1999) Macrophyte canopy structure and the success of an invasive marine bivalve. Oikos 84: 398–416.

Ricart AM, Dalmau A, Pérez M, Romero J (2015) Effects of landscape configuration on the exchange of materials in seagrass ecosystems. Marine Ecology Progress Series 532: 89–100.

Ries L, Sisk TD (2004) A predictive model of edge effects. Ecology 85: 2917–2926.

Rossi F, Colao E, Martinez MJ, Klein JC, Carcaillet F, Callier MD, de Wit R, Caro A (2013) Spatial distribution and nutritional requirements of the endosymbiont-bearing bivalve *Loripes lacteus* (sensu Poli, 1791) in a Mediterranean *Nanozostera noltii* (Hornemann) meadow. Journal of Experimental Marine Biology and Ecology 440: 108–115.

Roth BM, Rose KA, Rozas LP, Minello TJ (2008) Relative influence of habitat fragmentation and inundation on brown shrimp *Farfantepenaeus aztecus* production in northern Gulf of Mexico salt marshes. Marine Ecology Progress Series 359: 185–202.

Rountree RA, Able KW (2007) Spatial and temporal habitat use patterns for salt marsh nekton: implications for ecological functions. Aquatic Ecology 41: 25–45.

Rozas LP (1995) Hydroperiod and its influence on nekton use of the salt marsh: a pulsing ecosystem. Estuaries 18(4): 579–590.

Rozas LP, Minello TJ (1998) Nekton use of salt marsh, seagrass, and nonvegetated habitats in a south Texas (USA) estuary. Bulletin of Marine Science 63(3): 481–501.

Saintilan N, Wilson NC, Rogers K, Rajkaran A, Krauss KW (2014) Mangrove expansion and salt marsh decline at mangrove poleward limits. Global Change Biology 20(1): 147–157.

Salo T, Reusch TBH, Boström C (2015) Genotype-specific responses to light stress in eelgrass *Zostera marina*, a marine foundation plant. Marine Ecology Progress Series 519: 129–140.

Scott DB, Frail-Gauthier J, Mudie PJ (2014) Coastal Wetlands of the World: Geology, Ecology, Distribution and Applications. Cambridge University Press, New York, NY.

Sheaves M. Baker R, Nagelkerken I, Connolly RM (2015) True value of estuarine and coastal nurseries for fish: incorporating complexity and dynamics. Estuaries and Coasts 38: 401–414.

Short FT, Carruthers T, Dennison W, Waycott M (2007) Global seagrass distribution and diversity: A bioregional model. Journal of Experimental Marine Biology and Ecology 350: 3–20.

Silliman BR (1994) Salt marshes. Current Biology 24: 348–350 .

Sintes T, Marbá N, Duarte CM, Kendrick G (2005) Non-linear processes in seagrass colonization explained by simple clonal growth rules. Oikos 108: 165–175.

Smith TS, Hindell JS, Jenkins GP, Connolly RM (2008) Edge effects on fish associated with seagrass and sand patches. Marine Ecology Progress Series 359: 203–213.

Smith TS, Hindell JS, Jenkins GP, Connolly RM (2010) Seagrass patch size affects fish responses to edges. Journal of Animal Ecology 79: 275–281.

Smith TM, Hindell JS, Jenkins GP, Connolly RM, Keough MJ (2011) Edge effects in patchy seagrass landscapes: the role of predation in determining fish distribution. Journal of Experimental Marine Biology and Ecology 399(1): 8–16.

Smith TM, Jenkins GP, Hutchinson N (2012) Seagrass edge effects on fish assemblages in deep and shallow habitats. Estuarine, Coastal and Shelf Science 115: 291–299.

Tanner JE (2005) Edge effects on fauna in fragmented seagrass meadows. Austral Ecology 30: 210–218.

Tiner RW (2013) Tidal Wetlands Primer. University of Massachusetts Press, Amherst, MA.

Turner RE, Baustian JJ, Swenson EM, Spicer JS (2006) Wetland sedimentation from hurricanes Katrina and Rita. Science 314: 449–452.

Unsworth RKF, Cullen-Unsworth LC (2014) Biodiversity, ecosystem services and the conservation of seagrass meadows. In Maslo B & Lockwood JL (eds) (2014) Coastal Conservation. Cambridge University Press, Cambridge.

Valdemarsen T, Wendelboe K, Egelund JT, Kristensen E, Flindt M (2011) Burial of seeds and seedlings by the lugworm *Arenicola marina* hampers eelgrass (*Zostera marina*) recovery. Journal of Experimental Marine Biology and Ecology 410: 45–52.

Valentine JF, Heck Jr. KL (1999) Seagrass herbivory; evidence for the continued grazing of marine grasses. Marine Ecology Progress Series 176: 291–302.

Valle M, Chust G, del Campo A, Wisz MS, Olsen SM, Garmendia JM, Borja A (2014) Projecting future distribution of the seagrass *Zostera noltii* under global warming and sea level rise. Biological Conservation 170: 74–85.

Waycott M, Duarte CM, Carruthers TJB, Orth RJ, Dennison WC, Olyarnik S, Cardinale A, Fourqurean JW, Heck KL Jr., Hugher AR, Kendrivk GA, Kenworthy WJ, Short FT, Williams SL (2009) Accelerating loss of seagrasses across the globe threatens coastal ecosystems. Proceedings National Academy of Sciences 106: 12377–12381.

Webb, SR, and Kneib, RT (2002) Abundance and distribution of juvenile white shrimp *Litopenaeus setiferus* within a tidal marsh landscape. Marine Ecology Progress Series 232: 213–223.

Weinstein MP, Litvin SY (2016) Macro-restoration of tidal wetlands: a whole estuary approach. Ecological Restoration 34(1): 27–38.

Weslawski JM, Snelgrove PVR, Levin LA, Austen MC and others (2004) Marine sediment biota as providers of ecosystem goods and services. In Wall DH (ed.) Sustaining Biodiversity and Ecosystem Services in Soils and Sediments. Island Press, Washington, DC, pp. 73–98.

Whaley SD, Minello TJ (2002) The distribution of benthic infauna of a Texas salt marsh in relation to the marsh edge. Wetlands 22(4): 753–766.

Williams CB (1964) Patterns in the Balance of Nature. Academic Press, New York, NY.

Williams JA, Holt GJ, Robillard MM, Holt SA, Hensgen G, Stunz GW (2016) Seagrass fragmentation impacts recruitment dynamics of estuarine-dependent fish. Journal of Experimental Marine Biology and Ecology 479: 97–105.

Williams, PB, Orr MK, Garrity NJ (2002) Hydraulic geometry: a geomorphic design tool for tidal marsh channel evolution in wetland restoration projects. Restoration Ecology 10(3): 577–590.

Yamada K, Tanaka Y, Era T, Nakaoka M (2014) Environmental and spatial controls of macroinvertebrate functional assemblages in seagrass ecosystems along the Pacific coast of northern Japan. Global Ecology and Conservation 2: 47–61.

Yeager LA, Acevedo CL, Layman CA (2012) Effects of seascape context on condition, abundance, and secondary production of a coral reef fish, *Haemulon plumierii*. Marine Ecology Progress Series 462: 231–240.

Yeager LA, Layman CA, Allgeier JE (2011) Effects of habitat heterogeneity at multiple spatial scales on fish community assembly. Oecologia 167(1): 157–168.

Yeager LA, Keller DA, Burns TR, Pool AS, Fodrie FJ (2016) Threshold effects of habitat fragmentation on fish diversity at landscapes scales Ecology doi: 10.1002/ecy.1449.

Zeff ML (1999) Salt marsh tidal channel morphometry: applications for wetland creation and restoration. Restoration Ecology 7: 205–211.

6

Seascape Patch Dynamics

Emma L. Jackson, Rolando O. Santos-Corujo and Simon J. Pittman

6.1 Introduction

Our world is heterogeneous, patchy, at a range of scales in time and space (Wiens 1976), yet the majority of scientific models examining ecological relationships oversimplify complexity by averaging out values at a single focal scale (Grünbaum 2012). In contrast, landscape and seascape ecology embrace spatial heterogeneity and place temporal variability into a spatial context. The importance of placing greater focus on 'patchiness' was emphasized by Simon Levin during his Robert H. MacArthur Award Lecture in 1989, where he proposed that the key to understanding and predicting ecological pattern and the consequences of disturbance was to understand the mechanisms driving patch structure (Levin 1992).

What is a patch? Patches represent relatively discrete areas (spatial domain) or periods (temporal domain) of relatively homogeneous environmental conditions. In the broadest dictionary definition, a patch has been defined as 'a surface area differing from its surrounding' (Wiens 1976) or 'a bounded, connected discontinuity in a homogeneous reference background' (Levin & Paine 1974). When represented in maps, patches can differ in size, shape, boundary characteristics, composition, position in the seascape and duration and, as such, vary in quality and function for organisms. In the 1940s, the plant ecologist Alex Watt recognized the importance of patch context where 'each patch in this space-time mosaic is dependent on its neighbours and develops under conditions partly imposed by them' (Watt 1947). Through the patch-centred perspective of mainstream landscape ecology, patches form the basic elements or component of a landscape or seascape. Typically, a patch has been represented as a spatially discrete homogeneous area within a matrix (the most extensive and connected landscape element) or a component of a patch-mosaic (no one dominating element in the landscape) (Forman 1995). For more information on the structure and function of the matrix see Driscoll *et al.* (2013). Patch boundaries, or edges, are recognized as important structural attributes in landscape ecology and the study of the ecological consequences of different structural characteristics of boundaries is a major focus of scientific enquiry (Ries & Sisk 2004; Boström *et al.* 2011).

Patches shift in time and space and can result in complex ecological patterning. As such, patch dynamics are considered to be of profound importance in landscape ecology (Pickett & Thompson 1978). According to Pickett & Thompson (1978), three

Seascape Ecology, First Edition. Edited by Simon J. Pittman.
© 2018 John Wiley & Sons Ltd. Published 2018 by John Wiley & Sons Ltd.

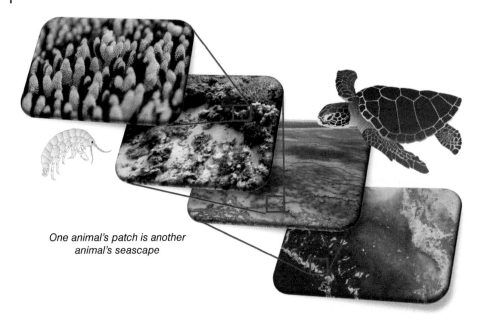

*One animal's patch is another
animal's seascape*

Figure 6.1 The perception and responses to a patch are organism specific, one animal's patch is another animal's seascape (adapted from Pittman 2013). Symbols courtesy of the Integration and Application Network, University of Maryland Center for Environmental Science (ian.umces.edu/symbols/, accessed 28 May 2017).

sets of phenomena and process define the dynamics of patches: (i) disturbance regime; (ii) within-patch species composition, population demographics, movements and organism geometry; and (iii) patch longevity. While a single patch at a single scale can be the focus of study, the modern concept of patch dynamics is based on the recognition that ecosystem heterogeneity occurs across multiple scales of time and space (*i.e.*, hierarchical patch dynamics) and can be represented mathematically in equations and through geometry (Levin 1992; Wu & Loucks 1995). For instance, imagine that a patch at any given scale has an internal structure that is a reflection of patchiness at finer scales and the mosaic containing that patch has a structure that is determined by patchiness at broader scales (Kotliar & Wiens 1990). Although all species live in patchy environments, in reality the definition is more complex still since the perception and responses to patch structure are organism specific (*e.g.*, shark or shrimp, plant or people, see Figure 6.1) and are linked to specific life history traits and influenced by past disturbance and colonization events (Wiens & Milne 1989; Girvetz & Greco 2007) and other environmental heterogeneity (Wiens 1976; Kotliar & Wiens 1990).

Taking the example of a vegetated seascape featuring seagrass beds, a patch could either be an unvegetated patch in a matrix of homogenous vegetation (also termed a gap or halo) or a vegetated patch in a matrix of unvegetated substrata (*e.g.*, sand). Alternatively, a patch can be created where a different species composition exists (*e.g.*, a patch reef or mangrove island). The temporal delineation of patch configurations within a seascape is explicitly related to the disturbance regime (frequency, severity and predictability) and temporal variability in colonization, competition and dieback.

Landscapes and seascapes are thus composed of hierarchically structured mosaics of patches recognizable and fluctuating across different spatial and temporal scales. The term 'shifting mosaics' was coined to describe the process of pattern change for terrestrial forest systems, whereby patches were observed changing as a result of the dynamic between growth and mortality. The 'shifting mosaic steady-state model' put forward by Bormann & Likens (1979) proposed that individual patches could be in different phases of succession, yet the proportions of those successional phases across the landscape remained constant. Thus, the shifting mosaic steady-state model recognized that dynamics occurring at one scale could produce a steady state at a different scale. Similarly, highly heterogeneous shifting mosaics have been identified and studied in marine systems (*i.e.*, algae, seagrasses, plankton) (Dayton *et al.* 1984; Holmquist 1997; Bell *et al.* 1999; Lévy *et al.* 2015; Lim *et al.* 2016). This more complex examination of patchiness has formed a major focus of landscape ecology and resource management (Bissonette 2012).

Studying patch dynamics is possible at a variety of scales. For example, Garrabou *et al.* (2002) and Teixidó *et al.* (2002) mapped and quantified changes in spatial patterning of sessile benthic fauna within $310\,\text{cm}^2$ and $1\,\text{m}^2$ photo quadrats respectively over a two year period using a set of spatial pattern metrics. Using similar techniques, Santos *et al.* (2016) mapped and quantified spatial patterning of seagrasses at coarser spatial resolution across far broader spatial scales (hundreds of metres to tens of kilometres) than Garrabou *et al.* (2002) and over a longer time series of images (71 years). Spatial change can be measured in different ways, for example two dimensional areal change in patch size (*e.g.*, Paling *et al.* 2008) and patch configurations across a seascape and the change in vertical structure (leaf length, rugosities) within a patch (*e.g.*, Ramage & Schiel 1999) and the change in composition of the community of organisms living within a patch (Terrados *et al.* 1998). In the past, this complexity and lack of standardized methodology have led to confusion when quantifying change, but we now have an increasing capacity to monitor, map and measure change remotely from space, air and water, and huge archives exist that have not yet been fully exploited for long-term change analysis (Garrabou *et al.* 1998; Manson *et al.* 2003).

In this chapter, we examine the evolution of the theory of seascape patch dynamics and the models that have attempted to simplify these dynamics. We also discuss the importance of scaling and identify the mechanisms influencing seascape patterns (with a particular focus on marine vegetated habitats such as seagrass) and the new spatial analytical approaches, technologies and tools available to monitor, map, measure and interpret spatial change.

6.2   From Patch Dynamics to Seascape Ecology

The recognition and study of the consequences of spatial patchiness on pattern and process in marine ecosystems is not new (Hutchinson 1953; Steele 1978). Early development of patch dynamic theory built on the study of succession in plant ecology and dynamic mosaics, focusing on the idea that patch distribution was uniform across time and space, leading to an equilibrium of patches within a landscape (White & Pickett 1985). This equilibrium concept was later identified as an exception rather than the norm (White & Pickett 1985) and it was proposed that an equilibrium landscape or seascape was only likely to occur in closed systems and where a feedback between communities and disturbances meant that susceptibility to a disturbance increased with time or where a patch was small relative to the area of the surrounding matrix habitat (White & Pickett 1985; Wu & Loucks 1995).

Ecological studies in marine systems (pelagic and benthic) contributed significantly to the early theoretical development of patch dynamics (Pickett & White 1985). Initial experimental studies examining patch dynamics focused on observing the response of a system to human and other drivers of disturbance through the creation of gaps. Levin & Paine (1974) developed a conceptual and mathematical model relating community structure (patch formation and spatial heterogeneity) to the level of environmental disturbance and applied it to an intertidal rocky shore population of mussels. The authors suggested that we view the community as a spatial and temporal mosaic of 'patches', which cannot be viewed as closed. Rather, they are part of an integrated 'patchwork', with individual patches constantly exchanging materials directly, or indirectly (Levin & Paine 1974). By experimentally removing mussel cover to create bare patches of varying size and positions and monitoring recolonization and by analysing the age structure of the whole mosaic, they identified different stages of patch recolonization according to different stages of succession. By treating a patch as the fundamental unit of community structure, recognizing intrapatch dynamics, and the stochasticity of recolonization events and environmental fluctuation, their model was one of the first to abandon static equilibria and include dynamic processes. Whilst this allowed comparisons of various geographic areas with different levels of disturbance, the model, by their own admission, did not consider specific dispersion patterns: for instance the influence of nearest-neighbour relationships on recolonization. In other words, it did not take a seascape approach.

Theory quickly moved from an examination of gaps to patches. Modelling focal habitat patch type (*e.g.*, seagrass) in a matrix of a less favourable habitat (*e.g.*, bare substrate) is considered analogous to oceanic islands and therefore assessments of patch dynamics have often taken an Island Biogeography Theory approach (IBT, MacArthur & Wilson 1967). IBT described patterns in species richness on islands as a function of island area and isolation (distance from the mainland) predicting several key relationships: (i) an equilibrium is reached when the extinction and invasion rate equalizes on an island or focal habitat patch; (ii) as patch area increases the rate of extinction decreases (larger patches support larger populations and are less likely to suffer local extinction); and (iii) isolation increases the distance from potential sources of colonizers thereby reducing new recruits. In the marine environment patches are considered to be more connected due to pelagic dispersal, although this is sometimes a function of the matrix. Metapopulation models (Hanski & Gilpin 1997) which focus on patchy habitats, emphasize the importance of dispersal between isolated patches, assuming a network of patches without a persistent mainland and focusing on the dynamics of usually single species. Like other early approaches to patch dynamics in population modelling (*i.e.*, patch occupancy models, source-sink dynamics and IBT), early metapopulation models had unrealistic assumptions (for example that all patches were similar in size, shape, composition and connectivity) and were unable to predict what was happening in nature (Fahrig 2007). To address this perceived oversimplification of spatial and temporal heterogeneity, researchers began integrating modern landscape ecology concepts into metapopulation ecology. This resulted in more spatially realistic models that allowed for different shapes, qualities and connectivity of patches and focused on determining which patches were most likely to go extinct or become colonized (Hanski 2001). Prior to the landmark study by Hanski (2001), detailed patch dynamic studies on kelp forests in the 1980s by Dayton *et al.* (1984) had already recognized the need for more spatially complex models.

Dayton *et al.* (1984) posed four hypotheses for the existence of distinct patches in nature. Firstly, niche separation and the idea that patches are due to local physical characteristics tolerated by only specific species. Secondly, dispersal potential results in patches being colonized by some species, but not others. Thirdly that the system is well mixed, with all species having equal chance of colonizing, but that different life histories determine patch composition. Finally, they hypothesized that beneficial (*e.g.,* mutualism) or deleterious (*e.g.,* competition, predation) interactions may enhance or prevent successful colonization. These hypotheses have since been supported through patch clearance and transplant experiments within many different seascapes including rocky shores (Levin & Paine 1974; Paine & Levin 1981; Sousa 1984a), seagrasses (Bell *et al.* 1999; Ramage & Schiel 1999) and saltmarsh (Shumway & Bertness 1994).

Change over time and space within a seascape is conceptualized by a spectrum of models (Table 6.1), all of which model change (from one cover type to another) in different landscape elements (patch type, class, cover type) over time. The models (Table 6.1) differ based on the level of aggregation (large or small number of patches, whole or distributional seascape models), whether continuous or discrete temporal data is available, as well as by the methods for defining a change in the state of a patch and the type of output (Weinstein & Shugart 1983). In general, the approach is to divide the area of interest into a grid of cells and predict change for each cell, to model change based on three components: (i) initial configuration; (ii) type of change (*e.g.,* extinction/colonization); and (iii) output configuration. Where continuous temporal data exists and there are a large number of elements, differential equations are used because the change in the number of elements is a result of a balance between patches moving from one type to another (distributional seascape models, see Table 6.1). If time is discrete (for example, an annual snapshot of a seascape or an acute disturbance event), then a difference equation is used based on the probability of a patch changing from one type to another at different times (Markov model). If there are only a few elements and patches change considerably over time, the common modelling approach is to define the dynamics of each patch with detailed models. Differential equation approaches do not incorporate the within-patch (*i.e.,* community or species) ecological response to changes to the environment (*e.g.,* through disturbance).

Models that do include response mechanisms at the element scale include yield models (*e.g.,* using regression to predict change in growth rate under different environmental conditions) and population models (which predict recruitment and extinction rates). But, both are very site specific and therefore difficult to scale up to a broader seascape. Gap replacement models, however, measure rates of change for specific species or communities within a patch type and the influence of environmental modifiers as probability functions. These models allow assessments of seascape changes due to regional disturbance, but still do not consider emergent properties relating to the pattern, juxtapositions and relationships within a seascape mosaic, like spatial seascape models do (Table 6.1). Exchanges between patches and with the matrix can have a significant influence on responses, for example the speed of colonization, susceptibility to disturbance and recoverability. The size and shape of patches also influence responses

Table 6.1 Different types of seascape models.

Model	Description	Examples	Issues/advantages
Whole seascape models	Model the change in value of a specific landscape variable in aggregate, for example the change in total number of patches.	Used as submodels in spatial models. Cunha *et al.*, 2005; Bell *et al.*, 1999	Does not incorporate patch spatial characteristics or dynamics and disturbance
Distributional seascape models	Model the distribution of a specific seascape variable, *e.g.* patch size distribution. Time is discrete time: Difference equation models (Markov chain models). Time is continuous: Differential equation models	Li & Yang (2015); Bell *et al.*, 1999; Cuhna *et al.*, 2005;	Based only on initial rates of change provided. Not sensitive to additional external disturbance.
Spatial seascape models	Model spatial location and configuration	Mosaic models; Element models (each element or patch type has its own submodel) Kendall, 2005; Ault *et al.*, 2003.	
Gap replacement model	Use patch colonization, growth and mortality rates in simulations	Yield models Population models Probabilistic cellular automaton Langmead & Shepherd, 2004; Vidondo *et al.*, 1997; Paine & Levin, 1981.	Can be scaled up to allow assessment of landscape scale changes due to local and regional disturbances.

and interconnections due to the influence of factors such as edge effects and ecosystem engineering. The ability to simultaneously model such interacting variables within a spatially explicit framework is characteristic of a landscape ecology approach. Similarly, a model that incorporates different temporal and spatial scales (*e.g.*, related to the age or size distributional properties of patches, or species composition) would thereby predict the essential dynamic properties of interest in a system.

6.3 Scale

A seascape can be defined conceptually and operationally as a hierarchy of patch mosaics across a range of temporal and spatial scales, the dynamics of which vary depending on perception (Wiens & Milne 1989; Kotliar & Wiens 1990; Pittman & McAlpine 2003). In order to investigate patch dynamics, patches must be defined

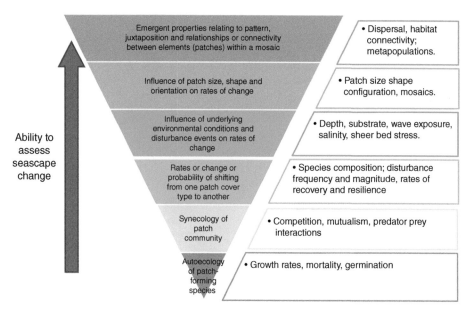

Figure 6.2 Our ability to assess seascape change increases as we consider processes occurring at different scales.

relative to the phenomenon under consideration. For example, in this chapter we focus on vegetated marine habitats with a patch defined as a structurally and visually discrete, relatively homogenous area of marine vegetation. Our ability to assess seascape change increases as we consider processes occurring at different scale (Figure 6.2). However, irrespective of the focal scale an investigation of patch dynamics benefits from an understanding of processes across multiple temporal and spatial scales. If the focal level is a patch of seagrass, the dynamics of that patch are influenced by climatic conditions operating at regional to global scales. For example, the dynamics of a patch of seagrass in the tropics will differ to that of a patch of seagrass in temperate regions, due to the higher frequency of disturbance and faster growth rates of the species of seagrass that dominate in the tropics (Kilminster *et al.*, 2015). Underlying substrate and depth could also be considered boundary conditions in a vertical plane. Another consideration is therefore dimensionality and how to measure and track dynamics of 3D patch structure. Most assessments of patch dynamics focus on two-dimensional 'flatscapes', but in reality habitat structure is of higher dimensionality (Pittman *et al.* 2010). Thus, the spatial patterns of seascape elements, ecological processes and responses of species to seascape structures vary with the scale at which they are observed (Levin 1992; Holland *et al.* 2004; Wu 2004). Multi-scale seascape approaches are important to describe and understand the spatial seascape structures and functioning since most patterns at the seascape level are spatially correlated and scale-dependent (Purkis *et al.* 2007; Boström *et al.* 2011; Santos *et al.* 2011). Therefore, understanding how scale influences observations is a critical aspect in seascape ecology (Sleeman *et al.* 2005; Bell *et al.* 2007).

Through time, a sampling regime that measures the extent of a patch of seagrass once a year may see little change, however seasonal measurements may indicate that at certain times of year the same patch may disappear completely, only to recover from seed or

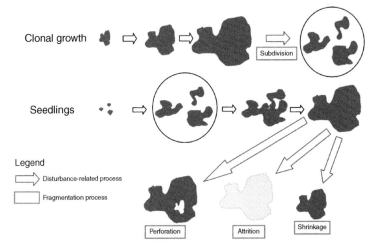

Figure 6.3 Fusion and fragmentation processes within a seascape. A depiction of how different processes of disturbance, colonization and clonal growth in a seagrass meadow can result in similar patch formations within a seascape and the four spatial processes involved in fragmentation: shrinkage, attrition, perforation and subdivision. *Source:* Adapted from Boström *et al.* (2011).

rhizome in the following summer months. A patch of marine vegetative habitat within a seascape may be a result of fragmentation of a larger patch or colonization processes (Duarte *et al.* 2006). Figure 6.3 illustrates, using seagrass meadows as an example, how each of these processes may result in a similar patch formation. Similarly, short term monitoring may indicate a decline over a period of a few years, however over longer periods this may be part of longer term dynamics driven by climatic conditions. In tropical Australia, the frequency and magnitude of extreme storms and flood events are correlated with the El Nino southern oscillation (ENSO). Above average rainfall, and more frequent storms and cyclones associated with La Niña events increase turbidity, decrease salinity and cause physical disturbance to the plants and seed banks, all resulting in large scale declines of seagrass (Coles *et al.* 2015; McKenna *et al.* 2015). During El Nino events, reduced storm frequency and lower rainfall have been correlated with increases in seagrass cover.

6.4   Factors Influencing Seascape Patchiness

Whilst understanding and incorporating scaling issues is imperative in understanding the dynamics of a seascape, the dynamics of a patch are influenced by internal and external factors. Internal factors encompass a species ability to resist and recover from disturbance (recolonization ability, growth rates and other life history traits) and external factors include those that create a disturbance (the frequency and magnitude of disturbance), or influence the rates of internal factors. As Duarte *et al.* (2006) pointed out, whilst the extension of a seagrass patch is governed by the ability of the plants rhizomes to elongate, there are no constraints when it comes to a patch receding or dying completely. Therefore, predicting patch dynamics becomes less about understanding the growth rates of the plants and more about the disturbances.

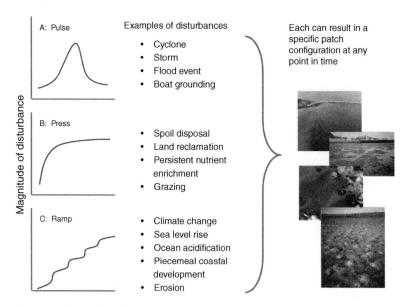

Figure 6.4 Examples of types of disturbance magnitudes, causes and examples of different resulting patch configurations using seagrass as an example. *Source:* Adapted from Lake (2000).

Disturbance has long been recognized as important in structuring marine communities (Dayton 1971; Sousa 1984b). White and Pickett (1985) defined disturbance as 'any relatively discrete event that disrupts the structure of an ecosystem, community, or population and changes resource availability or the physical environment.' Drivers of disturbance in seascapes are reviewed elsewhere (Wilson *et al.* 2006; Airoldi & Beck 2007; Gedan *et al.* 2009); here we discuss the types of disturbance and their ecological consequences, which depend on the frequency, duration and magnitude of the events. With regard to seascape patterning, the spatial pattern of the disturbance (*e.g.,* its geographical shape and extent) and the resilience characteristics of the seascape will be important. A major problem facing seascape ecologists is that very little is known about the historical ecology and spatial patterning of disturbance events. Three classes of disturbance events are relevant to seascape patch dynamics (Figure 6.4). These are pulse, press (*sensu* Bender *et al.* 1984) and ramp disturbances (Lake 2013). Pulsed (analogous to Type II perturbations, as defined by Sutherland 1981) represent low frequency, but high magnitude formative events (for example cyclones, flooding, or major storms). In contrast, a press disturbance (analogous to Type III perturbations, as defined by Sutherland 1981) is one that may occur quickly, but is then maintained at a constant level (an example could be burial or persistent pollution). Finally, a ramp disturbance involves a disturbance that gradually increases either continuously or levelling off after a long time (*e.g.,* global warming, the spread of invasive species, sedimentation, increases in bioturbation or creeping coastal development). There is inconsistency in the use of this terminology (see Glasby & Underwood 1996), mainly because some definitions refer only to the cause of the disturbance (Bender *et al.* 1984), the effect (Sutherland 1981), or cause and effect together (Glasby & Underwood, 1996). The third approach recognizes scale and the fact that the response is determined by the higher order and lower order influences. For example, a pulse disturbance such as a storm in a temperate persistent seagrass meadow (Kilminster *et al.* 2015) may result in a shift in seascape patch configuration (for example fragmentation) that persists for a long period consistent with a press disturbance. Additionally, gaps in the *Posidonia oceanica* seagrass meadows in Corsica caused by World War II bomb blasts were still evident in 1996 (Pasqualini *et al.* 1999). In contrast, a pulse pressure in tropical seagrass meadows where colonizing and opportunistic seagrass species dominate can exhibit fast recovery from seed banks (Rasheed *et al.* 2014).

An individual disturbance is unlikely to exist in isolation in an ecosystem, as often there will be a multitude of disturbances occurring simultaneously or in succession, interacting and modifying the impact of and recovery from other disturbances. Exposure to multiple interacting disturbances is a major challenge for scientists and managers, as is considering the capacity of different ecosystems to cope with disturbance, without switching to an alternative (and undesirable) state (resilience). The landscape ecology approach has helped through the understanding of disturbance regimes at multiple scales; a focus on feedbacks and uncovering the factors (internal and external) that mediate the resilience of landscapes. According to Turner (2010) three new insights have emerged from studies of disturbance and landscape dynamics: (i) even very large disturbances do not homogenize the landscape; rather, they create spatial heterogeneity, often at multiple scales; (ii) landscape equilibrium is scale-dependent (*i.e.*, steady state mosaics only applied in very specific cases) and is but one of a suite of dynamics that systems may exhibit (*e.g.*, nonequilibria and dynamic equilibria); (iii) conditions under which spatial heterogeneity influences ecological processes can be identified sometimes, but we still do not know when it is, or isn't, important.

Spatial patterning in the seascape, just like ecological organization of any ecosystem, emerges from the interaction of structures and processes operating at different scales (Kotliar & Wiens 1990; Peterson 2000). According to a hierarchical perspective, understanding patch dynamics across a range of scales is fundamental to understanding emergent ecosystem responses such as resilience, because resilience is influenced by spatial variables such as geographical position, functional connectivity and context (*e.g.*, juxtaposition with other habitats). The hierarchical patch dynamics concept states that changes in spatial patterns and processes at different scales both impact and are impacted by local system resilience (Wu & Loucks 1995). For example, seagrasses engineer their environment by binding sediments, buffering water currents and changing local biogeochemistry (van der Heide *et al.* 2011). Feedbacks of this nature mitigate the direct effect of fluctuations resulting in nonlinear relationships between an increasing pressure and ecosystem responses (Nyström *et al.* 2012), including spatial patterning. Fonseca *et al.* (2000) examined the relative importance of chronic versus extreme hydrodynamically mediated disturbance in the expression of seascape patterns of seagrasses by mapping spring and autumn coverage of seagrass in adjacent 1 m^2 grid cells from 1991 to 1995. In 1993, the sites experienced an extreme wind event and they found that sites with higher cover suffered less loss than those with lower initial cover. The result fuelled evidence from a previous study (Fonseca & Bell 1998) examining the relationship between physical setting and spatial patterns of seagrass, which suggested that below a critical threshold of 59% seagrasses were structurally incapable of withstanding the scouring effects of a storm disturbance. Fonseca & Bell (1998) proposed that this threshold effect fits with percolation theory, whereby below a cover value of 59% landscape elements occur as discrete patches and above it they join ('percolating clusters', With & Crist 1995). At this threshold, or tipping point, landscapes look like large patches joined by narrow corridors. These narrow corridors may be particularly vulnerable to acute physical disturbance (Fonseca *et al.* 2000), resulting in an abrupt change. Although few marine studies have done so, it may be possible to predict the likely change in cover, contiguity and fragmentation an additional disturbance may cause (Grober-Dunsmore *et al.* 2009).

6.5   Mapping and Quantifying Seascape Change

The quantification of spatio-temporal patterns are needed for evaluating change, understanding and predicting the effects of pattern on ecological processes, for analysing differences between communities and assessing the implications for system resilience. The quantification of seascape changes can be approached in three logical steps: (i) define research questions based on the nature of the system (*i.e.,* spatiotemporal scales, disturbance frequency, pattern characteristics); (ii) quantify and describe spatio-temporal patterns with maps and spatial pattern metrics; and (iii) build models to describe and forecast seascape dynamics (see Figure 6.5). The robustness and accuracy of seascape dynamic analyses will depend on the assumptions of the research questions and specifications and limits of the tools selected. Therefore, it is important to consider the advantages and disadvantages of mapping techniques and spatial pattern metrics with respect to the scale and nature of the research questions.

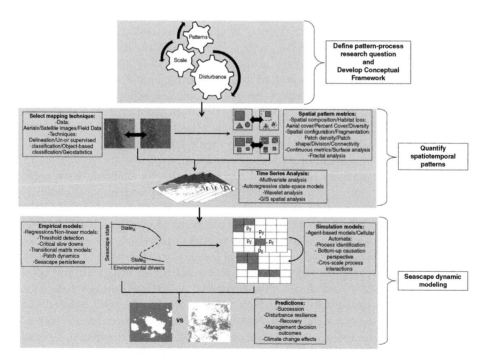

Figure 6.5 Example of a conceptual diagram showing the main steps in a study of seascape dynamics.

6.5.1 Habitat Mapping for Change Analysis

Habitat mapping provides a way to measure and investigate the spatial dynamics of seascapes at ecologically meaningful scales that are also operationally relevant for management actions (Santos *et al.* 2016; Lecours *et al.* 2015). Seascape ecology studies have used field mapping, remote sensing and geostatistical approaches to create broad-scale spatial mosaics of the benthic seascape (see Chapter 2 in this book). Scale can have a significant influence on the use of benthic habitat maps and therefore careful decisions must be made for the spatial, temporal and thematic resolution of the mapping. Ideally, these parameters should be determined by the research questions, the focal seascape patterning and relevant ecological processes, as well as availability of spatial data (Wedding *et al.* 2011; Lecours *et al.* 2015).

There is a growing interest in using remote sensing for submerged habitat mapping, since conventional field mapping techniques are usually time consuming and expensive to conduct over a continuum of scales (Mumby & Harborne 1999; Andréfouët 2008; Roelfsema *et al.* 2009). Seascape scale mapping with remote sensing has used aerial photographs, satellite images, radar and acoustic data (Sheppard *et al.* 1995; Silva *et al.* 2008; Hossain *et al.* 2015). Aerial photography has been used effectively to map coastal habitats (Robbins 1997; Hossain *et al.* 2015) because of the availability of a long time series and typically a high spatial resolution (<1 m) providing the opportunity for finer habitat discrimination. Despite limitations in spectral resolution, habitat delineation with aerial photography has proven successful for accurately mapping shallow coastal ecosystems (Pasqualini *et al.* 2001; Frederiksen *et al.* 2004a; Hernández-Cruz *et al.* 2006; Santos *et al.* 2011). Several studies have used archived aerial photographs to assess historical changes in seagrass meadows in relationship to environmental and anthropogenic disturbances (Frederiksen *et al.* 2004b; Hernández-Cruz *et al.* 2006; Santos *et al.* 2016). For example, in Puerto Rico, Hernández-Cruz *et al.* (2006) mapped seagrass patches from aerial imagery acquired between 1937 and 2000. Analyses revealed that the seagrass patchiness went through discrete episodes characterized by expansion in the number and spatial extent of small patches followed by an increase in patch size and agglomeration of small patches to form large homogeneous areas. Advances in digital cameras have now provided the advantage of obtaining aerials with more spatial accuracy and resolution and colour digital information (*e.g.*, RGB values), which are necessary for semiautomated image classifications.

Satellite and aerial imagery have proved highly effective for mapping submerged habitats, quantifying habitat cover, structure and complexity and physical-chemical properties of the water column over broad spatial extent (Andréfouët *et al.* 2003; Mumby *et al.* 2004; Lyons *et al.* 2010). Several platforms of medium (10–30 m pixel size – *e.g.*, Landsat) and high spatial resolution (1–4 m pixel size – *e.g.*, IKONOS, Quickbird, GeoEye, WorldView 3) multispectral imagery are commonly used to map submerged aquatic vegetation (SAV) habitats in coastal areas, mangrove and coral reef (Mumby *et al.* 2004; Andréfouët 2008; Hossain *et al.* 2015). There is a consensus that sensors with the greatest spatial resolution should provide the most detailed and accurate representation of coastal submerged habitats. However, there are advantages and limitations of each platform that depend on a tradeoff between the spatial, spectral and temporal resolutions across remote sensing techniques, field observations and cost-effective factors (Andréfouët *et al.* 2003).

Marine cartographers also use in water sensors to map seascape structure including unmanned aerial systems, autonomous underwater vehicles and boat-based systems to collect habitat data at higher spatial resolution (*i.e.,* submetre and centimetre-scale spatial resolution) (Klemas 2015; Chirayath & Earle 2016; Hedley *et al.* 2016). The development of miniaturized sensors, such as multi- and hyperspectral imagers and LIDAR sensors (light imaging, detection and ranging), adapted for unmanned autonomous systems has provided the flexibility to obtain both high spatial and spectral resolution timely data at a lower cost; thus, providing the opportunity to monitor seascape processes consistently over higher spatiotemporal resolutions (Klemas 2015). For example, Chirayath & Earle (2016) developed fluid lensing data acquisition, an experimental technology that removes wave distortion from images, to map at a centimetre-scale resolution coral reefs of Ofu Island (American Samoa) using multispectral imagery from a low-altitude unmanned aerial vehicle. Sonar data obtained from boat-based and autonomous underwater vehicles sonar-based systems have also been used to classify and map benthic habitats and sea-floor sediment characteristics (Erdey-Heydorn 2008; Hossain *et al.* 2015; Abadie *et al.* 2015; Hedley *et al.* 2016); especially in areas where the water column characteristics (*e.g.,* depth, turbidity) limits the use of remote sensing with aerials and satellite images. Broad-scale habitat maps based on different satellite platforms are usually created with semiauto-mated image classification techniques. Semiautomated image classification techniques identify spectral signatures representative of various habitat types and assign every image pixel to the closest habitat spectral signature. Pixel-based image classification techniques use either unsupervised or supervised training (see Chapter 2 in this book). These classification techniques quantify temporal changes in seagrass percentage cover using both medium and high spatial resolution satellite platforms (Mumby & Edwards 2002; Lyons *et al.* 2010; Roelfsema *et al.* 2014), exhibiting varying mapping accuracy (between 40 to 80%).

Supervised object-based classification has greatly improved accuracy compared to more commonly used un/supervised classification (Urbański *et al.* 2009; Blaschke 2010; Roelfsema *et al.* 2014). The segmentation of the images into objects with distinct properties (*e.g.,* tone, colour contrast, texture and shape) help distinguish the borders between different classes. One important aspect of an object-based classification is that it offers the advantage of mapping the hierarchical and scale-dependent structure of submerged habitats by providing image segmentation at multiple levels. For example, an object-based classification approach could be used to map SAV seascape at the meadow, bed or patch scale. Therefore, the object-based approach shows considerable promise for the interpretation of remote sensing data (Hossain *et al.* 2015) and should be implemented under different scenarios to study multiscale seascape changes of submerged habitats.

The combination of field surveys and geostatistical techniques offer the potential for efficient broad-scale mapping in marine environments where limited field-collected data is available and where the implementation of remote sensing techniques is limited due to physical constraints (*e.g.,* turbid water column, deep habitats, cloud cover) (Holmes *et al.* 2007; Kendrick *et al.* 2008; Barrell & Grant 2013). Geostatistical techniques are based on detecting, modelling and estimating spatial patterns in spatially correlated data (Fortin & Dale 2005). For example, methods of variography and kriging interpolations (*e.g.,* ordinary, simple and indicator kriging) estimate values at unmeasured points to form a continuous surface based on measured values and spatial relationships (*e.g.,* autocorrelation, distance, direction, adjacency).

Geostatistical and spatial distribution models have been applied to map the continuous distribution of seagrass density (Scardi *et al.* 2006; Leriche *et al.* 2011) and probability of occurrence (Holmes *et al.* 2007; March *et al.* 2013; Ooi *et al.* 2014). Some of these techniques can also map the distribution and spatial structure of submerged habitats as products of biotic and abiotic interactions across multiple scales, thus offering the advantage of identifying the scale at which spatial patterning of seascapes mostly occur (Kendrick *et al.* 2008; Barrell & Grant 2013). However, the accuracy of geostatistical maps is highly dependent on the number and distribution of field observations and the spatial heterogeneity of submerged habitats. In addition, depending on the geostatistical technique, the habitat spatial structure could not be analysed under a patch-mosaic model since patch boundaries are not readily apparent unless cutoff values are used to assign different habitat classes.

6.5.2 Characterization of Spatial Patterns

The ability to quantify seascape structure is a prerequisite for the assessment of seascape patch dynamics (Kupfer 2012). Regardless of the type of metric model used (*e.g.,* patch-mosaic or continuous metrics), the development of metrics within a seascape context are still necessary since they are the main tool for resource managers for assessing and monitoring changes in habitat structure and the effects on underlying ecological processes at broad scales (Wedding *et al.* 2011; Kupfer 2012).

Spatial data from habitat maps based on discrete and meaningful categories or elements (*i.e.,* patches) may illustrate mosaic patterns of varying configuration ranging from highly fragmented to continuous formations. Therefore, the spatial structure of habitats represented in thematic maps can be characterized using a landscape ecology approach based on a patch-mosaic model, in which landscapes (or seascapes) are conceptualize and analysed as mosaics of discrete patches (Forman 1995; Turner *et al.* 2001; McGarigal & Cushman 2002). In this approach, landscape elements are defined as discrete patches and habitat spatial patterns are described using spatial pattern metrics developed to quantify patch (*e.g.,* patch size, shape, isolation) and mosaic or landscape (*e.g.,* patch diversity, configuration and composition) level characteristics (Kupfer 2012). Interest in using spatial pattern metrics to describe spatial structure of marine habitats has increased over the years, but still more work is needed to understand which landscape ecology techniques best represent spatial structure of different seascapes (Sleeman *et al.* 2005; Boström *et al.* 2011).

Spatial-pattern metrics can be quantified and analysed at three levels of a spatial hierarchy: patch, class or habitat type and seascape (Turner *et al.* 2001; Botequilha *et al.* 2006). Patch-level metrics quantify the spatial characteristics of individual patches and tend to be aggregated and used to calculate landscape metrics at higher hierarchical levels (*e.g.,* class or landscape level metrics) (Turner *et al.* 2001; McGarigal & Cushman 2002). Class-level metrics quantify the spatial extent and arrangement of specific patch or habitat type. Landscape-level metrics encompass the overall patch spatial data within the full extent of the area of study (McGarigal & Cushman 2002; Botequilha *et al.* 2006) and are used to quantify the spatial structure of habitat mosaics over broad scales.

Spatial pattern metrics at the landscape and class level can provide useful information about two seascape structure categories: those that quantify the composition of the mosaic without reference to the spatial characteristics, and those that quantify configuration of seascape elements in the mosaic taking into account spatial attributes (McGarigal *et al.* 2002; Sleeman *et al.* 2005). Therefore, spatial pattern metrics quantifying seascape composition are used to quantify habitat loss, whereas metrics of configuration are used to assess and describe habitat fragmentation. Advances in computing, remote sensing and GIS technologies have increased the ability of researchers to acquire spatially explicit data and compute spatial pattern metrics to characterize spatial change associated with habitat loss and fragmentation (McGarigal *et al.* 2005).

6.5.2.1 Continuous Metrics and Surface Analysis

Even though the patch matrix model, in which spatial metrics are based on, has helped in the progress in understanding landscape and seascape spatial pattern-process interactions, it is unable to represent and quantify continuous spatial heterogeneity (McGarigal *et al.* 2009; Barrell & Grant 2013). In the process of classifying the seascape in discrete units or habitat patches, this classification homogenizes all internal heterogeneity, which may result in significant loss of information. When applying the patch mosaic model in practice, it is prudent to ask whether the magnitude of information loss is acceptable (McGarigal *et al.* 2009). Many terrestrial and marine studies are now quantifying continuous heterogeneity outside of the patch-matrix model, by applying topographical metrics and surface analysis on 3D landscape/seascape patterns (Hoechstetter *et al.* 2008; McGarigal *et al.* 2009; Wedding *et al.* 2011).

Morphometrics commonly used in geomorphology and industrial engineering to quantify surface features and complexity have performed well as predictors of fish diversity and species distributions across coral reef seascapes (Wedding *et al.* 2011). For example, Wedding & Friedlander (2008) and Pittman *et al.* (2009) derived depth, slope and rugosity metrics from 3D surface data to identify the best predictors of fish abundance and species richness. In addition, continuous morphometric data have been used to examine the effects of geologically forced seascape structure on kelp seascapes (Parnell 2015). However, surface metrics have not been applied to quantify and characterize changes in seascape patterns. Measurement on submerged aquatic vegetation such as canopy height and leaf area index can also provide further information on habitat structure, which could extend analysis of within-patch variability to three dimensions (Barrell & Grant 2013). Moving windows of predetermined size can be used to create continuous surfaces of spatial pattern metrics across large areas (Kupfer 2012; McGarigal *et al.* 2012). FRAGSTATS v4 software can compute a variety of surface metrics that describe the spatial structure of continuous surfaces or landscape gradients (McGarigal *et al.* 2012). Surface creation of spatial pattern metrics using moving windows could be useful for examining the effects of scale on habitat patterns (Riitters & Coulston 2005; Kupfer 2012). In addition, surface metrics or gradients of seascape patterns can be applied into habitat suitability models to relate spatial habitat structure to gradients of environmental parameters (*e.g.,* temperature, salinity, nutrients, light availability and depth) and predict seascape spatial changes/dynamics under different environmental scenarios.

6.5.2.2 Metrics, Scale and Sensitivity Analysis

With the availability of hundreds of spatial pattern metrics to quantify seascape composition and configuration, it is the scientist's task to determine which metrics provide useful information across multiple scales (Turner *et al.* 2001; Kelly *et al.* 2011) to select the most appropriate set of metrics particular for the research question and subjects of investigation (*e.g.,* habitat types, pattern processes and organism relationships). It is very important to be cautious looking at the results of metrics, specially within a multi-scale approach, since these are known to be sensitive to changes of scale (*e.g.,* change in resolution and/or extension) (Turner *et al.* 1989, 2001; Saura & Martinez-Millan 2001; Wu 2004). Thus, comparisons of spatial pattern metrics quantified at different scales may be inappropriate because results reflect scale-related errors or scale effects, rather than differences in seascape patterns (Turner *et al.* 2001).

Beside the spatial scale effects, it is important to select spatial pattern metrics with ecological merit that could provide linkages to functional properties of patches and seascape mosaics. Many of the spatial pattern metrics available quantify landscape structure and have a structural basis rather than a functional one. Thus, the ecological relevance of spatial pattern metrics should be evaluated *a priori* to facilitate meaningful interpretations of pattern-process relationships and describe seascape spatial dynamics within a mechanistic framework. For example, connectivity metrics could better describe change dynamics of seascape dependent on long dispersal characteristics; however, seascape patterns dependent on clonal dispersion of patches may be better quantify with subdivision, shape and edge metrics. Habitat maps are often associated with some level of error and uncertainty, which add another cautionary aspect when using spatial pattern metrics and interpreting seascape patterns (see Mapping Technique section above; Andréfouët 2008; Roelfsema *et al.* 2009; Wedding *et al.* 2011). The accuracy level of habitat maps and the thematic resolution could propagate through the results of spatial pattern metrics and seascape analyses (Huang *et al.* 2006; Shao & Wu 2008). Even acceptably low misclassification rates in the spatial data use in landscape analyses can be propagated and magnified into significant errors in metric values (Langford *et al.* 2006; Shao & Wu 2008; Wedding *et al.* 2011).

6.5.2.3 Quantifying Seascape Change

The combination of remotely sensed imagery and spatial pattern metrics to document change in marine and coastal habitats over time is becoming increasingly important as anthropogenic stresses significantly influence coastal systems (Santos *et al.* 2011; Wedding *et al.* 2011; Cuttriss *et al.* 2013). The availability of long archive of aerial and satellite images and freeware to calculate spatial pattern metrics (*e.g.,* FRAGSTATS, Q-Rule, Patch Analyst) offers the opportunity to study and detect broad-scale seascape changes in a cost-effective way. Perturbation to coastal and submerged habitats are most commonly quantified at fine spatial resolution by *in situ* observations and descriptions of patch characteristics (*e.g.,* SAV shoot/ramet density, percentage cover of small components such as coral colonies and seagrass blades) (Waycott *et al.* 2009; Lyons *et al.* 2013). However, there has been an increasing application of time series remote sensing analyses in combination with spatial pattern metrics to assess temporal changes of broad scale seascape properties (Cuttriss *et al.* 2013; Roelfsema *et al.* 2014; Santos *et al.* 2016). Seascape level studies using this time series approach often focus on quantifying changes in the areal extent and species composition (*i.e.,* studies of seascape composition change or habitat loss). Thus, most broad-scale mapping studies have used seascape composition metrics such as habitat area, percentage cover and diversity.

In Moreton Bay, Australia, time series of moderate and high spatial resolution imagery (*e.g.*, Landsat TM, Quickbird) were classified using pixel- and object-based classification techniques to quantify area long-term changes of seagrass patch types defined by species composition, percentage cover and biomass classes (Lyons *et al.* 2010, 2013; Roelfsema *et al.* 2014). Other studies have quantified seagrass cover change using stepwise correlation and change detection analysis of spectral information gathered by multispectral moderate spatial resolution imagery (see Hossain *et al.* 2015 for a list of examples). In addition, time series of historical aerial photographs have been used to study the spatiotemporal dynamic of seagrass meadow extent (Ferguson *et al.* 1993; Kendrick *et al.* 2002; Frederiksen *et al.* 2004a). Broad scale mapping and time series seascape studies like these have helped to understand how seascape composition characteristics and habitat loss are driven by disturbance events such as freshwater and nutrient loading, sedimentation, storms, ice scouring, fishing and disease outbreaks (Frederiksen *et al.* 2004a, 2004b; Duarte *et al.* 2006; Gullström *et al.* 2006; Lyons *et al.* 2013). For instance, Frederiksen *et al.* (2004b) used a time series of aerial photographs to characterized changes in eelgrass spatial distribution and acreage associated with wave dynamics, where exposed sites were composed of smaller and complex seagrass patches and presented the greatest spatial changes.

Many of the available examples have provided insight on the highly dynamic spatiotemporal nature of coastal submerged seascapes, which highlights the importance of integrating a collection of metrics and analysis techniques to capture different components of seascape change (Cunha & Santos 2009). Historically, few large-scale mapping and monitoring studies have considered the significance of seascape configuration dynamics and how this spatial transformation may influence the resilience of submerged habitat seascapes and population dynamics (Cunha & Santos 2009; Montefalcone *et al.* 2010; Cuttriss *et al.* 2013). By incorporating both spatiotemporal dynamics in areal extent and other measure of seascape composition and seascape configuration, seascape ecology studies could have a better quantification of both habitat loss and fragmentation and how these seascape changes respond to disturbances with distinct intensity and persistence.

Monitoring and assessing changes in seascape composition does not present a complete evaluation of the dynamics of submerged habitats as fragmentation can also influence the population dynamics of founding species (*i.e.*, habitat forming species such as seagrass, corals, macroalgae) (Vermaat *et al.* 2004; Bell 2006; Huntington *et al.* 2010). For example, studies that used configuration metrics in addition to areal extent identified an increase in the fragmentation state of submerged vegetation seascapes concomitant with different trends in habitat extent (*i.e.*, spatial extent or area increased, decreased or stay neutral) (Cunha *et al.* 2005; Cuttriss *et al.* 2013; Santos *et al.* 2016). Hernández-Cruz *et al.* (2006) observed an expansion of seagrass cover within an embayment, but the near shore portion of the study area influenced by effluent discharges revealed increases in patchiness and fragmentation. In Biscayne Bay (Miami, Florida, United States), studies have demonstrated how SAV seascape composition and configuration is spatially variable, where the interaction of habitat loss and fragmentation was associated to freshwater canals and the proximity to the shore (Santos *et al.* 2011, 2016).

Disturbance events and associated regime shifts may induce both habitat loss and fragmentation. The former is the reduction in the amount or the proportion of habitat space occupied by habitat patches within the seascape, while the latter refers to the breaking apart of continuous larger patches. Therefore, habitat loss refers to the reduction in cover and fragmentation to the changes in the spatial arrangement of the remaining habitat within the seascape (Fahrig 2003; McGarigal *et al.* 2005; Liao *et al.* 2013). Both of these aspects can have either independent or interactive effects on the resilience and persistence of macrophyte populations and influence faunal connectivity patterns among habitats, species diversity and ecological interactions (*e.g.,* competition, predation, foraging behaviour) (Irlandi & Crawford 1997; Fahrig 2003; Jackson *et al.* 2006; Boström *et al.* 2011). Fragmentation of seascapes is likely to have a greater effect on habitat persistence than changes in cover (Fonseca & Bell 1998; Sleeman *et al.* 2005; Duarte *et al.* 2006). Increasing fragmentation may hinder the persistence of coastal submerged habitats since the extension or elongation of existing patches can be a slow process and most often the probability of newly established patches to reach a large size is low (Vidondo *et al.* 1997; Kendrick *et al.* 2005). Evidence is illustrated in the common lack of recovery of propeller scars within *Thalassia testudinum* beds in Florida where rhizomes are not able to bridge the denuded gaps, which may persist for years or decades.

There are different alternative scenarios of seascape habitat loss and fragmentation associated with different seascape spatial disruptions (*e.g.,* perforation, dissection, sub-division, shrinkage and attrition) and tradeoffs between disturbance and succession processes (Forman 1995; McGarigal *et al.* 2005). Seascapes can exhibit an increase in habitat cover without a reduction in fragmentation if the habitat patches increase the above-ground biomass and a steadily expansion of large patches over time concentrated on one edge of the seascape, but fail to expand into denuded areas (illustrated by a downward progression along the seascape cover vertical axis in Figure 6.5). Conversely, seascapes can exhibit reduced cover without a change in fragmentation if remaining patches only 'thin out' in unfavourable environmental conditions. Under favourable conditions, habitat patches can increase in biomass/cover as well as expand into suitable habitat and fill out gaps among patches. When conditions are not consistently favourable, tradeoffs may lead to conflicting and dynamic patterns such as documented in different seascape studies (Frederiksen *et al.* 2004b; Cunha & Santos 2009; Cuttriss *et al.* 2013; Santos *et al.* 2016). Under extreme or persistent unfavourable conditions, this may lead to both habitat loss and fragmentation (illustrated by the black dashed line in Figure 6.6). Under this scenario, the habitat patches within the seascape are subject to: (i) perforation; (ii) subdivision; (iii) shrinkage; and (iv) attrition processes that lead to a gradual breakdown and formation of discrete fragments (Forman 1995). Based on a patch-focused analysis using GIS (Geographical Information Systems) methods and spatial pattern metrics, Li & Yang (2015) demonstrated how these four type of seascape change can occur simultaneously and at different rates, as well as during different periods.

Many of these metrics are strongly correlated, exhibit statistical interactions, present scaling problems and typically have strong relationships with the amount of habitat (*i.e.,* relationship with seascape composition metrics) (McGarigal *et al.* 2002; Bogaert 2003; Fahrig 2003; Li & Wu 2004). Consequently, the selection of spatial pattern

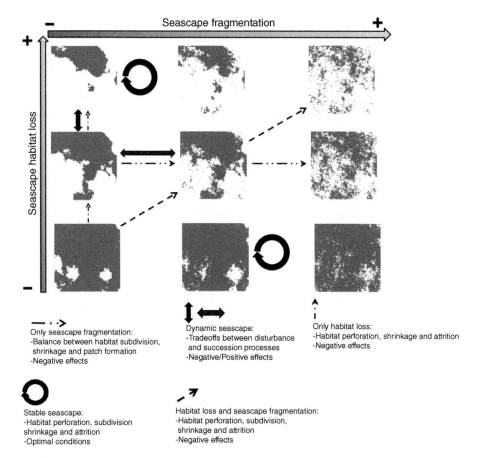

Figure 6.6 Conceptual diagram of the interaction of seascape habitat loss and fragmentation and general pathways of seascape pattern outcomes.

metrics quantifying seascape configuration and fragmentation should be well thought out to avoid multicollinearity statistical problems, increase the robustness of the analysis across multiple spatial scales, provide accurate connection between pattern and process and, more importantly, to separate the effects and processes of habitat loss from the processes of spatial configuration change and habitat fragmentation (Wu *et al.* 2002; Bogaert 2003; Li & Wu 2004; Sleeman *et al.* 2005). Once a temporal stack of aerial or satellite images are identified and processed and seascape composition and configuration are quantified with the proper set of spatial pattern metrics, various time-series analysis techniques should be considered to assess both temporal trends and spatiotemporal fluctuations of habitat loss and fragmentation. Ordination procedures (*e.g.,* principal component analysis, nonmetric multidimensional scaling, canonical correspondence analysis), a group of techniques that reduce multivariate data into fewer independent axes or variables that represent different gradients in the data, can be used to identify temporal trends and variability in seascape structure (Wedding *et al.* 2011). Santos *et al.* (2016) used a series of principal component analyses (PCAs) to identify how different sites changed from continuous to fragmented seascapes, and by combining the PCAs with a vector analysis the authors were able to illustrate how the temporal variability in seascape configuration was associated with distinct disturbance regimes. The PCA axes and other indices of seascape fragmentation could also be developed and used to simplify the contrast and interpretation of habitat loss and fragmentation trends. Similar approaches have been used previously to assess the effects of fragmentation on species diversity, probability of occurrence and abundance of terrestrial and marine species independent of habitat loss (McGarigal & McComb 1995; Trzcinski *et al.* 1999).

Fractal dimension indices derived from fractal theory have been applied by several studies to evaluate scaling behaviours and persistence patterns of seagrass seascapes (Cunha & Santos 2009; Irvine *et al.*, 2016). Fractal dimension indices are used to assess scale-free patterning in the form of power law patch-size distributions that help explore the concepts of scale and spatial variability in vegetation spatial patterns. For example, Cunha & Santos (2009) used the Korcak dimension and Hurst exponent, a fractal dimension index, to characterize the persistence of seagrass spatial patterns in relation to changes of a coastal inlet in Portugal. Using this approach, they found that seascapes with larger and nonfragmented seagrass areas showed higher persistence, but in contrast, more complex and highly fragmented landscapes dominated by many small areas were associated with a reduced persistence. In addition, different domains of scale identified across patch size ranges suggested a shift in generating processes by which seagrass seascapes meadows grow. The Korcak dimension and exponent have also been applied to provide estimates of patchiness and reforestation in terrestrial systems (Imre *et al.* 2011). Irvine *et al.* (2016) applied other scaling and spatial statistics to different seagrass seascapes in the Isles of Scilly, which suggested a significant degree of fractality in the vegetative community and determined localized banding patterns of seagrass patches.

GIS also offer the empirical platform to identify and illustrate spatial changes of seascapes across different time steps (*i.e.*, monthly, yearly or decadal changes). For example, using GIS, change detection analysis can be used to quantify the areas where habitat was either lost, gained, or stable between two consecutive time periods (Figure 6.7). This type of spatial change detection can assist in the localization of areas of management concern and assess the dynamics of spatial changes at the local scale. Various seascape studies have used this GIS approach to estimate an index of relative change (according to Frederiksen *et al.* 2004a) that describes the proportion of the total area covered by seagrass habitats that was either loss or gained from two consecutive sampling periods; thus providing, a measure of spatial variability that tend to be overlooked when the total area change is only considered in broad-scale studies (Frederiksen *et al.* 2004b; Ferreira *et al.* 2009; Santos *et al.* 2016).

Change detection studies have used satellite images to capture multitemporal trends over time of bottom-reflectance value representing different habitat classes. Tools such as LandTrendr (Landsat-based detection of trends in disturbance and recovery) can be used to detect both abrupt disturbance events and longer term stress-induced degradation that cause both habitat loss and fragmentation by estimating changes in pixel values across multiple years (Kennedy *et al.* 2010). Terrestrial ecology studies have used this tool to assess and characterize the spatial recovery, stability and mortality of forest landscapes after natural disturbance events (Kennedy *et al.* 2010; Senf *et al.* 2015). Other methods have been developed to assess landscape heterogeneity changes using pixel values of satellite images that could be also implemented in the studies of seascape

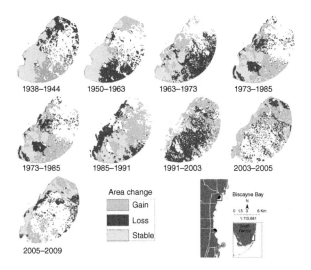

Figure 6.7 Illustration of spatiotemporal analysis of seascape change using GIS. Time step differences between areas where submerged habitat gain or loss cover can be used to estimate an index of relative change, and identify areas of concern.

spatial dynamics using broad-scale images (Li *et al.* 2015). There is a limited example of seascape studies using spectral indexes and trends to assess seascape spatiotemporal dynamics (see Gullström *et al.* 2006; Misbari & Hashim 2016).

6.5.4 Seascape Modelling

The complex and multiscale interaction of exogenous, including disturbance (*e.g.*, hurricane, sedimentation, ice scouring) and/or endogenous (*e.g.*, biotic interactions such as interspecies interaction) processes result in spatiotemporally heterogeneous and dynamic seascapes. The behaviour and effects of this complex interaction are constrained by the spatial pattern of the seascape but also act to shape that pattern (Perry & Enright 2006). Thus, there are strong feedbacks between seascape pattern and process (Turner *et al.* 2001). Due to this complex nature of seascape pattern-process relationships a range of modelling approaches should be employed to gain insights into the mechanisms underlying spatio-temporal patterns and forecast the dynamics and recovery times of seascapes under different disturbance scenarios.

Models used to study seascape-level habitat change can be classified into two broad categories: (i) empirical (*e.g.*, analytical and statistical models) and (ii) simulation models (Perry & Enright 2006; Schröder & Seppelt 2006; Reuter *et al.* 2010). Empirical models are based on systems of linear equations, regressions and/or correlations and are often used for pattern detection and description (*i.e.*, identify processes based on patterns by selecting models supposed to represent the underlying mechanisms and determining the model parameters) (Perry & Enright 2006; Schröder & Seppelt 2006). In contrast, simulation models are mechanistic models designed for process description by generating patterns that emerge under the assumption of a set of driving processes (Schröder & Seppelt 2006). In addition, this type of modelling has the potential to process changing interaction structures between a large number of heterogeneous components (Reuter *et al.* 2010).

There are various time series and statistical analysis techniques that have been used in environmental, fishery and atmospheric studies that can be adapted to assess seascape change dynamic when using data with complex behaviours (*e.g.,* nonstationarity, multicollinearity, nonlinear trends, scaling issues). The spatial structure of seascapes can be highly dynamic and variable across temporal scales (Frederiksen *et al.* 2004a; Duarte *et al.* 2006; Lyons *et al.* 2013), thus, time-series analysis techniques that control for nonstationarity (*i.e.,* when statistical properties change with time), such as dynamic factor analysis (DFA, Zuur *et al.* 2003) and wavelet analysis (WA, Cazelles *et al.* 2008), should be employed to assess temporal dynamics of spatial pattern metrics. DFA models time series as linear combinations of common trends, a level parameter and noise (Zuur *et al.* 2003). Thus, DFA can be used to detect underlying temporal trends across different seascapes and assess how trends in habitat loss and fragmentation interact across time. In contrast, WA is a technique that allows for the decomposition of the variance over time (Cazelles *et al.* 2008; Hidalgo *et al.* 2011). WA is currently gaining increasing attention in landscape studies to assess spatiotemporal patterns (Schröder & Seppelt 2006) and it has the potential to study temporal fluctuations in seascape structure across different scales and how these relate to climatic stressors, environmental oscillation and the frequency of anthropogenic disturbances.

Fragmentation and fractal indices, configuration and composition metrics can also be used as dependent variables in linear and nonlinear models (*e.g.,* generalized additive models, boosted regression trees) to characterize complex trends in habitat loss and fragmentation (*e.g.,* Irvine *et al.*, 2016). Moving windows of predetermined size can be used to create continuous surfaces of spatial pattern metrics across large areas (Kupfer 2012; McGarigal *et al.* 2012), which could be used as input in habitat suitablity models to relate spatial habitat structure to gradients of environmental parameters (*e.g.,* temperature, salinity, nutrients, light availability and depth) and predict seascape spatial changes/dynamics under different environmental scenarios using empirical and nonlinear statistical models. Such statistical models could also identify significant breakpoints in the time series, which may indicate regime shifts in seascape and disturbance structure and identify threshold responses to external drivers (Andersen *et al.* 2009; Lindegren *et al.* 2012). For example, using a regression with a cubic fit, Fonseca & Bell (1998) found a threshold of seagrass habitat amount that identified the seagrass seascapes vulnerable to high energy disturbance events (*e.g.,* storms, hurricanes, waves). Critical slowing down indicators (*i.e.,* critical transitions that arises in the vicinity of bifurcations or thresholds – see Scheffer *et al.* 2012 and Dakos *et al.* 2015 for details) can also be estimated from models to identify the tipping point between alternate seascape states, determine the resilience of seascapes under a range of disturbance regimes and provide a signal that forewarn a major shift in habitat composition and spatial configuration.

Seagrass patches of different classes and sizes have differing mortality, growth and fragmentation (Duarte & Sand-Jensen 1990; Kendrick *et al.* 2005) due to the differentiation in rhizome extension, anchoring capabilities and sediment deposition rate between patch sizes and seagrass species. Transition matrix (Markov) models frequently are used to assess landscape change (Perry & Enright 2006) and could be applied to a series of seascape maps to estimate the probabilities associated to patch growth, shrinkage,

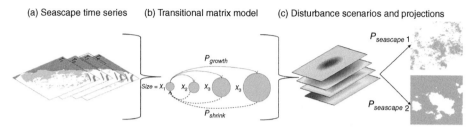

Figure 6.8 A transitional model approach to understand seascape dynamics. (a) A collection of seascape maps are stacked where the transition of different patch size classes (X_n) are follow to estimate change probabilities using (b) transitional matrix models. Then transitional models could be applied to (c) disturbance scenarios to project seascape changes.

mortality and fragmentation (Figure 6.3 and Figure 6.8). For example, different patch size classes can be identified in a seascape time series and follow the transition of patch classes across the time series to estimate the probabilities of patch dynamics (*e.g.*, growth, fragmentation, mortality) (Figure 6.8a and 6.8b). In turn, these transition probabilities could be used to simulate seascape change and predict the persistence of seascapes under different restoration, disturbances or climate change scenarios (Figure 6.8c). Similar approaches have been applied to study harvesting effects on sponge beds (Cropper & DiResta 1999), coral community dynamics and resilience to storm frequency (Lirman 2003; Tanner *et al.* 2009) and landscape changes in coastal areas (Lei *et al.* 2012).

Cellular automata (CA) and agent-based models (ABM) are two mechanistic model approaches with the potential to represent a wide range of ecological processes driving seascape patterns. These approaches have been widely used in landscape ecology studies addressing forest disturbance and succession dynamics, plant population spatial dynamics and coral interactions (see Perry and Enright 2006 for list of models developed in terrestrial studies; Schröder & Seppelt 2006; Reuter *et al.* 2010). CA and ABM can reproduce the emergence of complex pattern behaviour from simple rule sets iteratively applied to interacting cells on a lattice (in CA cases) and to individual and independent units (in ABM cases) (Reuter *et al.* 2010). Through agent-based models and other simulation approaches, studies have determined that patch growth by elongation and branching of horizontal rhizomes during a branching growth phase, compact growth phase by self-accelerating processes and coalescence of patches with large meadow formation are the ecological processes responsible for observed seagrass seascape patterns in temperate coastal system (Kendrick *et al.* 2005; Sintes *et al.* 2005). These models should be expanded to tropical system scenarios, and modified to understand how the importance and sensitivity of ecological processes change under different disturbance regimes and capture the persistence of seascapes that use different long dispersal strategies.

6.6   The Future of Seascape Dynamics Research

Marine ecosystems are becoming increasingly disrupted and altered due to human population growth concentrated in coastal areas (Orth *et al.* 2006; Valiela 2009; Waycott *et al.* 2009) and at a global scale due to the direct and indirect impacts of climate change (Hughes *et al.* 2005). Shifts in disturbance regimes (for example, increased storm frequency), biogeographical shifts in species distributions and composition (due to increased temperature change, sea level rise and ocean acidification) and increased direct human pressures causing habitat loss will result in increases in fragmentation, shifting mosaics, habitat losses/gains and coastal squeeze (low water mark migration due to sea level rise where high water migration is restricted by coastal defence). Taking into consideration that marine ecosystem services (*e.g.,* habitat provision, carbon sequestration, erosion protection, species persistence and diversity, Barbier *et al.* 2011) may be influenced by the spatial structure of habitats and processes such as fragmentation (Boström *et al.* 2011; Mizerek *et al.* 2011; Mitchell *et al.* 2015), the understanding, measurement and prediction of seascape patch dynamics will become a necessity for maintaining or restoring the resilience of seascapes and supporting human benefits we gain from them.

The spatial structure and patch dynamics of vegetated seascapes are factors that are commonly overlooked by managers when assessing the level of ecosystem function and services influenced by disturbance events (Mitchell *et al.* 2015; Ricart *et al.* 2015). For example, ignoring the spatial heterogeneity and spatial structure of seagrass and salt marsh habitats could have significant implications in the estimation of carbon sequestration and transport within these habitats (Macreadie *et al.* 2013; Ricart *et al.* 2015); therefore, hindering managers' capacity to understand how natural systems (and their restoration) could be used to mitigate CO_2 emissions and climate change. Shifting mosaics and restructuring of seagrass seascape through habitat loss and fragmentation can also have significant effects on organisms at higher trophic levels and the fish abundance that fuel important recreational and commercial fisheries (Jackson *et al.*, 2006; Mizerek *et al.* 2011; Mitchell *et al.* 2015). Detailed seascape maps and quantification of spatial characteristics (*e.g.,* seascape composition, configuration and connectivity metrics) should be employed more frequently to understand the functional aspects of marine ecosystems at relevant spatial scales and assess the related patterns – processes that feed back or erode ecosystem services and the resilient of marine habitats.

Studying seascape change using several landscape analytical approaches will provide the necessary tools to distinguish between natural patterns of disturbance and human driven disturbance and how these interact with the habitat resilience across multiple spatiotemporal scales. Understanding patch dynamics in the sea facilitates our ability to forecast and manage change; predict consequences, assess risk and identify causality. For instance, approaches that reconstruct and characterize the dynamics of previous disturbance regimes and the spatial structure of marine habitats could be used to identify thresholds in spatial patterns that either accelerate or buffer the effects of disturbance events. Research should be expanded to understand how the importance and sensitivity of ecological processes change under different disturbance regimes and capture the persistence of patches. Such information is the key to parameterizing the function components of seascape patch dynamic models. Thus, understanding and being able to model patch dynamics makes it possible to simulate dynamic seascapes under different management and disturbance scenarios (see Chapter 8 in this book).

Currently coastal and marine nature conservation management focuses on species or focal habitats, even in situations where time series of seascapes exist. For example, ecosystem health assessments may be based on the overall extent of a focal habitat and within focal patch changes in quality, which not only ignore natural patch dynamics, but obscure the larger scale connectivity and dynamics important to the health of the system and important indicators of tipping points and resilience thresholds. For instance, marine species ontogenetic and metapopulation processes may depend on habitat characteristics expressed at broad spatial scales and the habitat interconnectivity provided by distinct seascape arrangements (Mumby *et al.* 2004; Olds *et al.* 2012; Nagelkerken *et al.* 2015). Thus, restoration that focuses on restoring individual habitat types ignores the fact that protecting and restoring optimal seascape types based on ecological requirements of species and communities may be the true target.

This chapter described some of the existing tools and approaches to examining seascape patch dynamics that would facilitate a shift towards a more holistic and spatially explicit approach to ecosystem-based management (Pittman *et al.* 2011). However, further research is required not only to apply landscape modelling techniques to seascapes, but also to make approaches practical for natural resource management application. Unlike their counterpart terrestrial studies, changes over time of seascape structure have been overlooked in seascape ecology studies (Wedding *et al.* 2011). By integrating the quantification on change dynamics, seascape studies could provide a broader understanding of pattern-process interactions that influence the persistence and ecological integrity of marine habitats (Perry & Enright 2006; Wedding *et al.* 2011; Kendrick *et al.* 2012).

References

Abadie A, Gobert S, Bonacorsi M, Lejeune P, Pergent G, Pergent-Martini C (2015) Marine space ecology and seagrasses. Does patch type matter in *Posidonia oceanica* seascapes? Ecological Indicators 57: 435–446.

Airoldi L, Beck MW (2007) Loss, status and trends for coastal marine habitats of Europe. Oceanography and Marine Biology 45: 345–405.

Andersen T, Carstensen J, Hernandez-Garcia E, Duarte CM (2009) Ecological thresholds and regime shifts: approaches to identification. Trends in Ecology and Evolution 24: 49–57.

Andréfouët S (2008) Coral reef habitat mapping using remote sensing: a user vs producer perspective. Implications for research, management and capacity building. Journal of Spatial Science 53: 113–129.

Andréfouët S, Kramer P, Torres-Pulliza D, Joyce KE, Hochberg EJ, Garza-Pérez R, Mumby PJ, Riegl B, Yamano H, White WH (2003) Multi-site evaluation of IKONOS data for classification of tropical coral reef environments. Remote Sensing of Environment 88: 128–143.

Ault JS, Luo J, Wang JD (2003) A Spatial Ecosystem Model to Assess Spotted Seatrout Population Risks from Exploitation and Environmental Changes. CRC Press, Boca Raton, FL, pp. 267–296.

Barbier EB, Hacker SD, Kennedy C, Koch, EW Stier AC, Silliman BR (2011) The value of estuarine and coastal ecosystem services. Ecological Monographs 81(2): 169–193.

Barrell J, Grant J (2013) Detecting hot and cold spots in a seagrass landscape using local indicators of spatial association. Landscape Ecology 28: 2005–2018.

Bell SS (2006) Seagrasses and the metapopulation concept: developing a regional approach to the study of extinction, colonization, and dispersal. In Kritzer JP, Sale PF (eds) Marine Metapopulations. Academic Press, London.

Bell SS, Fonseca MS, Stafford NB (2007) Seagrass Ecology: New Contributions from a Landscape Perspective. Seagrasses: Biology, Ecology and Conservation. Springer, Dordrecht.

Bell SS, Robbins BD, Jensen SL (1999) Gap dynamics in a seagrass landscape. Ecosystems 2(6): 493–504.

Bender EA, Case TJ, Gilpin ME (1984) Perturbation experiments in community ecology: theory and practice. Ecology 65(1): 1–13.

Bissonette JA (2012) Wildlife and Landscape Ecology: Effects of Pattern and Scale. Springer Science and Business Media, New York, NY.

Blaschke T (2010) Object based image analysis for remote sensing. ISPRS Journal of Photogrammetry and Remote Sensing 65: 2–16.

Bogaert J (2003) Lack of agreement on fragmentation metrics blurs correspondence between fragmentation experiments and predicted effects. Ecology and Society 7(1): r6.

Bormann FH, Likens GE (1979) Catastrophic disturbance and the steady state in northern hardwood forests: A new look at the role of disturbance in the development of forest ecosystems suggests important implications for land-use policies. American Scientist 67(6): 660–669.

Boström C, Pittman SJ, Simenstad C, Kneib RT (2011) Seascape ecology of coastal biogenic habitats: advances, gaps, and challenges. Marine Ecology Progress Series 427: 191–217.

Botequilha A, Miller J, Ahern J, McGarigal K (2006) Measuring Landscapes. A planner's Handbook. Island Press, Washington.

Cazelles B, Chavez M, Berteaux D, Ménard F, Vik JO, Jenouvrier S, Stenseth NC (2008) Wavelet analysis of ecological time series. Oecologia 156: 287–304.

Chirayath V, Earle SA (2016) Drones that see through waves: preliminary results from airborne fluid lensing for centimetre-scale aquatic conservation. Aquatic Conservation: Marine and Freshwater Ecosystems 26: 237–250.

Coles RG, Rasheed MA, McKenzie LJ, Grech A, York PH, Sheaves M, McKenna S, Bryant C (2015) The Great Barrier Reef World Heritage Area seagrasses: Managing this iconic Australian ecosystem resource for the future. Estuarine, Coastal and Shelf Science 153: A1–A12.

Cropper WP, DiResta D (1999) Simulation of a Biscayne Bay, Florida commercial sponge population: effects of harvesting after Hurricane Andrew. Ecological Modelling 118: 1–15.

Cunha AH, Santos R (2009) The use of fractal geometry to determine the impact of inlet migration on the dynamics of a seagrass landscape. Estuarine, Coastal and Shelf Science 84: 584–590.

Cunha A, Santos R, Gaspar A, Bairros M (2005) Seagrass landscape-scale changes in response to disturbance created by the dynamics of barrier-islands: a case study from Ria Formosa (Southern Portugal). Estuarine, Coastal and Shelf Science 64: 636–644.

Cuttriss AK, Prince JB, Castley JG (2013) Seagrass communities in southern Moreton Bay, Australia: Coverage and fragmentation trends between 1987 and 2005. Aquatic Botany 108: 41–47.

Dakos V, Carpenter SR, van Nes EH, Scheffer M (2015) Resilience indicators: prospects and limitations for early warnings of regime shifts. Philosophical Transactions of the Royal Society of London B: Biological Sciences 370: 20130263.

Dayton PK (1971) Competition, disturbance and community organization: the provision and subsequent utilization of space in a rocky intertidal community. Ecological Monographs 41: 351–389.

Dayton PK, Currie V, Gerrodette T, Keller BD, Rosenthal R, Tresca DV (1984) Patch dynamics and stability of some California kelp communities. Ecological Monographs 54: 253–289.

Driscoll DA, Banks SC, Barton PS, Lindenmayer DB, Smith AL (2013) Conceptual domain of the matrix in fragmented landscapes. Trends in Ecology and Evolution 28(10): 605–613.

Duarte CM, Fourqurean JW, Krause-Jensen D, Olesen B (2006) Dynamics of seagrass stability and change. In Larkum AWD, Orth RJ, Duarte CM (eds) Seagrasses: Biology, Ecology and Conservation. Springer, Dordrecht.

Duarte CM, Sand-Jensen K (1990) Seagrass colonization: Biomass development and shoot demography in *Cymodocea nodosa* patches. Marine Ecology Progress Series 67(1): 93–103.

Erdey-Heydorn M (2008) An ArcGIS seabed characterization toolbox developed for investigating benthic habitats. Marine Geodesy 31: 318–358.

Fahrig L (2003) Effects of habitat fragmentation on biodiversity. Annual Review of Ecology, Evolution, and Systematics 34: 487–515.

Fahrig L (2007) Landscape heterogeneity and metapopulation dynamics. In Wu J, Hobbs RJ (eds) Key Topics and Perspectives in Landscape Ecology. Cambridge University Press, Cambridge, pp. 78–89.

Ferguson RL, Wood LL, Graham DB (1993) Monitoring spatial change in seagrass habitat with aerial photography. Photogrammetric Engineering and Remote Sensing 59: 1033–1038.

Ferreira M, Andrade F, Bandeira S, Cardoso P, Mendes RN, Paula J (2009) Analysis of cover change (1995–2005) of Tanzania/Mozambique trans-boundary mangroves using Landsat imagery. Aquatic Conservation: Marine and Freshwater Ecosystems 19: S38–S45.

Fonseca MS, Bell SS (1998) Influence of physical setting on seagrass landscapes near Beaufort, North Carolina, USA. Marine Ecology Progress Series 171: 109–121.

Fonseca M, Kenworthy W, Whitfield P (2000) Temporal dynamics of seagrass landscapes: a preliminary comparison of chronic and extreme disturbance events. Biologia Marina Mediterranea 7: 373–376.

Forman RT (1995) Land mosaics: the ecology of landscapes and regions. Cambridge University Press, Cambridge.

Fortin M-J, Dale MRT (2005) Spatial Analysis: A Guide for Ecologists. Cambridge University Press, Cambridge.

Frederiksen M, Krause-Jensen D, Holmer M, Laursen JS (2004a) Spatial and temporal variation in eelgrass (*Zostera marina*) landscapes: influence of physical setting. Aquatic Botany 78: 147–165.

Frederiksen M, Krause-Jensen D, Holmer M, Sund Laursen J (2004b) Long-term changes in area distribution of eelgrass (*Zostera marina*) in Danish coastal waters. Aquatic Botany 78: 167–181.

Garrabou J, Ballesteros E, Zabala M (2002) Structure and dynamics of north-western Mediterranean rocky benthic communities along a depth gradient. Estuarine, Coastal and Shelf Science 55(3): 493–508.

Garrabou J, Riera J, Zabala M (1998) Landscape pattern indices applied to Mediterranean subtidal rocky benthic communities. Landscape Ecology 13: 225–247.

Gedan KB, Silliman B, Bertness M (2009) Centuries of human-driven change in salt marsh ecosystems. Annual Review Marine Science 1: 117–141.

Girvetz EH, Greco SE (2007) How to define a patch: a spatial model for hierarchically delineating organism-specific habitat patches. Landscape Ecology 22: 1131–1142.

Glasby TM, Underwood A (1996) Sampling to differentiate between pulse and press perturbations. Environmental Monitoring and Assessment 42: 241–252.

Grober-Dunsmore R, Pittman SJ, Caldow C, Kendall MS, Frazer TK (2009) A landscape ecology approach for the study of ecological connectivity across tropical marine seascapes. In Nagelkerken, I (ed.) Ecological Connectivity among Tropical Coastal Ecosystems. Springer, Dordrecht, pp. 493–530.

Grünbaum D (2012) The logic of ecological patchiness. Interface Focus 2: 150–155.

Gullström M, Lundén B, Bodin M, Kangwe J, Öhman MC, Mtolera MS, Björk M (2006) Assessment of changes in the seagrass-dominated submerged vegetation of tropical Chwaka Bay (Zanzibar) using satellite remote sensing. Estuarine, Coastal and Shelf Science 67: 399–408.

Hanski I (2001) Spatially realistic theory of metapopulation ecology. Naturwissenschaften 88: 372–381.

Hanski I, Gilpin ME (1997) Metapopulation Biology: Ecology, Genetics, and Evolution, Vol 1. Academic Press, San Diego, CA.

Hedley JD, Roelfsema CM, Chollett I, Harborne AR, Heron SF, Weeks S, Skirving WJ, Strong AE, Eakin CM, Christensen TR, Ticzon V (2016) Remote sensing of coral reefs for monitoring and management: A review. Remote Sensing 8(2): 118.

Hernández-Cruz LR, Purkis SJ, Riegl BM (2006) Documenting decadal spatial changes in seagrass and *Acropora palmata* cover by aerial photography analysis in Vieques, Puerto Rico: 1937–2000. Bulletin of Marine Science 79: 401–414.

Hidalgo M, Rouyer T, Molinero JC, Massutí E, Moranta J, Guijarro B, Stenseth NC (2011) Synergistic effects of fishing-induced demographic changes and climate variation on fish population dynamics. Marine Ecology Progress Series 426: 1–12.

Hixon MA, Pacala SW, Sandin SA (2002) Population regulation: historical context and contemporary challenges of open vs. closed systems. Ecology 83: 1490–1508.

Hoechstetter S, Walz U, Dang L, Thinh N (2008) Effects of topography and surface roughness in analyses of landscape structure – a proposal to modify the existing set of landscape metrics. Landscape Online 3: 1–14.

Holland JD, Bert DG, Fahrig L (2004) Determining the spatial scale of species' response to habitat. BioScience 54: 227–233.

Holmes K, Van Niel K, Kendrick G, Radford B (2007) Probabilistic large-area mapping of seagrass species distributions. Aquatic Conservation: Marine and Freshwater Ecosystems 17: 385–407.

Holmquist JG (1997) Disturbance and gap formation in a marine benthic mosaic: influence of shifting macroalgal patches on seagrass structure and mobile invertebrates. Marine Ecology Progress Series 17: 121–130.

Hossain M, Bujang JS, Zakaria M, Hashim M (2015) The application of remote sensing to seagrass ecosystems: an overview and future research prospects. International Journal of Remote Sensing 36: 61–114.

Huang C, Geiger E, Kupfer J (2006) Sensitivity of landscape metrics to classification scheme. International Journal of Remote Sensing 27: 2927–2948.

Hughes TP, Bellwood DR, Folke C, Steneck RS, Wilson J (2005) New paradigms for supporting the resilience of marine ecosystems. Trends in Ecology and Evolution, 20(7): 380–386.

Huntington BE, Karnauskas M, Babcock EA, Lirman D (2010) Untangling natural seascape variation from marine reserve effects using a landscape approach. PLoS One 5: e12327.

Hutchinson GE (1953) The concept of pattern in ecology. Proceedings of the Academy of Natural Sciences of Philadelphia 105: 1–12.

Imre AR, Cseh D, Neteler M, Rocchini D (2011) Korcak dimension as a novel indicator of landscape fragmentation and re-forestation. Ecological Indicators 11: 1134–1138.

Irlandi E, Crawford M (1997) Habitat linkages: the effect of intertidal saltmarshes and adjacent subtidal habitats on abundance, movement, and growth of an estuarine fish. Oecologia 110: 222–230.

Irvine MA, Jackson EL, Kenyon EJ, Cook KJ, Keeling MJ, Bull JC (2016) Fractal measures of spatial pattern as a heuristic for return rate in vegetative systems. Royal Society Open Science 3(3): 150519.

Jackson EL, Attrill MJ, Jones MB (2006) Habitat characteristics and spatial arrangement affecting the diversity of fish and decapod assemblages of seagrass (*Zostera marina*) beds around the coast of Jersey (English Channel). Estuarine, Coastal and Shelf Science 68: 421–432.

Kelly M, Tuxen KA, Stralberg D (2011) Mapping changes to vegetation pattern in a restoring wetland: Finding pattern metrics that are consistent across spatial scale and time. Ecological Indicators 11: 263–273.

Kendall, MS (2005) A method for investigating seascape ecology of reef fish. In Proceedings of the Gulf and Caribbean Fisheries Institute 56: 1–11.

Kendrick GA, Aylward MJ, Hegge BJ, Cambridge ML, Hillman K, Wyllie A, Lord DA (2002) Changes in seagrass coverage in Cockburn Sound, Western Australia between 1967 and 1999. Aquatic Botany 73: 75–87.

Kendrick GA, Duarte CM, Marbà N (2005) Clonality in seagrasses, emergent properties and seagrass landscapes. Marine Ecology Progress Series 290: 291–296.

Kendrick GA, Holmes KW, Van Niel KP (2008) Multi-scale spatial patterns of three seagrass species with different growth dynamics. Ecography 31: 191–200.

Kendrick GA, Waycott M, Carruthers TJ, Cambridge ML, Hovey R, Krauss SL, Lavery PS, Les DH, Lowe RJ, i Vidal OM (2012) The central role of dispersal in the maintenance and persistence of seagrass populations. BioScience 62: 56–65.

Kennedy RE, Yang Z, Cohen WB (2010) Detecting trends in forest disturbance and recovery using yearly Landsat time series: 1. LandTrendr:Temporal segmentation algorithms. Remote Sensing of Environment 114: 2897–2910.

Kilminster K, McMahon K, Waycott M, Kendrick GA, Scanes P, McKenzie L, O'Brien KR, Lyons M, Ferguson A, Maxwell P (2015) Unravelling complexity in seagrass systems for management: Australia as a microcosm. Science of The Total Environment 534: 97–109.

Klemas VV (2015) Coastal and environmental remote sensing from unmanned aerial vehicles: An overview. Journal of Coastal Research 31: 1260–1267.

Kotliar NB, Wiens JA (1990) Multiple scales of patchiness and patch structure: a hierarchical framework for the study of heterogeneity. Oikos 59(2): 253–260.

Kupfer JA (2012) Landscape ecology and biogeography: Rethinking landscape metrics in a post-FRAGSTATS landscape. Progress in Physical Geography 36(3): 400–420.

Lake P (2000) Disturbance, patchiness, and diversity in streams. Journal of the North American Benthological Society 19: 573–592.

Lake P (2013) Resistance, resilience and restoration. Ecological Management and Restoration 14: 20–24.

Langford WT, Gergel SE, Dietterich TG, Cohen W (2006) Map misclassification can cause large errors in landscape pattern indices: examples from habitat fragmentation. Ecosystems 9: 474–488.

Langmead O, Sheppard C (2004) Coral reef community dynamics and disturbance: a simulation model. Ecological Modelling 175(3): 271–290.

Lecours V, Devillers R, Schneider DC, Lucieer VL, Brown CJ, Edinger EN (2015) Spatial scale and geographic context in benthic habitat mapping: review and future directions. Marine Ecology Progress Series 535: 259–284.

Lei S, Jinghai Z, Shaohong R, Yuanman H, Miao L (2012) Landscape pattern change prediction of Jinhu Coastal Area based on Logistic-CA-Markov Model. Advances in Information Sciences and Service Sciences 4(11): 1–10.

Leriche A, Boudouresque C-F, Monestiez P, Pasqualini V (2011) An improved method to monitor the health of seagrass meadows based on kriging. Aquatic Botany 95: 51–54.

Levin SA (1992) The problem of pattern and scale in ecology: the Robert H. MacArthur award lecture. Ecology 73: 1943–1967.

Levin SA, Paine RT (1974) Disturbance, patch formation, and community structure. Proceedings of the National Academy of Sciences 71: 2744–2747.

Lévy M, Jahn O, Dutkiewicz S, Follows MJ, d'Ovidio F (2015) The dynamical landscape of marine phytoplankton diversity. Journal of The Royal Society Interface 12(111): 20150481.

Li H, Wu J (2004) Use and misuse of landscape indices. Landscape Ecology 19: 389–399.

Li L, Chen Y, Xu T, Liu R, Shi K, Huang C (2015) Super-resolution mapping of wetland inundation from remote sensing imagery based on integration of back-propagation neural network and genetic algorithm. Remote Sensing of Environment 164: 142–154.

Li S, Yang B (2015) Introducing a new method for assessing spatially explicit processes of landscape fragmentation. Ecological Indicators 56: 116–124.

Liao J, Li Z, Hiebeler DE, El-Bana M, Deckmyn G, Nijs I (2013) Modelling plant population size and extinction thresholds from habitat loss and habitat fragmentation: Effects of neighbouring competition and dispersal strategy. Ecological Modelling 268: 9–17.

Lim IE, Wilson SK, Holmes TH, Noble MM, Fulton CJ (2016) Specialization within a shifting habitat mosaic underpins the seasonal abundance of a tropical fish. Ecosphere 7(2): art. e01212.

Lindegren M, Dakos V, Gröger JP, Gårdmark A, Kornilovs G, Otto SA, Möllmann C (2012) Early detection of ecosystem regime shifts: a multiple method evaluation for management application. PLoS One 7: e38410.

Lirman D (2003) A simulation model of the population dynamics of the branching coral Acropora palmata Effects of storm intensity and frequency. Ecological Modelling 161: 169–182.

Lyons MB, Phinn SR, Roelfsema CM (2010) Long Term Monitoring of Seagrass Distribution in Moreton Bay, Australia, from 1972–2010 using Landsat MSS, TM, ETM+. Proceedings of Geoscience and Remote Sensing Symposium (IGARSS), July 25–30 2010, Honolulu, HI, IEEE International, New York, NY.

Lyons MB, Roelfsema CM, Phinn SR (2013) Towards understanding temporal and spatial dynamics of seagrass landscapes using time-series remote sensing. Estuarine, Coastal and Shelf Science 120: 42–53.

MacArthur RH, Wilson EO (1967) The Theory of Island Biogeography: Monographs in Population Biology. Princeton University Press, Princeton, NJ.

Macreadie PI, Hughes AR, Kimbro DL (2013) Loss of 'blue carbon' from coastal salt marshes following habitat disturbance. PloS One 8(7): p.e69244.

Manson F, Loneragan N, Phinn S (2003) Spatial and temporal variation in distribution of mangroves in Moreton Bay, subtropical Australia: a comparison of pattern metrics and change detection analyses based on aerial photographs. Estuarine, Coastal and Shelf Science 57: 653–666.

March D, Alós J, Cabanellas-Reboredo M, Infantes E, Jordi A, Palmer M (2013) A Bayesian spatial approach for predicting seagrass occurrence. Estuarine, Coastal and Shelf Science 131: 206–212.

McGarigal K, Cushman SA (2002) Comparative evaluation of experimental approaches to the study of habitat fragmentation effects. Ecological Applications 12: 335–345.

McGarigal K, Cushman S, Maile N (2012) FRAGSTATS v4: Spatial Pattern Analysis Program for Categorical and Continuous Maps. Department of Environmental Conservation, University of Massachusetts, Amherst, MA.

McGarigal K, Cushman SA, Neel MC, Ene E (2002) FRAGSTATS: spatial pattern analysis program for categorical maps. Computer software program produced by the authors at the University of Massachusetts, Amherst, MA.

McGarigal K, Cushman S, Regan C (2005) Quantifying Terrestrial Habitat Loss and Fragmentation: A Protocol. US Department of Agriculture Technical Report, USDA, Rocky Mountain Region, Golden, CO.

McGarigal K, McComb WC (1995) Relationships between landscape structure and breeding birds in the Oregon Coast Range. Ecological Monographs 65(3): 235–260.

McGarigal K, Tagil S, Cushman SA (2009) Surface metrics: an alternative to patch metrics for the quantification of landscape structure. Landscape Ecology 24: 433–450.

McKenna S, Jarvis J, Sankey T, Reason C, Coles R, Rasheed M (2015) Declines of seagrasses in a tropical harbour, North Queensland, Australia, are not the result of a single event. Journal of Biosciences 40: 389–398.

Misbari S, Hashim M (2016) Change detection of submerged seagrass biomass in shallow coastal water. Remote Sensing 8(3): 200 doi: 10.3390/rs8030200.

Mitchell MG, Suarez-Castro AF, Martinez-Harms M, Maron M, McAlpine C, Gaston KJ, Johansen K, Rhodes JR (2015) Reframing landscape fragmentation's effects on ecosystem services. Trends in Ecology Evolution 30(4): 190–198.

Mizerek T, Regan HM, Hovel KA (2011) Seagrass habitat loss and fragmentation influence management strategies for a blue crab Callinectes sapidus fishery. Marine Ecology Progress Series 427: 247–257.

Montefalcone M, Parravicini V, Vacchi M, Albertelli G, Ferrari M, Morri C, Bianchi CN (2010) Human influence on seagrass habitat fragmentation in NW Mediterranean Sea. Estuarine, Coastal and Shelf Science 86: 292–298.

Mumby PJ, Edwards AJ (2002) Mapping marine environments with IKONOS imagery: enhanced spatial resolution can deliver greater thematic accuracy. Remote Sensing of Environment 82: 248–257.

Mumby PJ, Harborne AR (1999) Development of a systematic classification scheme of marine habitats to facilitate regional management and mapping of Caribbean coral reefs. Biological Conservation 88: 155–163.

Mumby PJ, Skirving W, Strong AE, Hardy JT, LeDrew EF, Hochberg EJ, Stumpf RP, David LT (2004) Remote sensing of coral reefs and their physical environment. Marine Pollution Bulletin 48: 219–228.

Nagelkerken I, Sheaves M, Baker R, Connolly RM (2015) The seascape nursery: a novel spatial approach to identify and manage nurseries for coastal marine fauna. Fish and Fisheries 16(2): 362–371.

Nyström M, Norström AV, Blenckner T, de la Torre-Castro M, Eklöf JS, Folke C, Österblom H, Steneck RS, Thyresson M, Troell M (2012) Confronting feedbacks of degraded marine ecosystems. Ecosystems 15: 695–710.

Olds AD, Connolly RM, Pitt KA, Maxwell PS (2012) Habitat connectivity improves reserve performance. Conservation Letters 5(1): 56–63.

Ooi JL, Van Niel KP, Kendrick GA, Holmes KW (2014) Spatial structure of seagrass suggests that size-dependent plant traits have a strong influence on the distribution and maintenance of tropical multispecies meadows. PloS One 9: e86782.

Orth RJ, Carruthers TJ, Dennison WC, Duarte CM, Fourqurean JW, Heck KL, Hughes AR, Kendrick GA, Kenworthy WJ, Olyarnik S (2006) A global crisis for seagrass ecosystems. Bioscience 56: 987–996.

Paine RT, Levin SA (1981) Intertidal landscapes: disturbance and the dynamics of pattern. Ecological Monographs 51: 145–178.

Paling E, Kobryn H, Humphreys G (2008) Assessing the extent of mangrove change caused by Cyclone Vance in the eastern Exmouth Gulf, northwestern Australia. Estuarine, Coastal and Shelf Science 77: 603–613.

Parnell PE (2015) The effects of seascape pattern on algal patch structure, sea urchin barrens, and ecological processes. Journal of Experimental Marine Biology and Ecology 465: 64–76.

Pasqualini V, Pergent-Martini C, Clabaut P, Marteel H, Pergent G (2001) Integration of aerial remote sensing, photogrammetry, and GIS technologies in seagrass mapping. Photogrammetric Engineering and Remote Sensing 67(1): 99–105.

Pasqualini V, Pergent-Martini C, Pergent G (1999) Environmental impact identification along the Corsican coast (Mediterranean Sea) using image processing. Aquatic Botany 65: 311–320.

Peterson GD (2000) Scaling ecological dynamics: self-organization, hierarchical structure, and ecological resilience. Climatic Change 44(3): 291–309.

Perry GL, Enright NJ (2006) Spatial modelling of vegetation change in dynamic landscapes: a review of methods and applications. Progress in Physical Geography 30: 47–72.

Pickett ST, Thompson JN (1978) Patch dynamics and the design of nature reserves. Biological Conservation 13(1): 27–37.

Pickett ST, White PS (eds) (1985) The Ecology of Natural Disturbance and Patch Dynamics. Academic Press, Orlando, FL.

Pittman SJ (2013) Seascape ecology: A new science for the spatial information age. Marine Scientist 44: 20–23.

Pittman SJ, Costa BM, Battista TA (2009) Using lidar bathymetry and boosted regression trees to predict the diversity and abundance of fish and corals. Journal of Coastal Research 25(6): 27–38 (special issue no.53: Costal Applications of Airborne Lidar remote sensing).

Pittman SJ, Costa B, Jeffrey CF, Caldow C (2010). Importance of seascape complexity for resilient fish habitat and sustainable fisheries. In Proceedings of the Gulf and Caribbean Fisheries Institute 2010 Nov, Vol. 63, pp. 420–426. Gulf and Caribbean Fisheries Institute, Inc., Fort Pierce, FL.

Pittman SJ, Kneib RT, Simenstad CA (2011) Practicing coastal seascape ecology. Marine Ecology Progress Series 427: 187–190.

Pittman SJ, McAlpine C (2003) Movements of marine fish and decapod crustaceans: process, theory and application. Advances in Marine Biology 44: 205–294.

Pittman SJ, McAlpine C, Pittman K (2004) Linking fish and prawns to their environment: a hierarchical landscape approach. Marine Ecology Progress Series 283: 233–254.

Purkis SJ, Kohler KE, Riegl BM, Rohmann SO (2007) The statistics of natural shapes in modern coral reef landscapes. The Journal of Geology 115: 493–508.

Ramage DL, Schiel DR (1999) Patch dynamics and response to disturbance of the seagrass Zostera novazelandica on intertidal platforms in southern New Zealand. Marine Ecology Progress Series 189: 275–288.

Rasheed MA, McKenna SA, Carter AB, Coles RG (2014) Contrasting recovery of shallow and deep water seagrass communities following climate associated losses in tropical north Queensland, Australia. Marine Pollution Bulletin 83: 491–499.

Reuter H, Jopp F, Blanco-Moreno JM, Damgaard C, Matsinos Y, DeAngelis DL (2010) Ecological hierarchies and self-organisation: Pattern analysis, modelling and process integration across scales. Basic and Applied Ecology 11: 572–581.

Ries L, Sisk TD (2004) A predictive model of edge effects. Ecology 85(11): 2917–2926.

Ricart AM, York PH, Rasheed MA, Pérez M, Romero J, Bryant C V, Macreadie PI (2015) Variability of sedimentary organic carbon in patchy seagrass landscapes. Marine Pollution Bulletin 100(1): 476–482.

Riitters KH, Coulston JW (2005) Hot spots of perforated forest in the eastern United States. Environmental Management 35: 483–492.

Robbins BD (1997) Quantifying temporal change in seagrass areal coverage: the use of GIS and low resolution aerial photography. Aquatic Botany 58: 259–267.

Roelfsema CM, Lyons M, Kovacs EM, Maxwell P, Saunders MI, Samper-Villarreal J, Phinn SR (2014) Multi-temporal mapping of seagrass cover, species and biomass: A semi-automated object based image analysis approach. Remote Sensing of Environment 150: 172–187.

Roelfsema C, Phinn S, Udy N, Maxwell P (2009) An integrated field and remote sensing approach for mapping seagrass cover, Moreton Bay, Australia. Journal of Spatial Science 54: 45–62.

Santos RO, Lirman D, Pittman SJ (2016) Long-term spatial dynamics in vegetated seascapes: fragmentation and habitat loss in a human-impacted subtropical lagoon. Marine Ecology 37(1): 200–214.

Santos RO, Lirman D, Serafy JE (2011) Quantifying freshwater-induced fragmentation of submerged aquatic vegetation communities using a multi-scale landscape ecology approach. Marine Ecology Progress Series 427: 233–246.

Saura S, Martinez-Millan J (2001) Sensitivity of landscape pattern metrics to map spatial extent. Photogrammetric Engineering and Remote Sensing 67: 1027–1036.

Scardi M, Chessa LA, Fresi E, Pais A, Serra S (2006) Optimizing interpolation of shoot density data from a Posidonia oceanica seagrass bed. Marine Ecology 27: 339–349.

Scheffer M, Carpenter SR, Lenton TM, Bascompte J, Brock W, Dakos V, Van De Koppel J, Van De Leemput IA, Levin SA, Van Nes EH (2012) Anticipating critical transitions. Science 338: 344–348.

Schröder B, Seppelt R (2006) Analysis of pattern–process interactions based on landscape models – overview, general concepts, and methodological issues. Ecological Modelling 199: 505–516.

Senf C, Leitão PJ, Pflugmacher D, van der Linden S, Hostert P (2015) Mapping land cover in complex Mediterranean landscapes using Landsat: Improved classification accuracies from integrating multi-seasonal and synthetic imagery. Remote Sensing of Environment 156: 527–536.

Shao G, Wu J (2008) On the accuracy of landscape pattern analysis using remote sensing data. Landscape Ecology 23: 505–511.

Sheppard C, Matheson K, Bythell J, Murphy P, Myers CB, Blake B (1995) Habitat mapping in the Caribbean for management and conservation: use and assessment of aerial photography. Aquatic Conservation: Marine and Freshwater Ecosystems 5: 277–298.

Shumway SW, Bertness MD (1994) Patch size effects on marsh plant secondary succession mechanisms. Ecology: 564–568.

Silva TS, Costa MP, Melack JM, Novo EM (2008) Remote sensing of aquatic vegetation: theory and applications. Environmental Monitoring and Assessment 140: 131–145.

Sintes T, Marba N, Duarte CM, Kendrick GA (2005) Nonlinear processes in seagrass colonisation explained by simple clonal growth rules. Oikos 108: 165–175.

Sleeman JC, Kendrick GA, Boggs GS, Hegge BJ (2005) Measuring fragmentation of seagrass landscapes: which indices are most appropriate for detecting change? Marine and Freshwater Research 56: 851–864.

Sousa WP (1984a) Intertidal mosaics: patch size, propagule availability, and spatially variable patterns of succession. Ecology 65: 1918–1935.

Sousa WP (1984b) The role of disturbance in natural communities. Annual Reviews of Ecology, Evolution and Systematics 15: 353–392.

Steele JH (1978) Spatial Pattern in Plankton Communities. NATO Conference Series IV: Marine Sciences. Proceedings of the NATO conference on marine biology held at Erice, Italy, November 13–27, 1977. Plenum Press, New York, NY.

Sutherland JP (1981) The fouling community at Beaufort, North Carolina: a study in stability. American Naturalist 118(4): 499–519.

Tanner JE, Hughes TP, Connell JH (2009) Community-level density dependence: an example from a shallow coral assemblage. Ecology 90: 506–516.

Teixidó N, Garrabou J, Arntz W (2002) Spatial pattern quantification of Antarctic benthic communities using landscape indices. Marine Ecology Progress Series 242: 1–14.

Terrados J, Duarte CM, Fortes MD, Borum J, Agawin NS, Bach S, Thampanya U, Kamp-Nielsen L, Kenworthy W, Geertz-Hansen O (1998) Changes in community structure and biomass of seagrass communities along gradients of siltation in SE Asia. Estuarine, Coastal and Shelf Science 46(5): 757–768.

Trzcinski MK, Fahrig L, Merriam G (1999) Independent effects of forest cover and fragmentation on the distribution of forest breeding birds. Ecological Applications 9(2): 586–593.

Turner MG (2010) Disturbance and landscape dynamics in a changing world. Ecology 91(10): 2833–2849.

Turner MG, Gardner RH, O'Neill RV (2001) Landscape Ecology in Theory and Practice: Pattern and Process. Springer Science & Business Media, New York, NY.

Turner MG, O'Neill RV, Gardner RH, Milne BT (1989) Effects of changing spatial scale on the analysis of landscape pattern. Landscape Ecology 3: 153–162.

Urbański J, Mazur A, Janas U (2009) Object-oriented classification of QuickBird data for mapping seagrass spatial structure. Oceanological and Hydrobiological Studies 38: 27–43.

Valiela I (2009) Global Coastal Change. Blackwell, Oxford.

van der Heide T, van Nes EH, van Katwijk MM, Olff H, Smolders AJ (2011) Positive feedbacks in seagrass ecosystems: Evidence from large-scale empirical data. PloS One 6: e16504.

Vermaat JE, Rollon RN, Lacap CDA, Billot C, Alberto F, Nacorda HM, Wiegman F, Terrados J (2004) Meadow fragmentation and reproductive output of the SE Asian seagrass *Enhalus acoroides.* Journal of Sea Research 52: 321–328.

Vidondo B, Duarte CM, Middelboe AL, Stefansen K, Lützen T, Nielsen SL (1997) Dynamics of a landscape mosaic: Size and age distributions, growth and demography of seagrass *Cymodocea nodosa* patches. Marine Ecology Progress Series 158: 131–138.

Watt AS (1947) Pattern and process in the plant community. Journal of Ecology 35(1/2): 1–22.

Waycott M, Duarte CM, Carruthers TJB, Orth RJ, Dennison WC, Olyarnik S, Calladine A, Fourqurean JW, Heck KLJ, Hughes AR, Kendrick GA, Kenworthy WJ, Short FT, Williams SL (2009) Accelerating loss of seagrasses across the globe threatens coastal ecosystems. Proceedings of the National Academy of Sciences 106: 12377–12381.

Wedding LM, Friedlander AM (2008) Determining the influence of seascape structure on coral reef fishes in Hawaii using a geospatial approach. Marine Geodesy 31: 246–266.

Wedding LM, Lepczyk CA, Pittman SJ, Friedlander AM, Jorgensen S (2011) Quantifying seascape structure: extending terrestrial spatial pattern metrics to the marine realm. Marine Ecology Progress Series 427: 219–232.

Weinstein D, Shugart H (1983) Ecological modelling of landscape dynamics. In Mooney HA, Godron M (eds) Disturbance and Ecosystems. Springer, Berlin, pp. 29–45.

White PS, Pickett S (1985) Natural disturbance and patch dynamics: an introduction. In Pickett STA, White PS (eds) The Ecology of Natural Disturbance and Patch Dynamics. Academic Press, San Diego, CA, pp. 3–13.

Wiens JA (1976) Population responses to patchy environments. Annual Review of Ecology and Systematics 81–120.

Wiens JA, Milne BT (1989) Scaling of 'landscapes' in landscape ecology, or, landscape ecology from a beetle's perspective. Landscape Ecology 3: 87–96.

Wilson SK, Graham NA, Pratchett MS, Jones GP, Polunin NV (2006) Multiple disturbances and the global degradation of coral reefs: are reef fishes at risk or resilient? Global Change Biology 12: 2220–2234.

With KA, Crist TO (1995) Critical thresholds to landscape structure. Ecology 76: 2446–2459.

Wu J (2004) Effects of changing scale on landscape pattern analysis: scaling relations. Landscape Ecology 19: 125–138.

Wu J, Loucks OL (1995) From balance of nature to hierarchical patch dynamics: a paradigm shift in ecology. Quarterly Review of Biology 70(4): 439–466.

Wu J, Shen W, Sun W, Tueller PT (2002) Empirical patterns of the effects of changing scale on landscape metrics. Landscape Ecology 17: 761–782.

Zuur A, Tuck I, Bailey N (2003) Dynamic factor analysis to estimate common trends in fisheries time series. Canadian Journal of Fisheries and Aquatic Sciences 60: 542–552.

7

Animal Movements through the Seascape: Integrating Movement Ecology with Seascape Ecology

Simon J. Pittman, Benjamin Davis and Rolando O. Santos-Corujo

7.1   Introduction

All around us animals are on the move. Even those that appear sedentary for much of their life will at some period in time exhibit movement that is often critical to their survival, be it during the early life stage through dispersal via wind, water, or even being carried by another organism. Some of the most spectacular and fascinating natural phenomena on earth are the mass movements of organisms such as salmon migrations from the ocean to distant upland rivers to spawn (Eiler *et al.* 2014); and whales, seabirds, seals, turtles and sharks ,which travel thousands of kilometres across oceans to complete their lifecycle; each individual exhibiting astonishing endurance and navigational precision (Hays & Scott 2013; Luschi 2013). For instance, several species of albatross and many other far-ranging seabirds, perform the impressive feat of returning to the same nest site year after year (Fisher 1971). Leatherback turtles have been recorded travelling distances of more than 11 000 kilometres between nesting and foraging sites across the Pacific Ocean (Benson *et al.* 2011), returning to nest at the very same beach where they were born. Despite their small size, some zooplankton swim a vertical distance of several thousand feet during regular day and night migrations to avoid predators in the deep during daytime and to search for food in shallower waters during nocturnal periods. These and many other marine animals schedule movements to coincide with specific patchiness in environmental conditions and biological patterning (*e.g.,* temperature, salinity, prey, predators, mates, habitat edges, *etc.*). Often dynamic spatial processes such as tides, winds and currents are used to enable safe and energy efficient movements between ocean spaces, but these same processes also impede movements. From an ecological perspective, the movement of animals is a fundamental ecological process that is central to understanding how organisms respond to seascape structure and dynamics (Wiens *et al.* 1993; Pittman & McAlpine 2003; Nathan *et al.* 2008). Movement patterns throughout the life cycle influence gene flow, social organisation, community structure, evolutionary processes and patterns of biodiversity (Moorcroft *et al.* 2006; Nathan *et al.* 2008). The way that animals use space and the timing of space use, together with their locomotion and navigational abilities, have been shaped by natural selection to maximise their fitness and survival (Nathan *et al.* 2008).

   People too have adapted their space-use patterns to coincide with movements of marine animals to maximise fitness and survival. Ancient indigenous coastal communities in Australia, Polynesia and elsewhere have long used their intimate ecological knowledge of marine animal movements to optimise success in fishing. In Queensland, Australia, migrating mullet were such an important seasonal phenomenon for the Quandamooka people that they timed their own migration to the coast from the mountains, to coincide with the spawning run. Fishermen of the Quandamooka people of Moreton Bay prepared for the migrating mullet (*Mugil cephalus*) entering the bay by using observations on a sequence of animal movements as spatial and temporal cues. The first cue for the beginning of the mullet migration was the migration of a terrestrial animal, the 'hairy grub', a species of moth caterpillar, which travel, linked in single file, to form a long chain. Next, the presence of sea eagles helped fishers to locate and track the schools of mullet as they entered the bay. Then fishers in the water would 'call' the dolphins by slapping the water with their spears and digging the spears in the sand. The dolphins, would then herd the fish into the nets and were rewarded with hand-fed fish (Neil 2002). This special way of fishing was celebrated in a social gathering, or corroboree, called *Bulka Booangun*. In this way, the legend of 'calling the dolphins' integrates spirituality with traditional ecological knowledge of animal movement and plays out as an interconnected sequence of phenomena in time and space linking intimately the movement of people with the movements of other species across landscape and seascape.

Knowledge of movement patterns is now becoming a crucial information requirement for modern marine management, particularly where there is a need to maintain, enhance or exploit seascape connectivity, anticipate biological invasion pathways, assess the ecological effects caused by disturbance events, protect important 'blue corridors', or to identify and map critical habitat for conservation planning (Crowder & Norse 2008; Foley *et al.* 2010; see also Chapter 9 in this book). Widespread modifications to the marine environment by human activities have resulted in fragmented seascapes, degraded and newly created habitat, geographical shifts in the spatial arrangement of habitat mosaics and disruptions or improvements to connectivity (Crook *et al.* 2015). These changes now provide impetus to gain a more reliable understanding of the spatial ecology of animal movements and the interaction with the surrounding seascape at a range of spatial scales. For instance, with warming oceans, marine animals are shifting geographically to maintain access to favourable conditions, which can involve extended movements to track food and relocate to cooler waters (Perry *et al.* 2005; Pinsky *et al.* 2013). Studies of the influence of seascape patterning on species movement rates in response to broad-scale environmental change will help to address questions related to species extinction risk and spatial adaptation to the dynamic distributions of fishes and their associated fisheries.

Furthermore, understanding the movement response to seascape structure at the level of the individual animal will allow us to better understand the individualistic responses that occur within a species, or within a specific life stage or gender. Sometimes the high variability and flexibility of behavioural traits can confound our attempts to generalise results from investigations of behavioural relationships in seascape ecology. Such knowledge is critical for understanding the ecology and evolution of species and their response to ecological change (Brooker *et al.* 2016). It was not until relatively recently that we started to understand how fish perceive the world around them and how animals' movements through space could affect the way they learn and remember spatial information (Patton & Braithwaite 2015). For example, field observations and experimental research in spatial cognition using mesocosms or microlandscapes has shown that fish create multidimensional mental representations (or mental maps) of the environment to aid in navigation (Burt de Perera *et al.* 2016). Although spatial learning in fish has been known for some time (Dodson 1988), such studies have been relatively fine scale and have rarely benefited from the integration of detailed maps of the seascape, thus limiting their application to marine environments.

Increasing scientific attention is being given to the how, why, where and when of animal movements, with important implications for our knowledge, exploitation and management of impacts to marine life from anthropogenic activities. From a seascape ecology perspective, however, we frame our questions on the ecological relevance of seascape structure. For example: How do animals respond to the shape of patches, edges and ecotones, gaps between focal habitat patches, topographic features on the seafloor, wave exposure gradients, plankton patch size and patch density? What are the mechanisms and biological and ecological implications of movement responses to the spatial structuring of seascapes? This chapter reviews some of the many tools and techniques for quantifying the spatial and temporal patterning of marine animal movements, with the purpose of encouraging a greater integration of movement ecology within seascape ecology. We recognise a shift in complexity of these tools to account for complex multidimensional movement patterns (*e.g.*, account for three- and four-dimensional space use) and also the inclusion of a wider range of sensors capturing physiological data on organisms as they navigate through seascapes, which greatly increases the realism in modelling animal-seascape interactions. Throughout we also identify some key knowledge gaps and present research ideas and challenges to guide future studies.

Understanding of the behavioural interactions between animals and spatial structure in the seascape is an important theme at the frontier of seascape ecology and is set to reveal important ecological insights that were not previously accessible with traditional conceptual models and techniques. A deepening of our spatially explicit understanding of marine animal movements and of the functional significance of these movements has major implications for marine ecology and marine stewardship.

7.1.1 Why Animal Movement is Central to Seascape Ecology

Organisms move through the seascape in response to the spatial structure they perceive (*e.g.*, thermoclines, prey patchiness, refuge spaces, *etc.*) and at the same time can function as creators of dynamic spatial structure through their physical presence and foraging activities. For example, feeding by stingrays, dugongs and green turtles influences the spatial heterogeneity of seagrass seascapes (Thayer *et al.* 1984; Townsend & Fonseca 1998). It is becoming clear through studies of both structural and functional connectivity that some spatial structures will facilitate movements to varying degree and some will inhibit or constrain movements (Grober-Dunsmore *et al.* 2009; Turgeon *et al.* 2010; Pittman & Olds 2015; Crooks *et al.* 2015; see also Chapters 5 and 9 in this book). What do different patterns of connectivity mean for the spatial distribution of biodiversity, fisheries potential, or the growth and survival of individuals?

In terrestrial systems, it has long been recognised that understanding the key flows and movements of animals, plants, material and energy across landscapes will help identify the optimal spatial arrangement of habitat patches to guide the prioritisation of conservation actions (Crooks & Sanjayan 2006). For example, corridors of favourable habitat structure are typically considered to facilitate connectivity between patches and this concept has been widely integrated into conservation planning, albeit with mixed success (Simberloff & Cox 1987; Anderson & Jenkins 2006). In marine systems, however, few attempts have been made to investigate how spatial arrangement of seascape patches influence animal movement and this limits our understanding of the function of habitat corridors or 'blue corridors' (but see also Chapters 9 and 10 in this book).

The theoretical framework of landscape ecology has incorporated animal movement dynamics as one of the main underlying mechanisms rooted in the interaction between spatially explicit patterns of habitat mosaics and ecological processes influencing reproduction, mortality, fitness and access to resources (Wiens 1976, 1989; Ryall & Fahrig 2006). Movement behaviour, specifically the scale of space use, is at the core of landscape ecology. The point is best highlighted through the concept of ecological neighbourhoods (Wright 1943; Southwood 1977; Addicott *et al.* 1987; Pittman & McAlpine 2003; Palumbi 2004); an extension of the concept of habitat, whereby habitat is scaled by the movements of individuals and movements are influenced by the patterning of the landscape (or seascape). Addicott *et al.* (1987) defined ecological neighbourhoods using three properties: an ecological process, a time scale appropriate to that process and an organism's activity or influence during that time period. As such they state that the ecological neighbourhood of an organism, for a given ecological process (such as movement), is the region within which that organism is active or has some influence during the appropriate period of time. The ecological neighbourhood space can be used as a sample unit area for quantifying the environment, or perhaps just as a technique for anchoring the study at an ecologically meaningful place and period of time (*e.g.*, daily home range, spawning migration, territoriality, *etc.*).

Seascape ecology, the marine equivalent of landscape ecology, with a focus on studying the geometric patterning in the marine environment and its ecological consequences, provides an appropriate conceptual and analytical framework to examine movement patterns (Pittman & McAlpine 2003). All seascapes are patchy at some scale or another and some patches are more accessible than others, or will offer greater quality for foraging or refuge from predators, depending on their location and biophysical characteristics. With this is mind, the landscape ecologist John Wiens encouraged us to visualise the heterogeneous landscape mosaic as a cost-benefit surface, with 'peaks' where benefits exceed costs and 'valleys' where costs exceed benefits, corresponding to high-quality and poor-quality patches (Wiens 1997). In fact, this perspective of organism-landscape interactions is analogous to the use of the terms 'prey landscapes' (*e.g.*, Sims *et al.* 2006) when referring to patchiness in food resources, or 'seascapes of fear' (Wirsing *et al.* 2008) to communicate the response of prey to the spatial distribution of predators across the seascape. In addition to predator-prey distributions, competitors and breeding partners will also be distributed heterogeneously in time and space. Thus, when animals navigate across a seascape, behavioural decisions are made that will usually result in a tradeoff to maximise growth and survival. As such, movement patterns are often highly predictable with regular high use areas being easily identifiable (hence patchiness), although exploratory excursions to relocate are also common, particularly when environmental changes make a seascape less optimal for growth and survival (Pittman & McAlpine 2003).

In terrestrial ecology, the integration of spatial pattern with animal movement has long been recognised (Wiens *et al.* 1993; Lima & Zollner 1996), with many examples in the *Journal of Ecology, Animal Ecology, Movement Ecology* and *Landscape Ecology*. Although interest in applying landscape ecology approaches for examining species-seascape pattern relationships in marine ecosystems has increased in recent years, investigation of the consequences of seascape structure for marine animal movements are among the most overlooked applications of landscape ecology (Grober-Dunsmore *et al.* 2009; Boström *et al.* 2011; see also Chapter 9 in this book). An exceptional example of applying the landscape ecology perspective to coastal seascapes is that of Irlandi & Crawford (1997) whom recognised that the juxtaposition of intertidal salt marshes and subtidal seagrasses can influence the movements and growth of coastal fishes through trophic transfer during tidal migrations. Using enclosure experiments, Irlandi & Crawford (1997) found that estuarine pinfish (*Lagodon rhomboides*) movements were greater where intertidal marsh edge existed in close proximity to subtidal seagrass compared with marsh edge adjacent to unvegetated subtidal. This seascape configuration resulted in greater abundance and growth rates in fishes moving across the continuously vegetated tidal corridor. In landscape ecology, this ecological boosting effect of spatial configuration is known as landscape complementation where resources in both patch types are not substitutable (organisms cannot forage elsewhere), or landscape supplementation where resources are substitutable (organism can forage elsewhere) (Dunning *et al.* 1992). Such studies linking spatial patterning to the key ecological process of movement and its consequences for organism biology are still rare in marine ecology (Olds *et al.* 2016). However, even though progress in this direction has been slow and largely constrained by data availability, we are now on the verge of some breakthroughs fuelled by advances in the design and application of technologies such as telemetry, biologging, spatial mapping and modelling.

7.1.2 Advances in Movement Ecology and its Application in Marine Systems

Technological advances in biotelemetry and the development of spatial statistical and visualisation approaches have increased our capacity to understand the dynamics of spatial activities of marine species such as foraging, homing navigation to nest or shelter sites and relocation for mating (Pittman & McAlpine 2003; Grober-Dunsmore *et al.* 2009; Block *et al.* 2011). The technological revolution in animal tracking and mapping combined with a greater spatial awareness in society has also fuelled the emergence of a new subdiscipline of ecology called movement ecology, which focuses on understanding the movement of organisms in the context of their internal states, traits, constraints and interactions among themselves and with the environment (*i.e.*, 'external factors') (Nathan *et al.* 2008). For many animals, particularly those that are highly mobile moving through spatially heterogeneous seascapes, the spatial structure of the surrounding environment will comprise some of these key external factors. For example, the complex process of movement will depend on feedback interactions between an individual and the surrounding environment such as the ability of an individual to sense its environment, remember landmarks, construct mental maps and process information, as well as biological, physical and chemical attributes of the seascape that can hinder or facilitate movement and influence the pattern of the movement pathway (Getz & Saltz 2008). Figure 7.1 expands the conceptual model of movement ecology by placing organism movement behaviour within its broader ecological context, with a focus on the relationship between movement patterns and the spatial patterning of surrounding seascape, as well as recognizing the consequences of these interactions for the condition of individuals.

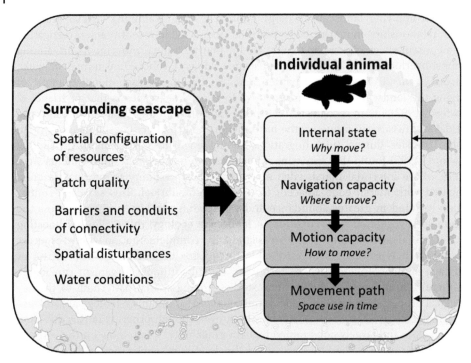

Figure 7.1 Conceptual model of movement ecology incorporating focal patterning from landscape ecology. Seascape structure and conditions can act as drivers for the internal state of individual animals (physiological and psychological states), can influence their navigation (mental maps, orientation), motion capacity (locomotory morphology/traits) and ultimately the movement process, which results in a specific pathway in time and space. *Source*: Adapted from Nathan *et al.* (2008).

7.1.3   Tracking and Mapping Capabilities

In the field of movement informatics, movement observations are spatiotemporal signals carrying information that can reveal underlying mechanisms driving the movement patterns (Nathan *et al.* 2008). Tracking of marine animals using a wide range of devices is now providing vast amounts of data allowing us to reconstruct movement pathways across a variety of spatiotemporal scales. With the most detailed locational information, it is possible to map a representation of the actual movement trajectory that can reveal behavioural decisions resulting from: foraging, responses to seafloor structure or water column properties, evasion of predators, or attraction to other organisms. Pathways can be quantified to measure characteristics such as path complexity using measures of the tortuosity, sinuosity and application of fractal indices (Turchin 1998). Emerging digital and communications technologies have refined our ability to measure movement at the resolution of fractions of seconds with concomitant spatial precision, while kinematical (*e.g.*, acceleration), physiological (*e.g.*, heart beat and temperature) and behavioural (*e.g.*, vocalisations) information are simultaneously recorded (Getz & Saltz 2008). This has significantly broadened the scope of questions that we can address in seascape-movement ecology and boosted the accuracy of evidence that we can collect to help answer these questions.

Two main forms of digital positioning technology are used to track animals in the wild, each with different advantages and disadvantages and suitability for different applications. GPS-based satellite tags provide virtually unlimited global coverage and are ideal for studying movements of some marine megafauna and avian species at a range of scales in time and space (Kuhn *et al*. 2009; Hussey *et al*. 2015). However, since tags need to be exposed to air to transmit signals, in the marine environment their application is restricted to studying far-ranging movements of animals that regularly surface (such as cetaceans and turtles). Acoustic telemetry systems on the other hand are capable of near-continuous tracking of animals underwater, operating via ultrasonic communications between animal-borne transmitter tags and receivers (Hussey *et al*. 2015). Animals can be acoustically tracked either manually, with tag transmissions detected by a boat-based hand-held hydrophone (Hitt *et al*. 2011a), or passively through a network of fixed location underwater receivers that continuously listen for tag transmissions (Jacoby *et al*. 2012). While the former is capable of high resolution positioning (on a scale of a few metres), it is only feasible to track one individual at a time and the duration of tracking is limited by constraints of a high labour demand (Papastamatiou *et al*. 2009; Hitt *et al*. 2011a). Conversely, passive acoustic tracking enables automated monitoring of multiple individuals over extended periods and, while it has historically been limited to coarser spatial resolutions ($>\sim$15 m) (Hussey *et al*. 2015), recent developments mean tag positions can triangulated between three or more receivers with high levels of precision ($\sim$2–5 m) (Espinoza *et al*. 2011; Biesinger *et al*. 2013; Dance & Rooker 2015).

It is largely because of the increasing precision and relatively low labour demands that passive acoustic telemetry is proving to be a powerful tool for quantifying detailed movement patterns in marine environments across a range of scales. Receivers can be strategically arranged, offering flexibility to address specific questions, and often reflect a tradeoff between areal coverage and precision (Heupel *et al*. 2006). Close clustering of receivers results in high overlap in listening range, and thus high-resolution positioning, enabling reconstructed movements to be more tightly associated with environmental structure on maps (Espinoza *et al*. 2011). Sparser distribution of receivers enables broader range movements to be detected, but with less accuracy. In some circumstances integration of high and low resolution receiver arrays is desirable, for instance, to look at detailed home-range movements and connectivity between patches when animals relocate their home ranges (*e.g.*, ontogenetic shifts) (Pittman *et al*. 2014). The suite of capabilities now possible with passive acoustic telemetry suitably mirrors the multiscale, multispecies framework that underpins the field of seascape ecology.

To better understand an animal's perception of its environment, digital tracking tags can be used in conjunction with integrated data loggers that record information on physiology, movement characteristics, or the environment it is moving within. Geolocated measures of physiological factors (*e.g.*, heart rate, body temperature) and movement characteristics (*e.g.*, acceleration, turning angle) can help to discern behavioural mechanisms underpinning movements and space use, such as foraging, prey avoidance, mating behaviours and movements between patches (Murchie *et al*. 2011; Hussey *et al*. 2015). Meanwhile, by getting the animal to act as a sensor of its own environment, more accurate information on details of temperature, salinity, depth, and light conditions can be used to explain patterns of movement (Hussey *et al*. 2015; Horodysky *et al*. 2015). Crucially, regular depth measurements (obtained through pressure loggers) can also be synchronised with geographical fixes, enabling animal positions to be located in three spatial dimensions, with massive implications for our potential to understand how marine animals interact with their environment.

7.2   Using Animal Movements to Scale Ecological Studies

It appears that the more we track and map marine animal movements the more we are discovering that many marine animals move across far greater spatial scales than was previously assumed while others show high site attachment. From the perspective of studying marine animal ecology, this new knowledge on marine animal movements is central to defining and delineating seascape patterning at functionally meaningful spatial and temporal scales. The absence of information, or the lack of continuous observation on the way animals use space through time, can all too often result in insufficient consideration of seascape context potentially resulting in misleading conclusions on the primary drivers of ecological patterns and processes (Addicott *et al.* 1987; Meetenmeyer 1989; Wiens 1989; Pittman & McAlpine 2003). Quantitative and spatially explicit information on animal movements can address several critical assumptions that exist in many ecological field studies. Movement data will help to select ecologically meaningful scales with which to define an ecological space, or area of interest, to guide the design of field data collection or data acquisition. With this in mind, movement data will also challenge the assumption of a single patch focus (*e.g.*, seagrasses or coral reefs), particularly for highly mobile organisms that use multiple patch types in daily home range excursions, for example, between foraging and resting areas, ontogenetic habitat shifts across mosaics of patch types and seasonal migrations for breeding (Pittman & McAlpine 2003).

Although it has been widely acknowledged in conceptual models that marine organisms are influenced by patterns and processes occurring at a range of scales of space, time and organisational complexity (Haury *et al.* 1978; Hatcher *et al.* 1987; Steele 1988, 1989; Barry & Dayton 1991; Holling 1992; Levin 1992; Marquet *et al.* 1993), few marine ecologists have designed studies that consider responses to habitat and resource structure across a range of spatial scales. In seascape ecology, we start with an assumption that marine animals are likely to respond to and be constrained by the composition and spatial arrangement of resources in a hierarchical way, as has been revealed for many terrestrial animals (Senft *et al.* 1987; Schaefer & Messier 1995; McAlpine *et al.* 1999; Rolstad *et al.* 2000). Even when a focal scale has been determined, we propose that measurements carried out at only a single scale cannot, by definition, incorporate important patterns and processes at scales above and below the focal scale and a single scale approach is therefore limited in ecology (Pittman & Olds 2015). Working at one scale is particularly inappropriate for studies of multispecies assemblages since species vary in their response to the environment due to functional differences related to dietary requirements, habitat specialisation, foraging tactics and body size (Betts *et al.* 2014). Similarly, ecological studies constrained to a single patch type when questions are addressing habitat use for species that use multiple patch types can also only offer an incomplete picture of animal ecology. As argued by Roughgarden *et al.* (1988), 'studies at only one of the habitats tell no more than half the story'. Therefore, if information on animal movement is not available, the assumption of single habitat use should be considered carefully or rejected entirely. If an assumption is to be made, then a multihabitat type assumption may be more suitable, thus allowing for the consideration of broader scale movement and potential linkages between component habitat types.

Some systems ecologists have proposed a hierarchical approach based on hierarchy theory to facilitate scale awareness and operational measures of scale (Wu & Loucks 1995). The hierarchical frameworks are by definition a simplified composite of complexity, which in reality varies across a space-time continuum, but would be less pragmatic for analytical purposes. Nevertheless, these spatial hierarchical constructs, allow one to focus on an event at a particular scale, while recognizing that there are other scales relevant to that event (Urban *et al.* 1987). For example, Pittman & McAlpine (2003) present a three-level spatial hierarchy (adapted from Allen & Starr 1982) whereby lower levels, L^{-1}, occupy less space and are characterised by processes operating at faster rates and finer time scales (Box 7.1). In contrast, higher levels, L^{+1}, are of broader temporal and spatial scales. Thus, there is no single correct scale or level for observations and ultimately, the appropriate scales will depend on a consideration of the questions asked, the organisms studied and the time period considered (Wiens 1989).

Fauchald & Tveraa (2006) tracked the movements of individual Antarctic petrel (*Thalassoica antarctica*) using satellite transmitters to quantify the spatial dynamics and spatial scales of their foraging behaviour. The authors understood that marine pelagic fish and krill are organised in a nested hierarchical structure (Stommel 1963; Haury *et al.* 1978; Fauchald 1999; Fauchald *et al.* 2000) and so analysed the bird's movements to see if they exhibited nested search strategies. The study of 36 birds used first-passage time (FPT), a scale-dependent measure of the animal's search effort at each point on its movement trajectory and showed that the birds did indeed exhibit a spatially nested search strategy to find food whereby they travelled faster over longer distances (>100 km scale) to locate large clusters of patches and then within those areas they concentrated their search at finer scales to locate smaller patches within which to feed. Over time, the broadest scale patches lasted weeks, while at finer scale patches of plankton disappeared or moved within days. Similarly, for basking sharks (*Cetorhinus maximus*) locating prey across the continental shelf of northwest Europe, movement tracks reconstructed from satellite-linked archival transmitters and linked to maps of zooplankton biomass suggested that sharks exhibited directed movement through a cross-shelf gradient of prey in order to locate the preferred high-density prey patches where they then exhibited a reduction in speed and high turning rates as they fed (Sims & Quayle 1998; Sims *et al.* 2006).

Box 7.1

For an example of how a hierarchical framework might be relevant to defining the environment for highly mobile animals, consider an assemblage of juvenile fish using a tidally dominated inshore area. In order to evade predators in deeper water and to forage amongst intertidal seagrasses and mangroves, the animals move back and forth with the flooding and ebbing tide through a mosaic of patches from the subtidal at low

(Continued)

Box 7.1 (Continued)

tide to the intertidal at high tide. A conceptual framework (Figure 7.2) can be constructed with the home range or ecological neighbourhood as the focal level (**L**) and intermediate in the hierarchy. The focal level is the level at which the phenomenon or process under study characteristically operates and is a functional part of a higher and lower level. At any single level, the consequences and significance can only be understood at a higher level(s) and the mechanistic explanation must be investigated at a lower level(s). The finer scale **L**$^{-1}$ components of the mosaic (*i.e.*, within-patch level) may consist of structure such as seagrass leaf length, epiphyte biomass and patch size and these may explain some of the fish distribution patterns found at high tide. For instance, some animals may have a preference for relatively large patches of long leaf seagrasses. This relationship, however, may not adequately explain the patterns at the broader scale of the home range, which may also include unvegetated subtidal areas where animals spend considerable time at low tide. At the extent of the home range, distributions may also be influenced by the spatial arrangement of patches of seagrasses and the relative proximity of seagrasses to complementary resources in adjacent mangroves and coral reefs (**L**$^{+1}$). Lower level explanations may be further lost at the **L**$^{+1}$ level, which would include the environment surrounding the home range (*i.e.*, where animal distributions and abundance respond to a suite of physico-chemical constraints such as a gradient in wave action, salinity, temperature, turbidity *etc*.). In this way, the intermediate level of the hierarchy is defined by the spatial activity patterns of the animal(s) within relevant time periods, thus anchoring the hierarchy to an ecologically meaningful scale in time and space.

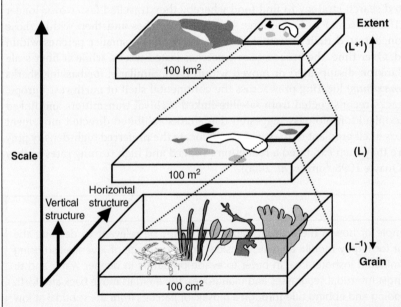

Figure 7.2 Relevance of a hierarchical framework to the definition of an environment for highly mobile animals.

7.2.1   Building Movement Scales into Conceptual and Operational Frameworks

Conceptual and operational frameworks for the study of movement responses to seascape structure have not yet emerged in the new field of behavioural seascape ecology and are rarely found in the terrestrial landscape ecology literature. An example of an operational framework for scaling the seascape using information on animal movements is shown in Figure 7.3. This framework was developed to guide decision making through a logical sequence of procedures when employing a landscape ecology approach to the study of connectivity in tropical marine seascapes. The framework consists of three components: (i) developing a conceptual framework; (ii) determining appropriate scales for analysis and (iii) conducting geospatial analyses scaled to the organism or process of interest. The operational framework also identifies critical feedback loops that emphasise the importance of iteratively adapting the framework as new information becomes available.

7.2.1.1   Component 1: Build a Conceptual Model
The conceptual framework will determine the data needs and feed into decision making with regard to data collection protocols. Ultimately, the type of ecological questions will determine appropriate scales. For instance, studies focused on understanding seascape connectivity throughout the daily home range (*i.e.,* routine foraging or territorial movements) may be conducted at different scales than studies focused at connectivity throughout the lifecycle (Pittman & McAlpine 2003). Where appropriate data are unobtainable, computer simulations can be used to explore connectivity and examine scale effects on species.

7.2.1.2   Component 2: Selecting Scale
The second step of the framework is the selection of an appropriate scale at which to conduct the study. Selecting appropriate scales (*i.e.,* refer to Schneider Chapter 4 for spatial scale definition and examples) for connectivity studies is important because ecological phenomena (spatial patterns processes, animal populations and materials) exhibit considerable variability in time and space. Likewise, environmental factors that influence ecological phenomena also exhibit significant spatiotemporal variability (*e.g.,* nutrient flows, current patterns, salinity gradients *etc.*). As a consequence, animals respond to environmental heterogeneity at different scales and in different ways (Johnson *et al.* 1992), which is linked to the way animals move throughout the seascape (*i.e.,* home ranges of highly mobile species compared with more sedentary species, or differences between life stages of the same species).

Therefore, to be ecologically meaningful, the extent of the seascape used for connectivity studies should be scaled to reflect the natural history and ecology of the organism or process being studied (Wiens & Milne 1989; Pittman & McAlpine 2003). This means that the size (spatial extent) and grain (resolution) of the seascape would differ among organisms. For reef fishes, organism-scaled home range size may require pilot studies that use tracking or fixed station sampling to determine temporal and spatial aspects of connected habitats (patch uses). However, the problem is even more complex for seascape studies of multispecies fish assemblages typical of coral reefs. The spatial extent of the seascape may need to be large enough to include all critical resource patches used by the organisms being studied including those used during different life phases.

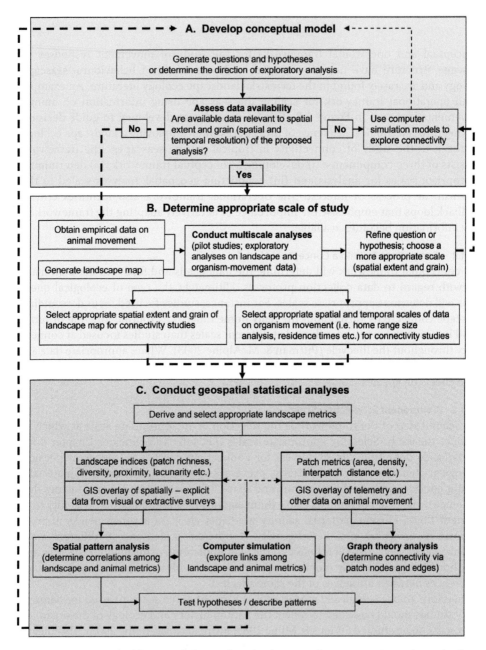

Figure 7.3 An operational framework that applies a landscape ecology perspective to the study of connectivity in tropical marine seascapes. The framework consists of three components: developing a conceptual framework that guides the study, determining appropriate scales for analysis and conducting geospatial analyses scaled to the organism of interest. Solid arrows indicate directional flow among the subcomponents. Broken arrows represent directional flow and important feedback loops among and within the three components. *Source*: Developed through personal communication with C. Jeffrey, NOAA Biogeography Branch.

For example, a review of movement patterns of 210 species of fishes associated with tropical coral reefs revealed that the scale of movements is influenced by a range of factors including body size, gender, behaviour, population density, habitat characteristics, season, tide and time of day, with some fish moving less than 0.1 kilometres and others tens to hundreds and even thousands of kilometres (Green *et al.* 2015).

7.2.1.3   Component 3: Tools Identification

The third component of the operational framework is the selection of the appropriate analytical tool for conducting the geospatial analysis. Tool selection is driven primarily by the questions asked and the type of data available (Calabrese & Fagan 2004). Spatially explicit visual interpretation or extracted data on species distributions are correlated with spatial pattern metrics to determine structural connectivity within seascapes, or potential connectivity if information on species movement is known. Alternatively, graph-theory analysis can be used to combine metrics such as interpatch distance with telemetry data to determine actual or functional connectivity. Information gained from spatial pattern analysis and graph theory could be further incorporated in computer simulation models to identify relationships among seascape structure and fauna, or to predict the influence of future changes in seascape structure on species distribution and abundance. However, as pointed out by Shamoun-Baranes *et al.* (2011), the steps needed to go from data collection to gain new ecological insight, such as data organisation, exploration, visualisation, quantification, inference and generalisation of movement data, can be extremely demanding. At every step of the process, great care must be taken to consider scale effects, errors associated with spatial data and the potential for propagation of errors through spatial analyses (Wedding *et al.* 2011).

7.3   Advances in the Visualisation and Quantification of Space-use Patterns

With the increasing spatiotemporal precision and size of animal tracking data sets, we now have the opportunity to capture and analyse data on how animals use space in ways that better approximate reality and offer novel insights into the spatial ecology of animals. However, challenges remain in finding effective ways to make sense of these vast and complex data sets. Here we describe several techniques and recent advances for the spatial analysis and mapping of organism space-use patterns.

7.3.1   Estimating and Mapping Utilisation Distributions

Typically, animal space-use is visualised as a utilisation distribution using kernel density estimation (KDE) techniques. This is a way of representing the probability of finding an animal in a given place at a given time through aggregating positional data, and is a commonly accepted means of estimating home-range dynamics. It works by placing probability decay kernels around individual location points and summing them into continuous density surfaces, essentially providing a heat map representing intensity of space use (Silverman 1986) (Figure 7.4). The 95% probability surface (the smallest area in which 95% of geographical fixes occur) is often taken to estimate the broader

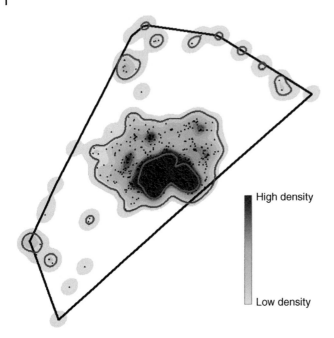

High density

Low density

Figure 7.4 Kernel density estimation with different colours representing iso-surfaces of different kernel densities. The black line represents the home range as defined by minimum convex polygon (MCP) techniques. Red contour lines are 95% kernel home range and the blue contour line is 50% core range. Black dots are GPS locations of an animal. *Source*: http://gis4geomorphology.com/home-range-kernel/ (accessed 25 May 2017).

home-range area, while the 50% surface is taken to represent the core area of use (Figure 7.3).

When overlaid onto benthic seascape maps, these surfaces can be used to infer preferences for different patch types and different seascape compositions and to assess variations in space use among individuals, species, or across time periods. Such techniques have previously been used to define and examine the home ranges of marine fishes (grunts and snapper) from manual tracking data on coral reefs (Hitt *et al.* 2011a, Hitt *et al.* 2011b) and black-tip reef sharks from passive acoustic networks in coastal waters (Heupel *et al.* 2004). Hitt *et al.* (2011a, b) quantified the size and shape of day and night space use patterns for several reef fish and then used that space to sample the patch-mosaic structure of the underlying benthic seascape providing for the first time a seascape context to help explain the movement ecology of two reef fish species. The stepwise spatial analysis process of quantifying individual space use and seascape structure is summarised in Figure 7.5.

KDE represents a significant advance on its predecessor, minimum convex polygons (MCPs), which involve drawing a simple boundary around the outer extent of animal movements, thus placing profound restrictions on geographic interpretation of space use. Although KDE techniques offer a means of visualizing relatively detailed structure of home ranges, they have historically neglected information contained in the serial correlation of points, by assuming position fixes are independent rather than connected by movements. By placing kernels over consecutive points to form segments or paths,

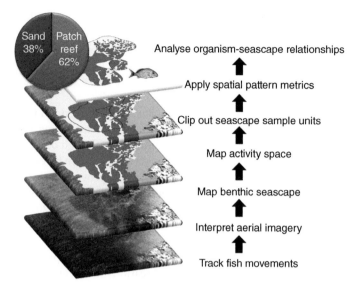

Figure 7.5 Stepwise spatial analysis to map and quantify benthic seascape patterning at a spatial extent defined using the diel activity space calculated from tracking an individual fish following the methods of Hitt *et al.* (2011a).

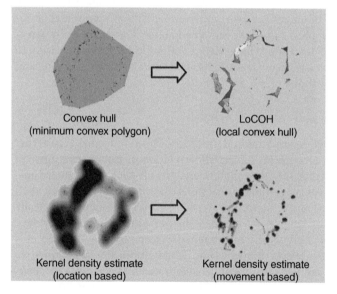

Figure 7.6 Differences in precision of utilisation distribution using four different home-range estimation techniques. *Source*: https://www.werc.usgs.gov/ProjectSubWebPage.aspx? SubWebPageID=1&ProjectID=258 (accessed 25 May 2017).

recent adaptations have allowed utilisation distributions to be represented more accurately and more precisely (Figure 7.6), enabling more detailed interpretation of space-use and animal-landscape interactions (Benhamou & Cornélis 2010). This produces what is known as a movement-based kernel density estimate (MKDE). Movement-based kernel density estimates can be further enhanced by modelling the uncertainty in location of the animal between paired fixes. This is often done with a Brownian decay function, which factors in a random walk model, constrained by the distance between points and animal velocity (Kranstauber *et al.* 2012; Demšar *et al.* 2015a). The effect is that the kernel becomes 'pinched' between consecutive data points (resembling a dumbbell), as the movement path between points is more likely to be direct than convoluted.

One criticism of KDE and MKDE techniques is that kernel surfaces often extend into areas that are not part of the animal's home range, as an artefact of smoothing the probability decay kernels placed around data points or segments. Even though obvious boundaries (such as lakes for terrestrial animals, land for marine animals) can be easily 'cut out' of surfaces (Benhamou & Cornélis 2010), KDE can ultimately provide misleading pictures of home ranges. To address this issue a nonparametric utilisation distribution visualisation tool, Local Convex Hulls (LoCoHs), was developed (Lyons *et al.* 2013). LoCoH involve drawing polygons around points closely clustered in space and time, with each polygon containing an equal number of points. These hulls are then sorted by the density of points within them and subsequently merged into density isopleths representing probability surfaces (*e.g.*, 50% isopleth, containing 50% of location fixes) (Figure 7.6). Since hulls hug the edges of position fixes, animal trajectories can be more reliably and tightly associated with landscape structure and sharp boundaries, which allows the user to examine the influence of habitat edges, connection corridors and other structural features on movement. In addition, adaptations to LoCoH and KDE algorithms have been developed to constrain home ranges using environmental boundaries and gradients based on known geographical range and suitable habitat. For example, Tarjan & Tinker (2016) developed the Permissible Home Range Estimation (PHRE) algorithm to restrict sea otter (*Enhydra lutris*) space-use estimates from overlapping terrestrial land and from extending too far offshore.

7.3.2 Analysing Spatiotemporal Utilisation Patterns

What if we want to observe spatiotemporal interactions in movement patterns? Utilisation distributions can account for temporal correlation of points, but they are still limited by being static; that is, they do not allow analysis of space use through time, obscuring potentially important space-time interactions that typify daily routines, such as movement between foraging and refuge areas. To address this issue, data has typically been split into discrete time windows, *e.g.* to compare day versus night utilisation distributions. For example, Hitt *et al.* (2011a) split movement data into predefined diel periods to investigate differences in daytime versus night-time habitat utilisation. This approach, however, can subsume more detailed movement behaviours such as ecologically meaningful changes in movement path complexity within each period (Hitt *et al.* 2011b).

Recently, enterprising collaborations between ecologists and information scientists have helped to create new techniques to visualise continuous space use through time, shedding light on patterns lost through static techniques (Demšar *et al.* 2015a). Here, data are represented in a volumetric space-time cube (STC), where the x and y axes represent two spatial dimensions and the z (vertical) axis time. In essence, this technique is analogous to multiple 2D kernel density maps stacked on top of one another on a continuous time axis. Generally, the bottom of the z axis represents midnight and the top the following midnight, such that the time-space trajectory for a single day is plotted in the

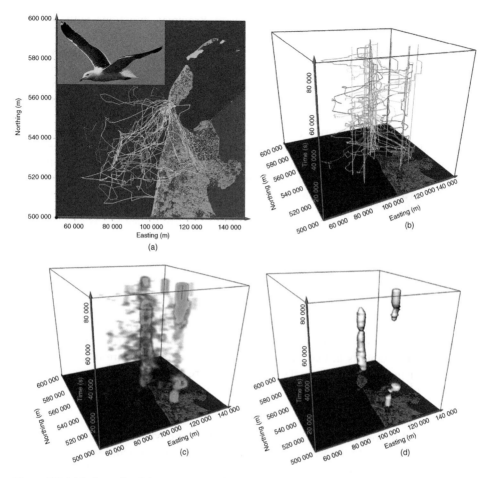

Figure 7.7 (a) Daily tracks of the movement of one lesser black-backed gull (*Larus fuscus*) collected over one month in the Netherlands; (b) Space-time cube of the same tracks with the *z*-axis showing time of day in seconds and ranging from midnight at the bottom to midnight at the top; (c) space-time kernel density estimate of the seagull's movements; (d) iso-surface with high intensity value to show two core areas of use separated by space and time. These areas are spatiotemporal hotspots and indicate previously unknown particularities in this gull's movement: consistently spending nights at a mainland location outside the nest and days in and around the nest. *Source*: Adapted from Demšar & van Loon (2013).

STC, before the trajectory for the following day is overprinted on the same space. To deal with visual clutter of multiple overlapping trajectories, space-time paths are smoothed into continuous density surfaces in similar ways to utilisation distributions, but with 3D kernels placed over whole trajectories, rather than segments or points (Demšar *et al.* 2015b). Space-time densities have previously been used to analyse daily movement patterns of a seagull (Figure 7.7), revealing two spatiotemporal hotspots separated by diffuse movements: a night-time nesting site based on an island and a mainland feeding site that the animal visited several times during the day (Demšar & van Loon 2013).

Rather than placing probability decay functions around whole trajectories, an alternative is to smooth the data by constructing probabilistic space-time prisms between consecutive position fixes. In a similar way to Brownian functions with MKDE, space-time prisms incorporate information on maximum velocity of the animal, to define the area of space potentially used between position fixes (Winter & Yin 2011; Downs *et al.* 2015), with probability weighted towards the estimated space-time path (*i.e.*, the trajectory connecting the sequence of points). This results in elliptical kernels that are greatest halfway between points and represent actual uncertainty about movement paths that is scaled by the animal's movement capabilities rather than an arbitrary decay function (as opposed to a parametric approach such as KDE). However, since space-time prisms are based on a single individual's movements, data from multiple days or individuals are not easily aggregated and are usually displayed separately (Downs *et al.* 2015). Furthermore, where depth or altitude is an important element of an animal's habitat, the time axis uses up a valuable third spatial dimension.

7.3.3 Visualizing Movement Patterns across Three Spatial Dimensions

Computer visualisations have emerged as powerful tools for exploring movements and space use, enabling analysts to identify generalised patterns from complex multidimensional datasets and draw insight through their tacit sense of space and knowledge of ecological phenomena (Andrienko & Andrienko 2013). The ability to represent complex space-use patterns through computer-based visualisations is continually progressing and while such tools have been primarily developed for GPS data on terrestrial animal or vehicle movements, they are equally applicable to marine organisms. By representing movements in two spatial dimensions (*x* and *y*), we are mapping only a simple planar portion of an animal's spatial use. In reality, an animal moves within a 3D geographic domain (*x*, *y* and z), thus compressing movements into two dimensions reduces our ability to link spatial behaviours with environmental heterogeneity and the movements of other animals (Tracey *et al.* 2014a). This is problematic where animals exhibit vertical movements through the water column and where the use of structures extending vertically into the water column (*e.g.*, coral, rocky reefs and seamounts) can form critical components of animal activity spaces. Accounting for depth (or altitude for birds) becomes particularly relevant when comparing space use patterns between species, whereby two animals may be located in the same general space (x, y) on the horizontal plane, but occupy very different water depths (z on the vertical dimension) and may therefore never interact with one another. Simpfendorfer *et al.* (2012) demonstrated this problem by comparing two-dimensional and three-dimensional kernel techniques to represent volumetric space use of two individual European eels (incorporating data from pressure sensors). The analysis revealed that 2D methods had significantly overestimated overlap in space use. More recently, 3D movement-based MKDEs have been implemented to visually analyse volumetric movement of animals in a 3D environment (including terrestrial, avian and marine species) (Tracey *et al.* 2014a, Tracey *et al.* 2014b). By representing movements in the same number of dimensions as they occur, these state-of-the-art techniques allow more biologically accurate interpretations of space use and promise novel ecological insights from telemetry data. Figure 7.8 shows the 3D MKDE for a dugong (*Dugong dugon*), which was tagged with an Argos satellite GPS tag and depth recorder and tracked for 41 days over seagrass beds in Hervey Bay, Australia (Tracey *et al.* 2014a). 3D models defined the extent of the home range movements and the proportion of the home range at different depths and can be linked to the spatial patterning of seagrasses; their primary food source.

(a)

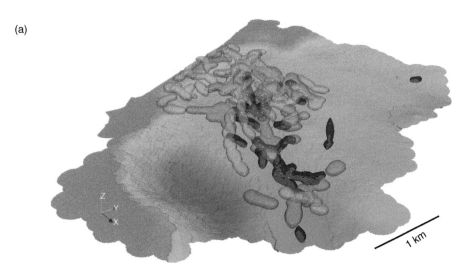

(b)

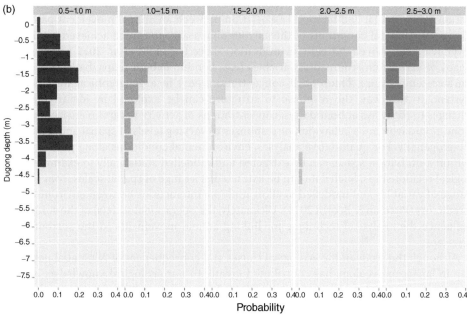

Figure 7.8 Three-dimensional movement-based kernel density estimates (3D MKDE) for an individual satellite tracked dugong overlain on 10 m resolution bathymetry. The probabilities representing 3D space use were mapped to 10 (*x*) × 10 (*y*) × 0.5 (*z* - depth) metre cubes (voxels). (a). The 99% contour volumes for 3D MKDEs based on locations when tidal heights ranged from 0.5–1.0 (red), 1.0–1.5 (orange), 1.5–2.0 (yellow), 2.0–2.5 (light green) and 2.5–3.0 (green) metres are shown. Based on the 3D MKDEs for each tidal height category, the probability that the dugong could have been at different water depths was computed into 0.5 m bins (b). The value on the *y*-axis is the upper depth value for each 0.5 m bin (*i.e.*, 0 indicates 0.0–0.5 m depth). *Source*: Reproduced from *PloS One* (Tracey *et al.* 2014a).

In another study, 3D analysis of Pacific leatherback turtle (*Dermochelys coriacea*) trajectories revealed that subsurface variables (*e.g.*, currents and prey) play an important role in shaping movement patterns such as changes in depth to locate patches of food, therefore explaining significant variance in geographic space use that would have been missed by typical 2D analyses (Schick *et al.* 2013).

We have discussed two different types of 3D model, one incorporating time as the third dimension and one incorporating a third spatial dimension. Now we face the challenge of finding ways to visually represent three spatial dimensions, as well as a continuous temporal dimension (*x, y, z & t*), in what would effectively be a four-dimensional model. The answer is likely to lie in animation of kernels through time and interactive manipulation of data, drawing upon tools that are used currently to monitor boat and aircraft traffic (Andrienko & Andrienko 2013).

7.4   Linking Animal Movement Patterns to Seascape Patterns

For preliminary exploration of seascape-movement patterns, a simple visual interpretation of the spatial and temporal association between an animal movement path and the environmental conditions in the surrounding seascape can be achieved by simply overlaying the movement path, or utilisation space (*e.g.*, Hitt *et al.* 2011 a and b), on maps of various environmental variables (*e.g.*, benthic habitat maps, bathymetry, sea surface temperature, chlorophyll concentration, *etc.*) in a geographical information system. In this way, the response to structure can be explored by identifying any changes to pathways at boundaries, concentrations of activity in certain environmental conditions, avoidance behaviour, or responses to varying levels of spatial and temporal heterogeneity. For quantitative analyses, environmental conditions along pathways, or within a utilisation space, can be extracted and modelled statistically to examine ecological relationships. Such data can be extracted at a range of spatial scales from a wide range of variables and modelled to allow multiple interactions between environmental predictors (Pittman & Brown 2011). Ultimately, the operational scale(s) of the study will influence the analytical technique for linking animal movements to the surrounding seascape.

Although the study of the ecological causes and consequences of 2D spatial patterning in landscape ecology has contributed greatly to our ecological knowledge, it also represents a considerable oversimplification of the true variability in structural complexity that exists in nature (McGarigal *et al.* 2009; Lausch *et al.* 2015). In contrast to the conventional 2D planar surfaces composed of discrete and internally homogenous patches, real seascapes exhibit multidimensional structural complexity at a range of spatial scales that is also functionally meaningful to organisms and ecological processes. Many terrestrial and marine studies are now quantifying continuous heterogeneity by applying topographical metrics and surface analysis to 3D terrains (Hoechstetter *et al.* 2008; McGarigal & Cushman 2005; McGarigal *et al.* 2009; Wedding *et al.* 2011). In turn, metrics designed to quantify surface features and topographic complexity have performed well as predictors of fish diversity and species distributions across coral reef seascapes (Wedding *et al.* 2011; Barrell & Grant 2013).

Spatially continuous representation of seascape patterns (spatial gradients) can be combined with novel surface characterisation of animal movement such as KUD, 3D and space-time density approaches to identify critical thresholds in animal-seascape relationships. Critical thresholds (*i.e.*, where a small change in the seascape results in abrupt changes in the movement state) can be identified using statistical nonlinear models and multivariate techniques that regress continuous seascape surfaces with movement surfaces (Large *et al.* 2015). For instance, the 3D movement-based kernel density estimator developed by Tracey *et al.* 2014 could be useful to understand the niche partitioning of species in the water column in relation to remotely sensed, continuously mapped seafloor structure and finer scale within-habitat structure (*e.g.*, reef rugosity, seagrass canopy high). Analytical tools that incorporate temporal variability such as time-geographic density estimation (Downs *et al.* 2011) and space-time density of trajectories (Demšar & Virrantaus 2010), could also be combined with surface metrics that quantify spatial patterning of the seascape to provide a dynamic context in the study of movement-seascape relationships.

To help guide the selection of analytical procedure we offer Figure 7.9, which recommends modelling techniques, data requirements and considerations and recommended reading when addressing a specific set of questions on animal-seascape spatial relationships. These techniques are also explained in more detail within the chapter.

7.4.1 Linking Individual Movement Trajectories to Seascape Structure

In addition to mapping and quantifying home range data, high resolution movement tracks typically provide sufficient information to classify movement phases into distinct ecological activities such as foraging, escaping or chasing, resting or migrating (*i.e.*, movement or behaviour states) (Turchin 1998; Nathan *et al.* 2008; Demšar *et al.* 2015a). The twist and turns of a movement path (path topology) and the spatial scales at which animals change between directed to random walks provide signals of drivers of movement and dispersal (Nams 2005, 2006). The ability to quantify speed, direction, turning angle and shape of track forms the basis of several models used in different landscape ecology studies (*e.g.*, fractal analysis, state-space models, random walks) to test hypotheses on the interaction between specific ecological activities and specific seascape characteristics (Pittman & McAlpine 2003; Schick *et al.* 2008). Analysis of path topology, the geometric properties of a trajectory, has received renewed interest with a growing number of metrics available for segmenting and characterizing movement paths (reviewed by Edelhoff *et al.* 2016; Gurarie *et al.* 2016).

From a foraging theory perspective, this analytical approach can be applied to assess how foraging activity is influenced by seascape patterning such as fragmentation that can increase interpatch distances, abundance of food and the energetic requirements to locate food (McIntyre & Wiens 1999). For example, Tilley *et al.* (2013) compared the orientation behaviour of acoustically tracked stingrays (*Dasyatis americana*) across a spatially heterogeneous lagoon with a computer-generated correlated random walk (CRW) model to measure the spatial scale at which directional bias occurred. The study determined that individual stingrays orientate at spatial scales up to 100 m as a result of experiential learning of depth and key topographical features such as patch reefs. The authors conclude that rays may use the distribution of patch reefs as a network of refuges, connected by pathways of potential foraging areas. The movement response to different oceanic 'windscapes' has also been examined. For northern gannets (*Morus bassanus*), Amélineau *et al.* (2014) monitored energy expenditure, flight patterns and

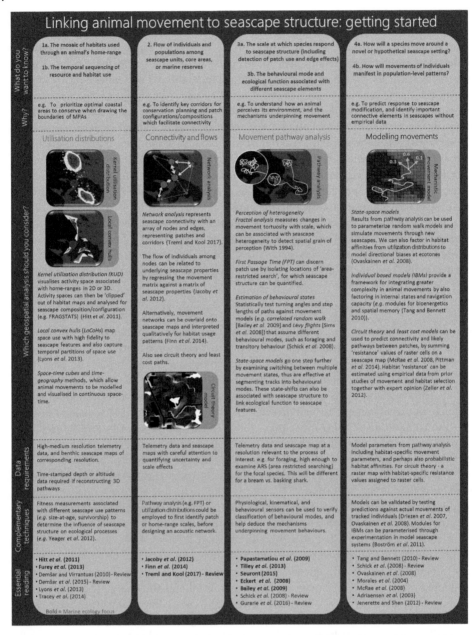

Figure 7.9 Summary and guidance for analytical tool selection for specific types of research questions when quantifying space use patterns and linking animal movement to seascape structure.

wind force and direction resulting in the identification of three behavioural states based on movement trajectories: a high tortuosity path at medium speed while foraging; a straight path at high speed while commuting; and a straight path at low speed while resting. Although wind force strongly shaped flight energy expenditure, gannets did not optimise their flight paths to avoid strong wind.

Spatial models such as hierarchical state-space models and individual-based models (IBMs) can also be used to quantify the spatial interactions between organisms and seascapes (Morales *et al.* 2004; Jonsen *et al.* 2005; Eckert *et al.* 2008; see also Chapter 8 in this book). By considering different types of random walk models this approach calculates the probabilities of an individual changing to a different movement state when environmental conditions change (Morales *et al.* 2004). Eckert *et al.* (2008) used hierarchical Bayesian state-space models with satellite telemetry data to predict movement pathways for juvenile loggerhead turtles (*Caretta caretta*) to quantitatively determine how oceanographic covariates influence movement states. The models indicated that some of the turtles were more likely to switch to intensive search behaviour (slower travel rate with higher turn angles) when encountering deeper waters where they are thought to be feeding. In contrast, larger individuals used ocean currents to perform directed movements (faster travel rate, lower turn angles) towards patchy ephemeral food resources.

The need to incorporate increased behavioural complexity in models of space-use was highlighted by Lima & Zollner (1996). In general, ecologists that have attempted to build realistic behaviour into models have obtained better fits when predicting movements and when linking movement behaviour to landscape patterns, including response to patch edges, gaps between patches and switches between behavioural states (Schick *et al.* 2008).

7.4.2 Individual Movement and Seascape Connectivity

In landscape ecology, functional connectivity is generally defined by the movement response of organisms to various structural features of the landscape (Schooley & Wiens 2003). In the sea too, measurements of movement can be used to test the permeability and connectivity of seascapes perceived by different marine species (Grober-Dunsmore *et al.* 2009; Dale & Fortin 2010). For instance, movement may be more difficult through some patch types than others and habitat edges may differ in their permeability or attraction (Ries *et al.* 2004). In addition to field observations and telemetry, landscape ecologists apply graph-theoretical approaches such as network analysis to examine connectivity. Network analysis can integrate seascape features (*e.g.*, patch type, size, isolation) with movement data to determine the paths that offer the least to highest resistance to movement, or to identify least-cost pathways (Grober-Dunsmore *et al.* 2009; Urban *et al.* 2009; Dale & Fortin 2010; see also Chapter 10 in this book). In addition, graph models provide the opportunity to test the hypothesis of nonlinear seascape connectivity as a function of increasing dispersal capacity (Urban *et al.* 2009).

Analysis of telemetry data with network analysis methods is increasing in ecology and has been applied effectively to several marine species (Jacoby *et al.* 2012; Dale & Fortin 2010; Finn *et al.* 2014). Finn *et al.* (2014) used network analysis derived from graph theory to visualise the direction of movement and number of movements between acoustic receivers (nodes) for individual fish (bonefish, barracuda and permit) across coral reef ecosystems off the island of Culebra, eastern Puerto Rico. Spatial movement graphs reflected individual behavioural differences with some fish exhibiting the pattern of a central place forager (bonefish), while others cruised along a territory (great barracuda and permit). The study was able to detect home ranges and site fidelity of different marine fish species in relation to distance from shore and habitat types. Such an approach could contribute to the quantification of interhabitat energetic flows and support marine management and restoration activities by incorporating individual movement variability into the design of marine protected area networks and ecological corridors (Grober-Dunsmore *et al.* 2009; Jacoby *et al.* 2012). Similarly, Jacoby *et al.* (2012) overlaid movement data of sharks (detected with an array of acoustic receivers) with spatially explicit information on seascape characteristics to demonstrate how network analysis could be used to predict how animal movements and their home range might be impacted by disturbances such as habitat loss. Lédée *et al.* (2015) compared kernel-based estimators of home range with a network analysis for acoustically tracked sharks in Queensland showing that although the two analyses provided similar results for estimates of core use areas, the network analysis was able to identify movement pathways that can be used to identify which corridors are most important for maintaining connectivity across the study region. Patch-based graphs, graphs that model the relationships among patches of habitat, show great promise in the operationalisation of graph-theoretic approaches in seascape ecology by measuring connectivity for focal species between patches of habitat across real seascapes. The use of network analysis in acoustic monitoring studies, especially within a seascape context, is still in its infancy and its utility will be determined through a broader range of marine applications addressing questions such as: Which areas of the seascape are connected? Which patches are most important for connectivity? What are the critical thresholds in species response to changing spatial structure? How does connectivity influence the spread of invasive species?

7.4.3 Linking Species Interactions and Physiology with Movements across Seascapes

Integrating species interaction and physiological measurements with movement data will help to disentangle the full spectrum of ecological consequences resulting from interaction with seascape patterning. For instance, species interactions are known to significantly affect space-use patterns across heterogeneous seascapes (Connolly & Hindell 2006; Boström *et al.* 2011), but are rarely considered in explanations of space-use patterns. As additional data across multiple years and seascape types are collected, it is expected that different movement analyses (*e.g.*, network analysis and time-geographic density estimation, 3D KUD) will be effective in assessing social and predator-prey interactions within and among species (Finn *et al.* 2014; Tracey *et al.* 2014a and b). For example, by combining a time-series of satellite tagged individual bull shark (*Carcharhinus leucas*) and tarpon (*Megalops atlanticus*), Hammerschlag *et al.* (2012) was able to determine how tarpon movement changed relative to the bull sharks' core area of habitat utilisation (quantified with KUD) and with respect to the spatial structural properties of their habitat. The majority of edge effect and fragmentation studies on predator-prey dynamics have concentrated mostly on direct predation effects (Creel & Christianson 2008). However, empirical studies have shown that risk effects can be significant and sometimes substantially larger than direct effects (Brown & Kotlier 2004; Heithaus *et al.* 2009). Since seascape properties can influence both predator and prey behaviour, it is essential to adapt movement ecology studies to test how changes in allocation time, vigilance, foraging, aggregation and movement patterns (*i.e.*, evidence predation risk) are influenced by spatial attributes of the seascape.

It is important to recognise that animal-seascape relationships will be mediated by nonspatial factors too, such as the health and condition of the individual or individualistic behaviour including social dynamics (Wilson *et al.* 2015). Currently, we know very little about how movement and animal-seascape interactions translate to individual fitness or the persistence of populations within metapopulations. One way forward is to integrate data from multiple techniques such as tracking telemetry, accelerometers and other biologgers, together with chemical tissue sampling and maps of seascape structure. In this way, movement ecology and seascape ecology studies can integrate knowledge to understand the ecological processes that influence an organism' fitness in relationship to the spatial attributes of seascapes. This suggests the augmentation of sequential positions with complementary data on the physiological (*e.g.*, hormonal, temperature) and behavioural (*e.g.*, feeding rates) state of the focal individual to explain how influential external factors (*e.g.*, seascape characteristics, species interaction) influence the overall fitness of individuals (Nathan *et al.* 2008; Brownscombe *et al.* 2014; Demšar *et al.* 2015a).

Triaxial accelerometers measure changes in velocity over time in three dimensions, which have helped to obtain estimates of fine-scale behaviour, such as foraging behaviour and energy expenditure in a wide range of species (Murchie *et al.* 2011; Brown *et al.* 2013; Demšar *et al.* 2015a). Using this approach, two studies on bonefish (*Albula spp.*) demonstrated how swimming and foraging behaviours and their energetic scope varied among individuals, over diel cycles and habitat types (Murchie *et al.* 2011; Brownscombe *et al.* 2014). These examples illustrate the potential of using accelerometers to develop bioenergetics models within a seascape ecology context and create seascape-fitness maps to identify optimal energetic seascape conditions. In addition, bio-loggers incorporating both location and environmental sensors that collect oceanographic parameters such as conductivity, temperature, depth and salinity could be used to assess the spatial structure of pelagic environments. Since pelagic seascapes are characterised by spatial and temporal discontinuities in energy and matter (Jelinski 2015), the influence of patterning in open waters can also be analysed using a landscape ecology approach and bio-loggers could help us to understand how pelagic species respond to oceanographic features, such as fronts, eddies, *etc.* In addition, video observations can be collected simultaneously with locational and accelerometer data that could serve as ground truthing for behaviour types and state-space dynamics (Demšar *et al.* 2015a). One example of this is the REMUS SharkCam developed by Woods Hole Oceanographic Institute (2015), an autonomous underwater vehicle equipped with video cameras and navigational and environmental loggers that enable it to locate and track the movement behaviour of large marine species such as the North Atlantic white shark (*Carcharodon carcharias*) (https://www.whoi.edu/main/remus-sharkcam, accessed 25 May 2017).

Furthermore, novel combinations of different measurements have been used to explain variation in movement behaviour across spatially heterogeneous seascapes. One example is the integrated analyses of animal movement pathways and stable isotope analysis (Layman *et al.* 2007a; Hammerschlag & Layman 2010). The relative positioning of organisms in a $\delta^{13}C$-$\delta^{15}N$ 2D space or $\delta^{13}C$-$\delta^{15}N$-$\delta^{34}S$ 3D space can reveal important aspects of trophic structure and niche space (Layman *et al.* 2007; Jackson *et al.* 2011), since different isotopes can provide proxies for the trophic position and ultimate sources of dietary carbon of organisms (Fry 2007). Novel quantitative metrics based on these representations of the niche (*i.e.*, 2D or 3D isotopic space) may be powerful tools to test ecological theory and study ecological responses to anthropogenic impacts such as habitat fragmentation (Layman *et al.* 2007; Schmidt *et al.* 2014). For example, different isotopic metrics such as isotopic range, niche width (*e.g.*, convex hull or ellipses), central tendency and niche overlap (for details Layman *et al.* 2007; Jackson *et al.* 2011) could be implemented in movement-seascape ecology studies to assess intraspecific tradeoffs in movement, foraging strategies and individual specialisation (*i.e.*, variation in resource use within the same species sex and age/stage class) in relation to seascape characteristics (Fodrie *et al.* 2015).

7.4.4 Experimental Seascapes to Investigate Animal Response to Seascape Patterns

Experimental approaches in seascape ecology can provide direct evidence of causation and have played an important role in linking patterns to processes in seascape ecology (Boström *et al.* 2011 and Chapter 5 in this book). Experiments can also provide validation data for models and field observations and generate hypotheses and parameter values to guide model development. Where controlled manipulative experiments are required to address questions about specific spatial configurations, then the options are likely to include either physical alteration of the seascape, or the use of artificial patches, such as artificial seagrass units, mangrove and patch reefs. Artificial patches have been applied successfully to examine the ecological consequences of seascape context, structural complexity, patch size, edge effects and patch proximity (*e.g.*, Jelbart *et al.* 2006; Walsh 1985; Nagelkerken & Faunce 2008; Yeager *et al.* 2012), but are typically limited in spatial scale to just a few metres or tens of metres; often considerably finer scales than the home ranges of the species under investigation. Although broad-scale manipulation of seascape structure may be unfeasible, or unethical for the purpose of experimental studies, we can make use of seascapes that have already been altered, or are about to be altered (*i.e.*, opportunistic experiments), to conduct studies on movement behaviour, both before and after the spatial configuration has been modified. These experiments could still reveal strong inference on causes driving the movement response of marine animals if we can also effectively account for variability in ecological characteristics among different seascapes. Replication, however, is often not possible due to the scale of the experiments and where broad scale structural changes have occurred, the opportunity is often a one-time event. A review of the challenges and progress in landscape experiments along with guidance on the ideal features of a manipulative experiment has been provided in the landscape ecology literature (*e.g.*, McGarigal & Cushman 2002; Jenerette & Shen 2012).

Wiens & Milne (1989) advocate using experimental microlandscapes to examine pattern-process linkages with the potential that such studies serve as a model of broader scale landscapes. There are several advantages of this microlandscape approach: (i) measurements may be taken with a level of detail that is difficult to attain at a broader scale; (ii) sample sizes may be greater, or sampling at a given intensity may provide a more accurate representation of the phenomenon being investigated; (iii) experimental manipulations may be conducted with relative ease and (iv) experiments or observations may be replicated over many plots or treatments with relative ease. Johnson *et al.* (1992) proposed that general ecological principles will emerge from experimental studies using microlandscapes (with appropriately scaled animals) that may be translated to other organisms and broader spatial scales. The biggest limitation is that scaling up can lead to erroneous conclusions and conducting microlandscape experiments at organism-relevant scales is challenging for many mobile marine species. However, this approach may be insightful for small-bodied individuals and those with a relatively small home range, or to understand how highly mobile species respond to structure in a specific part of their home range (edge response, crossing unsuitable patch types). Perhaps, the biggest challenge in scaling up movement patterns are the complexities of individual behaviour (Morales & Ellner 2002). However, the use of microseascapes to examine behavioural responses to seascape patterning in the way that landscape ecologists have done is extremely rare, yet this approach, combined with spatial modelling such as individual-based modelling (see Chapter 8 in this book) could provide strong inference on pattern-process relationships in seascapes.

7.4.5 Mechanistic Models

Even though the probabilistic models and empirical studies mentioned above do not necessarily allow one to reveal and disentangle the mechanistic underpinnings of movement directly, the statistics they provide could serve as assessment criteria for simulation models that do implement and combine different movement mechanisms (Mueller & Fagan 2008). Empirical studies can help in the parameterisation of individual-based simulations (*i.e.*, individual-based models IBMs) and increase the ability to identify multiple characteristic movement statistics as emergent responses to the complex spatial heterogeneity of seascapes and biological distributions (Mueller & Fagan 2008; Nathan *et al.* 2008; Schick *et al.* 2008). For example, an IBM has been applied to test if change in seascape composition and configuration influenced predator-prey interactions and cohort size for a group of settling juvenile blue crabs (*Callinectes sapidus*) (Hovel & Regan 2008; see also Chapter 8 in this book). IBM models incorporating different movement simulations and states can reveal which combination of spatial patterns, organism internal state, dispersal abilities facilitate or impede movement across seascapes (Morales *et al.* 2005; Vergara *et al.* 2015). Other examples in terrestrial studies have combined artificial neural networks with IBMs to model complex animal movement (Schick *et al.* 2008). Individual-based neural network algorithms are generally advantageous because the technique integrates qualitatively different input information can be used to explore how different alternative movement mechanism (*e.g.*, nonoriented, oriented, spatial memory) could induce a variation of emergent patterns under different seascape scenarios (Mueller & Fagan 2008; Nathan *et al.* 2008; Schick *et al.* 2008). Hence, the parameterisation of this type of model with identified and quantified behavioural responses to seascape structure could help determine how movement mechanisms interplay to influence the functional connectivity, dispersion and persistence of marine populations within spatially dynamic seascapes.

7.5 Implications of Animal-Seascape Understanding for Marine Stewardship

Advances in our understanding of animal-seascape relationships will have many benefits to ecosystem-based management including informing the design of both static and dynamic marine protected areas (Hyrenbach *et al.* 2000; Game *et al.* 2009; Pittman *et al.*

2014; Espinoza *et al.* 2015; Lea *et al.* 2016), building coherence into network design (Weeks *et al.* 2016), as well as incorporating key animal movement pathways, 'blue corridors', or 'blueways', into marine spatial planning (Martin *et al.* 2006; Pendoley *et al.* 2014; Brenner *et al.* 2016). It is widely acknowledged that connectivity among protected areas is an essential part of ecological coherence. In the Gulf of Mexico and in the Baltic region, migratory pathways have been identified to enhance spatial planning. For example, to understand migratory pathways across the Gulf of Mexico, identify priority spaces and relevant scales for management, The Nature Conservancy synthesised and mapped large volumes of satellite telemetry data for highly mobile species (marine fish, sea turtles, marine mammals and birds) (Brenner *et al.* 2016). The study identified partial migratory corridors and movement density within the corridors, occurrence hotspots, locations of aggregations and multispecies aggregations and areas of potential threat to migratory movements from human activities (Figure 7.10).

Furthermore, using animal movements to improve our scaling of habitat will have practical implications for the identification and mapping of habitat concepts important for legislated protection, such as essential fish habitat and critical habitat. Understanding functional connectivity between multiple patch types across the seascape can guide spatial prioritisation activities in conservation planning and ensure that decision makers are aware of the consequences of the potential disruptions to connectivity of environmental change. Fortunately, connectivity can also be facilitated by management actions such as prioritizing protection for connected seascapes (Nagelkerken *et al.* 2015; see also Chapter 9 in this book). Likewise, patterns of fish movements have application to

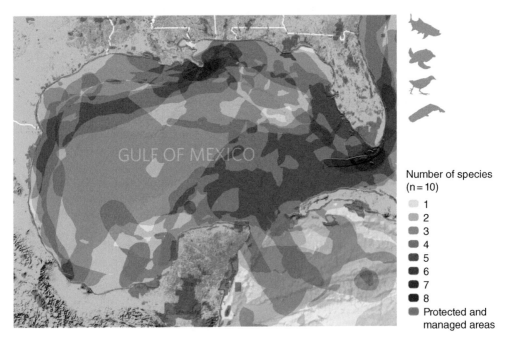

Figure 7.10 Overlay of marine species corridors for 10 marine species (representing fishes, sea turtles and marine mammals) and protected and management areas in the Gulf of Mexico. *Source*: Provided with permission by Jorge Brenner of The Nature Conservancy (Brenner *et al.* 2016).

the design and evaluation of habitat restoration and habitat creation projects for marine species. For example, knowledge of fish movement patterns can be used as a functional metric for assessing the performance of restoration projects for estuarine fish habitat (Freedman *et al.* 2016).

For ecology, we expect that many new insights into the relationships between marine animals and the seascape will emerge from the application of spatial technologies for both mapping individual animal movements and mapping the surrounding seascape structure. New developments in home range estimators, particularly those that incorporate time, network analyses using graph theoretic approaches, dynamic individual-based models combined with advances in animal tracking technologies and habitat mapping have greatly facilitated the quantification of spatial patterns in movement ecology. Many interesting and useful questions can be addressed through integration of movement ecology within seascape ecology, particularly studies that link seascape patterning to movement behaviour and the consequences for organism biology such as growth, energetics, condition and survival. Studies are needed to understand the behavioural movement responses to structural features, such as: corridors of favourable patch composition and gaps of unfavourable patch composition; movement responses to boundary structure; the use of seascape features as navigational aids; the importance of cognitive mapping of the seascape in marine animal movements; and spatial patterns that facilitate or restrict connectivity, which is required to complete the life cycle. Finally, it is worth recognizing that researchers studying moving objects, such as ships, cars or people, face similar challenges in data analysis, interpretation and visualisation suggesting that we may benefit from the review and application of approaches and tools from these other disciplines to advance our knowledge of the how, why, where and when of animal movements through the seascape.

References

Addicott JF, Aho JM, Antolin MF, Padilla DK, Richardson JS, Soluk DA (1987) Ecological neighbourhoods: scaling environmental patterns. Oikos 1: 340–346.

Adriaensen F, Chardon JP, De Blust G, Swinnen E, Villalba S, Gulinck H, Matthysen E (2003) The application of 'least-cost' modelling as a functional landscape model. Landscape and Urban Planning 64(4): 233–247.

Allen TFH, Starr TB (eds) (1982) Hierarchy: Perspectives for Ecological Complexity. University of Chicago Press, Chicago, IL.

Amélineau F, Péron C, Lescroël A, Authier M, Provost P, Grémillet D (2014) Windscape and tortuosity shape the flight costs of northern gannets. Journal of Experimental Biology 217(6): 876–885.

Anderson AB, Jenkins CN (2006) Applying nature's design: Corridors as a strategy for biological conservation. Columbia University Press, New York, NY.

Andrienko N, Andrienko G (2013) Visual analytics of movement: An overview of methods, tools and procedures. Information Visualization 12(1): 3–24.

Bailey H, Mate BR, Palacios DM, Irvine L, Bograd SJ, Costa DP (2009) Behavioural estimation of blue whale movements in the Northeast Pacific from state-space model analysis of satellite tracks. Endangered Species Research 10: 93–106.

Barrell J, Grant J (2013) Detecting hot and cold spots in a seagrass landscape using local indicators of spatial association. Landscape Ecology 28(10): 2005–2018.

Barry JP, Dayton PK (1991) Physical heterogeneity and the organization of marine communities. In Kolasa J. & Pickett ST (eds) Ecological heterogeneity. Springer, New York, NY, pp. 270–320.

Benhamou S, Cornélis D (2010) Incorporating movement behaviour and barriers to improve kernel home range space use estimates. The Journal of Wildlife Management 74: 1353–1360.

Benson SR, Eguchi T, Foley DG, Forney KA, Bailey H, Hitipeuw C, Samber BP, Tapilatu RF, Rei V, Ramohia P, Pita J (2011) Large-scale movements and high-use areas of western Pacific leatherback turtles, *Dermochelys coriacea*. Ecosphere 2(7): 1–27.

Betts MG, Fahrig L, Hadley AS, Halstead KE, Bowman J, Robinson WD, Wiens JA, Lindenmayer DB (2014) A species-centered approach for uncovering generalities in organism responses to habitat loss and fragmentation. Ecography 37(6): 517–527.

Biesinger Z, Bolker BM, Marcinek D, Grothues TM, Dobarro JA, Lindberg WJ (2013) Testing an autonomous acoustic telemetry positioning system for fine-scale space use in marine animals. Journal of Experimental Marine Biology and Ecology 448: 46–56.

Block BA, Jonsen ID, Jorgensen SJ, Winship AJ, Shaffer SA, Bograd SJ, Hazen EL, Foley DG, Breed GA, Harrison AL, Ganong JE (2011) Tracking apex marine predator movements in a dynamic ocean. Nature 475(7354): 86–90.

Boström C, Pittman SJ, Simenstad C, Kneib RT (2011) Seascape ecology of coastal biogenic habitats: advances, gaps, and challenges. Marine Ecology Progress Series 427: 191–217.

Brenner J, Voight C, Mehlman D (2016) Migratory Species in the Gulf of Mexico Large Marine Ecosystem: Pathways, Threats and Conservation. The Nature Conservancy, Arlington, VA.

Brooker RM, Feeney WE, White JR, Manassa RP, Johansen JL, Dixson DL (2016) Using insights from animal behaviour and behavioural ecology to inform marine conservation initiatives. Animal Behaviour 10.1016/j.anbehav.2016.03.012 (accessed 25 May 2017).

Brown DD, Kays R, Wikelski M, Wilson R, Klimley AP (2013) Observing the unwatchable through acceleration logging of animal behaviour. Animal Biotelemetry 1: 20.

Brown JS, Kotlier BP (2004) Hazardous duty pay and the foraging cost of predation. Ecology Letters 7(10): 999–1014.

Brownscombe JW, Gutowsky LF, Danylchuk AJ, Cooke SJ (2014) Foraging behaviour and activity of a marine benthivorous fish estimated using tri-axial accelerometer biologgers. Marine Ecology Progress Series 505: 241–251.

Burt de Perera TB, Holbrook RI, Davis V (2016) The representation of three-dimensional space in fish. Frontiers in Behavioural Neuroscience 10: 40.

Calabrese JM, Fagan WF (2004) A comparison-shopper's guide to connectivity metrics. Frontiers in Ecology and the Environment 2(10): 529–536.

Connolly RM, Hindell JS (2006) Review of nekton patterns and ecological processes in seagrass landscapes. Estuarine, Coastal and Shelf Science 68(3): 433–444.

Creel S, Christianson D (2008) Relationships between direct predation and risk effects. Trends in Ecology and Evolution 23(4): 194–201.

Crook DA, Lowe WH, Allendorf FW, Erős T, Finn DS, Gillanders BM, Hadwen WL, Harrod C, Hermoso V, Jennings S, Kilada RW (2015) Human effects on ecological connectivity in aquatic ecosystems: Integrating scientific approaches to support management and mitigation. Science of the Total Environment 534: 52–64.

Crooks KR, Sanjayan M (2006) Connectivity conservation: maintaining connections for nature. In Crooks KR, Sanjayan M (eds) Connectivity Conservation. Cambridge University Press, Cambridge, pp. 1–20.

Crowder L, Norse E (2008) Essential ecological insights for marine ecosystem-based management and marine spatial planning. Marine Policy 32(5): 772–778.

Dale MR, Fortin MJ (2010) From graphs to spatial graphs. Annual Review of Ecology, Evolution, and Systematics 41: 21–38.

Dance MA, Rooker JR (2015) Habitat-and bay-scale connectivity of sympatric fishes in an estuarine nursery. Estuarine, Coastal and Shelf Science 167: 447–457.

Demšar U, Buchin K, Cagnacci F, Safi K, Speckmann B, Van de Weghe N, Weiskopf D, Weibel R (2015a). Analysis and visualisation of movement: an interdisciplinary review. Movement Ecology 3(1): 5.

Demšar U, Buchin K, van Loon EE, Shamoun-Baranes J (2015b) Stacked space-time densities: a geovisualisation approach to explore dynamics of space use over time. Geoinformatica 19: 85–115.

Demšar U, van Loon E (2013) Visualising movement: the seagull. Significance 10: 40–42.

Demšar U, Virrantaus K (2010) Space-time density of trajectories: exploring spatio-temporal patterns in movement data. International Journal of Geographical Information Science 24(10): 1527–1542.

Dodson JJ (1988) The nature and role of learning in the orientation and migratory behaviour of fishes. Environmental Biology of Fishes 23(3): 161–182.

Downs JA, Horner MW, Tucker AD (2011) Time-geographic density estimation for home range analysis. Annals of GIS 17(3): 163–171.

Downs JA, Horner MW, Hyzer G, Lamb D, Loraamm R (2015) Voxel-based probabilistic space-time prisms for analysing animal movements and habitat use. International Journal of Geographical Information Science 28: 875–890.

Dunning JB, Danielson BJ, Pulliam HR (1992) Ecological processes that affect populations in complex landscapes. Oikos 1: 169–175.

Eckert SA, Moore JE, Dunn DC, van Buiten RS, Eckert KL, Halpin PN (2008) Modelling loggerhead turtle movement in the Mediterranean: importance of body size and oceanography. Ecological Applications 18(2): 290–308.

Edelhoff H, Signer J, Balkenhol N (2016) Path segmentation for beginners: an overview of current methods for detecting changes in animal movement patterns. Movement Ecology 4(1): 21.

Eiler JH, Masuda MM, Spencer TR, Driscoll RJ, Schreck CB (2014) Transactions of the American Fisheries Society 143(6): 1476–1507.

Espinoza M, Farrugia TJ, Webber DM, Smith F, Lowe CG (2011) Testing a new acoustic telemetry technique to quantify long-term, fine-scale movements of aquatic animals. Fisheries Research 108: 364–371.

Espinoza M, Lédée EJ, Simpfendorfer CA, Tobin AJ, Heupel MR (2015) Contrasting movements and connectivity of reef-associated sharks using acoustic telemetry: implications for management. Ecological Applications 25(8): 2101–2118.

Fauchald P (1999) Foraging in a hierarchical patch system. American Naturalist 153: 603–613.

Fauchald P, Erikstad KE, Skarsfjord H (2000) Scale-dependent predator-prey interactions: the hierarchical spatial distribution of seabirds and prey. Ecology 81: 773–783.

Fauchald P, Tveraa T (2006) Hierarchical patch dynamics and animal movement pattern. Oecologia 149(3): 383–395.

Finn JT, Brownscombe JW, Haak CR, Cooke SJ, Cormier R, Gagne T, Danylchuk AJ (2014) Applying network methods to acoustic telemetry data: modelling the movements of tropical marine fishes. Ecological Modelling 293: 139–149.

Fisher HI (1971) Experiments on homing in Laysan albatrosses, *Diomedea immutabilis*. The Condor 73(4): 389–400.

Fodrie FJ, Yeager LA, Grabowski JH, Layman CA, Sherwood GD, Kenworthy MD (2015) Measuring individuality in habitat use across complex landscapes: approaches, constraints, and implications for assessing resource specialization. Oecologia 178(1): 75–87.

Foley MM, Halpern BS, Micheli F, Armsby MH, Caldwell MR, Crain CM, Prahler E, Rohr N, Sivas D, Beck MW, Carr MH (2010) Guiding ecological principles for marine spatial planning. Marine Policy 34(5): 955–966.

Freedman RM, Espasandin C, Holcombe EF, Whitcraft CR, Allen BJ, Witting D, Lowe CG (2016) Using movements and habitat utilization as a functional metric of restoration for estuarine juvenile fish habitat. Marine and Coastal Fisheries 8(1): 361–373.

Furey NB, Dance MA, Rooker JR (2013) Fine-scale movements and habitat use of juvenile southern flounder *Paralichthys lethostigma* in an estuarine seascape. Journal of Fish Biology 82(5): 1469–1483.

Game ET, Grantham HS, Hobday AJ, Pressey RL, Lombard AT, Beckley LE, Gjerde K, Bustamante R, Possingham HP, Richardson AJ (2009) Pelagic protected areas: the missing dimension in ocean conservation. Trends in Ecology and Evolution 24: 360–369.

Getz WM, Saltz D (2008) A framework for generating and analyzing movement paths on ecological landscapes. Proceedings of the National Academy of Sciences 105(49): 19066–19071.

Green AL, Maypa AP, Almany GR, Rhodes KL, Weeks R, Abesamis RA, Gleason MG, Mumby PJ, White AT (2015) Larval dispersal and movement patterns of coral reef fishes, and implications for marine reserve network design. Biological Reviews 90(4): 1215–1247.

Grober-Dunsmore R, Pittman SJ, Caldow C, Kendall MS & Frazer TK (2009) A landscape ecology approach for the study of ecological connectivity across tropical marine seascapes. In Nagelkerken I (ed.) Ecological Connectivity among Tropical Coastal Ecosystems. Springer, Heidelberg, pp. 493–530.

Gurarie E, Bracis C, Delgado M, Meckley TD, Kojola I, Wagner CM (2016) What is the animal doing? Tools for exploring behavioural structure in animal movements. Journal of Animal Ecology 85(1): 69–84.

Hammerschlag N, Luo J, Irschick DJ, Ault JS (2012) A Comparison of spatial and movement patterns between sympatric predators: Bull Sharks (Carcharhinus leucas) and Atlantic Tarpon (Megalops atlanticus). PLoS ONE 7(9): e45958.

Hammerschlag-Peyer CM, Layman CA (2010) Intrapopulation variation in habitat use by two abundant coastal fish species. Marine Ecology Progress Series 415: 211–220.

Hatcher BG, Imberger J, Smith SV (1987) Scaling analysis of coral reef systems: an approach to problems of scale. Coral Reefs 5(4): 171–181.

Haury LR, McGowan JA, Wiebe PH (1978) Patterns and processes in the time-space scales of plankton distributions. In Steele JH (ed.) Spatial Pattern in Plankton Communities. Plenum Press, New York, NY.

Hays GC, Scott R (2013) Global patterns for upper ceilings on migration distance in sea turtles and comparisons with fish, birds and mammals. Functional Ecology 27(3): 748–756.

Heithaus MR, Wirsing AJ, Burkholder D, Thomson J, Dill LM (2009) Towards a predictive framework for predator risk effects: the interaction of landscape features and prey escape tactics. Journal of Animal Ecology 78(3): 556–562.

Heupel M, Semmens J, Hobday A (2006) Automated acoustic tracking of aquatic animals: scales, design and deployment of listening station arrays. Marine and Freshwater Research 57: 1–13.

Heupel MR, Simpfendorfer CA, Hueter RE (2004) Estimation of shark home ranges using passive monitoring techniques. Environmental Biology of Fishes 71: 135–142.

Hitt S, Pittman SJ, Brown KA (2011b). Tracking and mapping sun-synchronous migrations and diel space use patterns of *Haemulon sciurus* and *Lutjanus apodus* in the US Virgin Islands. Environmental Biology of Fishes 92(4): 525–538.

Hitt S, Pittman SJ, Nemeth RS (2011a) Diel movements of fishes linked to benthic seascape structure in a Caribbean coral reef ecosystem. Marine Ecology Progress Series 427: 275–291.

Hoechstetter S, Walz U, Dang LH, Thinh NX (2008) Effects of topography and surface roughness in analyses of landscape structure: a proposal to modify the existing set of landscape metrics. Landscape Online 3: 1–4.

Holling CS (1992) Cross-scale morphology, geometry, and dynamics of ecosystems. Ecological Monographs 62(4): 447–502.

Horodysky AZ, Cooke SJ, Brill RW (2015) Physiology in the service of fisheries science: why thinking mechanistically matters. Reviews in Fish Biology and Fisheries 25(3): 425–447.

Hovel KA, Regan HM (2008) Using an individual-based model to examine the roles of habitat fragmentation and behavior on predator–prey relationships in seagrass landscapes. Landscape Ecology 23(1): 75–89.

Hussey NE, Kessel ST, Aarestrup K, Cooke SJ, Cowley PD, Fisk AT, Harcourt RG, Holland KN, Iverson SJ, Kocik JF, Flemming JE (2015) Aquatic animal telemetry: a panoramic window into the underwater world. Science 348(6240): 1255642.

Hyrenbach KD, Forney KA, Dayton PK (2000) Marine protected areas and ocean basin management. Aquatic Conservation: Marine and Freshwater Ecosystems 10: 435–458.

Irlandi EA, Crawford MK (1997) Habitat linkages: the effect of intertidal saltmarshes and adjacent subtidal habitats on abundance, movement, and growth of an estuarine fish. Oecologia 110(2): 222–230.

Jackson AL, Inger R, Parnell AC, Bearhop S (2011) Comparing isotopic niche widths among and within communities: SIBER–Stable Isotope Bayesian Ellipses in R. Journal of Animal Ecology 80(3): 595–602.

Jacoby DM, Brooks EJ, Croft DP, Sims DW (2012) Developing a deeper understanding of animal movements and spatial dynamics through novel application of network analyses. Methods in Ecology and Evolution 3(3): 574–583.

Jelbart JE, Ross PM, Connolly RM (2006) Edge effects and patch size in seagrass landscapes: an experimental test using fish. Marine Ecology Progress Series 319: 93–102.

Jelinski DE (2015) On a landscape ecology of a harlequin environment: the marine landscape. Landscape Ecology 30: 1–6.

Jenerette GD, Shen W (2012) Experimental landscape ecology. Landscape Ecology 27(9): 1237–1248.

Jonsen ID, Flemming JM, Myers RA (2005) Robust state–space modeling of animal movement data. Ecology 86(11): 2874–2880.

Johnson AR, Wiens JA, Milne BT, Crist TO (1992) Animal movements and population dynamics in heterogeneous landscapes. Landscape Ecology 7(1): 63–75.

Kranstauber B, Kays R, LaPoint SD, Wikelski M, Safi K (2012) A dynamic Brownian bridge movement model to estimate utilization distributions for heterogeneous animal movement. Journal of Animal Ecology 81(4): 738–746.

Kuhn CE, Johnson DS, Ream RR, Gelatt TS (2009) Advances in the tracking of marine species: using GPS locations to evaluate satellite track data and a continuous-time movement model. Marine Ecology Progress Series 393: 97–109.

Large SI, Fay G, Friedland KD, Link JS (2015) Critical points in ecosystem responses to fishing and environmental pressures. Marine Ecology Progress Series 521: 1–7.

Lausch A, Schmidt A, Tischendorf L (2015) Data mining and linked open data: New perspectives for data analysis in environmental research. Ecological Modelling 295: 5–17.

Layman CA, Quattrochi JP, Peyer CM, Allgeier JE (2007) Niche width collapse in a resilient top predator following ecosystem fragmentation. Ecology Letters 10(10): 937–944.

Lea JS, Humphries NE, von Brandis RG, Clarke CR, Sims DW (2016) Acoustic telemetry and network analysis reveal the space use of multiple reef predators and enhance marine protected area design. Proceedings Royal Society B 283(1834) 20160717.

Lédée EJ, Heupel MR, Tobin AJ, Knip DM, Simpfendorfer CA (2015) A comparison between traditional kernel-based methods and network analysis: an example from two nearshore shark species. Animal Behaviour 103: 17–28.

Levin SA (1992) The problem of pattern and scale in ecology: The Robert H. MacArthur award lecture. Ecology 73(6): 1943–1967.

Lima SL, Zollner PA (1996) Towards a behavioural ecology of ecological landscapes. Trends in Ecology and Evolution 11(3): 131–135.

Luschi P (2013) Long-distance animal migrations in the oceanic environment: orientation and navigation correlates. ISRN Zoology Article ID 631839.

Lyons AJ, Turner WC, Getz WM (2013) Home range plus: a space-time characterization of movement over real landscapes. Movement Ecology 1: 1–14.

Marquet PA, Fortin MJ, Pineda J, Wallin DO, Clark J, Wu Y, Bollens S, Jacobi CM, Holt RD (1993) Ecological and evolutionary consequences of patchiness: a marine-terrestrial perspective. In Patch dynamics. Springer, Berlin, pp. 277–304.

Martin G, Makinen A, Andersson Å, Dinesen GE, Kotta J, Hansen J, Herkül K, Ockelmann KW, Nilsson P, Korpinen S (2006) Literature review of the 'Blue Corridors' concept and its applicability to the Baltic Sea. BALANCE project. http://www.balance-eu.org/ (accessed 25 May 2017).

McAlpine CA, Grigg GC, Mott JJ, Sharma P (1999) Influence of landscape structure on kangaroo abundance in a disturbed semi-arid woodland of Queensland. The Rangeland Journal 21(1): 104–134.

McGarigal K, Cushman SA (2002) Comparative evaluation of experimental approaches to the study of habitat fragmentation effects. Ecological Applications 12(2): 335–345.

McGarigal K, Cushman SA (2005) The gradient concept of landscape structure. In Wiens J, Moss M (eds) Issues and perspectives in landscape ecology. Cambridge University Press, Cambridge, pp 112–119.

McGarigal K, Tagil S, Cushman SA (2009) Surface metrics: an alternative to patch metrics for the quantification of landscape structure. Landscape Ecology 24(3): 433–450.

McIntyre NE, Wiens JA (1999) Interactions between landscape structure and animal behavior: the roles of heterogeneously distributed resources and food deprivation on movement patterns. Landscape Ecology 14(5): 437–447.

McRae BH, Dickson BG, Keitt TH, Shah VB (2008) Using circuit theory to model connectivity in ecology, evolution, and conservation. Ecology 89(10): 2712–2724.

Meentemeyer V (1989) Geographical perspectives of space, time and scale. Landscape Ecology 3: 163–173.

Moorcroft PR, Lewis MA, Crabtree RL (2006) Mechanistic home range models capture spatial patterns and dynamics of coyote territories in Yellowstone. Proceedings of the Royal Society of London B: Biological Sciences 273(1594): 1651–1659.

Morales JM, Ellner SP (2002) Scaling up animal movements in heterogeneous landscapes: the importance of behaviour. Ecology 83(8): 2240–2247.

Morales JM, Fortin D, Frair JL, Merrill EH (2005) Adaptive models for large herbivore movements in heterogeneous landscapes. Landscape Ecology 20(3): 301–316.

Morales JM, Haydon DT, Frair J, Holsinger KE, Fryxell JM (2004) Extracting more out of relocation data: building movement models as mixtures of random walks. Ecology 85(9): 2436–2445.

Mueller T, Fagan WF (2008) Search and navigation in dynamic environments: from individual behaviours to population distributions. Oikos 117(5): 654–664.

Murchie KJ, Cooke SJ, Danylchuk AJ, Danylchuk SE, Goldberg TL, Suski CD, Philipp DP (2011) Thermal biology of bonefish (*Albula vulpes*) in Bahamian coastal waters and tidal creeks: an integrated laboratory and field study. Journal of Thermal Biology 36(1): 38–48.

Nagelkerken I, Faunce CH (2008) What makes mangroves attractive to fish? Use of artificial units to test the influence of water depth, cross-shelf location, and presence of root structure. Estuarine, Coastal and Shelf Science 79(3): 559–565.

Nagelkerken I, Sheaves M, Baker R, Connolly RM (2015) The seascape nursery: a novel spatial approach to identify and manage nurseries for coastal marine fauna. Fish and Fisheries 16(2): 362–371.

Nams V (2005) Using animal movement paths to measure response to spatial scale. Oecologia 143: 179–188.

Nams V (2006) Detecting oriented movement of animals. Animal Behaviour 72(5): 1197–1203.

Nathan R, Getz WM, Revilla E, Holyoak M, Kadmon R, Saltz D, Smouse PE (2008) A movement ecology paradigm for unifying organismal movement research. Proceedings of the National Academy of Sciences 105(49): 19052–19059.

Neil DT (2002) Cooperative fishing interactions between Aboriginal Australians and dolphins in eastern Australia. Anthrozoos 15: 3–18.

Olds AD, Connolly RM, Pitt KA, Maxwell PS (2012) Habitat connectivity improves reserve performance. Conservation Letters 5(1): 56–63.

Olds AD, Connolly RM, Pitt KA, Pittman SJ, Maxwell PS, Huijbers CM, Moore BR, Albert S, Rissik D, Babcock RC, Schlacher TA (2016) Quantifying the conservation value of seascape connectivity: a global synthesis. Global Ecology and Biogeography 25(1): 3–15.

Ovaskainen O, Luoto M, Ikonen I, Rekola H, Meyke E, Kuussaari M (2008) An empirical test of a diffusion model: predicting clouded apollo movements in a novel environment. The American Naturalist 171(5): 610–619.

Palumbi SR (2004) Marine reserves and ocean neighbourhoods: the spatial scale of marine populations and their management. Annual Review of Environment and Resources 29: 31–68.

Papastamatiou YP, Lowe CG, Caselle JE, Friedlander AM (2009) Scale-dependent effects of habitat on movements and path structure of reef sharks at a predator-dominated atoll. Ecology 90: 996–1008.

Patton BW, Braithwaite VA (2015) Changing tides: ecological and historical perspectives on fish cognition. Wiley Interdisciplinary Reviews: Cognitive Science 6(2): 159–176.

Pendoley KL, Schofield G, Whittock PA, Ierodiaconou D, Hays GC (2014) Protected species use of a coastal marine migratory corridor connecting marine protected areas. Marine Biology 161(6): 1455–1466.

Perry AL, Low PJ, Ellis JR, Reynolds JD (2005) Climate change and distribution shifts in marine fishes Science 308(5730): 1912–1915.

Pinsky ML, Worm B, Fogarty MJ, Sarmiento JL, Levin SA (2013) Marine taxa track local climate velocities. Science 341(6151): 1239–1242.

Pittman SJ, Brown KA (2011) Multi-scale approach for predicting fish species distributions across coral reef seascapes. PloS One 6(5): e20583.

Pittman SJ, McAlpine CA (2003) Movements of marine fish and decapod crustaceans: process, theory and application. Advances in Marine Biology 44(1): 205–294.

Pittman SJ, Monaco ME, Friedlander AM, Legare B, Nemeth RS, Kendall MS, Poti M, Clark RD, Wedding LM, Caldow C (2014) Fish with chips: Tracking reef fish movements to evaluate size and connectivity of Caribbean marine protected areas. PloS One 9: e96028.

Pittman SJ, Olds AD (2015) Seascape ecology of fishes on coral reefs. In Mora C (ed.) Ecology of Fishes on Coral Reefs. Cambridge University Press, Cambridge, pp. 274–282.

Ries L, Fletcher Jr RJ, Battin J, Sisk TD (2004) Ecological responses to habitat edges: mechanisms, models, and variability explained. Annual Review of Ecology, Evolution and Systematics 35: 491–522.

Rolstad J, Løken B, Rolstad E (2000) Habitat selection as a hierarchical spatial process: the green woodpecker at the northern edge of its distribution range. Oecologia 124(1): 116–129.

Roughgarden J, Gaines S, Possingham H (1988) Recruitment dynamics in complex life cycles. Proceedings of the National Academy of Sciences 85: 7418.

Ryall KL, Fahrig L (2006) Response of predators to loss and fragmentation of prey habitat: a review of theory. Ecology 87(5): 1086–1093.

Schaefer JA, Messier F (1995) Habitat selection as a hierarchy: the spatial scales of winter foraging by muskoxen. Ecography 18(4): 333–344.

Schick RS, Loarie SR, Colchero F, Best BD, Boustany A, Conde DA, Halpin PN, Joppa LN, McClellan CM, Clark JS (2008) Understanding movement data and movement processes: current and emerging directions. Ecology Letters 11(12): 1338–1350.

Schick RS, Roberts JJ, Eckert SA, Halpin PN, Bailey H, Chai F, Shi L, Clark JS (2013) Pelagic movements of pacific leatherback turtles (*Dermochelys coriacea*) highlight the role of prey and ocean currents. Movement Ecology 1: 11.

Schmidt DJ, Crook DA, Macdonald JI, Huey JA, Zampatti BP, Chilcott S, Raadik TA, Hughes JM (2014) Migration history and stock structure of two putatively diadromous teleost fishes, as determined by genetic and otolith chemistry analyses. Freshwater Science 33: 193–206.

Schooley RL, Wiens JA (2003) Finding habitat patches and directional connectivity. Oikos 102: 559–570.

Senft RL, Coughenour MB, Bailey DW, Rittenhouse LR, Sala OE, Swift DM (1987) Large herbivore foraging and ecological hierarchies. BioScience 37(11): 789–799.

Seuront L (2015) On uses, misuses and potential abuses of fractal analysis in zooplankton behavioural studies: A review, a critique and a few recommendations. Physica A: Statistical Mechanics and its Applications 432: 410–434.

Shamoun-Baranes J, van Loon EE, Purves RS, Speckmann B, Weiskopf D, Camphuysen CJ (2011) Analysis and visualization of animal movement. Biology Letters: rsbl20110764.

Silverman BW (1986) Density Estimation for Statistics and Data Analysis. CRC Press, Boca Raton, FL.

Simberloff D, Cox J (1987) Consequences and costs of conservation corridors. Conservation Biology 1(1): 63–71.

Simpfendorfer CA, Olsen EM, Heupel MR, Moland E (2012) Three-dimensional kernel utilization distributions improve estimates of space use in aquatic animals. Canadian Journal of Fisheries and Aquatic Sciences 69: 565–572.

Sims DW (2010) Tracking and analysis techniques for understanding free-ranging shark movements and behaviour. In Carrier JC, Musick JA & Heithaus MR (eds) Sharks and their relatives II: biodiversity, adaptive physiology, and conservation. CRC Press, Boca Raton, FL, pp. 351–392.

Sims DW, Quayle VA (1998) Selective foraging behaviour of basking sharks on zooplankton in a small-scale front. Nature 393: 460–464.

Sims DW, Southall EJ, Humphries NE, Hays GC, Bradshaw CJ, Pitchford JW, James A, Ahmed MZ, Brierley AS, Hindell MA, Morritt D (2008) Scaling laws of marine predator search behaviour. Nature 451(7182): 1098–1102.

Sims DW, Witt MJ, Richardson AJ, Southall EJ, Metcalfe JD (2006) Encounter success of free-ranging marine predator movements across a dynamic prey landscape. Proceedings of the Royal Society of London B: Biological Sciences 273(1591): 1195–1201.

Southwood TR (1977) Habitat, the templet for ecological strategies? The Journal of Animal Ecology 46: 337–365.

Steele JH (1988). Scale selection for biodynamic theories. In Rothschild BJ (ed.) Toward a Theory on Biological-Physical Interactions in the World Ocean. Proceedings of the NATO Advanced Research Workshop, Castéra-Verduzan, France, 1–5 June 1987. Kluwer Academic Publishers, Dordrecht, pp. 513–526.

Steele JH (1989) The ocean 'landscape'. Landscape Ecology 3(3): 185–192.

Stommel H (1963) Varieties of oceanographic experience. Science 139: 572–576.

Tang W, Bennett DA (2010) Agent-based modelling of animal movement: A review. Geography Compass 4(7): 682–700.

Tarjan LM, Tinker MT (2016) Permissible Home Range Estimation (PHRE) in Restricted Habitats: A new algorithm and an evaluation for sea otters. PLoS One 11(3): e0150547.

Thayer GW, Bjorndal KA, Ogden JC, Williams SL, Zieman JC (1984) Role of larger herbivores in seagrass communities. Estuaries 7(4): 351–376.

Tilley A, López-Angarita J, Turner JR (2013) Effects of scale and habitat distribution on the movement of the southern stingray *Dasyatis americana* on a Caribbean atoll. Marine Ecology Progress Series 482: 169–179.

Townsend EC, Fonseca MS (1998) Bioturbation as a potential mechanism influencing spatial heterogeneity of North Carolina seagrass beds. Marine Ecology Progress Series 169: 123–132.

Tracey JA, Sheppard JK, Lockwood GK, Chourasia A, Tatineni M, Fisher RN, Sinkovits RS (2014b) Efficient 3D Movement-based Kernel Density Estimator and Application to Wildlife Ecology. Proceedings of the 2014 Annual Conference on Extreme Science and Engineering Discovery Environment. Article no. 14. Association for Computing Machinery, New York, NY.

Tracey JA, Sheppard J, Zhu J, Wei F, Swaisgood RR, Fisher RN (2014a) Movement-based estimation and visualization of space use in 3D for wildlife ecology and conservation. PloS One 9: e101205.

Turchin P (1998) Quantitative Analysis of Movement: Measuring and Modelling Population Redistribution in Animals and Plants. Sinauer Associates Publishers, Sunderland, MA, p. 396.

Turgeon K, Robillard A, Grégoire J, Duclos V, Kramer DL (2010) Functional connectivity from a reef fish perspective: behavioural tactics for moving in a fragmented landscape. Ecology 91(11): 3332–3342.

Urban DL, O'Neill RV, Shugart HH (1987) Landscape ecology. BioScience 37: 119–127.

Urban DL, Minor ES, Treml EA, Schick RS (2009) Graph models of habitat mosaics. Ecology Letters 12(3): 260–273.

Vergara PM, Saura S, Pérez-Hernández CG, Soto GE (2015) Hierarchical spatial decisions in fragmented landscapes: Modelling the foraging movements of woodpeckers. Ecological Modelling 300: 114–122.

Walsh WJ (1985) Reef fish community dynamics on small artificial reefs: the influence of isolation, habitat structure, and biogeography. Bulletin of Marine Science 36(2): 357–376.

Wedding LM, Lepczyk CA, Pittman SJ, Friedlander AM, Jorgensen S (2011) Quantifying seascape structure: extending terrestrial spatial pattern metrics to the marine realm. Marine Ecology Progress Series 427: 219–232.

Weeks R, Green AL, Joseph E, Peterson N, Terk E (2016) Using reef fish movement to inform marine reserve design. Journal of Applied Ecology 54: 145–152.

Wiens JA (1976) Population responses to patchy environments. Annual Review of Ecology and Systematics 7(1): 81–120.

Wiens JA (1989) Spatial scaling in ecology. Functional Ecology 3(4): 385–397.

Wiens JA (1997) Metapopulation dynamics and landscape ecology. In Hanski I & Gaggiotti O. Metapopulation Biology: Ecology, Genetics, and Evolution. Academic Press, San Diego, CA, pp. 43–62.

Wiens JA, Milne BT (1989) Scaling of 'landscapes' in landscape ecology, or, landscape ecology from a beetle's perspective. Landscape Ecology 3(2): 87–96.

Wiens JA, Stenseth NC, Van Horne B, Ims RA (1993) Ecological mechanisms and landscape ecology. Oikos 1: 369–380.

Wilson AD, Brownscombe JW, Krause J, Krause S, Gutowsky LF, Brooks EJ, Cooke SJ (2015) Integrating network analysis, sensor tags, and observation to understand shark ecology and behaviour. Behavioural Ecology 26(6): 1577–1586.

Winter S, Yin Z-C (2011) The elements of probabilistic time geography. Geoinformatica 15: 417–434.

Wirsing AJ, Heithaus MR, Frid A, Dill LM (2008) Seascapes of fear: evaluating sublethal predator effects experienced and generated by marine mammals. Marine Mammal Science 24(1): 1–5.

With KA (1994) Using fractal analysis to assess how species perceive landscape structure. Landscape Ecology 9: 25–36.

Wright S (1943) Isolation by distance. Genetics 28(2): 114–138.

Wu J, Loucks OL (1995) From balance of nature to hierarchical patch dynamics: a paradigm shift in ecology. Quarterly Review of Biology 1: 439–466.

Yeager LA, Acevedo CL Layman CA (2012) Effects of seascape context on condition, abundance, and secondary production of a coral reef fish, *Haemulon plumierii*. Marine Ecology Progress Series 462: 231–240.

Zeller KA, McGarigal K, Whiteley AR (2012) Estimating landscape resistance to movement: a review. Landscape Ecology 27(6): 777–797.

8

Using Individual-based Models to Explore Seascape Ecology
Kevin A. Hovel and Helen M. Regan

8.1   Introduction

Individual-based models (IBMs) simulate the behavior of individual organisms (such as movements) and they are used to determine how the adaptive traits of individuals dictate the emergent properties of populations, communities and ecosystems. These models can readily be applied to the study of ecological interactions across a range of spatial scales because the effects of habitat structure on organisms are ultimately manifested through the decisions that individuals make as they interact with their environment (Wiens 1989; Morris 2003). For this reason, IBMs are useful tools to use in conjunction with empirical experiments and observations when determining how seascape structure affects ecological processes. Our goals in this chapter are to introduce the reader to the basic concepts involved in IBMs, to discuss the advantages of using IBMs for studying the principles of seascape ecology and to demonstrate their utility (and limitations) for studying processes at a range of spatial scales. We provide examples of IBMs that have been successfully used to predict and simulate the effects of habitat structure on animal behavior and how this translates to broader temporal and spatial scales in marine systems. We conclude by advocating for greater use of IBMs in seascape ecology and marine spatial planning and by discussing the most promising applications for IBMs for studying seascape ecology.

8.1.1   What are IBMs?

Spatial ecological models recognize the role of habitat in the distribution and abundance of species. Indeed, the notion of distribution that underpins ecology is inherently spatial. Spatial models acknowledge that species rely on habitats, which are heterogeneous (Hastings 1990; Pickett & Cadenasso 1995; Tilman & Kareiva 1998). Heterogeneity in habitat attributes can have differential effects on an organism's survival, reproduction, behavior and interactions with conspecifics and other organisms. These not only determine the fate of individual organisms, but can have profound effects on populations, communities and ecosystems. Spatial processes have long been recognized as critically important in understanding ecological processes, giving rise to numerous ways of representing space and its influence on organisms, populations and communities (Perry & Bond 2005). These range from simple metapopulation models with patches

Seascape Ecology, First Edition. Edited by Simon J. Pittman.

that can become extinct and be recolonized by a single species, to landscape models that represent communities of interacting species and complex ecosystem processes. With greater emphasis on ecological synthesis and the increase in software sophistication and computing power, ecological models have increasingly and explicitly included spatial components (Petrovskii & Petrovskaya 2012). This trend has occurred alongside the recognition that space matters in practical conservation planning (Margules & Pressey 2000) as conservation science and management have shifted away from a single species to focus on spatial planning for multiple species, biodiversity and ecosystem function.

Spatial simulation models attempt to represent or mimic important underlying ecological processes and relationships in the system under investigation. While there are many different types of spatial simulation models used to answer a broad range of ecological questions, in this chapter we focus on spatial individual-based models (IBMs) in which organisms explicitly occupy a spatial location in each time step of the model (Soares-Filho *et al.* 2002). The location of an organism can be a set of coordinates in continuous space, or more often a cell within a gridded seascape. Each cell is assigned a set of attributes (Figure 8.1) and cells can interact within a specified neighborhood according to a set of rules (Durrett & Levin 1994). Cell attributes can include species- and individual-specific vital rates, such as births and deaths, habitat type and quality, and rules that define organismal behavior. For example, a cell may include features that provide shelter for an organism, thus lowering mortality via predation; could be part of a nursery habitat, increasing survival of juveniles; may contain unsuitable habitat, raising the probability of movement from the cell in the next time step; or, could provide suitable habitat for more than one species, thus invoking competition (Grimm & Railsback 2005). Furthermore, habitat attributes may change with time depending on ecosystem or physical processes or the biotic activities and interactions occurring

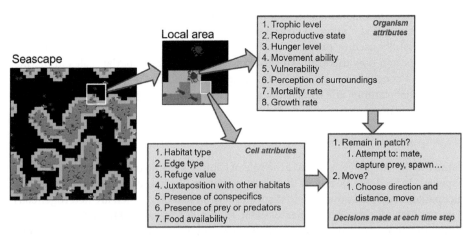

Figure 8.1 The basic components of a seascape ecology IBM, based loosely on the blue crab – seagrass model of Hovel & Regan (2008) created using NetLogo software. The seascape is composed of cells that represent seagrass habitat (gray: patch interior; blue: patch edge) and nonseagrass habitat (black). The seascape is populated by prey (blue 'bugs'), mesopredators (red crabs) and top predators (fish). Each cell is endowed with attributes that, in combination with attributes of individual organisms, influence decisions made by organisms in each model time step. It is possible to include many more attributes.

in the cell (*e.g.,* consumer-resource interactions) or in neighboring cells (*e.g.,* based on seascape composition and habitat configuration).

Individual-based models differ from statistical models that analyze observed correlations and ecological relationships or use likelihood techniques to fit models to data. They also differ from mechanistic models that describe ecological dynamics via differential equations, in that these types of models translate the system into a mathematical abstraction that is then solved using mathematical analytical techniques (DeAngelis & Grimm 2014). While these types of mathematical abstractions have proved enormously useful and enlightening in ecological theory they are limited in the number of processes and details that can be included. Conversely, IBMs acknowledge that individuals are the building blocks of ecological systems and they construct a system based on individual entities that interact with the landscape/seascape; hence, IBMs take a 'bottom-up' approach to modeling ecological processes and dynamics (Grimm & Railsback 2005). In the context of spatial ecological modeling, IBMs represent species and habitat attributes, ecological processes and behaviors, and project these through space and time with a set of rules or equations in each time step. Hence, ecological details that are individually and/or independently represented in the model can synthesize into emergent properties of the ecological system.

Individual-based models have been used to address ecological questions for around four decades. Early examples of IBMs largely focused on plant ecology, particularly forest succession (reviewed in Buggman 2001). Extensions to animal populations initially focused on fish populations, particularly the recruitment, survival and growth of young-of-the-year freshwater fish (DeAngelis & Mooij 2005). Examples of additional terrestrial applications of IBMs have included the effects of fire and disease on long-lived plants (Regan *et al.* 2011), fragmentation on animal movement (Bélisle & Desrochers 2002), fire and succession on ungulates (Turner *et al.* 1994) and hydrology and habitat change on the Cape Sable sea sparrow and snail kites (Mooij *et al.* 2002; Elderd & Nott 2008). Marine-based IBMs have investigated the effects of hydro- and tropho-dynamics on pelagic fish (Werner *et al.* 2001a; Parada *et al.* 2003) and, more recently, the effects of spatial heterogeneity on ecological processes at across the seascape (Werner *et al.* 2001b; Hovel & Regan 2008). In the next section we describe the advantages of using IBMs for studying seascape ecology and provide several examples of how this has been successfully accomplished for a variety of marine taxa and ecosystems.

8.2 Why use IBMs to Study Seascape Ecology?

8.2.1 The Effects of Habitat Structure on Populations are Consequences of Organismal Behavior

The primary advantage of using IBMs to study seascape ecology is that they allow tests of how linkages between habitat structure and organismal behavior influence populations and communities. IBMs take a 'bottom-up' approach to modeling the effects of the environment on organisms in that they explore how emergent properties of populations and communities are driven by the simulated behavior of individuals (Uchmanski & Grimm 1996; Grimm & Railsback 2005; Parrott *et al.* 2012). IBMs recognize that individuals are not identical to one another; each organism acts adaptively to maximize its fitness.

The adaptive nature of individuals depends on how they respond to their environment and, collectively, the decisions made by organisms in response to environmental stimuli result in patterns we see at higher levels of ecological organization: the growth or decline of populations, the spatial distribution of organisms and the structure of communities.

Individual-based models are well suited to the study of seascape ecology because habitat structure at a range of spatial scales can have profound effects on organismal behavior, which in turn can drive population dynamics and affect the strength of ecological interactions. Habitat structure influences a variety of processes that can affect an organism's fitness, including foraging efficiency, predator avoidance, competition and the ability to find mates. Habitat selection, the process whereby organisms continually select which parts of a habitat to occupy (Grimm & Railsback 2005), is one fundamental behavior that drives population dynamics and is strongly influenced by habitat structure at a variety of scales. Not surprisingly, this type of movement forms a basis for many IBMs examining effects of seascape structure on populations. The ability or willingness of motile organisms to move through the seascape may depend heavily on habitat cover, patch dispersion and the size of habitat patches (Schadt *et al.* 2002; Hitt *et al.* 2011; Claydon *et al.* 2015), as well as the quality of habitat within patches (which may affect resource availability; Haas *et al.* 2004). The strength of intraspecies and interspecies interactions depends on habitat selection. For instance, the ability of conspecifics to detect each other and their probability of encounter (which may influence the degree of competition and opportunities for reproduction) may depend on routes that organisms use to move across the seascape (Coates *et al.* 2013). Additionally, the ability of prey and predator species to detect each other and their responses after detection may depend on habitat features, such as patch size and patch connectivity (Hovel & Regan 2008). Organisms vulnerable to predators may avoid (or move from) small patches, patch edges, or seascapes with low connectivity if predators are more abundant or more efficient hunters in those locations. For some predators, patch edges may provide easier access to prey and opportunities to forage in different habitat types (Gates & Gysel 1978; Paton 1994).

Dispersal, a one-time movement to a new location, is another kind of movement that has been incorporated into IBMs operating at broad spatial scales. In marine systems, dispersal often takes place in the larval and postlarval stages of marine organisms. Seascape structure may influence dispersal by affecting the reproductive behavior of adults (*e.g.,* selection of spawning sites), by dictating patterns of larval dispersal, or by influencing settlement rates of larvae and postlarvae that are transitioning from the plankton to benthic habitats (Dytham 2003; Dytham & Simpson 2007; Kjelland *et al.* 2015). For species with complex life histories (*i.e.,* species that have a dispersive larval stage), IBMs offer the advantage of being able to track larvae and to test how different patterns of abiotic features of the environment and organismal behaviors, affect key ecological processes, such as recruitment, linkages between isolated populations, and gene flow (Daewel *et al.* 2015). Individual-based models that include components of seascape ecology can even be applied to sessile (nonmoving) species because most move (up to hundreds of kilometers) in their larval stages (*e.g.,* Munguia *et al.* 2011).

The field of terrestrial landscape ecology has a long history of using landscape metrics to explore and understand key features of the landscape that facilitate or inhibit movement including quantification of boundary permeability, habitat contiguity, habitat fragmentation and matrix habitat composition (Turner 1990; Cushman *et al.* 2008). Individual-based models can be employed to use this information to examine path tortuosity for individuals in both terrestrial and marine landscapes (Austin *et al.* 2004) to better inform spatial conservation planning. Though the effects of seascape structure on habitat selection and dispersal may be fundamental drivers of population dynamics (Butler *et al.* 2005), the effects of structure on these behaviors have been explored to a limited degree in real seascapes. This may be due to the difficulty of determining habitat preference and monitoring the movement of animals, particularly at broad spatial scales in the sea. Testing for habitat preferences requires manipulative experiments that control for confounding effects and that eliminate alternative explanations for why animals exhibit particular distributions or occupy particular habitats (Underwood *et al.* 2004). Testing how seascape features influence movement requires marking or identifying individuals in order to track them (*e.g.*, Lowe *et al.* 2003; Fodrie *et al.* 2015), which can be labor intensive and expensive, or requires recapturing marked individuals at a later time, which may be difficult (Pittman & McAlpine 2003). In contrast, testing how linkages between habitat structure and organismal behavior drive population dynamics may be much more tractable using IBMs, in which simulated organisms can be instilled with complex behaviors based on benefits and costs of making certain choices (Haas *et al.* 2004), or based on other aspects of their environment, such as the behavior of conspecifics, competitors, or predators (Grimm & Railsback 2005). If experimental data on behaviors are available, they can be used to increase the realism of IBMs by assigning probabilities of organisms making particular choices, or assigning consequences (*e.g.*, for survival or growth) for moving into and occupying different habitat types or seascape components. Individual-based models that incorporate realistic probabilities of particular behaviors can then be used to test how organisms respond to a variety of interacting factors, the influence of these responses at the population level and their output compared to available empirical data to assess realism (Dunstan & Johnson 2005; Claydon *et al.* 2015). This may be effective even when relatively little data from real systems are available (Kramer-Schadt *et al.* 2004).

A pair of studies conducted on abalone in California kelp forests (Coates *et al.* 2013; Coates & Hovel 2014) illustrates how IBMs can be parameterized with real data and can be used to predict the effects of behavior on ecological processes across the seascape. California abalone populations have been severely depleted due to overfishing and disease and low densities of remaining individuals may limit the reproductive success of these broadcast spawning species, unless a behavioral mechanism (*e.g.*, conspecific attraction) exists to bring them in close proximity when releasing gametes. In addition to conspecifics, seascape features of rocky reefs, including the distribution and juxta-position of rock and sand patches and the distribution of shelters (crevices in rocky habitat), may influence abalone movement and habitat use. In a field study conducted off the coast of San Diego, California, Coates *et al.* (2013) used acoustic telemetry to track pink abalone (*Haliotis corrugata*) moving through the kelp forest (*Macrocystis pyrifera*), with the goal of determining abalone home range and whether abalone main-tain spawning aggregations that enhance reproductive success (Figure 8.2). Abalone were placed into high density aggregations and then acoustically tracked for up to 1.5 years to determine whether individuals would move toward each other via conspe-cific attraction to maintain high densities necessary for suitable rates of fertilization for population growth and recovery. Though most individual abalone exhibited small home

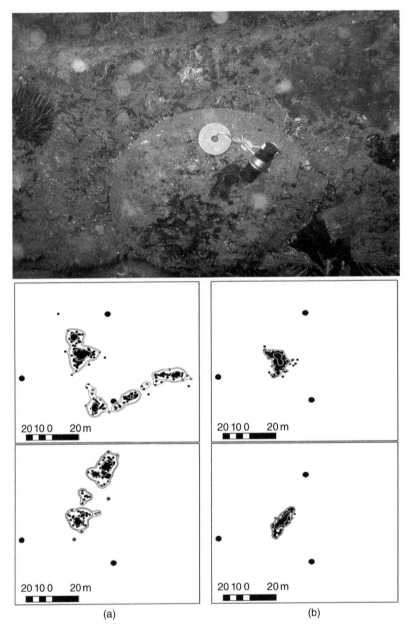

Figure 8.2 (a) An individual pink abalone (*Haliotis corrugata*) equipped with an acoustic transmitter that allows the position of the animal to be determined over time steps of minutes (photo credit: Julia Coates). (b) Home ranges of four example abalone tracked in Coates *et al*. (2013). Dots represent discrete positions in the seascape and gray lines represent statistically determined home ranges. Large dots arranged in a triangle represent the position of stationary acoustic transmitters used to triangulate the position of each animal at each time point.

ranges and repeated movements to a home shelter, high density aggregations were not maintained. Data on movement behavior in relation to conspecifics and seascape features, as well as data on population density and abalone fertilization success, were subsequently used to parameterize an IBM that modeled the effects of population density, movement behavior and spawning behavior (synchronous or asynchronous) on reproductive success in simulated seascapes (Coates & Hovel 2014). Tracking data and measurements of abalone home ranges were used to create habitat selection rules for individual abalone based on several variables: attraction to and position of conspecifics, remaining in suitable habitat, readiness to spawn, and maintaining a small home range centered on a home shelter (Figure 8.3). The effectiveness of these rules in simulating abalone movement was validated by comparing movement produced by these rules with movement behaviors observed for real abalone in the field study. The study found that home ranges (ranging from 84–1006 m^2) and movement via conspecific attraction could overcome asynchronous spawning by individuals (asynchronous spawning is more likely at low population densities where chemical signaling among individuals is reduced) to result in greater fertilization success. Unfortunately, fertilization rates under the best scenarios were not high enough to lead to the conclusion that movement is likely to mitigate for the negative effects of low population densities on abalone reproductive success and population recovery.

8.2.2 IBMs Allow for Extensive Manipulation of Seascapes

Another advantage of using IBMs to study seascape ecology is that in a simulation, seascapes can be manipulated to a degree that is impossible or unethical for many naturally occurring systems. Though marine field experiments have revealed a great deal about the effects of habitat spatial structure on populations and communities, they are generally limited to relatively fine spatial and temporal scales and low levels of replication. Many field-based studies on seascape ecology actually take place at the patch scale (within a single patch or patch type), due to the difficulty of replicating experiments or surveys at broader scales (Fahrig 2003). Moreover, most field-based studies on the effects of seascape structure utilize existing, prefragmented habitats as a proxy for habitat loss and fragmentation, because creating experimental treatments at broader scales that are ecologically more meaningful may simply not be possible or would require an unacceptable level of habitat destruction. Using existing, naturally occurring seascapes to test hypotheses, however, ignores the environmental factors that covary with or cause patterns of patchiness. For instance, shallow seagrass habitats, in which much of the empirical work on seascape structure has been done, are fragmented by waves and currents. Patchy areas of seagrass seascapes therefore often experience different levels of hydrodynamic activity, as well as sediment characteristics, compared to continuous areas (Fonseca & Bell 1998). Moreover, because vegetated habitats consist of living organisms that are affected by their environment, seagrasses and other biogenic habitats exhibit a high degree of variability in structural complexity (*e.g.*, density, length, or biomass of shoots, or other structural elements) within and among patches (Irlandi 1996; Hovel & Lipcius 2002; Boström *et al.* 2011). Hydrodynamics, structural complexity and other environmental features that covary with seascape structure may have strong effects on behavior, survival and growth of organisms (Irlandi 1996; Hovel *et al.* 2002; Peterson *et al.* 2004) and these covarying factors therefore complicate the interpretation of experiments designed to test for effects of seascape structure on organisms

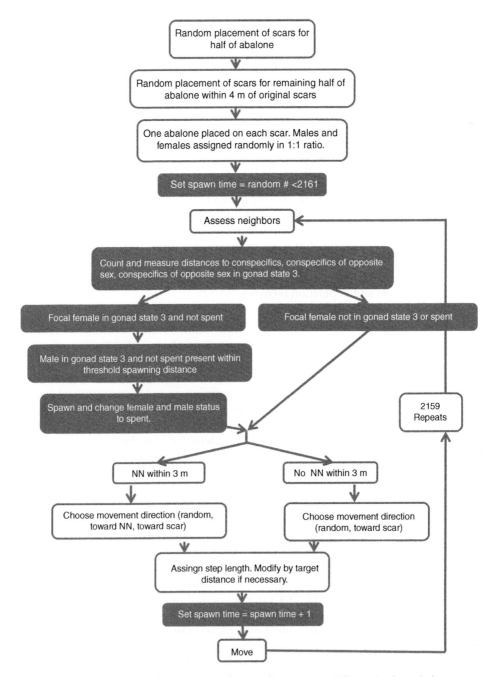

Figure 8.3 Model procedures for the pink abalone (*Haliotis corrugata*) IBM testing how abalone movement influences spawning success (Coates and Hovel 2014). Shelters ('scars') and abalone are placed in the seascape and abalone make movement decisions at each time step based on an assessment of their surroundings and readiness to spawn. NN = nearest neighbor.

that use these habitats. Models have the obvious advantages of allowing the elimination of confounding among environmental factors and allowing replication across space and time at any scale. Covarying factors can be ignored, or (perhaps more realistically) they can be varied systematically along with seascape structure to test for their relative and interactive effects.

An IBM produced by Hovel & Regan (2008) illustrates how IBMs can be used to control for covariation and confounding effects in seascape studies. The underlying goal of this model was to test how seagrass seascape structure interacts with predator and prey behavior to dictate prey population size. An important ecosystem function of seagrasses is to provide nursery habitat to juvenile fishes and invertebrates (*e.g.*, crabs, shrimp, lobsters, snails and bivalves; Williams & Heck 2001). The complex structure of nursery habitats provides vulnerable juveniles with enhanced refuge from predation and opportunities for foraging. However, assessments of the effects of habitat structure on predation risk for small organisms are difficult to produce and are often limited to quantifying relative effects of different levels of habitat structure on predator-prey interactions using techniques like tethering (*e.g.*, Heck & Thoman 1981; Pile *et al.* 1996; Eggleston *et al.* 2005; Hovel & Fonseca 2005). In this widely used technique, prey were fixed to short lengths of monofilament line, nylon line, or other strong material and attached to the substratum so that the organism can later be relocated to determine if the organism has been consumed by a predator. Tethering restricts all but very fine-scale prey movement and behavior and therefore inflates estimates of prey mortality by limiting the ability of prey to attempt to escape from predator attacks (Zimmer-Faust *et al.* 1994). Tethering therefore cannot estimate true organismal mortality rates, but is very useful for establishing relative mortality rates among experimental treatments if it can be established that there is no treatment-specific bias due to the technique (Peterson & Black 1994). Across the seascape, this means that the restriction of prey due to tethering must not inflate mortality rates more steeply in a smaller patch than in a larger patch. However, prey organisms generally are reluctant to move out of seagrass patches into unvegetated areas that offer little refuge, creating the plausible scenario that prey escape attempts from predators may be more successful in larger patches, which permit more extensive movement away from predators, than in smaller patches where escape distances may be limited. Speaking more generally, it is possible, if not expected, that seascape structure influences the behavior of prey being hunted by predators, which may not be fully incorporated in tethering experiments.

The major objectives in Hovel & Regan's (2008) study were to: (i) create simulated seascapes that closely resemble real seagrass seascapes, but to remove confounding factors due to hydrodynamics and structural complexity; (ii) test whether effects of seagrass patchiness on prey survival differed between prey allowed to move versus prey tethered in place; and (iii) test how different movement strategies for predators and prey (random movement versus directed hunting) and different settlement patterns for prey influenced the effects of seagrass patchiness on prey survival. The model, written in Net-Logo (Wilenski 1999), measured the survival rates for juvenile blue crabs (*Callinectes sapidus*), a well-studied and economically important fishery species in the southeastern US. Because abundance, distribution and relative mortality rates of juvenile and adult blue crabs in seagrass habitat have all been quantified in previous studies, the IBM could be parameterized with reasonable densities and habitat preferences for prey (juvenile blue crabs) and their predators (larger blue crabs; blue crabs are highly cannibalistic).

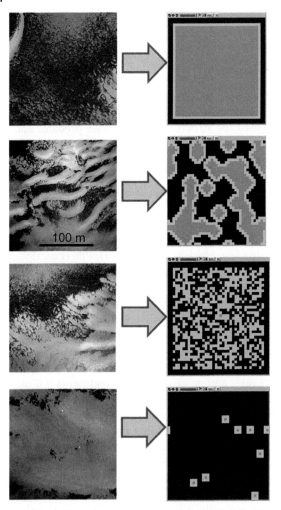

Figure 8.4 Left: aerial photographs of sections of an eelgrass (*Zostera marina*) seascape in the lower Chesapeake Bay, Virginia, United States, illustrating different patchiness regimes. From Hovel & Lipcius (2001). Right: simulated eelgrass seascapes created in NetLogo. Colored areas represent seagrass (gray: patch interior; blue: patch edge) and black represents unvegetated sediment. *Source:* From Hovel & Regan (2008).

Using aerial photographs of naturally occurring seagrass seascapes in the lower Chesapeake Bay, VA, four simulated seascapes were created that varied both in the degree of patchiness and amount of seagrass present (Figure 8.4). The 'continuous seagrass' seascape type was a large expanse (30 000 m²) of unfragmented seagrass; the 'large patch' seascape type represented large patches (thousands of square meters) of seagrass created from current scouring and isolated by at least several meters of unvegetated sediment; the 'small patch' seascape type represented small (tens of square meters) but highly connected patches of seagrass, which may be created by waves and currents, digging predators, propeller scarring and fishing practices; and the 'very small patch' seascape type represented small (ca. <1 m²) seagrass patches isolated by large distances of unvegetated sediment that are often found in areas of high hydrodynamic activity (Fonseca & Bell 1998). By modeling simulated seascapes based on these naturally occurring patterns of seagrass patchiness, we were able to represent increasing levels of habitat fragmentation and habitat loss in a landscape of constant size, but also to standardize for the confounding factors of hydrodynamic activity and seagrass structural complexity.

After structuring the seascape patterns, we populated each seascape type with three trophic levels: juvenile blue crabs, larger blue crabs that consume juvenile blue crabs and top-level predators (large fishes). Juvenile blue crabs 'settled' into seagrass habitat using one of three routines: along patch edges, within patch interiors, or randomly in seagrass. We compared survival rates for juvenile blue crabs tethered in place versus juvenile blue crabs allowed to move throughout the seascape within each of the four seascape types. Each organism type moved throughout the seascape based on hierarchical sets of rules; juvenile blue crabs prioritized minimizing encounter rates with larger blue crabs, larger blue crabs prioritized minimizing encounter rates with fishes, and fishes either hunted via random walk or were given the ability to detect and move toward their prey. We compared output between models in which larger blue crabs hunted randomly versus when they could detect juvenile blue crabs and move toward them. Juvenile blue crabs were forced to remain in seagrass habitat, whereas larger blue crabs were given a low probability of being outside of seagrass habitat, which allowed them to more rapidly move through the seascape, but came with the cost of increased vulnerability to fish predators. Fishes were assumed invulnerable to predation and were allowed to rapidly move anywhere in the seascape.

The primary finding from our study was that the effects of seascape structure on juvenile blue crab survival depended on their movement ability. Tethered juvenile blue crabs exhibited higher survival in the most fragmented seascapes (very small patches $< 1\,\mathrm{m}^2$) and lowest survival in continuous seagrass. This counterintuitive pattern matched results from field tethering trials with juvenile blue crabs and likely arose from the reluctance of larger blue crabs (predators of juveniles) to forage in highly fragmented seascapes, due to high predation risk from top predators. In contrast, mobile juvenile blue crabs exhibited low survival in very small patches and high survival in continuous seagrass, because they were able to effectively escape approaching large blue crabs by swimming away into seagrass habitat, an option not available to mobile crabs in small patches. Other aspects of prey and predator behavior also played into patterns of juvenile blue crab survival: larger blue crabs exhibiting directed hunting resulted in reduced juvenile blue crab survival across all seascape types. In contrast, crab settlement behavior had little effect on model outcomes. To add realism to our model, we have now added variation in structural complexity (*e.g.*, seagrass shoot density) among seascape types, as well as growth functions for juvenile organisms and are using this expanded version of the IBM to test how these additional factors help drive relationships between habitat structure and nursery habitat function. We are parameterizing this model using survey data on patterns of structural complexity in seagrass seascapes, which we collected over multiple seasons using sonar and SCUBA. Functions for the effects of structural complexity on foraging behavior and habitat preferences in these simulations are being parameterized using the results of laboratory experiments assessing foraging efficiency and habitat preferences for mesopredatory fish (Lannin & Hovel 2011; Tait & Hovel 2012), as well as growth rates from field caging experiments.

8.2.3 IBMs can be Used to Test for Ecological Effects of Habitat Configuration versus Habitat Amount

Another aspect of covariation in seascape structure exists between habitat amount and habitat configuration (Fahrig 2003). In her review of the effects of habitat loss on biodiversity, Fahrig (2003) distinguished between habitat amount and habitat fragmentation *per se*, which is the breaking apart of habitat into small, isolated patches. As habitat is lost, the configuration of remaining habitat pieces often changes as well and seascapes with low cover typically consist of isolated patches and small patches with a high proportion of edge habitat (Andrén 1994). Both habitat amount and habitat configuration can influence population dynamics (Estavillo *et al.* 2013), predator-prey interactions (Paton 1994) and biodiversity (Fahrig 2003). Testing for relative effects of habitat amount versus habitat configuration is difficult in real seascapes, where the two factors tend to go hand in hand. Collectively, the relatively small number of field studies that have focused on this issue suggest that habitat amount has stronger effects on populations and communities than does habitat configuration (Fahrig 2003). However, it is likely that these two aspects of structural change act synergistically to dictate ecological patterns at low levels of habitat cover (Andrén 1994; Estavillo *et al.* 2013). This interaction may be a primary driver of threshold effects in landscapes with high amounts of habitat loss (Fahrig 2002). Threshold effects constitute 'tipping points' beyond which further loss of habitat rapidly decreases survival, abundance, or diversity. From a conservation standpoint, potential thresholds are important to identify because they influence whether alteration of habitat configuration (and how much) can be used to rehabilitate populations and communities (Fahrig 2002).

Individual-based models are effective instruments for testing for relative effects of habitat amount and habitat configuration on ecological processes because any level of covariation in habitat amount and configuration is possible to simulate and this can be done in the absence of the processes (whether natural or human induced) that create seascape patterns. An example of this comes from a study on brown shrimp (*Farfantepenaeus aztecus*) survival in salt marsh habitat (Haas *et al.* 2004). Salt marsh habitats vary in configuration, particularly in the amount of edge habitat, which in field experiments has been shown to influence brown shrimp abundance and survival (Minello *et al.* 1994). Juvenile brown shrimp that use salt marshes as nursery habitat remain close to the interface between marsh vegetation and open water, where they find suitable foraging opportunities. As marshes break apart and begin to be degraded (*e.g.,* via coastal development) they acquire more edge habitat before losing much habitat cover, but the amount of edge in the seascape then drops with high levels of habitat loss (Browder *et al.* 1985). Haas *et al.* (2004) created four simulated seascapes (referred to as a marshscape) based on aerial imagery from coastal Louisiana, with the four marshscapes representing four steps in the process of marsh disintegration (Figure 8.5). Because the amount of edge habitat increases and then decreases as marsh disintegration progresses, the simulations can be considered a 2 × 2 factorial design with the amount of edge as one factor and overall habitat amount as the other factor. In the model, juvenile brown shrimp settled into the marsh as postlarvae and were then tracked until they either died or reached a size at which they no longer required nursery habitat and left the marsh for other habitats. Shrimp selected habitat at each time step and either moved to, or remained in, habitats that promoted growth and survival. Shrimp size and habitat type influenced rates of movement as well as the probability of predator-induced mortality. Using this IBM, Haas *et al.* (2004) found that shrimp survival depended more on habitat configuration than on habitat amount: marshscapes with more edge habitat resulted in higher shrimp survival, whereas marshes with high amounts of habitat had only 1–2% more survival than marshes with low habitat amount. Marshes with higher amounts of edge permitted juvenile brown shrimp to remain in high quality edge habitat, reduced their

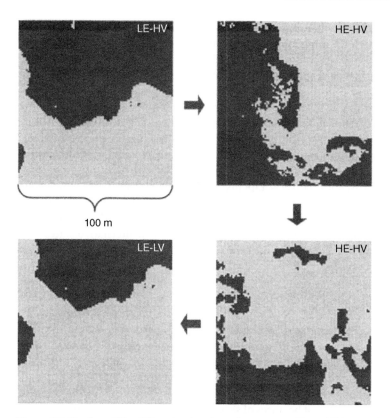

Figure 8.5 The four 100 × 100 m simulated marshscapes used in an IBM to test how marsh seascape configuration influences the survival and growth of the brown shrimp *Farfantepenaeus aztecus*. Gray is water and black represents vegetation. The four maps represent snapshots in a simplified continuum of marsh disintegration and different combinations of habitat configuration and cover: LE-HV = little edge and high amount of vegetation; HE-HV = high edge and high amount of vegetation; HE-LV = high edge and low amount of vegetation; LE-LV = little edge and low amount of vegetation. *Source:* From Haas *et al.* (2004).

need to move at each time step and reduced their local densities, all of which increased shrimp survival and density-dependent growth. Unlike the findings of Hovel & Regan's (2008) study on blue crabs in seagrass habitat, altering movement rules (*e.g.*, random selection of destination cells versus selection based on perceived value of cells) did not change the relationship between marshscape configuration and survival. Random movement did, however, reduce overall brown shrimp survival. The importance of marsh habitat configuration for brown shrimp production was corroborated in a subsequent study that also employed an IBM (Roth *et al.* 2008). The authors found that seawater inundation had stronger effects on export of shrimp from salt marshes than did habitat fragmentation, but that marsh habitat configuration had strong influences on production within a single inundation regime.

8.2.4 IBMs Allow Tests of How Seascape Change Influences Ecological Processes

Modeling also makes it much easier to test how *variation* in seascape structure influences individuals, populations and communities. Real seascapes are not static, but vary in spatial structure as environmental forces shape the distribution and configuration of habitat-forming species. Such changes are accommodated by the flexibility of IBMs (Butler *et al.* 2001). Seagrasses and other vegetated habitats (*e.g.*, kelp forests) are good examples of seascapes that can rapidly change and radically differ in structure over short time scales (see Chapter 6 in this book). For instance, in Chesapeake Bay, seascapes composed of eelgrass differ in structure from early to late summer as seasonally warming water temperatures defoliate eelgrass shoots and benthic predators disturb eelgrass via bioturbation (Hovel & Lipcius 2002). Structural differences include a loss of some patches, fragmentation of remaining patches and reduced structural complexity within patches. These changes profoundly influenced the vulnerability of juvenile blue crabs to predators in these seascapes (Hovel & Lipcius 2001, 2002). However, this conclusion is tempered by the fact that these studies were a snapshot in time and did not test for changes to seascape structure over multiple years. In IBMs, simulated seascapes can change in a multitude of ways and cycle back and forth between alternative states and encompass long time scales that may be necessary for testing how changes to seascapes influence ecological processes.

Butler and colleagues (Butler *et al.* 2001, 2005) have created multiple versions of a spatially explicit IBM to study how changes to benthic seascapes in the Florida Keys affect Caribbean spiny lobster (*Panulirus argus*) population dynamics and fishery yield. Caribbean spiny lobsters are an abundant resident of shallow hard bottom, reef and seagrass habitats in the Caribbean, and they support Florida's most valuable fishery. Their studies were prompted by large changes to the Florida Keys seascape, which forms a nursery habitat for *P. argus*. In this region, blooms of cyanobacteria in 1991–1992 caused a large-scale die off of sponges, which form the primary shelter-providing habitat for juvenile lobsters, and thousands of hectares of seagrass (which also provide habitat) have been lost due to declining water quality. Despite these cascading disturbances to benthic seascapes, the population of spiny lobsters in Florida *increased* during this period (Butler *et al.* 2005). Butler *et al.* (2001, 2005) created a spatially explicit IBM to evaluate several hypotheses that may explain the resilience of spiny lobsters to these perturbations. The model simulated a large seascape (a stretch of the Florida Keys region about 200 km long; Figure 8.6) and involved monthly settlement of lobster postlarvae to this seascape, which was composed of discrete cells (1000 m^2 in size). Each cell was designated as one type of habitat (seagrass or hard bottom), based on field surveys and GIS habitat data acquired in earlier studies. Hard bottom cells were given a particular carrying capacity for lobsters based on the dominant type of shelter providing habitat, including sponges, solution holes and corals. Lobsters periodically settled into the seascape and then exhibited growth, mortality and habitat selection based on functions derived from empirically derived data (*e.g.*, laboratory growth experiments, tethering experiments on relative mortality and field- and lab-based habitat selection experiments). Movement among habitat types and seascape cells was a key function in the model. The probability that lobsters would move from different types of shelter and between adjacent cells in the seascape was a function of lobster size and the availability of shelter in their current cell. This function was parameterized using field mark-recapture data for over 500 individual lobsters.

After creating their baseline IBM, Butler *et al.* (2005) ran simulations to predict the potential impact of cyanobacterial blooms on spiny lobster recruitment. This was done by comparing model runs without cyanobacterial blooms to model runs

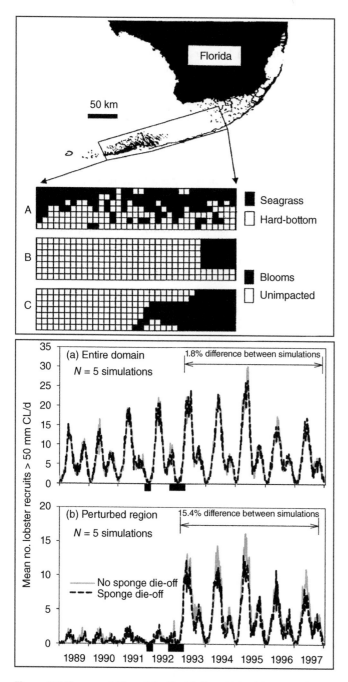

Figure 8.6 Top panel: Map of the Florida Keys, United States and (A) the simulated seascape composed of seagrass and hard bottom habitat used by Caribbean spiny lobsters (*Panulirus argus*). (B) and (C) represent two simulated cyanobacterial blooms that caused the die off of habitat-forming sponges. Bottom panel: results from the Caribbean spiny lobster IBM comparing lobster abundance between seascapes affected by, or not affected by sponge die off. The percentage difference in postbloom lobster abundance between seascapes affected by blooms or not affected by blooms is shown. The two black bars on the *x*-axes represent the periods when the algal blooms occurred. Upper panel represents lobster abundance in the entire Florida Keys region and lower panel represents lobster abundance only in the areas of cyanobacterial blooms. *Source:* From Butler *et al*. (2005).

that involved a daily reduction in sponge carrying capacity. Ten-year simulations revealed only localized effects on lobsters; effects of reductions in sponge habitat due to cyanobacterial blooms had little effect on lobster recruitment at the scale of the Florida Keys (Figure 8.6). The model revealed that resilience of lobsters to this major change to the seascape arose from higher than average postlarval supply that occurred in the years before the blooms, but also from increased movement of lobsters out of the impacted cells and subsequent use of alternative habitat types. Their model results were corroborated by comparing them to the results of field surveys for lobsters that took place during and after the blooms. These independent data were not used to formulate the model, but showed a remarkable correspondence to lobster behavior from simulations: real lobsters reduced their use of sponges and increased their use of alternative habitat types following cyanobacterial blooms. Habitat switching resulted in resiliency of the lobster population even though the model incorporated a cost (albeit a relatively small one) in terms of increased risk of mortality for moving outside of shelter and searching for alternative shelter.

Another illustration of the ability of IBMs to incorporate and test for effects of seascape change on organisms comes from benthic habitats of Long Island Sound (Munguia *et al.* 2011). The goal of this study was to use simulation modeling to explore how changes to coastal seascapes in eastern Long Island Sound influence the distributions of common benthic species with contrasting dispersal strategies. Similar to many other coastlines, seascapes in this heavily populated area are rapidly changing as humans homogenize benthic habitat by installing man-made, hard-substrate structures (*e.g.*, rip-rap that slows coastal erosion, but provides additional, homogeneous habitat). To test how varying life history traits (*i.e.*, different larval durations) interact with seascape structure, Munguia *et al.* (2011) coupled a hydrodynamic model of Long Island Sound with an IBM that provided the ability to manipulate life history traits. The hydrodynamic model dictated patterns of larval distribution for a pool of propagules available to settle into simulated benthic seascapes. The IBM simulated the distribution and abundance of four species that can dominate benthic habitat in this system and that represent alternative stable states: blue mussels (*Mytilus edulis*) with long-lived larvae (~2 weeks), and three species with very short larval durations of 1 d or less: the native bryozoan *Schizoporella errata*, the native solitary ascidian *Styleta clava* and the invasive colonial ascidian *Diplosoma listerianum*. To vary seascape structure, the authors contrasted heterogeneous seascapes (patches of habitat, such as seagrass and hard substrate, mixed with patches of uninhabitable soft substrate) with homogeneous seascapes, in which the coastline consists only of hard substrate simulating the prevalence of man-made structures along the entire coastline. The model tracked individual larvae as they dispersed in the water column and then settled into suitable habitat. Each time step in the model determined the survival, direction and distance traveled by larvae and the survival and reproduction of adults (Munguia *et al.* 2011).

The IBM revealed an interactive effect of life history traits and seascape structure on organismal density in Long Island Sound, resulting in a switch from domination by native species to domination by invasive species, as has been observed for rocky habitat in New England. Heterogeneous (naturally occurring) seascapes were dominated by native blue mussels, the species with the greatest dispersal ability, which allowed colonization of distant patches of suitable substrate. In contrast, the solitary ascidian

S. clava did poorly in homogeneous habitats due to intermediate dispersal ability and poor competitive ability compared to invasive colonial ascidians, such as *D. listerianum*. Under the scenario in which humans altered the coastal zone by creating homogeneous hard substrate, space was dominated by the invasive *D. listerianum*, which broods larvae giving them the ability to settle immediately after release and which outcompetes other species for space. Given that coastlines in this area are being developed at a rate of 16–58% per year, these results have implications for the fate of native versus non-native species, as well as biodiversity, as mussels constitute a foundation species that promotes species richness on hard substrate.

8.2.5 IBMs Allow the Coupling of Processes Operating over Different Scales

Individual-based models can easily incorporate processes that occur across a range of spatial scales. Though patterns of habitat distribution across seascapes may vary widely within a habitat or region and have large effects on organisms, these patterns and processes do not occur in isolation from those in other habitats or at other scales. The IBMs of Munguia *et al.* (2011) and Butler *et al.* (2005) described above illustrate how simulations of seascape structure can be coupled with simulations of larval dispersal and settlement (see also Kjelland *et al.* 2015). In these coupled models, processes that operate at regional scales (larval transport) dictate the input of individuals whose fates subsequently are modeled across benthic seascapes. Whereas Munguia *et al.* (2011) tracked individual larvae dispersing through Long Island Sound using an IBM, other studies (*e.g.*, Kjelland *et al.* 2015) used hydrodynamic models to estimate the overall transport success of larvae to suitable habitat, which thereafter are modeled as individual juveniles and adults. Using this approach, Kjelland *et al.* (2015) found that the long-term persistence of market oysters (the eastern oyster *Crassostrea virginica*) and the fishery they support in Chesapeake Bay depended strongly on the spatial position of isolated oyster reefs at the seascape scale, which influenced recruitment and connectivity among reefs. Kjelland *et al.* (2015) further coupled models incorporating oyster life history with models of harvesting pressure and harvest regime to suggest which scenarios of oyster exploitation resulted in long-term persistence of the fishery. Population and fishery persistence was only likely under a management scenario that incorporated selective, rotational harvests on individual reefs.

Most IBMs are structured within a two-dimensional horizontal landscape. However, for seascapes consideration of depth can be just as important for understanding ecological processes as is horizontal space. Few applications of IBMs in ecology include depth, however competition for light has been represented as a three dimensional process in IBMs of forest and savanna dynamics (Simioni *et al.* 2000; Bugmann 2001). While rare, examples of three-dimensional marine IBMs do exist in the literature; Hermann *et al.* (2001) linked a three-dimensional hydrodynamic model with an IBM to investigate the effects of spatial heterogeneity of environmental variables such as temperature, irradiance, salinity, turbulence, mixed layer depth, currents and plankton dynamics on walleye pollock individuals. The considerable literature on three-dimensional IBMs in other fields such as engineering, medicine, computing and environmental and social sciences indicates broad scope for application of three dimensions in IBMs for marine settings.

Individual-based models not only allow organisms to be studied over a variety of spatial scales, but over a variety of temporal scales as well. IBMs work by manipulating (and measuring the status of) individuals at each time step in a simulation. Time steps used in IBMs can vary from seconds to much longer periods, but typically are short because IBMs often explicitly simulate organismal behavior. However, IBMs can extend simulations much farther in time than normally is possible in field or laboratory-based studies (*e.g.*, the 10-year simulations of spiny lobster abundance created by Butler *et al.* 2005). The IBMs of Dytham (2003) and Dytham & Simpson (2007) are good examples of how models incorporating dispersal and seascape structure can be used to study long-term processes. Their models of coral reef fish populations involve linkages between dispersive larval stages and juvenile and adult stages that inhabit coral reef seascapes. However, they also link processes occurring over the short term (time steps of 10 minutes) with evolutionary processes operating across generations. Dytham & Simpson (2007) tested how mortality rates for fish larvae interacted with coral reef seascape structure to dictate the evolution of fish dispersal strategies. Across four seascape types, which represented different levels of reef connectivity, they varied mortality rates for larvae that remained on coral reefs (as opposed to dispersing off reefs) and monitored overall population sizes and the evolution of dispersal strategies over 180 generations. Evolved dispersal strategies were highly seascape specific; contiguous reef arrangements generated a strategy of low dispersal, even when on-reef mortality was high, because fish could disperse off reefs to avoid some predation but then settle locally. In contrast, seascapes consisting of more isolated reefs selected for high rates of larval movement to enable dispersal among reefs. Extreme isolation of reefs resulted in population extinctions when larval mortality levels on reefs were high because reefs were too far apart to sustain dispersal among reefs. Dytham and Simpson (2007) noted that their random walk-based modeling of dispersal discounted more complex behaviors that larvae can use for orientation and homing and that future studies would benefit from more knowledge of the sensory landscape that larval stages are exposed to when undergoing dispersal.

8.3   Data for Parameterizing Seascape Ecology IBMs

8.3.1   Parameterization

Any IBM, even the simplest one, includes at least several parameters that must be defined in some way for the IBM to function. Parameters may define the behaviors exhibited by animals (*e.g.*, how far an animal moves in a given time step, their habitat preference, or their ability to detect other animals), the size of populations of organisms interacting in the model, rates of settlement and recruitment, growth rates, or the composition and configuration of the seascape (*e.g.*, the relative value of different habitats, habitat cover, or the size and spacing of patches), just to name a few examples. In most IBMs, some parameters are well known and may be easy to assign values to, whereas others are not. This leads to questions such as: how many parameters must be well defined in a model for the IBM to be useful? How important is it to have empirical data that can be used to define model parameters? And, is it OK to have parameters that are simply guesses?

Though the answers to these questions depend to some degree on the purpose of the IBM in question, it is encouraging to know that IBMs can be useful even when relatively few parameters are well defined (Kramer-Schadt *et al.* 2004) and educated guesses at parameter values are not necessarily problematic. The Caribbean spiny lobster models of Butler *et al.* (2001, 2005), described above, are examples of well-parameterized IBMs that involve a variety of submodels in which lobsters settle to habitat, select habitat and move among model cells in the seascape. The authors were able to parameterize these behaviors and other factors, such as lobster mortality rates, by using previously collected movement and survival data on this extensively studied, economically valuable species. Nonetheless, some behaviors were parameterized based on intuition. For those factors for which few data are initially available to parameterize an IBM, parameters may be assigned a range of plausible values that are then narrowed down after matching ecological patterns produced by the model to patterns from the real world ('inverse modeling'; Grimm & Railsback 2005). For instance, Butler *et al.* (2005) ran simulations with lobster movement rates that varied over two orders of magnitude and narrowed movement rates to those that produced behaviors similar to those observed for real lobsters. Likewise, sensitivity analyses can determine which parameters do not have large effects on model outcomes, or can be used to arrive at appropriate parameter ranges. A full description of these techniques is beyond the scope of this chapter, but can be found in Grimm & Railsback (2005).

8.3.2   Movement and Habitat Selection

Models addressing seascape ecology have a variety of goals, but many incorporate habitat selection by organisms. Animals may move through different habitat types and through and around habitat patches. Movement can also include the dispersion and transport of larvae and settlement of larvae to habitat. Thus, broad-scale and fine-scale movement in relation to habitat type and other environmental factors (*e.g.*, predator presence, currents, shelter, or food levels) are typically integral components of IBMs and data on these behaviors can be very useful when constructing an IBM. For some organisms, moment-to-moment and long-term movement behaviors may be measurable using technological advances, such as acoustic tags, which permit active or passive tracking of animals underwater, or satellite tags that transmit animal positions when animals are at the water's surface (Pittman & McAlpine 2003). This technology has been applied to organisms such as fishes (*e.g.*, Lowe *et al.* 2003), lobsters (Butler *et al.* 2005; Withy-Allen & Hovel 2013), seabirds (Louzao *et al.* 2011) and turtles (Makowski *et al.* 2006). The pink abalone IBM described above (Coates & Hovel 2014) involved submodels for abalone movement that were parameterized using data from dozens of acoustically tracked abalone (Coates *et al.* 2013).

In many instances, few data will be available that can be used to structure movement routines in an IBM, or to parameterize movement behaviors. Field measurements of animal movement are rarely available for small, cryptic organisms residing in complex habitat, or for larvae. Thus, fine-scale movement may be the most speculative part of many IBMs (Haas *et al.* 2004). One strategy for constructing movement routines for marine organisms is to use results from field or laboratory studies on habitat preferences. If done correctly (see cautions and guidelines expressed by Underwood *et al.* 2004), these data may be used to parameterize habitat selection behavior in IBMs, based on assumptions that organisms will move toward or tend to remain in preferred habitat types. For instance, Claydon *et al.* (2015) parameterized habitat selection rules for their IBM on rainbow parrotfish (*Scarus guacamaia*) distribution by using data on rainbow parrotfish abundance and habitat use from extensive (over 65 linear km) snorkeling surveys in coral and mangrove habitat. We also advocate that researchers not only survey animals for habitat associations and derive experimental results for habitat selection, but also take opportunities to *observe* their study species when conducting experiments or surveys. Our models involving organismal movement in seagrass habitat have been greatly informed by observing real organisms as we conducted habitat selection experiments. For instance, in our habitat selection experiments with juvenile giant kelpfish (*Heterostichus rostratus*), a small fish that resides in seagrass habitat, fish spent the vast majority of their time remaining cryptic in dense, highly structured seagrass habitat (relative to low density habitat), but occasionally made quick hunting forays into low density habitat where foraging on invertebrate prey was easier (Tait & Hovel 2012). Only by watching fish in these experiments did we realize that fish use both habitat types, but for different purposes and for different durations.

8.3.3   Seascape and Habitat Structure

With the development of GIS and remote sensing, relevant data with which to structure simulated seascapes is now widely available, even for underwater habitats. The distribution, extent and persistence of coastal habitats (*e.g.*, seagrasses, kelp forests, coral reefs and salt marshes) are now routinely measured using sonar, satellite imagery, laser altimetry and aerial photography (see Chapter 2 in this book). These data often are publicly available. IBMs can incorporate actual habitat maps derived from these surveys (Breckling *et al.* 2005; Brown *et al.* 2005), or can involve simulated seascapes that are created in the program itself to match habitat spatial characteristics, such as mean patch size, nearest neighbor distances and juxtaposition of habitat types. Of course, a great advantage of using IBMs for seascape ecology is the ability to test how organisms respond to aspects of seascape structure that do not currently exist, but might exist due to habitat loss or fragmentation, climate change, or species introductions.

One consideration when using real spatial data is resolution: what is the size of the smallest unit of measurement and how does this correspond to the behavior of organisms that are interacting in the seascape? Spatial data may be collected over a great range of resolutions, from less than a square meter to square kilometers, depending on the habitat type, instrument type and budget. Resolution defines the minimum area over which habitat is defined ('grain'), which may correspond to the size of a habitat cell in an IBM. Cell size should be carefully considered in any IBM. If organisms make decisions based on characteristics of each cell they encounter, which is the basis of most grid-based IBMs, it is necessary to demonstrate that spatial variation and effects can be ignored over areas less than the cell size (Grimm & Railsback 2005). For instance, if spatial data are only available at a scale of square kilometers, but organismal behavior depends heavily on spatial factors that vary at the scale of meters, it will be necessary to add fine-scale spatial variability to the seascape by incorporating smaller cells in the IBM. For many shallow water habitats, data on fine-scale habitat structure (how habitat varies *within individual patches* that create the seascape) can be acquired using field-based sampling techniques. However, aligning movement rules in IBMs appropriately with the spatial resolution of the grid cell and the chosen time step (often annual) remains an ongoing challenge.

8.3.4 Other Factors

Patterns of larval and postlarval settlement may be strongly influenced by seascape structure, which in turn can influence the areas of the seascape in which juveniles may be found. Patterns can arise from the interplay of hydrodynamics and habitat structure. For instance, currents carrying propagules slow from the edge to the interior of patches and may be attenuated quickly in complex, dense habitat (Peterson *et al.* 2004). This can create 'settlement shadows' in which juveniles are found primarily along the edges of habitat patches (Haas *et al.* 2004). A similar edge effect can be created via habitat selection: for some species, first contact with any suitable habitat promotes settlement behavior. Because mortality rates may be particularly high for newly settled juveniles and because habitat edges may be areas of high predator density, patterns of simulated settlement and recruitment in IBMs may have a large effect on model outcomes.

8.4 Challenges and Future Directions in Using IBMs to Explore Seascapes

Though IBMs have great promise for testing seascape-related hypotheses and we hope to see them used more frequently in the future, they are not without their challenges and limitations. Some concerns pertain to IBMs in general; often these involve aspects of complexity and how IBMs are parameterized. IBMs have been characterized as 'data hungry' and some may be deterred from using IBMs for analysing seascape effects due to the perception that large amounts of data are necessary to parameterize models, particularly for aspects of behavior, such as movement rates among patches, which can be difficult to acquire. Another common concern for IBMs is *error propagation*, in which the effects of small changes in parameter values are magnified by the model's interactions, particularly when models are complex and include many factors (Grimm & Railsback 2005). IBMs use probabilities and rules to represent behaviors and while movement might be sufficiently well studied to provide a cogent rule in the model, the probability of movement is often a guess, particularly for species with large home ranges or cryptic behavior or those that are difficult to observe. Error propagation can occur when many such guesses are included in the model and uncertainty compounds to such an extent to deem the predictive capacity of the model unreliable. Fortunately, these concerns are being addressed with the development of field techniques used to acquire data on movement and seascape structure and the development of more sophisticated statistical techniques for parameterizing model functions (see Chapter 9 in Grimm & Railsback 2005). One of the major ways in which the effects of parameter uncertainty on model output are studied in ecological models is through sensitivity analyses. In these analyses, parameters are perturbed individually, in subsets, or altogether to determine how model output would change if the parameter(s) in question were in error. A challenge for IBMs is the sheer number of parameters and their reliance on spatial data, which may also be uncertain. Global sensitivity analyses, where all parameters are perturbed simultaneously, are the recommended procedure as they reveal the effects of dependencies between multiple uncertain parameters (Saltelli *et al.* 2008; Langford *et al.* 2011). However, these types of sensitivity analyses can be onerous and are rarely conducted for IBMs. Future directions in the application of IBMs to seascapes should include recommendations for best practices in conducting sensitivity and uncertainty analyses that can highlight opportunities to reduce uncertainty through data collection and further research. Protocols have been developed for describing, analysing and reducing uncertainty in IBMs (Grimm *et al.* 2005, 2006) and these need to be systematically applied as a matter of course. At present they rarely are (although see Topping *et al.* 2010 for an exception).

At the outset, consideration should be given to what IBMs can reveal about ecological dynamics in seascapes that other simpler and less data intensive models might be better suited for. If the goal is to reveal higher level emergent properties from known processes at the individual level then IBMs fit the bill (Breckling *et al.* 2005). In model construction, one must keep in mind the underlying theories to be tested and avoid constructing models so complex or specific that it would be easier to simply study nature (Grimm & Railsback 2005). The challenge lies in ascertaining what is known at the individual level. While some factors, such as survival and births, may be well studied, behaviors, such as movement, are notoriously difficult to represent faithfully, particularly for seascapes (Patterson *et al.* 2007). Accurate representation of behavior, particularly in the face of changes, such as habitat fragmentation or degradation, presents the greatest challenge for application of IBMs in seascapes.

Data acquisition for IBM construction also remains one of the most challenging aspects of the application of IBMs to seascapes, irrespective of uncertainty in available data. This is particularly the case for parameters on behavior, such as movement, habitat selection and predation choices, where reliable and meaningful information requires many hours of observation over very large scales. Some of this can be achieved with the use of new technology, though the logistics and expense of such data collection make seascapes a more challenging environment to work in than terrestrial landscapes. For this reason, laboratories and mesocosms present mechanisms for gaining data through experiments (Wirsing *et al.* 2008; Yu *et al.* 2011). However, these do not necessarily yield information at the appropriate scale (*e.g.,* wide ranging species can rarely be observed in the laboratory or in mesocosms) and variables observed in controlled environments may not mimic those in the field. For instance, as described above, tethering generally inflates mortality rates above those that would occur for mobile organisms. Therefore, opportunities should be sought to close the gap between manipulated experiments and observations of behaviors and other variables in the field. In particular, it will be important to ascertain which parameters and variables have the most influence over model results and are the most important in answering the research or management question.

Though we have identified several challenges for IBMs applied to seascape ecology, we also note that one of the most useful applications of IBMs is to explore the *potential* effects of some factor of interest on ecological interactions, with the purpose of formulating hypotheses (rather than testing them) based on the results. A perfectly legitimate use of IBMs is to explore the ways in which unmeasured factors may influence behavior and populations, which can be thought of as the development of a sophisticated conceptual model. For instance, IBMs, with their spatially explicit representation of different habitat features, can be used to optimize habitat restoration proposals when the goal is to facilitate movement or maximize population viability, which cannot be readily tested empirically at the planning stage. Likewise, IBMs can be used to anticipate the potential effects of disease transmission or spread of invasive species for pre-emptive planning and management. While these practical applications are not restricted to IBM frameworks, IBMs provide an ideal format for testing conservation management plans particularly when individual behavior is a key consideration. The model may become more realistic or accurate as empirical data are collected and revised models can refine hypotheses and lead to new ones. When setting out to explore ecological phenomena, the creation of an IBM can be useful for organizing thoughts, directing experimentation and identifying major gaps in data. A comprehensive research program to improve the application of IBMs to seascapes might involve the combination and coordination of individual-based modeling and empirical data collection; in an iterative fashion, a model may be used to construct appropriate hypotheses, which are tested with real organisms and these data are then plugged back into the model to modify hypotheses and to further develop an understanding of ecological patterns and processes (Grimm *et al.* 2005). In doing so, the reliability and realism of IBMs can be improved in ways most meaningful to the research question.

Finally, a relatively new frontier for the application of IBMs in seascapes is in conservation management. Animal behavior, the representation of which sets IBMs apart from other more aggregated modeling frameworks used in conservation ecology, has recently been explored in the literature as important in evaluating the impacts of threats and in informing conservation action, particularly in the areas of population viability analysis, effects of genetic, demographic and environmental stochasticity on populations, habitat fragmentation, reserve connectivity and captive breeding and reintroductions (Caro 1998). With the development of frameworks to explicitly include behavior in conservation management decisions (Berger-Tal *et al.* 2011) and well developed software tools with which to build IBMs (Railsback *et al.* 2006), a road map exists to provide pertinent information on how behavior in the face of environmental change can inform conservation decision making. Like many other modeling frameworks, IBMs can also be linked with a range of other models that represent environmental dynamics, such as models of global warming and increased carbon dioxide (Botkin *et al.* 2007), hydrodynamics (Rochette *et al.* 2012) and socio-economic processes (Heckbert *et al.* 2010). In such coupled modeling frameworks feedbacks are represented between environmental processes and the fate of organisms. This gives IBMs even greater scope for answering questions about complex systems that span multiple disciplines. Furthermore, the tendency of IBMs to avoid abstraction can make communication of the model and its results more palatable and comprehensible to managers. Hence, IBMs, with their spatial explicitness, intuitive rules and intended mimicry of the environment and its processes, are well poised to serve not only the understanding of ecology in seascapes, but also to inform policy on how to manage seascapes in the face of threats and stressors.

References

Andrén H (1994) Effects of habitat fragmentation on birds and mammals in landscapes with different proportions of suitable habitat: a review. Oikos 1: 355–366.

Austin D, Bowen WD, McMillan JI (2004) Intraspecific variation in movement patterns: modelling individual behaviour in a large marine predator. Oikos 105: 15–30.

Bélisle M, Desrochers A (2002) Gap-crossing decisions by forest birds: an empirical basis for parameterizing spatially-explicit, individual-based models. Landscape Ecology 17: 219–231.

Berger-Tal O, Polak T, Oron A, Lubin Y, Kotler BP, Saltz D (2011) Integrating animal behaviour and conservation biology: a conceptual framework. Behavioural Ecology 22: 236–239.

Boström C, Pittman SJ, Simenstad C, Kneib RT (2011) Seascape ecology of coastal biogenic habitats: advances, gaps, and challenges. Marine Ecology Progress Series 427: 191–217.

Botkin DB, Saxe H, Araújo MB, Betts R, Bradshaw RHW, Cedhagen T, Chesson P, Dawson TP, Etterson JP, Faith DP, Ferrier S, Guisan A, Skjoldborg Hansen A, Hilbert DW, Loehle C, Margules C, New M, Sobel MJ, Stockwell DRB (2007) Forecasting the effects of global warming on biodiversity. BioScience 57: 227–236.

Breckling B, Müller F, Reuter H, Hölker F, Fränzle O (2005) Emergent properties in individual-based ecological models – introducing case studies in an ecosystem research context. Ecological Modelling 186: 376–388.

Browder JA, Bartley HA, Davis KS (1985) A probabilistic model of the relationship between marshland-water interface and marsh disintegration. Ecological Modelling 29: 245–260.

Brown DG, Riolo R, Robinson DT, North M, Rand W (2005) Spatial process and data models: Toward integration of agent-based models and GIS. Journal of Geographical Systems 7: 25–47.

Bugmann H (2001) A review of forest gap models. Climatic Change 51: 259–305.

Butler MJ, Dolan T, Herrnkind WF, Hunt J (2001) Modelling the effect of spatial variation in postlarval supply and habitat structure on recruitment of Caribbean spiny lobster. Marine and Freshwater Research 52: 1243–1252.

Butler MJ, Dolan III TW, Hunt JH, Rose KA, Herrnkind WF (2005) Recruitment in degraded marine habitats: a spatially explicit, individual-based model for spiny lobster. Ecological Applications 15: 902–918.

Caro T (1998) Behavioural Ecology and Conservation Biology. Oxford University Press, New York, NY.

Claydon JAB, Calosso MC, De Leo GA, Peachey RBG (2015) Spatial and demographic consequences of nursery-dependence in reef fishes: an empirical and simulation study. Marine Ecology Progress Series 525: 171–183.

Coates JH, Hovel KA (2014) Incorporating movement and reproductive asynchrony into a simulation model of fertilization success for a marine broadcast spawner. Ecological Modelling 283: 8–18.

Coates JH, Hovel KA, Butler JL, Klimley AP, Morgan SG (2013) Movement and home range of pink abalone *Haliotis corrugata*: implications for restoration and population recovery. Marine Ecology Progress Series 486: 189–201.

Cushman SA, McGarigal K, Neel MC (2008) Parsimony in landscape metrics: Strength, universality, and consistency. Ecological Indicators 8(5): 691–703.

Daewel U, Schrum C, Gupta AK (2015) The predictive potential of early life stage individual-based models (IBMs): an example for Atlantic cod *Gadus morhua* in the North Sea. Marine Ecology Progress Series 534: 199–219.

DeAngelis DL, Grimm V (2014) Individual-based models in ecology after four decades. F1000 Prime Reports 6: 39.

DeAngelis DL, Mooij WM (2005) Individual-based modelling of ecological and evolutionary processes. 1. Annual Review of Ecology Evolution and Systematics 36: 147–168.

Dunstan PK, Johnson CR (2005) Predicting global dynamics from local interactions: individual-based models predict complex features of marine epibenthic communities. Ecological Modelling 186: 221–233.

Durrett R, Levin SA (1994) Stochastic spatial models: A user's guide to ecological applications. Philosophical Transactions of the Royal Society of London B: Biological Sciences 343(1305): 329–350.

Dytham C (2003) How landscapes affect the evolution of dispersal behaviour in reef fishes: results from an individual-based model. Journal of Fish Biology 63 (supplement A): 213–225.

Dytham C, Simpson SD (2007) Elevated mortality of fish larvae on coral reefs drives the evolution of larval movement patterns. Marine Ecology Progress Series 346: 255–264.

Eggleston DB, Bell GW, Amavisca AD (2005) Interactive effects of episodic hypoxia and cannibalism on juvenile blue crab mortality. Journal of Experimental Marine Biology and Ecology 325: 18–26.

Elderd BD, Nott MP (2008) Hydrology, habitat change and population demography: an individual-based model for the endangered Cape Sable seaside sparrow *Ammodramus maritimus mirabilis*. Journal of Applied Ecology 45: 258–268.

Estavillo C, Pardini R, da Rocha PLB (2013) Forest loss and the biodiversity threshold: an evaluation considering species habitat requirements and the use of matrix habitats. PLoS One 8:e82369.

Fahrig L (2002) Effect of habitat fragmentation on the extinction threshold: a synthesis. Ecological Applications 12: 346–353.

Fahrig L (2003) Effects of habitat fragmentation on biodiversity. Annual Review of Ecology Evolution and Systematics 34: 487–515.

Fodrie FJ, Yeager LA, Grabowski JH, Layman CA, Sherwood GD, Kenworthy MD (2015) Measuring individuality in habitat use across complex landscapes: approaches, constraints, and implications for assessing resource specialization. Oecologia 178(1): 75–87.

Fonseca MS, Bell SS (1998) Influence of physical setting on seagrass landscapes near Beaufort, North Carolina, USA. Marine Ecology Progress Series 171: 109–121.

Gates JE, Gysel LW (1978) Avian nest dispersion and fledgling success in field-forest ecotones. Ecology 59: 871–883.

Grimm V, Berger U, Bastiansen F, Eliassen S, Ginot V, Giske J, Goss-Custard J, Grand T, Heinz SK, Huse G, Huth A, Jepsen JU, Jørgensen C, Mooij WM, Müller B, Pe'er G, Piou C, Railsback SF, Robbins AM, Robbins MM, Rossmanith E, Rüger N, Strand E, Souissi S, Stillman RA, Vabø R, Visser U, DeAngelis DL (2006) A standard protocol for describing individual-based and agent-based models. Ecological Modelling 198: 115–126.

Grimm V, Railsback SF (2005) Individual-based modelling and ecology. Princeton University Press, Princeton, NJ.

Grimm V, Revilla E, Berger U, Jeltsch F, Mooij WM, Railsback SF, Thulke H-H, Weiner J, Wiegand T, DeAngelis DL (2005) Pattern-oriented modelling of agent-based complex systems: lessons from ecology. Science 310: 987–991.

Haas HL, Rose KA, Fry B, Minello TJ, Rozas LP (2004) Brown shrimp on the edge: linking habitat to survival using and individual-based simulation model. Ecological Applications 14: 1232–1247.

Hastings A (1990) Spatial heterogeneity and ecological models. Ecology 71: 426–428.

Heck KL, Jr., Thoman TA (1981) Experiments on predator-prey interactions in vegetated aquatic habitats. Journal of Experimental Marine Biology and Ecology 53: 125–134.

Heckbert S, Baynes T, Reeson, A (2010) Agent-based modelling in ecological economics. Annals of the New York Academy of Sciences 1185: 39–53.

Hermann A, Hinckley S, Megrey B, Napp J (2001) Applied and theoretical considerations for constructing spatially explicit individual-based models of marine larval fish that include multiple trophic levels. ICES Journal of Marine Science 58: 1030–1041.

Hitt S, Pittman SJ, Nemeth RS (2011) Diel movements of fishes linked to benthic seascape structure in a Caribbean coral reef ecosystem. Marine Ecology Progress Series 427: 275–291.

Hovel KA, Fonseca MS (2005) Influence of seagrass landscape structure on the juvenile blue crab habitat-survival function. Marine Ecology Progress Series 300: 170–191.

Hovel KA, Fonseca MS, Myer DL, Kenworthy WJ, Whitfield PE (2002) Effects of seagrass landscape structure, structural complexity and hydrodynamic regime on macrofaunal densities in North Carolina seagrass beds. Marine Ecology Progress Series 243: 11–24.

Hovel KA, Lipcius RN (2001) Habitat fragmentation in a seagrass landscape: patch size and complexity control blue crab survival. Ecology 82: 1814–1829.

Hovel KA, Lipcius RN (2002) Effects of seagrass habitat fragmentation on juvenile blue crab survival and abundance. Journal of Experimental Marine Biology and Ecology 271: 75–98.

Hovel KA, Regan HM (2008) Using an individual-based model to examine the roles of habitat fragmentation and behaviour on predator-prey relationships in seagrass landscapes. Landscape Ecology 23: 75–89.

Irlandi EA (1996) The effects of seagrass patch size and energy regime on growth of a suspension-feeding bivalve. Journal of Marine Research 54: 161–185.

Kjelland ME, Piercy CD, Lackey T, Swannack TM (2015) An integrated modelling approach for elucidating the effects of different management strategies on Chesapeake Bay oyster metapopulation dynamics. Ecological Modelling 308: 45–62.

Kramer-Schadt S, Revilla E, Weigand T, Breitenmoser U (2004) Fragmented landscapes, road mortality and patch connectivity: modelling influences on the dispersal of Eurasian lynx. Journal of Applied Ecology 41: 711–723.

Langford WT, Gordon A, Bastin L, Bekessy SA, White MD, Newell G (2011) Raising the bar for systematic conservation planning. Trends in Ecology and Evolution 26: 634–640.

Lannin R, Hovel KA (2011) Variable prey density modifies the effects of seagrass habitat structure on predator–prey interactions. Marine Ecology Progress Series 442: 59–70.

Louzao M, Pinaud D, Péron C, Delord K, Wiegand T, Weimerskirch H (2011) Conserving pelagic habitats: seascape modelling of an oceanic top predator. Journal of Applied Ecology 48: 121–132.

Lowe CG, Topping DT, Cartamil DP, Papastamatiou YP (2003) Movement patterns, home range, and habitat utilization of adult kelp bass *Paralabrax clathratus* in a temperate no-take marine reserve. Marine Ecology Progress Series 256: 205–216.

Makowski C, Seminoff JA, Salmon M (2006) Home range and habitat use of juvenile Atlantic green turtles (*Chelonia mydas* L.) on shallow reef habitats in Palm Beach, Florida, USA. Marine Biology 148: 1167–1179.

Margules CR, Pressey RL (2000) Systematic conservation planning. Nature 405: 243–253.

Minello TJ, Zimmerman RJ, Medina R (1994) The importance of edge for natant macrofauna in a created salt marsh. Wetlands 14: 184–198.

Mooij WM, Bennetts RE, Kitchens WM, DeAngelis DL (2002) Exploring the effect of drought extent and interval on the Florida snail kite: interplay between spatial and temporal scales. Ecological Modelling 149: 25–39.

Morris DW (2003) Toward an ecological synthesis: a case for habitat selection. Oecologia 136: 1–3.

Munguia P, Osman RW, Hamilton J, Whitlach R, Zajac R (2011) Changes in habitat heterogeneity alter marine sessile benthic communities. Ecological Applications 21: 925–935.

Parada C, van der Lingen CD, Mullon C, Penven P (2003) Modelling the effect of buoyancy on the transport of anchovy (*Engraulis capensis*) eggs from spawning to nursery grounds in the southern Benguela: an IBM approach. Fisheries Oceanography 12: 170–184.

Parrott L, Chion C, Gonzales R, Latombe G (2012) Agents, individuals, and networks: modelling methods to inform natural resource management in regional landscapes. Ecology and Society 17: 32.

Paton PWC (1994) The effect of edge on avian nest success: how strong is the evidence? Conservation Biology 8: 17–26.

Patterson TA, Thomas L, Wilcox C, Ovaskainen O, Matthiopoulos J (2007) State-space models of individual animal movement. Trends in Ecology and Evolution 23: 87–94.

Perry GLW, Bond NR (2005) Spatial population models for animals. In Wainright J, Mulligan M (eds) Environmental Modelling: Finding Simplicity in Complexity (1st edition), John Wiley & Sons, Ltd., Chichester.

Peterson CH, Black R (1994) An experimentalist's challenge: when artifacts of intervention interact with treatments. Marine Ecology Progress Series 111: 289–297.

Peterson CH, Leuttich RA, Jr., Micheli F, Skilleter GA (2004) Attenuation of water flow inside seagrass canopies of differing structure. Marine Ecology Progress Series 268: 81–92.

Petrovskii S, Petrovskaya N (2012) Computational ecology as an emerging science. Interface Focus 2: 241–254.

Pickett, STA, Cadenasso ML (1995) Landscape ecology: spatial heterogeneity in ecological systems. Science 269: 331–334.

Pile AJ, Lipcius RN, van Montfrans J, Orth RJ (1996) Density-dependent settler-recruit-juvenile relationships in blue crabs. Ecological Monographs 66(3): 277–300.

Pittman SJ, McAlpine CA (2003) Movements of marine fish and decapod crustaceans: process, theory and application. Advances in Marine Biology 44: 205–294.

Railsback SF, Lytinen SL, Jackson SK (2006) Agent-based simulation platforms: review and development recommendations. Simulation 82: 609–623.

Regan HM, Keith DA, Regan TJ, Tozer MG, Tootell N (2011) Fire management to combat disease: turning interactions between threats into conservation management. Oecologia 167: 873–882.

Rochette S, Huret M, Rivot E, Le Pape O (2012) Coupling hydrodynamic and individual-based models to simulate long-term larval supply to coastal nursery areas. Fisheries Oceanography 21: 229–242.

Roth BM, Rose KA, Rozas LP, Minello TJ (2008) Relative influence of habitat fragmentation and inundation on brown shrimp (*Farfantepenaeus aztecus*) production in northern Gulf of Mexico salt marshes. Marine Ecology Progress Series 359: 185–202.

x

Saltelli A, Ratto M, Andres T, Campolongo F, Cariboni J, Gatelli D, Saisana M, Tarantola S (2008) Global Sensitivity Analysis: The Primer. John Wiley & Sons, Ltd, Chichester.

Schadt S, Knauer F, Kaczensky P, Revilla E, Wiegand T, Trepl L (2002) Rule-based assessment of suitable habitat and patch connectivity for the Eurasian lynx. Ecological Applications 12: 1469–1483.

Simioni G, Le Roux X, Gignoux J, Sinoquet H (2000) Treegrass: a 3D, process-based model for simulating plant interactions in tree–grass ecosystems. Ecological Modelling 131: 47–63.

Soares-Filho BS, Cerqueira GC, Pennachin CL (2002) DINAMICA – a stochastic cellular automata model designed to simulate the landscape dynamics in an Amazonian colonization frontier. Ecological Modelling 154: 217–235.

Tait KJ, Hovel KA (2012) Do predation risk and food availability modify prey and mesopredator microhabitat selection in eelgrass (Zostera marina) habitat? Journal of Experimental Marine Biology and Ecology 426–427: 60–67.

Tilman D, Kareiva P (eds) (1998) Spatial Ecology: The Role of Space in Population Dynamics and Interspecific Interactions. Princeton University Press, Princeton, NJ.

Topping CJ, Høye TT, Olesen CR (2010) Opening the black box: Development, testing and documentation of a mechanistically rich agent-based model. Ecological Modelling 221: 245–255.

Turner MG (1990) Spatial and temporal analysis of landscape patterns. Landscape Ecology 4: 21–30.

Turner MG, Wu Y, Wallace LL, Romme WH, Brenkert A (1994) Simulating winter interactions among ungulates, vegetation, and fire in northern Yellowstone Park. Ecological Applications 4: 472–496.

Uchmański J, Grimm V (1996) Individual-based modelling in ecology: what makes the difference? Trends Ecological Evolution 11: 437–441.

Underwood AJ, Chapman MG, Crowe TP (2004) Identifying and understanding ecological preferences for habitat or prey. Journal of Experimental Marine Biology and Ecology 300: 161–187.

Werner FE, MacKenzie BR, Perry RI, Lough RG, Naimie CE, Blanton BO, Quinlan JA (2001a) Larval trophodynamics, turbulence and drift on Georges Bank: a sensitivity analysis of cod and haddock. Scientia Marina 65: 99–115.

Werner FE, Quinlan JA, Lough RG, Lynch DR (2001b) Spatially-explicit individual based modelling of marine populations: A review of the advances in the 1990s. Sarsia 86: 411–421.

Wiens JA (1989) Spatial scaling in ecology. Functional Ecology 3: 385–397.

Wilensky U (1999) NetLogo (and NetLogo User Manual). Center for Connected Learning and Computer-Based Modeling, Northwestern University, Evanston, IL.

Williams SL, Heck KL, Jr. (2001) Seagrass community ecology. In Bertness MD, Gaines SD, Hay ME (eds) Marine Community Ecology. Sinauer Associates Inc., Sunderland, MA, pp. 317–337.

Wirsing AJ, Heithaus MR, Frid A, Dill LM (2008) Seascapes of fear: evaluating sublethal predator effects experienced and generated by marine mammals. Marine Mammal Science 24: 1–15.

Withy-Allen KR, Hovel KA (2013) California spiny lobster (Panulirus interruptus) movement behaviour and habitat use: implications for the effectiveness of marine protected areas. Marine and Freshwater Research 64: 359–371.

Yu PC, Matson PG, Martz TR, Hofmann GE (2011) The ocean acidification seascape and its relationship to the performance of calcifying marine invertebrates: Laboratory experiments on the development of urchin larvae framed by environmentally-relevant pCO2/pH. Journal of Experimental Marine Biology and Ecology 400: 288–295.

Zimmer-Faust RK, Feilder DR, Heck KL, Jr., Coen LD, Morgan SG (1994) Effects of tethering on predatory escape by juvenile blue crabs. Marine Ecology Progress Series 111: 299–303.

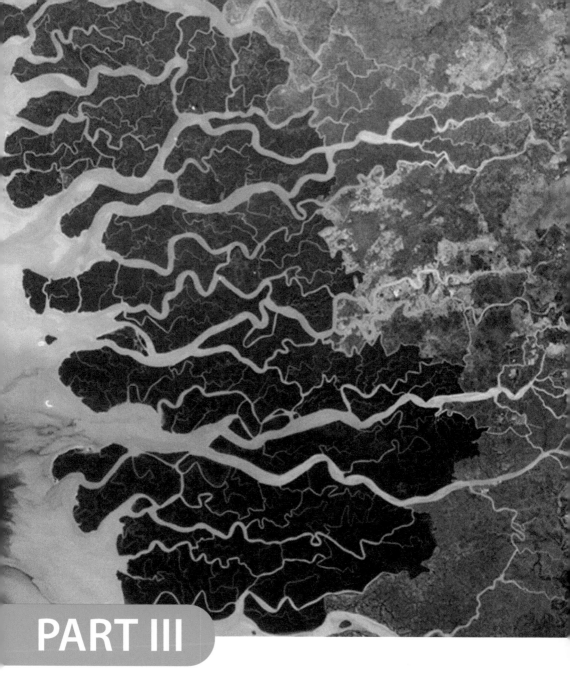

PART III
Seascape Connectivity

9

Connectivity in Coastal Seascapes

Andrew D. Olds, Ivan Nagelkerken, Chantal M. Huijbers, Ben L. Gilby, Simon J. Pittman and Thomas A. Schlacher

9.1 Introduction

Movements of organisms, energy and nutrients link populations, habitats, food webs and ecosystems across the boundaries of seascapes and landscapes (Loreau *et al.* 2003; Lundberg & Moberg 2003). In the sea, the magnitude and direction of connections are shaped by topography, geological history and oceanography and the spatial arrangement of ecosystems (Cowen *et al.* 2007; Grober-Dunsmore *et al.* 2009). The ecological importance of these spatial linkages is widely recognized (reviewed by Nagelkerken 2009; Sheaves 2009; Kool *et al.* 2013) and their effects on distribution, movement and biology have been reported for decades (Odum 1968; Parrish 1989). Connectivity is also important for the recovery of populations, assemblages and ecosystems from disturbance (Cumming 2011; Bernhardt & Leslie 2013) and, consequently, has become an important consideration in marine conservation and fisheries management (Magris *et al.* 2014; Beger *et al.* 2015).

Connectivity can be defined and quantified in many different ways and operates at all levels of ecological organization (Grober-Dunsmore *et al.* 2009; Boström *et al.* 2011) (Table 9.1). Widely reported spatial linkages (Figure 9.1) that illustrate the multiple ecological effects of connectivity include, ontogenetic movements of fishes between coastal ecosystems (Nagelkerken *et al.* 2015); migration of humpback whales between polar feeding grounds and tropical breeding areas (Rosenbaum *et al.* 2014); transfer of carbon from sea to land by raptors foraging along beaches (Huijbers *et al.* 2015c); stranding of marine plant material (*e.g.*, kelp, seagrass) on shorelines (Hyndes *et al.* 2014); dispersal of propagules in currents (Lee *et al.* 2014b); and advection of nutrients into coastal waters from estuaries (Schlacher & Connolly 2009).

9.2 Global Synthesis of Connectivity Research

To understand and quantify characteristics of global research on connectivity we compiled a database of all peer-reviewed studies that reported effects of connectivity in the sea and recorded: (i) the types of connectivity assessed; (ii) region of study; (iii) focal

Seascape Ecology, First Edition. Edited by Simon J. Pittman.
© 2018 John Wiley & Sons Ltd. Published 2018 by John Wiley & Sons Ltd.

Table 9.1 Common terms relating to connectivity that are adopted in studies of movement biology and seascape ecology. Definitions, examples and citations to relevant published studies are provided for each term.

Terms	Definition and examples	Relevant study
Connectivity		
How connectivity is quantified:		
Actual	Measured directly (*e.g.*, fish movement; telemetry).	Pittman *et al.* (2014)
Potential	Inferred from distribution and spatial properties of seascapes, with some data on mobility (*e.g.*, coral dispersal; graph theory).	Treml *et al.* (2008)
Structural	Inferred from distribution and spatial properties of seascapes (*e.g.*, fish abundance; spatial pattern metrics).	Meynecke *et al.* (2008)
The scale of measurement:		
Population	Among individuals in different places (*e.g.*, fish genetics).	Cowen *et al.* (2007)
Habitat	Among different patches of the same habitat (*e.g.*, coral abundance).	Huntington *et al.* (2010)
Ecosystem	Among different habitats (*e.g.*, invertebrate abundance & movement).	Micheli & Peterson (1999)
Seascape	Among habitats and ecosystems across seascapes (*e.g.*, fish abundance and ecological functions).	Berkström *et al.* (2012)
What is being measured:		
Genetic	Spatial effects of adults on the DNA of juveniles from elsewhere (*e.g.*, fish genetics).	Harrison *et al.* (2012)
Trophic	Spatial effects of one ecosystem on food web structure in another (*e.g.*, fish diet and foraging ecology).	Davis *et al.* (2014b)
Functional	Spatial effects of one ecosystem on ecological functions in another (*e.g.*, fish abundance and herbivory).	Olds *et al.* (2012)
Ontogenetic	Movement of organisms from one ecosystem to another with age as their resource requirements change (*e.g.*, fish migration).	Nagelkerken *et al.* (2012)
Terms used interchangeably with connectivity:		
Movement	Nonpermanent movement of organisms between different places.	Sheaves (2005)
Migration	Permanent movement of organisms from one place to another.	Gillanders *et al.* (2003)
Dispersal	Nondirectional movement of larvae from spawning locations.	Jones *et al.* (2005)
Habitat linkages	See habitat connectivity.	Irlandi & Crawford (1997)

(Continued)

Table 9.1 (Continued)

Terms	Definition and examples	Relevant study
Trophic linkages	See trophic connectivity.	Valentine *et al.* (2008)
Trophic transfer	See trophic connectivity.	Heck *et al.* (2008)
Trophic relay	Spatial effects of inshore ecosystems on offshore food webs through a series of predator-prey interactions and migration (see trophic connectivity).	Kneib (1997)
Mobile link	Organisms that move among habitats in seascapes and link food webs and ecological functions (see trophic and functional connectivity).	Lundberg & Moberg (2003)
Spatial subsidy	See trophic and functional connectivity.	Vanderklift & Wernberg (2008)
Spatial coupling	See trophic and functional connectivity.	Schlacher & Connolly (2009)

seascapes; and (iv) organisms examined (Figure 9.2). The Scopus and ISI Web of Knowledge databases were searched using keywords related to connectivity (connect* OR link* OR move*) AND seascape ecology (seascape OR landscape AND ecology) AND marine ecosystems (marine OR ocean OR sea). This search yielded 592 studies.

9.2.1   Research Theme

Most research on connectivity has focused on examining ecological effects on animals, or the functions animals perform in linking ecosystems (500 studies) (Figure 9.2). The predominance of animal connectivity studies is evident in the several published reviews on the topic (*e.g.,* Grober-Dunsmore *et al.* 2009; Hazen *et al.* 2012; Hays & Scott 2013; Pittman & Olds 2015). Recurrent themes are effects on distribution (Randall 1965; Nagelkerken *et al.* 2000b), movement (Irlandi & Crawford 1997; Pittman *et al.* 2014) and the ecological functions performed by mobile species (Micheli & Peterson 1999; Olds *et al.* 2012). Connectivity is also critical for the dispersal of nonliving matter (124 studies), carbon (97) and plants (45) (see reviews by Bouillon *et al.* 2008; Heck *et al.* 2008; Hyndes *et al.* 2014; McMahon *et al.* 2015); many studies that focus on these aspects of connectivity also report on effects on animals. For example, animals are commonly affected by the movement of sediments and plant materials (*e.g.,* Vanderklift & Wernberg 2008; Schlacher & Connolly 2009) and are also responsible for transporting carbon and nutrients among ecosystems (*e.g.,* Kneib 1997; Layman *et al.* 2013) (Figure 9.3).

There is strong bias, however, in the taxonomic identity of marine and coastal animals examined, with the vast majority of all connectivity studies focussing on fish (359) (reviewed by Nagelkerken 2009; Sheaves 2009; Berkström *et al.* 2012; Olds *et al.* 2016). Conversely, our searches indicated fewer studies that assessed the effects of connectivity on corals (165), crustaceans (128), birds (126), mammals (121), molluscs (82), reptiles (74) and echinoderms (56) (*e.g.,* Treml *et al.* 2008; Hays & Scott 2013; Buelow & Sheaves 2015) (Figure 9.2). By contrast, the effect of seascape characteristics on plant dispersal (*i.e.,* through seed or propagule dispersal) is rarely evaluated (Figure 9.2) (but see Reynolds *et al.* 2013; McMahon *et al.* 2015; Van der Stocken *et al.* 2015). Quantification

Figure 9.1 Spatial linkages that illustrate diversity in the form and function of connectivity effects. Fish move between tropical coastal ecosystems with ontogenetic development (a). Humpback whales migrate between polar feeding grounds and tropical breeding areas (b). Raptors forage on fish stranded on sandy beaches and relay carbon from sea to land (c). Detached kelp washes ashore on exposed beaches where it provides food for invertebrates (d). Mangrove propagules disperse out to sea in coastal currents (e). Flood plumes from rivers transport sediments and nutrients in coastal waters (f). *Source*: Photographs by A. Olds, B. Gratwicke (CC BY-NC 2.0), J. Udy, T. Schlacher.

of the biological properties and benefits of connectivity for many invertebrates and marine plants may, therefore, improve our understanding of the ecology of these populations and enhance prospects for their conservation (McMahon *et al.* 2015; Olds *et al.* 2016).

9.2.2 Geographical Distribution

Research on connectivity is geographically widespread (Figure 9.2). It is mostly conducted in Australia (159 studies), North America (155 studies), South America (94), or in shallow tropical and subtropical seas (*e.g.*, Mediterranean Sea, 134 studies; Caribbean Sea, 114 studies; Oceania, 62 studies). The global distribution of connectivity studies

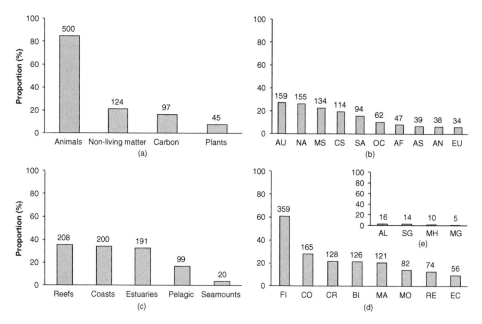

Figure 9.2 Summary of global research on connectivity in marine ecosystems showing variation in the: types of connectivity assessed (a); regions of study (b); focal seascapes (c); and organisms examined (d). The number of studies conducted is displayed for each: type, region, seascape and organism and bars show this figure as a proportion of all connectivity studies. Regions: AU, Australia; NA, North America; MS, Mediterranean Sea; CS, Caribbean Sea; SA, South America; OC, Oceania; AF, Africa; AS, Asia and Indonesia; AN, Antarctica; and EU, Europe. Organisms: FI, fish; CO, coral; CR, crustaceans; BI, bivalves; MA, mammals; MO, molluscs; RE, reptiles; EC, echinoderms; AL, algae; SG, seagrass; MH, marsh; and MG, mangroves.

may reflect the location, history and ecological interests of marine laboratories, or social and economic opportunities for research (sensu Grober-Dunsmore *et al.* 2009; Boström *et al.* 2011; Olds *et al.* 2016). In addition to these four regions, numerous studies have examined aspects of connectivity in Africa (47), Asia (39), Antarctica (38) and Europe (outside the Mediterranean; 34) (Figure 9.2) (*e.g.*, Unsworth *et al.* 2008; Krumme 2009; Berkström *et al.* 2012; Kimirei *et al.* 2013b).

Most connectivity research has been done in shallow coastal seascapes (<30 m depth), particularly those that contain reefs (coral and rocky) (208), inshore shelf habitats (30–200 m) (200) and estuaries (191) (*e.g.*, Gillanders *et al.* 2003; Nagelkerken 2009; Sheaves 2009). By contrast, research on connectivity is less common in offshore pelagic waters (99) (Gaspar *et al.* 2006) and in deeper seascapes, such as seamounts (20) (Schlacher *et al.* 2010) (Figure 9.2). There have also been few studies of connectivity at (or across) the interface between sea and land (but see Reef *et al.* 2014; Huijbers *et al.* 2015c); this result is particularly surprising for sandy beaches, which are easy to access, have strong biological links with other ecosystems and account for over 70% of the global coastline (Schlacher *et al.* 2015b). Opportunities for novel research on connectivity are, therefore, likely awaiting examination in the open ocean and deep sea, across coastal interfaces and among many seascapes that are commonly studied in isolation (*e.g.*, estuaries, surf zones) (Olds *et al.* 2016).

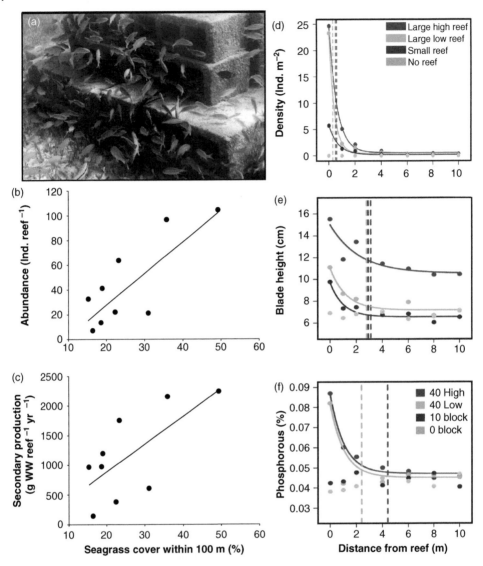

Figure 9.3 Connectivity effects on fish and seagrass ecosystems in the Bahamas. The abundance (b) and secondary production (c) of white grunts (*Haemulon plumierii*) on artificial reefs is positively correlated with seagrass cover. Fish density declines rapidly with distance from reefs (d), as does the height of seagrass (e), and the phosphorous content of seagrass leaves (f). *Source*: Adapted from Yeager *et al.* (2012) and Layman *et al.* (2013). Photograph by C. Layman.

9.2.3   Biological and Functional Consequences

Physical forces (*e.g.*, tides, currents, wind, waves) transport plants and nonliving matter across seascapes (Hyndes *et al.* 2014; McMahon *et al.* 2015). By contrast, changes in the biological requirements of animals drive active movements and migration for reproduction, colonization, or to maximise growth and reduce mortality (Lundberg & Moberg

2003; Nagelkerken *et al.* 2015). This movement has consequences for the structure of metapopulations and the functioning of metaecosystems (*i.e.*, populations and ecosystems that are biologically linked, but occur in different places) (Loreau *et al.* 2003). It can promote genetic diversity and fitness in metapopulations and provides the mechanism for their recovery following disturbance (Cowen *et al.* 2007; Kool *et al.* 2013). For ecosystems, this movement links ecological processes (*e.g.*, herbivory, predation, recruitment), food webs and nutrient cycling among habitats of the same type (*e.g.*, patches of seagrass) and those of different types (*e.g.*, coral reefs and mangrove forests) (Lundberg & Moberg 2003; Olds *et al.* 2016). This means that ecosystems do not exist in isolation and that their ecological condition can depend on the strength of functional linkages among habitats (Cumming 2011). For example, white (*Haemulon plumierii*) and bluestriped (*H. sciurus*) grunts shelter on reefs during the day and feed in seagrass at night. Grunt abundance is often positively correlated with the cover of seagrass nearby (Hitt *et al.* 2011; Yeager *et al.* 2012) (Figure 9.3). On reefs in the Bahamas, grunt excrement adds fertilizing nutrients to the seascape, which promotes the growth of seagrass plants near reefs (Layman *et al.* 2013) (Figure 9.3).

Functional linkages of ecological processes across seascapes are also pivotal for enabling ecosystems to resist, or recover, from disturbance: two key examples include, herbivory, which is performed by herbivorous fish that migrate between seagrass and corals and prevents both ecosystems from being overgrown by algae in eutrophic areas (Pages *et al.* 2014); and recruitment, which enables corals to recolonize reefs after perturbation (Olds *et al.* 2012). Functional connectivity is thus a central component of ecological resilience and may underpin the capacity of many ecosystems to cope with disturbance (Cumming 2011; Bernhardt & Leslie 2013). This hypothesis is widely recognized and is supported by modelling studies (Mumby & Hastings 2008), but it has not yet been tested empirically.

9.2.4   Connectivity is Scale Dependent

The spatial properties of seascapes and the dispersal capabilities of individuals and species are often scale dependent and these traits interact to structure the level of connectivity among populations and between ecosystems (Pittman & McAlpine 2003; Sheaves 2005; Nagelkerken *et al.* 2015). The spatial arrangement and accessibility of habitats in many seascapes is controlled by the interplay of bathymetry and hydrodynamic forces, with greater connectivity occurring for some species (*e.g.*, for tropical fishes) where tidal ranges are small (Igulu *et al.* 2014), or where strong currents and large tides transport animals and plants over large distances (Baker *et al.* 2013; McMahon *et al.* 2015). Many marine vertebrates (*i.e.*, birds, mammals, reptiles) are also capable of migrating vast distances, but do so infrequently (Battley *et al.* 2012; Putman *et al.* 2014; Rosenbaum *et al.* 2014). Animals can move further by travelling with currents and wind, but body size places strong energetic and biomechanical constraints on their movements; effects that are consistent across most marine taxa (Hein *et al.* 2011) and, which mean that larger animals can thus usually move greater distances. Dispersal capabilities also vary among life-history stages, with the larvae or juveniles of fishes and the young of turtles capable of dispersing large distances in ocean currents (Hays & Scott 2013; Green *et al.* 2014). To study connectivity, it is therefore, necessary to have a clear understanding of the movement capabilities of organisms, which are relevant to the life-history stage and seascape of interest (Pittman & McAlpine 2003; Martin *et al.* 2015). With recent advances in technology, this information is increasingly becoming available, or more easily attainable, for many species (*e.g.*, Hein *et al.* 2011; Hays & Scott 2013; Green *et al.* 2014).

9.3   Quantifying Connectivity: Advances in Key Tools and Techniques

Several tools are available to quantify connectivity among populations and ecosystems in coastal seascapes (Leis *et al.* 2011; Kool *et al.* 2013). This includes equipment for direct tracking of animal movement (*i.e.*, artificial tags, telemetry) (see reviews by Gillanders 2009; Hazen *et al.* 2012) and approaches for tracing and reconstructing dispersal histories using indirect methods (*i.e.*, genetics, ecogeochemical markers, isotopic and elemental ratios in animal tissues) (see reviews by Fry 2006; McMahon *et al.* 2013; Riginos & Liggins 2013). Empirical data on movement and dispersal can be linked with benthic habitat maps and circulation models (using graph theory and numerical modelling software) to predict connectivity among metapopulations or metaecosystems and identify priority locations for conservation (Finn *et al.* 2014; Thomas *et al.* 2014; Beger *et al.* 2015). Analytical approaches for quantifying dispersal and modelling potential connectivity are reviewed by Hovel & Regan (Chapter 8 in this book); Treml & Kool (Chapter 10 in this book) and Pittman *et al.* (Chapter 7 in this book). Here, we focus on recent advances in techniques (*i.e.*, tags, ecogeochemical markers, genetics) that are used to quantify connectivity by measuring animal movement.

9.3.1   Tags and Telemetry

Artificial tags, either externally attached to an organism, or internally embedded, have been used for decades to track animal movement (Hazen *et al.* 2012). Advances in tag technology are rapidly enhancing our knowledge of animal dispersal at local (*e.g.*, fish movement among habitats; Pittman *et al.* 2014), regional (*e.g.*, turtle migrations between ecosystems; Hays & Scott 2013) and global (*e.g.*, shorebird migrations among continents; Runge *et al.* 2014) scales. Reductions in the cost and size of tags, together with increases in battery life, are responsible for expansions in the geographic scope and temporal scale of tagging studies (Hazen *et al.* 2012; McMahon *et al.* 2013) and have markedly improved our capacity to measure actual connectivity in seascapes (Pittman & McAlpine 2003; Pittman *et al.* 2014).

Satellite tags, archival tags and popup satellite archival tags (PSAT) log data on the movement of animals and the environments they inhabit and can be used to track movements of large marine animals (*e.g.*, bluefin tuna, turtles, whales) over vast distances (Musyl *et al.* 2011). The development of mini-PSAT tags has allowed researchers to track movements of juvenile tuna as small as 100 mm and revealed seasonal patterns of habitat use that connect coastal seas along the entire east coast of north America (Galuardi & Lutcavage 2012). Acoustic tags transmit data on an animal's location at regular intervals; this is logged by acoustic receivers that are within range of the signal via telemetry (Hazen *et al.* 2012). Telemetry is typically used to provide detailed spatial information on how tagged animals move among habitats in coastal waters (Pittman *et al.* 2014).

Recent reductions in the size of acoustic tags (now available at 11 mm in length) mean that telemetry is now viable for a wide range of species and life stages. For example, telemetry was used recently to track movements of small juvenile fish (<110 mm) between nursery and adult habitats (Huijbers *et al.* 2015a). While smaller tags have facilitated tracking for a wider range of species, the battery life of artificial tags remains problematic and this precludes the tracking of individuals across multiple life stages. Studying connectivity often necessitates examining how animals use their environment throughout their lives and this usually requires reconstruction of movement histories from ecogeochemical or genetic markers (McMahon *et al.* 2013).

9.3.2   Ecogeochemical Markers

Otoliths are calcified structures in the heads of teleost fishes that grow continuously throughout life and remain chemically inert once formed (Campana 1999). Their chemistry is mainly derived from surrounding waters and thus preserves a lifelong record of a fish's environment, which can be used as a natural tag to discriminate fish stocks from different environments (Elsdon *et al.* 2008). This method has great promise for tracing fish migration among seascapes, however, few studies have attempted to reconstruct movements and quantify actual connectivity (Trueman *et al.* 2012). This is mainly because of issues with the temporal stability of otolith chemistry and difficulties in measuring small quantities of material derived from juvenile otoliths. To use otolith chemistry as a tool to investigate actual movement, a few basic steps need to be met (Gillanders 2009): (i) obtain otoliths from fish of known origin and ensure that differences exist between groups; (ii) validate the reliability of the otolith chemistry by quantifying temporal stability and spatial differences; (iii) assign fish of unknown origin to known groups based on maximum-likelihood estimation of their otolith chemistry. At the broadest scale, otolith chemistry has been used to quantify connectivity between bluefin tuna populations across the Atlantic Ocean (Rooker *et al.* 2008). It has also proven useful for quantifying connectivity at smaller scales; among coastal habitats (*e.g.*, mangroves, seagrass beds, coral reefs) and between juvenile nurseries and offshore adult habitats (Huijbers *et al.* 2013; Kimirei *et al.* 2013a). McMahon *et al.* (2012) used a novel analysis of carbon isotopes in otolith amino acids to trace fish migration and show high connectivity between coastal wetlands and shallow coral reefs. Analysis of otolith nitrogen content is also a promising avenue for future research and may provide a valuable tracer of anthropogenic impacts on fish migration, but this approach has not yet been applied in a marine setting (Grønkjær *et al.* 2013). Because the organic matrix of otoliths represents only a very small fraction (<10%) of their mass, broader application of these later approaches is impeded by technical challenges in how material is extracted from otoliths. Future technological advances will overcome these operational challenges and bring greater precision to studies that use ecogeochemical markers to quantify connectivity in coastal seascapes.

9.3.3   Genetics

Genetic markers, such as microsatellites, are popular tools for resolving structure in metapopulations; they can be used to determine the spatial origins of individual animals and to identify their parents (through parentage analysis) (reviewed by Kool *et al.* 2013). Genetic approaches are favoured over other methods for quantifying connectivity because genes are encoded in all animal tissues, which means that organisms do not need to be tagged (Andreou *et al.* 2012) and DNA samples can be collected without having to kill or capture individuals (Taberlet *et al.* 1999). Riginos & Liggins (2013) provide an in-depth review of the many ways that genetics can be used to measure seascape connectivity; thus, here, we examine how genetic analyses can be combined with ecogeochemical markers to improve our knowledge of connectivity in coastal seascapes. As a result of the longevity of many fish species, adult populations often comprise multiple age cohorts, which make it possible to compare genetic diversity on a cohort-by-cohort basis and thus contrast adults with juveniles (Leis *et al.* 2011). In combination with linking adult and juvenile residency areas through otolith chemistry, seascape connectivity can be measured at both ecological and evolutionary timescales. While this integrative approach shows great potential, few studies have used it thus far (Selkoe *et al.* 2008). Bradbury *et al.* (2008) used both genetics and otolith chemistry to examine dispersal of juvenile smelt between rivers and coastal bays. While their results showed genetic differentiation between spatially separated (>25 km) estuaries, a mixture of individuals from different spawning locations was observed in estuaries in closer proximity (<25 km). This combined approach is gaining traction in studies that focus on connectivity between rivers and estuaries (Woods *et al.* 2010; Schmidt *et al.* 2014), but it is yet to be applied in marine seascapes.

Technological advances in the tools that are used to quantify animal movement (*i.e.*, tags, ecogeochemical markers, genetics), together with new computational approaches for modelling connectivity (see Chapters 7 and 10 in this book), will improve our understanding of how animal movements maintain connections in seascapes. Individual tools can be used to describe movement patterns and define spatial linkages, but the greatest progress will be made by studies that integrate multiple and complementary tools to quantify the spatial and temporal scales over which animal movements connect populations and ecosystems (*e.g.*, Harrison *et al.* 2012; McMahon *et al.* 2012).

9.4 Application of Seascape Connectivity to Coastal Seascapes: Focal Topics

9.4.1 Focal Topic 1: Fish Movements Connecting Tropical Coastal Seascapes

Dispersal and movement form an intricate part of the life cycle of many marine organisms and are key structuring forces of metapopulations and food webs (Shima *et al.* 2010; Lipcius & Ralph 2011). Most marine organisms have a dispersive larval stage and multiple life stages are often completed in different ecosystems (Werner 1988; Lindeman *et al.* 2000; Pittman & McAlpine 2003). Hence, throughout ontogeny, movement and dispersal link habitats and ecosystems (Kinlan *et al.* 2005) and influence survival, growth, species interactions, habitat use and individual fitness (Grol *et al.* 2011; Rudolf & Rasmussen 2013).

Postsettlement movements connect ecosystems through day-night and tidal changes in habitat use and through ontogenetic and spawning migrations (Krumme 2009; Nagelkerken 2009; Berkström *et al.* 2012). These movement types have been widely studied for a range of species, but often from vastly different ecological perspectives. Many studies focus on individuals, using natural or artificial tags to understand the history, or final destination, of movement (*e.g.*, using otolith microchemistry, acoustic tags, or tissue stable isotopes). By contrast, many other studies focus on the environments fish inhabit, trying to understand how and why different ecosystems harbour different species, densities and size classes and infer how these are connected through fish migration. Movements of individuals (*e.g.*, home range movements) are often studied to quantify connections among populations or ecosystems or evaluate the function of seascape structure, whereas studies with an environmental focus typically seek to investigate the value of specific habitats as nurseries for juvenile fishes. To better understand how animals use seascapes and how ecosystems are connected through fish movements requires better integration of research on animal movement and habitat use. Nagelkerken *et al.* (2015) proposed a 'seascape nursery' approach for studying fish movement and habitat use; the authors postulated that we must now focus on identifying recruitment and postsettlement density hotspots, consecutive life stage habitats, movement corridors and home ranges that encompass multiple habitats to understand how seascapes structure fish behaviour and enhance the productivity of offshore populations.

The spatial patterning of ecosystems, geomorphology, environmental factors and species interactions play a tremendous role in structuring habitat use and the fitness of individuals across seascapes (Sheaves 2009; Harborne *et al.* 2015; Gilby *et al.* 2016), but these attributes can also be rather difficult to quantify. Every seascape is likely to be unique because of different combinations or characteristics of the attributes above; there is also behavioural plasticity in how different individuals of a particular species use habitats (Nagelkerken *et al.* 2008; Hitt *et al.* 2011). Consequently, we must seek to identify general patterns in population structure, ecological functioning and productivity that operate across seascapes. Following are some examples that typify the complexity of interacting factors that characterize coastal seascapes and their use by juvenile fishes, for three types of connectivity: day-night, tidal and ontogenetic.

Day-night changeovers in fish communities are common in tropical waters with high water clarity (Collette & Talbot 1972; Krumme 2009). Structurally complex habitats, such as rocky reefs and mangroves, are used during the day as shelter sites by nocturnally active species and vice versa for diurnal species at night (Ogden & Ehrlich 1977; Nagelkerken *et al.* 2000a). Changeover in fish communities typically occurs during dusk and dawn in a relatively structured way in terms of species and size classes (McFarland *et al.* 1979). This is an example of functional complementarity, with seascapes being used as resources by different fish communities at different times (Nagelkerken *et al.* 2006). Because fishes show homing behaviour to their resting sites (Dorenbosch *et al.* 2004; Verweij & Nagelkerken 2007), the distance fish can move into the surrounding habitats to forage is a function of their swimming capabilities and tidal amplitude. As a result, the position of resting sites can affect the composition and functioning of surrounding seascapes through the activities of the fish they harbour. For example, herbivorous fishes feed close to their patch-reef shelter sites to reduce predation risk and create grazing 'halos' in seagrass beds around these reefs (Ogden *et al.* 1973) that are visible from space (Madin *et al.* 2011). Furthermore, herbivory is often higher in the centre of seagrass patches than at the edges where predation risk is elevated (Pages *et al.* 2014). During inactive, resting periods, fish can enhance coral growth on patch reefs through defecation (Meyer *et al.* 1983, but see Burkepile *et al.* 2013). Similarly, nutrients excreted by fish during their feeding movements can enhance seagrass growth (Allgeier *et al.* 2013;

Layman *et al.* 2013) (Figure 9.3), while grazing on seagrass leaves leads to higher seagrass production rates (Christianen *et al.* 2012). Hence, the position of shelter and feeding habitats within seascapes modifies the spatial dynamics of day-night habitat shifts for fishes. We still know very little about these processes, but it is clear that they can link the ecological functioning and productivity of habitats across seascapes.

Tidal connectivity occurs on a daily basis and plays a greater role in the ecology of fishes in macrotidal than microtidal seascapes (Krumme 2009). Tidal exchange is a process that enables organisms to: (i) gain access to otherwise inaccessible feeding areas at high tide (Lugendo *et al.* 2007; Giarrizzo *et al.* 2010); (ii) move away with ingression of larger predators during the rising tide (Rypel *et al.* 2007; Lee *et al.* 2014a); (iii) travel between pools, creeks, channels, floodplains, or other habitats and relocate their home range to potentially more profitable areas (Skov *et al.* 2011; Baguette *et al.* 2013). The importance of these functions relies on abiotic conditions, geomorphology and the spatial setting of seascapes (Davis *et al.* 2014a). Interactions between tide and the day-night cycles (Castellanos-Galindo & Krumme 2013) and among species (*e.g.*, predator-prey relationships, Helfman 1986) are also important. Habitats that are closer to (deeper) channels or estuary mouths are more likely to see ingression of larger predators and pose increased predation risk (Baker & Sheaves 2005; Sridharan & Namboothri 2015). In these areas, tidal movements to distant and shallower habitats with fewer predators might be very important. On the other hand, shallower inshore areas with fewer predators pose a greater risk of stranding during ebbing tides (Krumme *et al.* 2004). Predation risk is not, however, the only factor shaping tidal connectivity, with fish movements also occurring in response to species environmental tolerances (*e.g.*, salinity, turbidity, oxygen content, water temperature; Blaber 2000) and food availability (Hammerschlag *et al.* 2010). Clearly, there are areas within seascapes that are favourable in terms of the combined effects of optimal environmental factors, food supply, shelter availability and predator abundance (Fodrie *et al.* 2009; Yeager *et al.* 2012; Kimirei *et al.* 2015) and such areas may form hotspots of productivity and diversity (Nagelkerken *et al.* 2015). To improve the spatial conservation and management of seascapes we must examine how the spatial patterning of seascapes interacts with abiotic (*e.g.*, currents, tides, water quality) and biotic (*e.g.*, food, shelter, predation risk) factors to form hotspots of animal diversity and productivity.

Ontogenetic connectivity (Table 9.1) usually occurs over broader temporal and spatial scales than diel and tidal connectivity (kilometres versus tens of kilometres) (Grober-Dunsmore *et al.* 2009; Huijbers *et al.* 2013; Olds *et al.* 2013). The best studied case is probably that of ontogenetic movements by fishes and crustaceans from juvenile inshore habitats to offshore adult habitats (Able 2005; Adams *et al.* 2006; Nagelkerken 2009) and habitats are often discussed in the context of providing nurseries for juvenile animals (nursery role hypothesis, Beck *et al.* 2001). Some species use intermediate habitats while moving from settlement to adult habitats and one Caribbean fish species (*Haemulon flavolineatum*), for example, can use up to five consecutive habitats (Grol *et al.* 2014). Ontogenetic changes between inshore habitats, such as mangroves, patch reefs, seagrasses and macroalgal beds, have traditionally been linked to provision of better food or shelter resources, or fewer predators compared to adult habitats (Parrish 1989). The food and predator hypotheses have recently been challenged; with studies in the Caribbean (Grol *et al.* 2014) and western Indian Ocean (Kimirei *et al.* 2013b) showing that adult (coral reef) habitat actually provides more food sources for juvenile fishes than their nursery habitats. Predation risk is, however, greater on coral reefs, suggesting that predator avoidance may be the primary function of inshore nursery habitats. This contention is supported by a recent meta-analysis which showed that few fish species actually depend on mangrove carbon for their diet (Igulu *et al.* 2013). Some nursery habitats can, however, also harbour abundant predators (Sheaves 2001; Dorenbosch *et al.* 2009). Ontogenetic movement across seascapes is not only driven by external factors, but also by the phenotype and internal state of individuals (*e.g.*, morphology, physiology, behaviour) (Clobert *et al.* 2009). Because seascape setting can play a large role in terms of structuring the spatial distribution and abundance of both predators and food in nursery habitats (Kimirei *et al.* 2015), it also likely contributes to the tradeoffs that fish face when deciding which nursery sites to occupy (Dahlgren & Eggleston 2000; Halpin 2000; Grol *et al.* 2011). Studies of seascape connectivity often link patterns in fish populations to the spatial distribution of environmental features, however, these effects are rarely connected with potential underlying mechanisms (*e.g.*, changes in resource availability, risk or animal behaviour) and this is now a key challenge for advancing this field.

9.4.2   Focal Topic 2: Connectivity across the Land-Sea Interface

The global ocean's boundaries with the continents form one of the planet's most distinctive and longest interfaces, connecting the sea with the land. These regions are of immense and irreplaceable significance to humankind – historically, culturally and economically. From an ecological perspective, the land-sea boundaries constitute distinct ecotones containing high biological and geochemical interactions, *i.e.* hotspots (Coupland *et al.* 2007).

A defining feature, arguably the defining feature, of the land-sea boundaries are connections (see Chapter 11 in this book). These connections functionally couple elements of the landscape and their ecosystems, particularly the tripartite system of dunes, shores and the nearshore marine zone. What makes connections in this coupled system particularly strong, persistent and functionally significant are four main characteristics: (i) boundaries are very long in relation to habitat areas (*i.e.*, large perimeter to area ratios); (ii) boundaries between habitats and food webs are highly permeable, presenting no significant impediments to the flow of matter and organisms; (iii) several vectors operate to transport matter and disperse organisms (*e.g.*, wind, waves, tides); and (iv) organisms in receiving environments have evolved to efficiently process incoming organic matter and nutrients (Schlacher *et al.* 2008, 2015b).

On many shorelines, inputs of marine matter dominate the flows between landscape elements and food webs. This material consists mostly of dislodged marine algae and seagrass, carcasses of marine animals and smaller organic detrital particles (Colombini & Chelazzi 2003; Krumhansl & Scheibling 2012). Inputs of plant material – 'wrack' – are a pivotal process that has major consequences in terms of: (i) enhancing the abundance, biomass and diversity of animal assemblages associated with habitat formed by wrack (Rodil *et al.* 2015); (ii) subsidizing shore-based food webs with carbon and other nutrients (Lastra *et al.* 2015); and (iii) providing the substrate for nutrient remineralization and nitrogen fixation (Barreiro *et al.* 2013; Hamersley *et al.* 2015).

More animals die from disease and other causes than from direct kills by predators, making animal carcasses an abundant and widespread resource in most food webs

(Wilson & Wolkovich 2011). Carcasses of marine animals regularly wash up on shores, subsidizing consumers in these systems with a highly nutritional source of 'necromass' (Sikes & Slowik 2010). Traditionally, connections involving such 'beach carrion' have been neglected in food-web studies of marine shores (but see Polis & Hurd 1996). Recent work has, however, shown that carrion connections are important for a diversity of scavengers on marine shores (Schlacher *et al.* 2013a, 2013b). These scavengers encompass species of marine provenance primarily associated with the shore habitat itself (*e.g.*, ghost crabs) and include a diverse suite of vertebrate scavengers with strong terrestrial affinities (*e.g.*, raptors, foxes, coyotes, Tasmanian devils, reptiles; Brown *et al.* 2015; Huijbers *et al.* 2015b, 2015c; Schlacher *et al.* 2015a).

Whilst fluxes from the ocean to the continents are often the energetically dominant connections at land-sea interfaces (particularly on small islands with low *in situ* productivity; Polis & Hurd 1995), connections from the land to the shore equally operate, albeit to a smaller extent. Prominent examples of terrestrial-marine linkages are the spillover of vertebrate consumers feeding on marine shores and the export of nutrients and organic matter from catchments. Terrestrial carnivores and raptors that extend their foraging ranges to marine shores are surprisingly diverse, widespread and abundant (*e.g.*, Carlton & Hodder 2003); their feeding activity across habitat boundaries functionally links food webs in abutting terrestrial and marine habitats and may constitute a biological vector for moving marine nutrients upland (sensu Havik *et al.* 2014). Conversely, terrestrial nutrients are assimilated into beach-food webs when discharge plumes from estuaries impinge on beaches bordering estuarine inlets (Schlacher & Connolly 2009; see also Chapter 11 in this book).

A recurring contention in beach ecology is the concept that animal species are structured as metapopulations (Defeo & McLachlan 2005). This hypothesis requires significant longshore connections to operate (*i.e.*, dispersal of larvae or adults parallel to the shoreline in addition to well-documented cross-shore connections). Supporting evidence for long-distance dispersal along coasts can be obtained from observed range extensions of beach species (Schoeman *et al.* 2015) and population genetics of beach invertebrates (Bezuidenhout *et al.* 2014). Other long-range connections in which sandy shorelines are focal nodes are nesting of marine turtles on coastal dunes interspersed by movements at the scale of ocean basins by individuals between nesting events (Harris *et al.* 2015).

Connections at land-sea boundaries have frequently been contextualized in terms of their consequences for consumers and food webs in recipient systems, the types of vectors that cause the exchange of nutrients and matter across boundaries, or the effect of cross-boundary coupling on nutrient stoichiometry (Polis *et al.* 1997; Schade *et al.* 2005; Hyndes *et al.* 2014; Sitters *et al.* 2015). For example, that marine plant carbon stranded on beaches is readily assimilated by invertebrates has been conclusively shown (Mellbrand *et al.* 2011). Similarly, beach scavengers respond rapidly to inputs of marine carrion leading to quick distributional shifts in consumer populations following food falls (Schlacher *et al.* 2013a, 2013b). These results suggest a significant role of connections in these systems, but the full range of biological and ecological effects of cross-boundary flows remains to be demonstrated. Thus, in our view, future research may fruitfully target the role of connectivity in shaping population dynamics and resilience, individual fitness and predator-prey dynamics in assemblages inhabiting linked surf-beach-dune habitats (research priorities 1–7, Box 9.1). Exchanges in this system also operate via a broad spatial spectrum of connections, from highly localized inputs of food parcels (centimetres to metres) to ocean-basin linkages via El Niño southern oscillation. We also predict that answers to the question '*What spatial scale is 'ecologically relevant?*' will to a large extent depend on the ecological neighbourhood, home range, foraging ambits and life-history characteristic of species and individuals. Thus, an important focus of future connectivity work at the land-sea boundary will be to identify which factors (*e.g.*, species behaviour and mobility, spatial positioning of seascapes) shape the spatial dimensions and significance of connectivity (research priorities 2–5, Box 9.1).

Box 9.1 Priority questions for research on seascape connectivity in coastal marine ecosystems. Cited studies provide examples of potential approaches for examining each priority question

Priority research questions

1) **Organisms:** Connectivity effects on marine vertebrates have been studied widely. How widespread are connectivity effects on invertebrates and marine plants, what biological properties of seascapes influence their distributions or biological traits (*e.g.*, body size, physiological condition) and what benefits do linkages confer to populations? (*e.g.*, Van der Stocken *et al.* 2015).

2) **Ecosystems:** To what extent does connectivity structure populations in the open ocean and deep sea, across coastal interfaces and among seascapes that are commonly studied in isolation (*e.g.*, estuaries, surf zones of beaches)? (*e.g.*, Schlacher *et al.* 2010).

3) **Scale:** How does species movement biology interact with seascape composition to modify the scale of connectivity effects among coastal ecosystems? (*e.g.*, Pittman *et al.* 2014).

4) **Functions:** How widespread and prominent are connectivity effects on ecological processes (*e.g.*, herbivory, predation) and to what extent does this link the ecological functioning and productivity of habitats across seascapes? (*e.g.*, Pages *et al.* 2014).

5) **Resilience:** How does connectivity modify the resistance of ecological entities (from populations to ecosystems) to disturbance and affect their recovery from perturbation? (*e.g.*, Mumby & Hastings 2008).

6) **Mechanisms:** To what extent does the spatial patterning of seascapes structure the ecological mechanisms (*e.g.*, resource availability, predation risk, reproduction) that underpin animal movements among ecosystems? (*e.g.*, Kimirei *et al.* 2013b).

7) **Hotspots:** How does habitat heterogeneity interact with abiotic (*e.g.*, currents, tides, water quality) and biotic (*e.g.*, food, shelter, predation risk) factors to form hotspots of animal diversity and productivity in seascapes? (*e.g.*, Nagelkerken *et al.* 2015).

8) **Conservation:** In what seascapes and over what scales do connectivity effects on populations, assemblages or ecosystems most improve the impact of management actions and what connections are most critical for promoting biodiversity, ecosystem functioning and productivity? (*e.g.*, Olds *et al.* 2016).

9.5 Integrating Connectivity into Marine Spatial Planning

Connectivity is an important consideration in marine conservation, restoration and fisheries management because movements of organisms and matter promote diversity, resilience and productivity in metapopulations and metaecosystems (Loreau *et al.* 2003; Rudnick *et al.* 2012; Baguette *et al.* 2013). Spatial linkages are also pivotal for facilitating recovery from disturbance (Cumming 2011; Bernhardt & Leslie 2013). Recovery

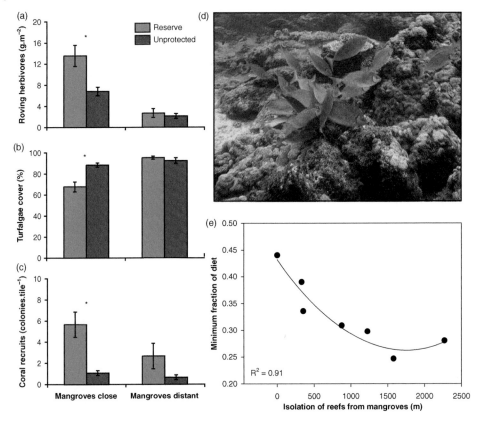

Figure 9.4 Combined effects of connectivity and seascape protection on fish, ecosystem functioning and food webs in eastern Australia. Herbivorous rabbitfish (*Siganus fuscescens*) are most abundant on coral reefs that are both near mangroves and protected in marine reserves (a); these fish feed on algae, which is less abundant on protected reefs near mangroves (b), which promotes coral recruitment (c). Rabbitfish feed on algae from both reefs (d) and mangroves and the contribution of mangrove carbon to their diet declines with reef isolation (e). Adapted from Olds *et al.* (2012) and Davis *et al.* (2014b). *Source*: Photograph by A. Olds.

occurs through the arrival of individuals or propagules from other populations (*i.e.*, recolonization effects) (Kool *et al.* 2013; Magris *et al.* 2014), or through animal migrations that link ecological functions among ecosystems (*i.e.*, mobile link effects; Lundberg & Moberg 2003; Olds *et al.* 2016).

Examples of how connectivity can improve marine spatial planning include: (i) optimizing the design of marine protected areas to link populations and functions among ecosystems (Olds *et al.* 2012; see also Chapter 14 in this book) (Figure 9.4), connect reserves in conservation networks (Pittman *et al.* 2014), or promote movement of harvested species across reserve boundaries (*i.e.*, spillover) (Harrison *et al.* 2012); (ii) enhancing performance of restoration projects by positioning sites to maximize natural recruitment (McMahon *et al.* 2015), or link habitat patches in fragmented landscapes (Reynolds *et al.* 2013); (iii) promoting conservation of migratory species by identifying and protecting nodes of critical habitat (*i.e.*, feeding, resting, breeding sites) or blue corridors on their dispersal paths (Pendoley *et al.* 2014; Fuentes *et al.* 2015) (Figure 9.5), an exercise that is particularly challenging for species that cross international boundaries (Runge *et al.* 2014); (iv) identifying seascape nurseries in coastal waters, which provide key habitats for juvenile fishes from offshore populations and function as hotspots for diversity and productivity (Nagelkerken *et al.* 2015); (v) improving fisheries management for species that are harvested as multiple stocks to ensure continued productivity across entire metapopulations (Moore & Simpfendorfer 2014); and (vi) identifying and conserving refugia for species that move to higher latitudes, or deeper cooler waters, with the thermal impacts of climate change (Makino *et al.* 2014). The sheer diversity in the spatial effects of connectivity and operational impediments to its quantification have, however, limited the scope and effectiveness with which connectivity has been integrated into conservation planning and spatial management (Moilanen *et al.* 2009; Rudnick *et al.* 2012; Magris *et al.* 2014).

Despite the importance of connectivity for marine spatial planning being widely appreciated (Lubchenco & Grorud-Colvert 2015), most marine reserve networks either fail to incorporate connectivity at all, or were designed using only qualitative estimates of its effects, which are often overly simplistic and therefore not likely to be particularly effective (Bernhardt & Leslie 2013; Magris *et al.* 2014). This is because until recently it has been difficult to validate models of connectivity with empirical data on the actual dispersal capabilities of organisms (Kool *et al.* 2013), or the scale over which ecological functions are linked among ecosystems (Hyndes *et al.* 2014).

The purported benefits of connectivity for marine spatial planning are also typically not considered by studies that quantify reserve effectiveness, restoration success, or fisheries productivity, which means we are unable to evaluate the role of connectivity effects and refine design parameters accordingly (Magris *et al.* 2014; Nagelkerken *et al.* 2015; Olds *et al.* 2016). To address this knowledge gap we must determine how connectivity affects the outcomes of management actions on populations, assemblages and ecosystems (Figure 9.4) and this requires empirical data on the spatial scale of connectivity effects and the seascapes in which they occur (Olds *et al.* 2016).

Human activities are, however, also directly altering the processes that link populations and ecosystems across seascapes (Saunders *et al.* 2015). Overharvesting is widespread in marine systems and has impacted populations of many marine animals (*e.g.*, fish, turtles, seals, whales, Jackson *et al.* 2001). Capturing large numbers of mobile organisms reduces the level of connectivity among populations of targeted species and may curtail the roles these animals play in linking ecological functions among ecosystems (Valentine *et al.* 2008; Estes *et al.* 2011). Animal migration is also impacted by engineering works that form barriers in seascapes (*e.g.*, sea walls, levees, bridges, Layman *et al.* 2007) and by coastal development that impinges on critical habitats along the routes of migratory species (*e.g.*, Murray *et al.* 2014) (Figure 9.5). These impacts likely limit the spatial and ecological extent of connectivity in seascapes and may therefore reduce the resilience and productivity of populations and ecosystems (Nagelkerken *et al.* 2015; Olds *et al.* 2016).

Many different tools and techniques are available for quantifying connectivity (Gillanders 2009; Riginos & Liggins 2013; Hyndes *et al.* 2014) and numerous software packages (*e.g.*, MARXAN, Zonation, ResNet) have been modified so that spatial linkages can be incorporated in marine spatial planning (Mumby 2006; Kool *et al.* 2013; Beger *et al.* 2015). The challenge is to bring these two complimentary fields of research together to

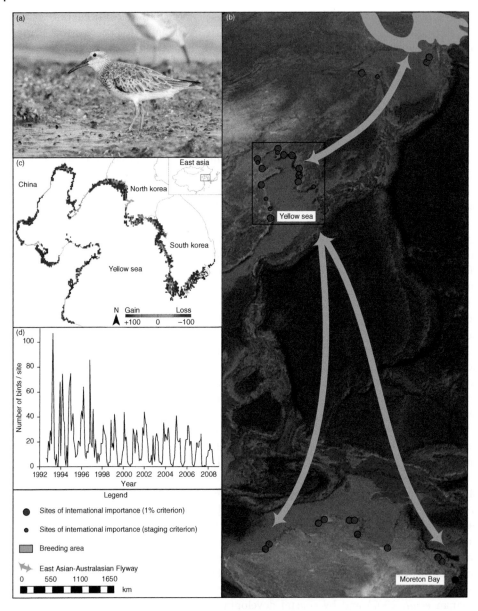

Figure 9.5 Great knots (*Calidris tenuirostris*) (a) migrate annually between feeding areas in Australia and breeding areas in eastern Russia (b). They depend on the tidal flats of the Yellow Sea as the sole resting site on this migration, but 65% of these flats have been lost to land reclamation since 1950 (c). Loss of this critical resting habitat has led to declines in great knot abundance in Moreton Bay (d), and across Australia (Bamford *et al.* 2008; Wilson *et al.* 2012; Murray *et al.* 2014). *Source*: Photograph by J. Harrison (CC BY-SA 3.0).

ensure that management decisions and conservation plans are based on empirical data on the scale of actual dispersal in seascapes and not qualitative estimates of connectivity (Magris *et al.* 2014; Olds *et al.* 2016).

To provide for comprehensive representations of connectivity in marine spatial planning we must: (i) determine the spatial scale over which connectivity affects species or ecosystems of interest (Hein *et al.* 2011; Martin *et al.* 2015); (ii) incorporate dispersal data for key connections that are pertinent to planning objectives (Beger *et al.* 2015; Magris *et al.* 2015); (iii) decide on a suitable planning approach that permits integration of the type/s of connectivity of interest (Kool *et al.* 2013; Olds *et al.* 2016); (iv) prioritize spatial configurations of management actions (*e.g.*, location, size and spacing of marine reserves) that match the dispersal capabilities of focal species (Green *et al.* 2014) and capture key nodes and corridors on the dispersal routes travelled by migratory species (Runge *et al.* 2014) (Figure 9.5); (v) determine whether prioritizing connectivity for one species or ecosystem results in tradeoffs with the spatial requirements of others, or conflicts with other planning considerations (*e.g.*, representation, habitat quality, ecosystem functioning) (Hodgson *et al.* 2011; Magris *et al.* 2014); and (vi) evaluate effectiveness of management actions and test whether connectivity improves outcomes for populations, assemblages or ecosystems (Harrison *et al.* 2012; Olds *et al.* 2016) (Figure 9.4).

Four recent studies have explored how integrating connectivity into marine spatial planning affects the position and configuration of conservation areas in reserve networks (Beger *et al.* 2010; Edwards *et al.* 2010; Beger *et al.* 2015; Magris *et al.* 2015). All focussed on conservation planning for coral reefs, but illustrate approaches that are likely to be more widely applicable in other seascapes. Beger *et al.* (2010) studied larval dispersal (modelled using pelagic larval duration, PLD) of coral trout among reefs on the Great Barrier Reef, Australia. Edwards *et al.* (2010) examined ontogenetic migration of fishes from mangroves to coral reefs (quantified as maximum dispersal distance) on the Belize Barrier Reef. Magris *et al.* (2015) modelled larval dispersal of four species (two corals and two fish) with varying PLDs among reefs in Brazilian seascapes. Beger *et al.* (2015) incorporated two types of connectivity for three different groups of species into conservation plans for reefs in the Coral Triangle. Dispersal of coral trout and sea cucumber larvae among reefs was modelled with PLD data and key nodes and pathways on the migration routes of sea turtles were quantified with telemetry data. These studies demonstrate that incorporating connectivity into the design of marine reserve networks results in prioritization of different sites for conservation to those that are selected on the basis of representation and habitat quality (Beger *et al.* 2010; Edwards *et al.* 2010). Their findings also suggest that complementary objectives and conservation plans may be needed to provide equitable representations of connectivity for species with differing dispersal capabilities (Beger *et al.* 2015; Magris *et al.* 2015).

9.6   Conclusions and Future Research Priorities

The importance of connectivity for populations, assemblages and ecosystems is widely appreciated and the published literature is dominated by studies on vertebrates conducted in shallow parts of the oceans located in tropical and subtropical regions. Less appears to be known about how connectivity structures invertebrate and plant populations (research priority 1, Box 9.1), or assemblages in the open ocean, deep sea, or land-sea interfaces (research priority 2, Box 9.1). The spatial patterning of seascapes likely affects the behaviour and movement biology of all species, but how this modifies the scale of connectivity effects is not well known. Recent technological advances in tools and techniques for quantifying movement and dispersal mean that it is becoming easier to measure connectivity. With empirical data on how seascapes affect animal migration and plant dispersal, we can refine estimates of the scale of connectivity effects and use these to improve the accuracy of spatial models for management of populations and ecosystems (research priority 3, Box 9.1).

Animal movements link ecological processes (*e.g.*, herbivory, predation) among ecosystems, but we do not know to what extent this functional connectivity links ecosystem functioning across seascapes (research priority 4, Box 9.1). Functional connectivity is believed to be pivotal for ecological resilience, but it is not clear whether the effects of connectivity on ecological processes actually translate into tangible benefits for how ecosystems cope with disturbance. To better integrate resilience theory into spatial ecology we need to know how connectivity shapes the relationship between disturbance intensity and frequency and the responses of ecological entities to these disturbance events (research priority 5, Box 9.1).

Studies of seascape connectivity often link patterns in animal populations to the spatial distribution of environmental features, however, these effects are rarely connected with potential underlying mechanisms (*e.g.*, changes in resource availability, risk or animal behaviour) and this is now a key challenge for advancing connectivity research (research priority 6, Box 9.1). To improve the spatial conservation and management of coastal ecosystems we must also examine how the spatial patterning of seascapes interacts with abiotic (*e.g.*, currents, tides, water quality) and biotic (*e.g.*, food, shelter, predation risk) factors to form hotspots of animal diversity and productivity (research priority 7, Box 9.1).

Connectivity is an important consideration for marine spatial planning, but it is currently rather poorly integrated into conservation, restoration and fisheries management. To address this knowledge gap we must determine how connectivity affects the outcomes of management actions on populations, assemblages and ecosystems and this requires empirical data on the spatial scale of connectivity effects and the seascapes in which they occur (research priority 8, Box 9.1). Many tools and techniques are available for quantifying connectivity and numerous computational approaches have been developed to allow incorporation of multiple different types of connectivity into spatial planning processes (see Chapters 7 and 10 in this book). We outlined a stepped approach for integrating these two complimentary fields of research and suggest that real benefits for conservation will occur where the planning process incorporates empirical data on the scale of actual dispersal in seascapes and where monitoring is conducted to determine how connectivity affects the outcomes of management actions.

References

Able KW (2005) A re-examination of fish estuarine dependence: evidence for connectivity between estuarine and ocean habitats. Estuarine, Coastal and Shelf Science 64: 5–17.

Adams AJ, Dahlgren CP, Kellison GT, Kendall MS, Layman CA, Ley JA, Nagelkerken I, Serafy JE (2006) Nursery function of tropical back-reef systems. Marine Ecology Progress Series 318: 287–301.

Allgeier JE, Yeager LA, Layman CA (2013) Consumers regulate nutrient limitation regimes and primary production in seagrass ecosystems. Ecology 94: 521–529.

Andreou D, Vacquie-Garcia J, Cucherousset J, Blanchet S, Gozlan RE, Loot G (2012) Individual genetic tagging for teleosts: an empirical validation and a guideline for ecologists. Journal of Fish Biology 80: 181–194.

Baguette M, Blanchet S, Legrand D, Stevens VM, Turlure C (2013) Individual dispersal, landscape connectivity and ecological networks. Biological Reviews 88: 310–326.

Baker R, Fry B, Rozas LP, Minello TJ (2013) Hydrodynamic regulation of salt marsh contributions to aquatic food webs. Marine Ecology Progress Series 490: 37–52.

Baker R, Sheaves M (2005) Redefining the piscivore assemblage of shallow estuarine nursery habitats. Marine Ecology Progress Series 291: 197–213.

Bamford M, Watkins D, Bancroft W, Tischler G, Wahl J (2008) Migratory shorebirds of the east Asian–Australian flyway: population estimates and internationally important sites. Wetlands International, Canberra, Australia.

Barreiro F, Gómez M, López J, Lastra M, de la Huz R (2013) Coupling between macroalgal inputs and nutrients outcrop in exposed sandy beaches. Hydrobiologia 700: 73–84.

Battley PF, Warnock N, Tibbitts TL, Gill RE, Piersma T, Hassell CJ, Douglas DC, Mulcahy DM, Gartrell BD, Schuckard R, Melville DS, Riegen AC (2012) Contrasting extreme long-distance migration patterns in bar-tailed godwits *Limosa lapponica*. Journal of Avian Biology 43: 21–32.

Beck MW, Heck KL, Jr., Able KW, Childers DL, Eggleston DB, Gillanders BM, Halpern B, Hays CG, Hoshino K, Minello TJ, Orth RJ, Sheridan PF, Weinstein MP (2001) The identification, conservation, and management of estuarine and marine nurseries for fish and invertebrates. Bioscience 51: 633–641.

Beger M, Linke S, Watts M, Game E, Treml E, Ball I, Possingham HP (2010) Incorporating asymmetric connectivity into spatial decision making for conservation. Conservation Letters 3: 359–368.

Beger M, McGowan J, Treml EA, Green AL, White AT, Wolff NH, Klein CJ, Mumby PJ, Possingham HP (2015) Integrating regional conservation priorities for multiple objectives into national policy. Nature Communications 6: 8208.

Berkström C, Gullström M, Lindborg R, Mwandya AW, Yahya SA, Kautsky N, Nyström M (2012) Exploring 'knowns' and 'unknowns' in tropical seascape connectivity: a review with insights from east African coral reefs. Estuarine, Coastal and Shelf Science 107: 1–21.

Bernhardt JR, Leslie HM (2013) Resilience to climate change in coastal marine ecosystems. Annual Review of Marine Science 5: 371–392.

Bezuidenhout K, Nel R, Hauser L (2014) Demographic history, marker variability and genetic differentiation in sandy beach fauna: What is the meaning of low FSTs? Estuarine, Coastal and Shelf Science 150: 120–124.

Blaber SJM (2000) Fish faunas and communities In Blaber SJM (ed.) Tropical Estuarine Fishes: Ecology, Exploitation, and Conservation. Blackwell Science, Malden.

Boström C, Pittman SJ, Simenstad C, Kneib RT (2011) Seascape ecology of coastal biogenic habitats: advances, gaps, and challenges. Marine Ecology Progress Series 427: 191–217.

Bouillon S, Connolly RM, Lee SY (2008) Organic matter exchange and cycling in mangrove ecosystems: recent insights from stable isotope studies. Journal of Sea Research 59: 44–58.

Bradbury IR, Campana SE, Bentzen P (2008) Estimating contemporary early life-history dispersal in an estuarine fish: integrating molecular and otolith elemental approaches. Molecular Ecology 17: 1438–1450.

Brown MB, Schlacher TA, Schoeman DS, Weston MA, Huijbers CM, Olds AD, Connolly RM (2015) Invasive carnivores alter ecological function and enhance complementarity in scavenger assemblages on ocean beaches. Ecology 96: 2715–2725.

Buelow C, Sheaves M (2015) A birds-eye view of biological connectivity in mangrove systems. Estuarine, Coastal and Shelf Science 152: 33–43.

Burkepile DE, Allgeier JE, Shantz AA, Pritchard CE, Lemoine NP, Bhatti LH, Layman CA (2013) Nutrient supply from fishes facilitates macroalgae and suppresses corals in a Caribbean coral reef ecosystem. Scientific Reports 3, doi: 10.1038/srep01493.

Campana SE (1999) Chemistry and composition of fish otoliths: pathways, mechanisms and applications. Marine Ecology Progress Series 188: 263–297.

Carlton JT, Hodder J (2003) Maritime mammals: terrestrial mammals as consumers in marine intertidal communities. Marine Ecology Progress Series 256: 271–286.

Castellanos-Galindo GA, Krumme U (2013) Tidal, diel and seasonal effects on intertidal mangrove fish in a high-rainfall area of the tropical eastern Pacific. Marine Ecology Progress Series 494: 249–265.

Christianen MJA, Govers LL, Bouma TJ (2012) Marine megaherbivore grazing may increase seagrass tolerance to high nutrient loads. Journal of Ecology 100: 546–560.

Clobert J, Le Galliard J-F, Cote J, Meylan S, Massot M (2009) Informed dispersal, heterogeneity in animal dispersal syndromes and the dynamics of spatially structured populations. Ecology Letters 12: 197–209.

Collette BB, Talbot FH (1972) Activity patterns of coral reef fishes with emphasis on nocturnal-diurnal changeover. Bulletin of the Natural History Museum of Los Angeles County 14: 98–124.

Colombini I, Chelazzi L (2003) Influence of marine allochthonous input on sandy beach communities. Oceanography and Marine Biology An Annual Review 41: 115–159.

Coupland GT, Duarte CM, Walker DI (2007) High metabolic rates in beach cast communities. Ecosystems 10: 1341–1350.

Cowen RK, Gawarkiewicz G, Pineda J, Thorrold SR, Werner FE (2007) Population connectivity in marine systems: an overview. Oceanography 20: 14–21.

Cumming GS (2011) Spatial resilience: integrating landscape ecology, resilience, and sustainability. Landscape Ecology 26: 899–909.

Dahlgren CP, Eggleston DB (2000) Ecological processes underlying ontogenetic habitat shifts in a coral reef fish. Ecology 81: 2227–2240.

Davis B, Baker R, Sheaves M (2014a) Seascape and metacommunity processes regulate fish assemblage structure in coastal wetlands. Marine Ecology Progress Series 500: 187–202.

Davis JP, Pitt KA, Fry B, Olds AD, Connolly RM (2014b) Seascape-scale trophic links for fish on inshore coral reefs. Coral Reefs 33: 897–907.

Defeo O, McLachlan A (2005) Patterns, processes and regulatory mechanisms in sandy beach macrofauna: a multi-scale analysis. Marine Ecology Progress Series 295: 1–20.

Dorenbosch M, Grol MGG, de Groene A, van Der Velde G, Nagelkerken I (2009) Piscivore assemblages and predation pressure affect relative safety of some back-reef habitats for juvenile fish in a Caribbean bay. Marine Ecology Progress Series 379: 181–196.

Dorenbosch M, Verweij MC, Nagelkerken I, Jiddawi N, Van der Velde G (2004) Homing and daytime tidal movements of juvenile snappers (Lutjanidae) between shallow-water nursery habitats in Zanzibar, western Indian Ocean. Environmental Biology of Fishes 70: 203–209.

Edwards HJ, Elliott IA, Pressey RL, Mumby PJ (2010) Incorporating ontogenetic dispersal, ecological processes and conservation zoning into reserve design. Biological Conservation 143: 457–470.

Elsdon TS, Wells BK, Campana SE, Gillanders BM, Jones CM, Limburg KE, Secor DH, Thorrold SR, Walther BD (2008) Otolith chemistry to describe movements and life-history parameters of fishes: hypotheses, assumptions, limitations and inferences. Oceanography and Marine Biology: An Annual Review 46: 297–330.

Estes JA, Terborgh J, Brashares JS, Power ME, Berger J, Bond WJ, Carpenter SR, Essington TE, Holt RD, Jackson JBC, Marquis RJ, Oksanen L, Oksanen T, Paine RT, Pikitch EK, Ripple WJ, Sandin SA, Scheffer M, Schoener TW, Shurin JB, Sinclair ARE, Soule ME, Virtanen R, Wardle DA (2011) Trophic downgrading of planet earth. Science 333: 301–306.

Finn JT, Brownscombe JW, Haak CR, Cooke SJ, Cormier R, Gagne T, Danylchuk AJ (2014) Applying network methods to acoustic telemetry data: modelling the movements of tropical marine fishes. Ecological Modelling 294: 139–149.

Fodrie FJ, Levin LA, Lucas AJ (2009) Use of population fitness to evaluate the nursery function of juvenile habitats. Marine Ecology Progress Series 385: 39–49.

Fry B (2006) Stable isotope ecology. Springer, New York, NY.

Fuentes MMPB, Chambers L, Chin A, Dann P, Dobbs K, Marsh H, Poloczanska ES, Maison K, Turner M, Pressey RL (2015) Adaptive management of marine mega-fauna in a changing climate. Mitigation and Adaptation Strategies for Global Change 21(2): 209–224.

Galuardi B, Lutcavage M (2012) Dispersal routes and habitat utilization of juvenile Atlantic bluefin tuna, *Thunnus thynnus*, tracked with mini PSAT and archival tags. PLoS One 7: e37829.

Gaspar P, Georges JY, Fossette S, Lenoble A, Ferraroli S, Le Maho Y (2006) Marine animal behaviour: Neglecting ocean currents can lead us up the wrong track. Proceedings of the Royal Society B: Biological Sciences 273: 2697–2702.

Giarrizzo T, Krumme U, Wosniok W (2010) Size-structured migration and feeding patterns in the banded puffer fish *Colomesus psittacus* (Tetraodontidae) from north Brazilian mangrove creeks. Marine Ecology Progress Series 419: 157–170.

Gilby BL, Tibbetts IR, Olds AD, Maxwell PS, Stevens T (2016) Seascape context and predators override water quality effects on inshore coral reef fish communities. Coral Reefs 35(3): 979–990.

Gillanders BM (2009) Tools for studying biological marine ecosystem interactions – natural and artificial tags. In Nagelkerken I (ed.) Ecological Connectivity among Tropical Coastal Ecosystems. Springer, Heidelberg.

Gillanders BM, Able KW, Brown JA, Eggleston DB, Sheridan PF (2003) Evidence of connectivity between juvenile and adult habitats for mobile marine fauna: an important component of nurseries. Marine Ecology Progress Series 247: 281–295.

Green AL, Maypa AP, Almany GR, Rhodes KL, Weeks R, Abesamis RA, Gleason MG, Mumby PJ, White AT (2014) Larval dispersal and movement patterns of coral reef fishes, and implications for marine reserve network design. Biological Reviews 90: 1215–1247.

Grober-Dunsmore R, Pittman SJ, Caldow C, Kendall MS, Frazer TK (2009) A landscape ecology approach for the study of ecological connectivity across tropical marine seascapes. In Nagelkerken I (ed.) Ecological Connectivity among Tropical Coastal Ecosystems. Springer, Heidelberg.

Grol M, Nagelkerken I, Rypel AL, Layman CA (2011) Simple ecological trade-offs give rise to emergent cross-ecosystem distributions of a coral reef fish. Oecologia 165: 79–88.

Grol M, Rypel AL, Nagelkerken I (2014) Growth potential and predation risk drive ontogenetic shifts among nursery habitats in a coral reef fish. Marine Ecology Progress Series 502: 229–244.

Grønkjær P, Pedersen JB, Ankjærø TT, Kjeldsen H, Heinemeier J, Steingrund P, Nielsen JM, Christensen JT (2013) Stable N and C isotopes in the organic matrix of fish otoliths: validation of a new approach for studying spatial and temporal changes in the trophic structure of aquatic ecosystems. Canadian Journal of Fisheries and Aquatic Sciences 70: 143–146.

Halpin PM (2000) Habitat use by an intertidal salt-marsh fish: trade-offs between predation and growth. Marine Ecology Progress Series 198: 203–214.

Hamersley MR, Sohm JA, Burns JA, Capone DG (2015) Nitrogen fixation associated with the decomposition of the giant kelp *Macrocystis pyrifera*. Aquatic Botany 125: 57–63.

Hammerschlag N, Heithaus MR, Serafy JE (2010) Influence of predation risk and food supply on nocturnal fish foraging distributions along a mangrove-seagrass ecotone. Marine Ecology Progress Series 414: 223–235.

Harborne AR, Nagelkerken I, Wolff NH, Bozec Y-M, Dorenbosch M, Grol MGG, Mumby PJ (2015) Direct and indirect effects of nursery habitats on coral-reef fish assemblages, grazing pressure, and benthic dynamics. Oikos 125(7): 957–967.

Harris LR, Nel R, Oosthuizen H, Meÿer M, Kotze D, Anders D, McCue S, Bachoo S (2015) Paper-efficient multi-species conservation and management are not always field-effective: The status and future of western Indian Ocean leatherbacks. Biological Conservation 191: 383–390.

Harrison HB, Williamson DH, Evans RD, Almany GR, Thorrold SR, Russ GR, Feldheim KA, Herwerden Lv, Planes S, Srinivasan M, Berumen ML, Jones GP (2012) Larval export from marine reserves and the recruitment benefit for fish and fisheries. Current Biology 22: 1023–1028.

Havik G, Catenazzi A, Holmgren M (2014) Seabird nutrient subsidies benefit non-nitrogen fixing trees and alter species composition in South American coastal dry forests. PLoS One 9: e86381.

Hays GC, Scott R (2013) Global patterns for upper ceilings on migration distance in sea turtles and comparisons with fish, birds and mammals. Functional Ecology 27: 748–756.

Hazen EL, Maxwell SM, Bailey H, Bograd SJ, Hamann M, Gaspar P, Godley BJ, Shillinger GL (2012) Ontogeny in marine tagging and tracking science: technologies and data gaps. Marine Ecology Progress Series 457: 221–240.

Heck KL, Jr., Carruthers TJB, Duarte CM, Hughes AR, Kendrick G, Orth RJ, Williams SW (2008) Trophic transfers from seagrass meadows subsidize diverse marine and terrestrial consumers. Ecosystems 11: 1198–1210.

Hein AM, Hou C, Gillooly JF (2011) Energetic and biomechanical constraints on animal migration distance. Ecology Letters 15: 104–110.

Helfman GS (1986) Fish behaviour by day, night and twilight. In Pitcher TJ (ed.) The Behaviour of Teleost Fishes. Croom Helm, London.

Hitt S, Pittman SJ, Nemeth RS (2011) Diel movements of fishes linked to benthic seascape structure in a Caribbean coral reef ecosystem. Marine Ecology Progress Series 427: 275–291.

Hodgson JA, Moilanen A, Wintle BA, Thomas CD (2011) Habitat area, quality and connectivity: striking the balance for efficient conservation. Journal of Applied Ecology 48: 148–152.

Huijbers CM, Nagelkerken I, Debrot AO, Jongejans E (2013) Geographic coupling of juvenile and adult habitat shapes spatial population dynamics of a coral reef fish. Ecology 94: 1859–1870.

Huijbers CM, Nagelkerken I, Layman CA (2015a) Fish movement from nursery bays to coral reefs: a matter of size? Hydrobiologia 750: 89–101.

Huijbers CM, Schlacher TA, McVeigh RR, Schoeman DS, Olds AD, Brown MB, Ekanayake KB, Weston MA, Connolly RM (2015b) Functional replacement across species pools of vertebrate scavengers separated at a continental scale maintains an ecosystem function. Functional Ecology 30(6): 998–1005.

Huijbers CM, Schlacher TA, Schoeman DS, Olds AD, Weston MW, Connolly RM (2015c) Limited functional redundancy in vertebrate scavenger guilds fails to compensate for the loss of raptors from urbanized sandy beaches. Diversiy and Distributions 21: 55–63.

Huntington BE, Karnauskas M, Babcock EA, Lirman D (2010) Untangling natural seascape variation from marine reserve effects using a landscape approach. PLoS ONE 5: e12327.

Hyndes GA, Nagelkerken I, McLeod RJ, Connolly RM, Lavery PS, Vanderklift MA (2014) Mechanisms and ecological role of carbon transfer within coastal seascapes. Biological Reviews 89: 232–254.

Igulu MM, Nagelkerken I, Dorenbosch M, Grol MGG, Harborne AR, Kimirei IA, Mumby PJ, Olds AD, Mgaya YD (2014) Mangrove habitat use by juvenile reef fish: meta-analysis reveals that tidal regime matters more than biogeographic region. PLoS One 9: e114715.

Igulu MM, Nagelkerken I, van der Velde G, Mgaya YD (2013) Mangrove fish production is largely fuelled by external food sources: a stable isotope analysis of fishes at the individual, species, and community levels from across the globe. Ecosystems 16: 1336–1352.

Irlandi EA, Crawford MK (1997) Habitat linkages: the effect of intertidal saltmarshes and adjacent subtidal habitats on abundance, movement, and growth of an estuarine fish. Oecologia 110: 222–230.

Jackson JBC, Kirby MX, Berger WH, Bjorndal KA, Botsford LW, Bourque BJ, Bradbury RH, Cooke R, Erlandson J, Estes JA, Hughes TP, Kidwell S, Lange CB, Lenihan HS, Pandolfi JM, Peterson CH, Steneck RS, Tegner MJ, Warner RR (2001) Historical overfishing and the recent collapse of coastal ecosystems. Science 293: 629–638.

Jones GP, Planes S, Thorrold SR (2005) Coral reef fish larvae settle close to home. Current Biology 15: 1314–1318.

Kimirei IA, Nagelkerken I, Mgaya YD, Huijbers CM (2013a) The mangrove nursery paradigm revisited: otolith stable isotopes support nursery-to-reef movements by Indo-Pacific fishes. PLoS One 8: e66320.

Kimirei IA, Nagelkerken I, Slooter N, Gonzalez ET, Huijbers CM, Mgaya YD, Rypel AL (2015) Demography of fish populations reveals new challenges in appraising juvenile habitat values. Marine Ecology Progress Series 518: 225–237.

Kimirei IA, Nagelkerken I, Trommelen M, Blankers P, van Hoytema N, Hoeijmakers D, Huijbers CM, Mgaya YD, Rypel AL (2013b) What drives ontogenetic niche shifts of fishes in coral reef ecosystems? Ecosystems 16: 783–796.

Kinlan BP, Gaines SD, Lester SE (2005) Propagule dispersal and the scales of marine community process. Diversiy and Distributions 11: 139–148.

Kneib R (1997) The role of tidal marshes in the ecology of estuarine nekton. Oceanography and Marine Biology Annual Review 35: 163–220.

Kool JT, Moilanen A, Treml EA (2013) Population connectivity: recent advances and new perspectives. Landscape Ecology 28: 165–185.

Krumhansl KA, Scheibling RE (2012) Production and fate of kelp detritus. Marine Ecology Progress Series 467: 281–302.

Krumme U (2009) Diel and tidal movements by fish and decapods linking tropical coastal ecosystems. In Nagelkerken I (ed.) Ecological Connectivity Among Tropical Coastal Ecosystems. Springer, Heidelberg.

Krumme U, Saint-Paul U, Rosenthal H (2004) Tidal and diel changes in the structure of a nekton assemblage in small intertidal mangrove creeks in northern Brazil. Aquatic Living Resources 17: 215–229.

Lastra M, López J, Neves G (2015) Algal decay, temperature and body size influencing trophic behaviour of wrack consumers in sandy beaches. Marine Biology 162: 221–233.

Layman CA, Allgeier JE, Yeager LA, Stoner EW (2013) Thresholds of ecosystem response to nutrient enrichment from fish aggregations. Ecology 94: 530–536.

Layman CA, Quattrochi JP, Peyer CM, Allgeier JE (2007) Niche width collapse in a resilient top predator following ecosystem fragmentation. Ecology Letters 10: 937–944.

Lee C-L, Huang Y-H, Chung C-Y, Lin H-J (2014a) Tidal variation in fish assemblages and trophic structures in tropical Indo-Pacific seagrass beds. Zoological Studies 53: 56.

Lee SY, Primavera JH, Dahdouh-Guebas F, McKee K, Bosire JO, Cannicci S, Diele K, Fromard F, Koedam N, Marchand C, Mendelssohn I, Mukherjee N, Record S (2014b) Ecological role and services of tropical mangrove ecosystems: a reassessment. Global Ecology and Biogeography 23: 726–743.

Leis JM, Van Herwerden L, Patterson H (2011) Estimating connectivity in marine fish populations: what works best? Oceanography and Marine Biology: An Annual Review 49: 193–234.

Lindeman KC, Pugliese R, Waugh GT, Ault JS (2000) Developmental patterns within a multispecies reef fishery: management applications for essential fish habitats and protected areas. Bulletin of Marine Science 66: 929–956.

Lipcius RN, Ralph GM (2011) Evidence of source-sink dynamics in marine and estuarine species. In Liu J, Hill V, Morzillo AT, Wiens JA (eds) Sources, sinks and sustainability. Cambrigde University Press, Cambridge.

Loreau M, Mouquet N, Holt RD (2003) Meta-ecosystems: a theoretical framework for a spatial ecosystem ecology. Ecology Letters 6: 673–679.

Lubchenco J, Groud-Colvert K (2015) Making waves: The science and politics of ocean protection. Science 350: 382–382.

Lugendo BR, Nagelkerken I, Kruitwagen G, van der Velde G, Mgaya YD (2007) Relative importance of mangroves as feeding habitats for fishes: a comparison between mangrove habitats with different settings. Bulletin of Marine Science 80: 497–512.

Lundberg J, Moberg F (2003) Mobile link organisms and ecosystem functioning: implications for ecosystem resilience and management. Ecosystems 6: 87–98.

Madin EMP, Madin JS, Booth DJ (2011) Landscape of fear visible from space. Scientific Reports 1: 14.

Magris RA, Pressey RL, Weeks R, Ban NC (2014) Integrating connectivity and climate change into marine conservation planning. Biological Conservation 170: 207–221.

Magris RA, Treml EA, Pressey RL, Weeks R (2015) Integrating multiple species connectivity and habitat quality into conservation planning for coral reefs. Ecography 39(7): 649–664.

Makino A, Yamano H, Beger M, Klein CJ, Yara Y, Possingham HP (2014) Spatiotemporal marine conservation planning to support high-latitude coral range expansion under climate change. Diversity and Distributions 20: 859–871.

Martin TSH, Olds AD, Pitt KA, Johnston AB, Butler IR, Maxwell PS, Connolly RM (2015) Effective protection of fish on inshore coral reefs depends on the scale of mangrove-reef connectivity. Marine Ecology Progress Series 527: 157–165.

McFarland WN, Ogden JC, Lythgoe JN (1979) The influence of light on the twilight migrations of grunts. Environmental Biology of Fishes 4: 9–22.

McMahon K, Berumen ML, Thorrold SR (2012) Linking habitat mosaics and connectivity in a coral reef seascape. Proceedings of the National Academy of Science, USA 109: 15372–15376.

McMahon K, Dijk Kv, Ruiz-Montoya L, Kendrick GA, Krauss S, Waycott M, Verduin J, Lowe R, Statton J, Eloise B, Duarte C (2015) The movement ecology of seagrasses. Proceedings of the Royal Society B 281: 20140878.

McMahon KW, Hamady LL, Thorrold SR (2013) A review of ecogeochemistry approaches to estimating movements of marine animals. Limnology and Oceanography 58: 697–714.

Mellbrand K, Lavery PS, Hyndes G, Hambäck PA (2011) Linking Land and Sea: Different Pathways for Marine Subsidies. Ecosystems 14: 732–744.

Meyer JL, Schultz ET, Helfman GS (1983) Fish schools: an asset to corals. Science 220: 1047–1049.

Meynecke J-O, Lee SY, Duke NC (2008) Linking spatial metrics and fish catch reveals the importance of coastal wetland connectivity to inshore fisheries in Queensland, Australia. Biological Conservation 141: 981–996.

Micheli F, Peterson CH (1999) Estuarine vegetated habitats as corridors for predator movements. Conservation Biology 13: 869–881.

Moilanen A, Wilson KA, Possingham HP (2009) Spatial Conservation Prioritization: Quantitative Methods and Computational Tools. Oxford University Press, Oxford.

Moore BR, Simpfendorfer CA (2014) Assessing connectivity of a tropical estuarine teleost through otolith elemental profiles. Marine Ecology Progress Series 501: 225–238.

Mumby PJ (2006) Connectivity of reef fish between mangroves and coral reefs: algorithms for the design of marine resreves at seascape scales. Biological Conservation 128: 215–222.

Mumby PJ, Hastings A (2008) The impact of ecosystem connectivity on coral reef resilience. Journal of Applied Ecology 45: 854–862.

Murray NJ, Clemens RS, Phinn SR, Possingham HP, Fuller RA (2014) Tracking the rapid loss of tidal wetlands in the Yellow Sea. Frontiers in Ecology and the Environment 12: 267–272.

Musyl MK, Domeier ML, Nasby-Lucas N, Brill RW, McNaughton LM, Swimmer JY, Lutcavage MS, Wilson SG, Galuardi B, Liddle JB (2011) Performance of pop-up satellite archival tags. Marine Ecology Progress Series 433: 1–28.

Nagelkerken I (2009) Evaluation of nursery function of mangroves and seagrass beds for tropical decapods and reef fishes: patterns and underlying mechanisms. In Nagelkerken I (ed.) Ecological Connectivity Among Tropical Coastal Ecosystems. Springer, Heidelberg.

Nagelkerken I, Bothwell J, Nemeth RS, Pitt JM, Van der Velde G (2008) Interlinkage between Caribbean coral reefs and seagrass beds through feeding migrations by grunts (Haemulidae) depends on habitat accessibility. Marine Ecology Progress Series 368: 155–164.

Nagelkerken I, Dorenbosch M, Verberk WCEP, Cocheret de la Moriniere E, van Der Velde G (2000a) Day-night shifts of fishes between shallow-water biotopes of a Caribbean bay, with emphasis on the nocturnal feeding of Haemulidae and Lutjanidae. Marine Ecology Progress Series 194: 55–64.

Nagelkerken I, Dorenbosch M, Verberk WCEP, Cocheret de la Moriniere E, van Der Velde G (2000b) Importance of shallow-water biotopes of a Caribbean bay for juvenile coral reef fishes: patterns in biotope association, community structure and spatial distribution. Marine Ecology Progress Series 202: 175–192.

Nagelkerken I, Grol MGG, Mumby PJ (2012) Effects of marine reserves versus nursery habitat availability on structure of reef fish communities. PLoS One 7: e36906.

Nagelkerken I, Sheaves M, Baker R, Connolly RM (2015) The seascape nursery: a novel spatial approach to identify and manage nurseries for coastal marine fauna. Fish and Fisheries 16: 362–371.

Nagelkerken I, van der Velde G, Verberk WCEP, Dorenbosch M (2006) Segregation along multiple resource axes in a tropical seagrass fish community. Marine Ecology Progress Series 308: 78–89.

Odum EP (1968) Energy flow in ecosystems: a historical review. American Zoologist 8: 11–18.

Ogden JC, Brown RA, Salesky N (1973) Grazing by the echinoid *Diadema antillarum* Philippi: formation of halos around West Indian patch reefs. Science 182: 715–717.

Ogden JC, Ehrlich PR (1977) The behaviour of heterotypic resting schools of juvenile grunts (Pomadasyidae). Marine Biology 42: 273–280.

Olds AD, Albert S, Maxwell PS, Pitt KA, Connolly RM (2013) Mangrove-reef connectivity promotes the functioning of marine reserves across the western Pacific. Global Ecology and Biogeography 22: 1040–1049.

Olds AD, Connolly RM, Pitt KA, Pittman SJ, Maxwell PS, Huijbers CM, Moore BR, Albert S, Rissik D, Babcock RC, Schlacher TA (2016) Quantifying the conservation value of seascape connectivity: a global synthesis. Global Ecology and Biogeography 25: 3–15.

Olds AD, Pitt KA, Maxwell PS, Connolly RM (2012) Synergistic effects of reserves and connectivity on ecological resilience. Journal of Applied Ecology 49: 1195–1203.

Pages JF, Gera A, Romero J, Alcoverro T (2014) Matrix composition and patch edges influence plant–herbivore interactions in marine landscapes. Functional Ecology 28: 1440–1448.

Parrish JD (1989) Fish communities of interacting shallow-water habitats in tropical oceanic regions. Marine Ecology Progress Series 58: 143–160.

Pendoley KL, Schofield G, Wittock PA, Lerodiaconou D, Hays GC (2014) Protected species use of a coastal marine migratory corridor connecting marine protected areas. Marine Biology 161: 1455–1466.

Pittman SJ, McAlpine CA (2003) Movements of marine fish and decapod crustaceans: process, theory and application. Advances in Marine Biology 44: 205–294.

Pittman SJ, Monaco ME, Friedlander AM, Legare B, Nemeth RS, Kendall MS, Poti M, Clark RD, Wedding LM, Caldow C (2014) Fish with chips: tracking reef fish movements to evaluate size and connectivity of caribbean marine protected areas. PLoS One 9: e96028.

Pittman SJ, Olds AD (2015) Seascape ecology of fishes on coral reefs. In Mora C (ed.) Ecology of fishes on coral reefs. Cambridge University Press, Cambridge.

Polis GA, Anderson WB, Holt RD (1997) Toward an integration of landscape and food web ecology: The dynamics of spatially subsidized food webs. Annual Review of Ecology and Systematics 28: 289–316.

Polis GA, Hurd SD (1995) Extraordinarily high spider densities on islands – flow of energy from the marine to terrestrial food webs and the absence of predation. Proceedings of the National Academy of Sciences of the United States of America 92: 4382–4386.

Polis GA, Hurd SD (1996) Linking marine and terrestrial food webs: allochthonous input from the ocean supports high secondary productivity on small islands and coastal land communities. American Naturalist 147: 396–423.

Putman NF, Scanlan MM, Billman EJ, O'Neil JP, Couture RB, Quinn TP, Lohmann KJ, Noakes DL (2014) An inherited magnetic map guides ocean navigation in juvenile Pacific salmon. Current Biology 24(4): 446–450.

Randall JE (1965) Grazing effect on sea grasses by herbivorous reef fishes in the West Indies. Ecology 46: 255–260.

Reef R, Feller IC, Lovelock CE (2014) Mammalian herbivores in Australia transport nutrients from terrestrial to marine ecosystems via mangroves. Journal of Tropical Ecology 30: 179–188.

Reynolds LK, Waycott M, McGlathery KJ, Silliman B (2013) Restoration recovers population structure and landscape genetic connectivity in a dispersal-limited ecosystem. Journal of Ecology 101: 1288–1297.

Riginos C, Liggins L (2013) Seascape genetics: populations, individuals, and genes marooned and adrift. Geography Compass 7: 197–216.

Rodil IF, Olabarria C, Lastra M, Arenas F (2015) Combined effects of wrack identity and solar radiation on associated beach macrofaunal assemblages. Marine Ecology Progress Series 531: 167–178.

Rooker JR, Secor DH, De Metrio G, Schloesser R, Block BA, Neilson JD (2008) Natal homing and connectivity in Atlantic bluefin tuna populations. Science 322: 742–744.

Rosenbaum HC, Maxwell SM, Kershaw F, Mate B (2014) Long-range movement of humpback whales and their overlap with anthropogenic activity in the South Atlantic Ocean. Conservation Biology 28: 604–615.

Rudnick DA, Ryan SJ, Beier P, Cushman SA, Dieffenbach F, Epps CW, Gerber LR, Hartter J, Jenness JS, Kintsch J, Merelender AM, Perkl RM, Preziosi DV, Trombulak SC (2012) The role of landscape connectivity in planning and implementing conservation and restoration priorities. Issues in Ecology 16: 1–20.

Rudolf VHW, Rasmussen NL (2013) Ontogenetic functional diversity: Size structure of a keystone predator drives functioning of a complex ecosystem. Ecology 94: 1046–1056.

Runge CA, Martin TG, Possingham HP, Williis SG, Fuller RA (2014) Conserving mobile species. Frontiers in Ecology and the Environment 12: 395–402.

Rypel AL, Layman CA, Arrington DA (2007) Water depth modifies relative predation risk for a motile fish taxon in Bahamian tidal creeks. Estuaries and Coasts 30: 518–525.

Saunders MI, Brown CJ, Foley MM, Febria CM, Albright R, Mehling MG, Kavanaugh MT, Burfeind DD (2015) Human impacts on connectivity in marine and freshwater ecosystems assessed using graph theory: a review. Marine and Freshwater Research 67(3): 277–290.

Schade JD, Espeleta JF, Klausmeier CA, McGroddy ME, Thomas SA, Zhang L (2005) A conceptual framework for ecosystem stoichiometry: balancing resource supply and demand. Oikos 109: 40–51.

Schlacher TA, Connolly RM (2009) Land–ocean coupling of carbon and nitrogen fluxes on sandy beaches. Ecosystems 12: 311–321.

Schlacher TA, Rowden A, Dower J, Consalvey M (2010) Seamount science scales undersea mountains: new research and outlook. Marine Ecology 31: 1–13.

Schlacher TA, Schoeman DS, Dugan JE, Lastra M, Jones A, Scapini F, McLachlan A (2008) Sandy beach ecosystems: key features, sampling issues, management challenges and climate change impacts. Marine Ecology 29 (S1): 70–90.

Schlacher TA, Strydom S, Connolly RM (2013a) Multiple scavengers respond rapidly to pulsed carrion resources at the land–ocean interface. Acta Oecologica 48: 7–12.

Schlacher TA, Strydom S, Connolly RM, Schoeman D (2013b) Donor-control of scavenging food webs at the land-ocean interface. PLoS One 8: e68221.

Schlacher TA, Weston MA, Lynn D, Schoeman DS, Huijbers CM, Olds AD, Masters S, Connolly RM (2015a) Conservation gone to the dogs: when canids rule the beach in small coastal reserves. Biodiversity and Conservation 24: 493–509.

Schlacher TA, Weston MA, Schoeman DS, Olds AD, Huijbers CM, Connolly RM (2015b) Golden opportunities: a horizon scan to expand sandy beach ecology. Estuarine, Coastal and Shelf Science 157: 1–6.

Schmidt DJ, Crook DA, Macdonald JI, Huey JA, Zampatti BP, Chilcott S, Raadik TA, Hughes JM (2014) Migration history and stock structure of two putatively diadromous teleost fishes, as determined by genetic and otolith chemistry analyses. Freshwater Science 33: 193–206.

Schoeman DS, Schlacher TA, Jones AR, Murray A, Huijbers CM, Olds AD, Connolly RM (2015) Edging along a warming coast: a range extension for a common sandy beach crab. PLoS One 10: e0141976.

Selkoe KA, Henzler CM, Gaines SD (2008) Seascape genetics and the spatial ecology of marine populations. Fish and Fisheries 9: 363–377.

Sheaves M (2001) Are there really few piscivorous fishes in shallow estuarine habitats? Marine Ecology Progress Series 222: 279–290.

Sheaves M (2005) Nature and consequences of biological connectivity in mangrove systems. Marine Ecology Progress Series 302: 293–305.

Sheaves M (2009) Consequences of ecological connectivity: the coastal ecosystem mosaic. Marine Ecology Progress Series 391: 107–115.

Shima JS, Noonburg EG, Phillips NE (2010) Life history and matrix heterogeneity interact to shape metapopulation connectivity in spatially structured environments. Ecology 91: 1212–1224.

Sikes DS, Slowik J (2010) Terrestrial arthropods of pre- and post-eruption Kasatochi Island, Alaska, 2008–2009: a shift from a plant-based to a necromass-based food web. Arctic, Antarctic, and Alpine Research 42: 297–305.

Sitters J, Atkinson CL, Guelzow N, Kelly P, Sullivan LL (2015) Spatial stoichiometry: Cross-ecosystem material flows and their impact on recipient ecosystems and organisms. Oikos 124: 920–930.

Skov C, Baktoft H, Bronmark C, Chapman B, Hansson LA, Nilsson PA (2011) Sizing up your enemy: individual predation vulnerability predicts migratory probability. Proceedings of the Royal Society B 278: 1414–1418.

Sridharan B, Namboothri N (2015) Factors affecting distribution of fish within a tidally drained mangrove forest in the Andaman and Nicobar Islands, India. Wetlands Ecology and Management 23: 909–920.

Taberlet P, Waits LP, Luikart G (1999) Noninvasive genetic sampling: look before you leap. Trends in Ecology & Evolution 14: 323–327.

Thomas CJ, Lambrechts J, Wolanski E, Traag VA, Blondel VD, Deleersnijder E, Hanert E (2014) Numerical modelling and graph theory tools to study ecological connectivity in the Great Barrier Reef. Ecological Modelling 272: 160–174.

Treml EA, Halpin PN, Urban DL, Pratson LF (2008) Modelling population connectivity by ocean currents, a graph-theoretic approach for marine conservation. Landscape Ecology 23: 19–36.

Trueman CN, MacKenzie KM, Palmer MR (2012) Identifying migrations in marine fishes through stable-isotope analysis. Journal of Fish Biology 81: 826–847.

Unsworth RKF, De León PS, Garrard SL, Jompa J, Smith DJ, Bell JJ (2008) High connectivity of Indo-Pacific seagrass fish assemblages with mangrove and coral reef habitats. Marine Ecology Progress Series 353: 213–224.

Valentine JF, Heck KL, Jr., Blackmon D, Goecker ME, Christian J, Kroutil RM, Peterson BJ, Vanderklift MA, Kirsch KD, Beck M (2008) Exploited species impacts on trophic linkages along reef-seagrass interfaces in the Florida keys. Ecological Applications 18: 1501–1515.

Vanderklift MA, Wernberg T (2008) Detached kelps from distant sources are a food subsidy for sea urchins. Oecologia 157: 327–335.

Van der Stocken T, De Ryck D, Vanschoenwinkel B, Deboelpaep E, Bouma TJ, Dahdouh-Guebas F, Koedam N (2015) Impact of landscape structure on propagule dispersal in mangrove forests. Marine Ecology Progress Series 524: 95–106.

Verweij MC, Nagelkerken I (2007) Short and long-term movement and site fidelity of juvenile Haemulidae in back-reef habitats of a Caribbean embayment. Hydrobiologia 592: 257–270.

Werner EE (1988) Size, scaling and the evolution of complex life cycles. In Ebenman B, Persson L (eds) Size-Structured Populations. Springer, New York, NY.

Wilson EE, Wolkovich EM (2011) Scavenging: How carnivores and carrion structure communities. Trends in Ecology and Evolution 26: 129–135.

Wilson HB, Kendall BE, Fuller RA, Milton DA, Possingham HP (2012) Analyzing variability and the rate of decline of migratory shorebirds in Moreton Bay, Australia. Conservation Biology 25: 758–766.

Woods RJ, Macdonald JI, Crook DA, Schmidt DJ, Hughes JM (2010) Contemporary and historical patterns of connectivity among populations of an inland river fish species inferred from genetics and otolith chemistry. Canadian Journal of Fisheries and Aquatic Sciences 67: 1098–1115.

Yeager LA, Acevedo CL, Layman CA (2012) Effects of seascape context on condition, abundance, and secondary production of a coral reef fish, *Haemulon plumierii*. Marine Ecology Progress Series 462: 231–240.

10

Networks for Quantifying and Analysing Seascape Connectivity

Eric A. Treml and Johnathan Kool

10.1   Introduction

In the 2000s, the use of graph (or network) theory became a popular tool in the study of seascape connectivity (Galpern *et al*. 2011; Treml *et al*. 2008; Urban *et al*. 2009). Graph theory is a branch of mathematics concerned with functional connections between discrete objects (Harary 1969) and is commonly used to explore and quantify the spatial and temporal patterns in real-world networks (*e.g.*, social networks, transportation systems, computer circuits, Internet). The use of graph-theoretic approaches has also increased significantly in quantifying and studying population dynamics (Kool *et al*. 2013) and in setting priorities in conservation applications (Beger *et al*. 2015; Magris *et al*. 2016; Minor & Urban 2008; Rayfield *et al*. 2011). Modelling populations or habitat mosaics (nodes in a graph) and the relationships among them (links in a graph) as a network structure (Figure 10.1) makes it possible to leverage the broad range of theories, algorithms and visualisation techniques associated with graph theory, although the specific manner in which they are applied depends on the seascape features and ecological processes represented by the structural elements of the graphs. How the features and ecological processes are defined, in turn, will depend on the research question of interest and the available data.

There are many different ways in which marine connectivity can be characterised. Indeed, the term 'connectivity' is somewhat ambiguous, context dependent and often applied to a particular spatial and temporal scale. Connectivity may refer to the structural or physical relationships among habitat patches, to a process such as dispersal or migration, or to a resultant spatiotemporal pattern. More specifically, population connectivity may refer to the biophysical functional relationships or the likelihood of individual movement among subpopulations over several generations (*e.g.*, demographic connectivity), or gene flow over the last millennia (*e.g.*, genetic or evolutionary connectivity). In this case, the linkages among subpopulations or habitat patches depend on the dispersal or movement capacity of the species, the structural topology of the habitat and the physical environment (*e.g.*, temperature, ocean currents, predators) between habitat patches (often called the matrix). The distinction lies in identifying the key physical and biological factors that are important to the connectivity question. As a result, the network and empirical approaches used to quantify and analyse marine population connectivity are highly diversified.

Seascape Ecology, First Edition. Edited by Simon J. Pittman.
© 2018 John Wiley & Sons Ltd. Published 2018 by John Wiley & Sons Ltd.

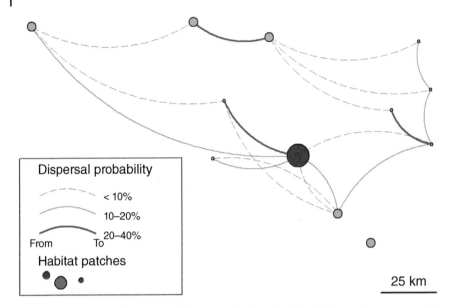

Figure 10.1 An example marine habitat network where functional linkages (edges or arcs) depict directional dispersal among discrete habitat patches (nodes). Direction in the linkages is implied by following arcs in a clockwise direction. Here, habitat area is illustrated by the relative size of the nodes and the colour represents whether the habitat is protected (blue) or unprotected (green).

To help characterise the various forms of marine connectivity, we highlight three broad connectivity typologies: structural, functional and realised. Structural connectivity incorporates information on the physical attributes of the seascape only. Within this typology, we recognise isolation-by-distance processes where the location, size, shape and separation of habitat patches or subpopulations are the focus, ignoring biological attributes such as dispersal (Calabrese & Fagan 2004). Functional connectivity incorporates additional information on the individual's dispersal or migration capacity. Functional connectivity may be characterised as 'potential connectivity' and may include information on reproduction, behavioural decisions, presettlement and postsettlement mortality and recruitment dynamics (Treml *et al.* 2015). Finally, realised connectivity refers to patterns derived from empirical data, either through direct observations, inferred from tracking or tagging data, or estimated from genetic data (often with embedded models of mutation and assumptions regarding population sizes, *etc.*). These three connectivity typologies are not necessarily mutually exclusive and therefore differences across typologies are often subtle. Clearly these differences, along with their assumptions, should be considered when exploring questions related to connectivity and when interpreting connectivity models and any inferred conclusions (Tischendorf & Fahrig 2000).

In this chapter, we introduce the basic elements of network theory and the process of network construction and analysis. We also discuss the empirical approaches for measuring connectivity, the insights from classic networks and the pitfalls of applying a graph-theoretical approach in the study of seascape ecology. To demonstrate the utility of the approach to address real world questions about seascape connectivity, we present a case study illustrating a network approach for mapping and analysing coral reef connectivity for the Hawaiian islands.

10.1.1   Structural Connectivity

The principal concept underpinning structural connectivity relates to the size, shape and physical arrangement of habitat patches and consequently their distances from one another. Quantifying the spatial structure of a seascape is often a necessary first step in the analysis of marine population connectivity. For example, the location and arrangement of habitat patches (*e.g.,* linear, clustered, random, isolated), or the minimum and mean distance among patch edges are aspects of the structural connectivity of a system. Several approaches have been developed and are widely used to quantify the structure of a seascape, often borrowing from the field of landscape ecology (Turner *et al.* 2001). For example, FRAGSTATS (McGarigal & Marks 1994) has been developed as a general-purpose tool to quantify aspects of landscape structure, including: patch cohesion index, connectance index and nearest neighbour metrics. Nearest neighbour geographic distance is often used as a proxy where movement is largely constrained by Euclidean (or over-water) distance alone. This may adequately represent broad geographic patterns of connectivity, but lacks the biophysical nuances of more advanced measures. As a result, quantifying population connectivity based purely on Euclidean or over-water distance is often criticised as it overlooks important seascape and biological attributes. For instance, by using simulated data to compare how several isolation measures predicted animal dispersal, Bender *et al.* (2003) showed that when patch size and shape were variable, nearest neighbour geographic distance was largely ineffective at predicting immigration rates.

In addition to geographic distance, connectivity estimates can also be based on the environmental similarity among populations (by calculating environmental distances rather than geographic distances), thereby including important environment parameters related to habitat suitability, which are lacking in simple distance-based estimates. Habitat suitability may be based on environmental parameters such as depth, mean net primary productivity and temperature variability, or habitat attributes such as patch size, perimeter and shape index. Since environmental distances do not directly incorporate information on the dispersal ability of individuals, these metrics assume that the connectivity is driven in large part by habitat suitability.

The structural connectivity of a seascape may also include the physical influence of ocean currents. Structural-based flow from a habitat patch upstream to a second patch downstream will be much shorter than flow from the downstream patch to the upstream habitat in the opposite direction. Incorporating these physical flows has been termed hydrological connectivity (Pringle 2003) and may be quantified in terms of the time it takes to flow between habitat patches (ignoring the biology of organisms). Hydrological (or hydrodynamic) connectivity can be modelled through Lagrangian, or Eulerian simulation methods, which incorporate ocean current flow fields and passive dispersers (Treml *et al.* 2008; Kool *et al.* 2011; Jönsson & Watson 2016). The addition of biological attributes to this hydrodynamic connectivity allows for the quantification of spatial and temporal dynamics of marine population connectivity and shifts the investigation to the functional relationships between habitat patches.

10.1.2 Functional Connectivity

Measurements of connectivity using purely physical or geographic attributes do not incorporate fundamental ecological processes that influence population connectivity, such as: reproduction, dispersal parameters, demographic processes, postsettlement mortality, or animal behaviour (Bélisle 2005; Treml *et al.* 2015). In contrast, functional connectivity incorporates species-level characteristics (*e.g.,* environmental tolerances and life history), population-specific attributes (*e.g.,* size, reproductive output) and / or individual behaviour (*e.g.,* vertical positioning, swimming behaviour). As a result, functional connectivity is a broad categorisation and can range from very simplistic estimates of potential connectivity to realistic models of effective connectivity that attempt to encapsulate the full suite of processes resulting in realised connectivity (next section). Potential connectivity may combine information on seascape structure with simple parameterisation for the species-specific dispersal abilities. Unlike spatially implicit approaches such as those based on geographic distance or environmental similarity, potential connectivity incorporates some individual biological attributes such as generalised survivorship during dispersal or maximum dispersal distances. Potential connectivity can also be estimated by developing movement resistance surfaces or cost surfaces to calculate distance-by-resistance estimates (McRae 2006). The potential connectivity among habitat patches can be defined by a distance-based dispersal kernel, or a function describing the relationship between distance travelled and the probability of arrival (Urban & Keitt 2001). This arrival probability can then be truncated at some critical distance indicating which connections may be meaningful for the study (*e.g.,* to focus on strong, demographically significant connections). Finally, a suite of functional connectivity models exists, incorporating particle tracking algorithms in flow fields where individual particles (representing individual disperses) have biological attributes such as behaviour, mortality, mobility and habitat preferences (see Chapter 8 in this book).

10.1.3 Realised Connectivity

Realised connectivity, sometimes referred to as 'actual' connectivity, includes the empirically based and often spatially explicit movement data of individuals through the seascape; it incorporates direct observations (or data-derived estimates) of emigration, dispersal rates and pathways, and immigration and thus provides direct, evidence-based estimates of population connectivity (Calabrese & Fagan 2004). There are numerous ways in which to measure realised connectivity, such as tracking animal movements, conducting mark-release-recapture or mass mark-recapture studies, or using a variety of genetic techniques. Importantly, these methods often require significant effort and resources, particularly when conducted at a regional level and each metric has its advantages and disadvantages. Furthermore, these measures rarely provide the opportunity to reconstruct individual movement pathways as they typically rely solely on linkages between sources and destinations only. For example, although tracking animal movements provides direct and precise estimates of realised connectivity, it typically provides only short-term data (several seasons or years) based on a limited number of individuals and is an inherently difficult and expensive undertaking, particularly with small or cryptic species. Although mark-recapture methods overcome some of these issues and have the additional benefit of providing information on population size, this technique is only appropriate if the wild and captured population sizes are sufficiently large. In marine environments, chemical signatures within the ear bones (otoliths) of fish provide an additional means of identifying source-destination population pairs (Neubauer *et al.* 2010). Genetic techniques have also been used to assess single and multigenerational connectivity among populations as well, although drawing inferences from these approaches requires models and associated assumptions relating to mutation rates, demographic and social structure and the mechanism of genetic exchange (Selkoe *et al.* 2008, 2016).

10.2   Network Models of Connectivity: Representing Pattern and Process

Once the biophysical context of the connectivity question is defined and the goals of the analysis have been articulated, the next step is to develop the most appropriate network model to represent the system's connectivity. Before this can be accomplished effectively, the spatial and temporal extent and resolution of the study, the biological attributes and the environment or habitat context all need to be clearly defined. Care should be taken to clearly outline the model assumptions and the appropriate or intended use of the results. Once the network model of marine population connectivity has been constructed, it will represent the key patterns and processes of the system, allowing for efficient graph analysis to quantify emergent connectivity patterns.

10.2.1   Defining Nodes and Links

Formally, a graph is a set of nodes (for definitions, see Table 10.1), which may be connected by some set of links (in landscape ecology; see Urban & Keitt 2001; Urban *et al.*

Table 10.1   Common network terms often applied to ecological and population networks. Also see Galpern *et al.* 2011; Urban *et al.* 2009; Minor & Urban (2008) for additional descriptions.

Network term	Definitions	Example and significance
Node (vertex)	Unique elements within a network that may be connected by links if they are associated in some way.	Nodes may represent suitable habitat patches, islands, subpopulations, or spawning sites.
Link (edge or arc)	A feature representing a functional connection between two nodes. Edge is synonymous with link and these can be binary (present or absent) or weighted (to represent strength or distance) and/or directional (referred to as arcs).	Links may represent distance, adjacency, dispersal or movement probability, flow of individuals, gene flow, similarity, or influence.
Degree	Total number of edges adjoining a node. Out-degree is the number of edges leaving a node. In-degree is the number of edges coming into a node.	The potential accessibility and influence of a source habitat patch to its downstream neighbours. Patches with a low in-degree may be more vulnerable to extinction if upstream sources are removed. Habitat with high out-degree values may be important source populations.

(Continued)

Table 10.1 (Continued)

Network term	Definitions	Example and significance
Hub	A node connected to many other nodes, thereby characterised as having a high degree.	In fragmented seascapes, conservation efforts may focus on protecting strong hubs, which have a greater capacity to rescue more downstream patches (*e.g.*, fished zones or other unprotected sites).
Path	Any sequence of edges connecting nodes through a network and in which no node is visited more than once.	Possible multi-generational dispersal stepping-stones through which genes move over time. If the network represents individual migrations or movements, the path may represent a potential corridor.
Betweenness centrality	The relative proportion of shortest paths between all possible node pairs that pass through the focal node or link of interest.	Habitat patches that may serve as important stepping-stones and highlights routes that may be important dispersal corridors.
Closeness centrality	The inverse of the average path distances from a node and to all other nodes in the network.	Habitat patches that occupy a central position in the network due to their proximity to other habitat patches.
Neighbourhood	The set of nodes that are within a given number of links (*i.e.*, order) from a focal node. The second-order neighbourhood includes nodes that are reachable in two or fewer steps.	Neighbourhood statistics are useful for quantifying the local importance of nodes. For example, a node's local influence could be quickly assessed by comparing that population's size or health to the nearest (1st order) neighbours.
Component	A connected subgraph or group of connected nodes for which a path (either direct or indirect) exists between all nodes.	Patches in the same component are mutually reachable. Conversely, there is no apparent movement between different components, implying some level of isolation.
Clustering coefficient	Average fraction of a focal node's immediate neighbours that are also connected as neighbours.	Highly clustered nodes may facilitate local dispersal and feedbacks. These nodes may be more resilient to localized disturbances due to redundant dispersal pathways within the local neighbourhood.
Cut node	A node whose removal breaks the graph into different components.	Cut nodes may be critically important for the cohesion of a network. If central cut nodes are removed or damaged, the broad network connectivity may be severely impacted.
Order	Total number of nodes in a graph.	The number of unique patches in the habitat network.
Degree distribution	The frequency distribution of node degrees for the entire network.	A skewed distribution due to the presence of several high-degree nodes (*i.e.*, hubs) may be indicative of some robustness in the network.

2009) (Figure 10.1). Nodes are used to represent habitat areas, subpopulations, or sites of interest in the connectivity analysis. In graph models of a marine population connectivity network, the minimum data structure for the nodes is a list of unique identification codes (*e.g.*, site names or numbers). It is, however, also useful to store real-world location data (*e.g.*, longitude and latitude) and other site attributes such as habitat size, quality, maximum population size and protected status. Although the majority of network analyses are independent of these other node attributes, they can be used to reselect, subset, or target network measures for specific types or suites of sites. The real-world feature that the nodes in a network represent may be point locations (*e.g.*, mass spawning sites or individual seamounts), areas of habitat (*e.g.*, reef patches, estuaries, or tuna breeding grounds), or volumes (*e.g.*, nearshore zone to 200 m). The geometry of the habitat patch or population will have implications for how links can be quantified. For example, with points, the pairwise distance calculations are clear and efficient, but may not accurately reflect actual distances between features if the site of interest is more appropriately represented as an area of habitat.

The structural, functional, or realised connectivity between habitat patches (nodes) is represented in the graph model as links (or edges) joining two habitat patches (Urban *et al.* 2009). Edges may represent binary adjacencies (*i.e.*, presence or absence of connectivity), be weighted to represent geographic distances or the likelihood of dispersal and be directional to illustrate asymmetric flow from one site to another. At one extreme, structural connectivity among habitat patches may be calculated by finding the Euclidean distance between sites. This may be modified to calculate a simple dispersal probability approximated by a negative-exponential decay function with respect to distance (Urban & Keitt 2001). At the other extreme, a very realistic functional connectivity estimate can be derived from spatially explicit biophysical models of larval dispersal incorporating spawning characteristics, larval behaviour, mortality, postsettlement survival, swimming and sensing capacities and temporally varying three-dimensional ocean currents (*e.g.*, Treml *et al.* 2015). Although these complex models may be difficult to validate, new techniques and technologies are continually being developed to provide new streams of information.

Regardless of how connectivity is defined or derived, the fundamental graph data structure for edges is the adjacency matrix or the edge list. The adjacency matrix is constructed where each element in the matrix, A_{ij}, represents the connectivity between the source site, i (row) and the destination site, j (column). For a binary adjacency matrix, only ones and zeroes are present, but it may be asymmetric ($A_{ij} \neq A_{ji}$) due to directed linkages. Similarly, the adjacency matrix may be weighted, consisting of probabilities, or fluxes or flow of individuals between sites, or the likelihood that an individual found in a destination site came from a particular source site (*i.e.*, a migration matrix). During graph construction, a threshold edge weight may be chosen to simplify the network structure to remove rare or weak connections or to identify only those linkages that are strong and demographically significant.

Depending on research goals and the objective of the model, networks can be either static or dynamic (Jacoby & Freeman 2016). Static networks are those where a single edge value represents linkages among nodes, often where dispersal or movement data have been pooled across time (*e.g.*, many spawning events or several decades), creating a directed and weighted adjacency matrix. These networks are important in illustrating how the seascape generally drives and shapes connectivity patterns. Conversely, dynamic networks explicitly incorporate a temporal dimension by including repeated dispersal or migration events through time (Jacoby & Freeman 2016). In dynamic networks, the directional movement between nodes is accompanied by a measure of time relating to previous and succeeding movement events and thus these models can reveal fundamental behavioural characteristics of the taxa and seascape of interest. Defining or determining the characteristics of the linkages (existence, weight, direction, persistence and dynamics) among habitat patches is a very time consuming aspect of constructing representative network models of marine population connectivity.

10.3 Modelling Marine Population Connectivity

Generating empirical data on larval dispersal and recruitment (*i.e.,* the network linkages) is difficult, as the small size and the potential distances covered during propagule development makes tracking an individual from a reproductive (source) site to a destination or settlement site and to monitor subsequent survival, inherently complex (but see Almany *et al.* 2007; Jones *et al.* 2005). Furthermore, population connectivity in marine systems approximated by the dispersal potential of larvae or propagules alone may be too simplistic, as the physical dynamics of the ocean environment (*e.g.,* currents, temperature, light) can interact with biological aspects of the disperser (*e.g.,* behaviour, growth and condition), all of which fluctuate over a broad range of spatial and temporal scales influencing dispersal pathways and connectivity outcomes (Cowen & Sponaugle 2009; Liggins *et al.* 2013).

Despite growing research interest in the biological and physical drivers of dispersal and recruitment patterns in marine systems, there is still little empirically based evidence on how individual biological parameters interact with environmental factors in the dispersal phase to influence dispersal for any particular species. Therefore, biophysical models are increasingly being used to estimate dispersal outcomes and population connectivity in marine landscapes (Cowen *et al.* 2007).

Broadly speaking, for marine systems there are four primary components required to define marine population connectivity and, thus, four main questions that need to be addressed to appropriately quantify marine connectivity: (i) How important are local population demographics at the source site, (ii) What biological traits of the disperser can significantly influence dispersal pathways and trajectories, (iii) How does the physical environment influence dispersal potential and (iv) What postsettlement processes at the destination site contribute to recruitment. A model aimed at predicting or understanding connectivity, or testing questions relating to observed connectivity, or metapopulation dynamics, should strive to include all four components. These biophysical marine connectivity models make it possible to ask specific questions and make quantifiable and testable predictions about populations, systems, or processes that are difficult to observe directly.

Individual-based biophysical models of marine connectivity (see Chapter 6), simulate population connectivity by modelling the movement of discrete individual organisms, each with a set of attributes and behaviours (*e.g.,* spatial location, physiological traits, behavioural traits) that can change or adapt through time, allowing each individual to respond independently to interactions with each other and with their environment (Grimm & Railsback 2005; North *et al.* 2009). Individual-based models are bottom-up models, where population-level connectivity is dependent on the interactions and behaviours of autonomous individuals (DeAngelis & Grimm 2014). Duration of dispersal (Siegel *et al.* 2003), behaviour during dispersal (Irisson *et al.* 2010; Irisson *et al.* 2015; Murphy *et al.* 2011), growth (Lett *et al.* 2010) and survival (Burgess *et al.* 2012; Connolly & Baird 2010) can all be used to characterise individual-level behaviour used in these models. Additional processes related to realised connectivity can also be accommodated by estimates of habitat quality at the settlement site, incorporating postsettlement processes such as density dependence, growth, reproduction and survival, with the goal of generating more realistic estimates of local recruitment and established connections among populations (Liggins *et al.* 2013). Irrespective of the detailed modelling approach used to quantify the functional connectivity among habitat patches or populations, these connectivity estimates are efficiently summarised as a single adjacency matrix (or suite of matrices across spawning events or years).

10.3.1   Empirical Estimates of Marine Population Connectivity

A strong understanding of species' life histories, individual dispersal behaviour and physiology, environmental parameters, ocean currents and how all these factors interact is required to help develop a more comprehensive and accurate view of marine population connectivity. Only with this information can we ultimately aim to fully understand and predict population persistence and adaptation to climate change, as well as effectively manage marine populations and design marine protected areas. It is important to consider appropriate spatial and temporal sampling strategies that are relevant to key biological attributes to avoid adopting and generalizing misleading patterns (Liggins *et al.* 2013). Here, we describe various methods – both direct and indirect – used for obtaining empirical data on marine population connectivity, the latest applications of these approaches and briefly discuss their strengths and weaknesses.

Direct methods of measuring connectivity require some form of observation, marking or tagging to assess the extent of an organism's dispersal. Mark-recapture methods are often useful in assessing marine connectivity by providing rough estimates of population size and movement patterns. Even conventional external tags can provide important insights when applied at appropriate spatial scales and allow the estimation of demographic drivers that may influence population connectivity such as growth and survival. For example, by using external T–bar anchor tags to mark cod in sheltered and open habitats over a period of five years, Rogers *et al.* (2014) found that recruitment synchrony and connectivity between populations was driven by habitat structure. Marking of this kind, however, only delineates the spatial scale of individual movements between source and destination sites and, while useful for generating connectivity estimates, it cannot reveal specific movement pathways, including fine-scale habitat use or movement corridors. Newer tagging technologies such as acoustic and satellite tags do allow the analysis of movement pathways and have recently been used to monitor a range of organisms from small coral reef fishes (Pittman *et al.* 2014) to bluefin tuna (Galuardi *et al.* 2010). Although these technologies provide opportunities to track larger and highly mobile species, cost and logistics are a consideration and size requirements prevent transmitting tags from being used to track small dispersing larvae. In general, the use of tagging for connectivity estimates and monitoring of younger life stages is not common or broadly feasible (Hazen *et al.* 2012).

Mass mark recapture is an alternative form of tagging that uses hard parts of fish and invertebrate species (otoliths and statoliths, respectively) to embed an identifiable stable isotope tag in a cohort of individuals. This isotopic tag may be passed from parent to progeny, allowing for dispersing larvae to be tagged in advance by introducing the enriched isotope to the mother (Almany *et al.* 2007; Jones *et al.* 1999). However, studying multiple patches and dispersal events within a marine metapopulation using this method can be logistically challenging (Botsford *et al.* 2009; Cowen & Sponaugle 2009), leading to complications when identifying the source-destination pairs in collected individuals. To help address this challenge, it is possible to use microchemical signatures absorbed naturally into calcified structures from the surrounding water as a natural form of tagging. In this approach, comparing the microchemical characteristics of different habitat patches with those of larval calcified structures can reveal patterns of source-destination movement in a similar fashion to marking fish otoliths with stable isotopes. This has been used to describe patterns of self-recruitment and provided evidence for behaviourally moderated dispersal, as opposed to passive, neutrally buoyant particles (Swearer *et al.* 1999). Microchemistry is also being used to characterise larval dispersal 'histories', whereby broad larval behaviour strategies and larval quality can be investigated (Shima and Swearer 2009). However, this approach relies on distinct differences in the chemical properties between sampled locations and individuals remaining sedentary long enough to uptake identifiable elemental signatures. Consequently, the spatial resolution possible is often more coarse than other direct methods. Additionally, the challenges associated with estimating connectivity across a marine seascape of multiple populations using only a subset of potential source locations can be significant.

Gene flow and genetic diversity are also important for a population's adaptive capacity and long-term persistence. Consequently, assessing 'genetic connectivity' through analysing gene flow and genetic differentiation offers valuable insights into understanding realised linkages among populations and habitat patches, albeit indirectly. Obtaining empirical data on gene flow is therefore critical to quantifying long-term connectivity and assessing the consequences of dispersal at longer time scales across multiple generations (Lowe & Allendorf 2010). By collecting individuals and using genetic markers, researchers can trace genetic movements between habitat patches and use these estimates to quantify genetic connectivity among populations. Similar to direct mark-recapture methods, genotypic data can be compared to assign individuals to a specific subpopulation of origin or their parents to directly estimate dispersal kernels (Jackson *et al.* 2014). Genotyping provides very accurate estimates of connectivity if samples are comprehensive and combined with the appropriate spatial data (Jones *et al.* 2005; Planes *et al.* 2009), although it is difficult to achieve this across realistic scales or for nonmodel species. In practice, the capacity to collect enough samples, particularly among large populations, is a limitation to these methods, as one must sample a significant proportion of the adult population as well as the new recruits over the entire potential dispersal range to reliably quantify connectivity (Botsford *et al.* 2009). Despite this, developments in next generation genome sequencing will see the increased use of single nucleotide polymorphisms analysis (SNPs), allowing comparisons between individuals and populations at the finest genetic resolution possible (*e.g.,* Riginos *et al.* 2016). Genome scanning technology is now capable of identifying 'outlier' loci, or loci with extreme levels of differentiation affected by selection due to localised adaptation (Gagnaire *et al.* 2015). The ability to refine the assignment of individuals to populations and habitats is already being put to work to improve fisheries management and biodiversity conservation (Bradbury *et al.* 2013; Funk *et al.* 2012). With this finer spatial and temporal resolution in genetic data now becoming available, we are quickly improving our ability to approximate marine population connectivity across scales and taxa.

10.4   Network Analysis of Marine Population Connectivity

Having established the marine population connectivity context and having defined the nature of the graph's nodes and edges to form the connectivity network, the next stage is to use the suite of graph-theoretic statistics, algorithms and analytical methods to study aspects of the system (for tools see Table 10.2). These can be broadly classified

Table 10.2   Descriptions of the tools used for the development of habitat networks and for network analysis. References are provided for additional information.

Tool	Description	Examples of use in marine connectivity
Conefor: Quantifying the importance of habitat patches and links for landscape connectivity **URL**: http://www.conefor.org/ (accessed 25 May 2017) **Reference**: Saura & Torné (2009)	A stand-alone Windows-based computer program designed to quantify the importance of habitat patches and links in maintaining connectivity with the intent of prioritizing sites for ecological connectivity (previously known as Conefor Sensinode). A plugin exists for both QGIS and ArcGIS. The analysis is performed on the basic graph structure with additional functionality for 'spatial graphs.'	Espinoza *et al.* (2015)
Connectivity modelling system (CMS): multiscale stochastic Lagrangian framework **URL**: https://github.com/beatrixparis/connectivity-modelling-system (accessed 25 May 2017) **Reference:** Paris *et al.* (2013)	This Fortran based open-source probabilistic connectivity modelling program couples hydrodynamic data with a biologically based Lagrangian framework to track particle displacement based on ocean physics and larval behaviour.	Wood *et al.* (2014) Kough *et al.* (2013)
FRAGSTATS: Spatial Pattern Analysis Program for Categorical Maps **URL**: http://www.umass.edu/landeco/research/fragstats/fragstats.html (accessed 25 May 2017) **Reference**: McGarigal *et al.* (2002)	A stand-alone Windows-based computer program designed to compute a wide variety of landscape metrics (*e.g.*, structural connectivity) for categorical (habitat) map patterns. Data and metrics are based on the patch-corridor-matrix model of habitat. A wide diversity of raster (gridded) data are accepted as input, including classic Geographic Information System data.	Anadón *et al.* (2011) Meynecke *et al.* (2007)

(Continued)

Table 10.2 (Continued)

Tool	Description	Examples of use in marine connectivity
igraph: the network analysis package. URL: http://igraph.org/ (accessed 25 May 2017) **Reference:** Csardi & Nepusz (2006)	igraph is an open source and freely available collection of network analysis tools applicable to a wide variety of network models. igraph functionality can be accessed through R, Python and C/C++. igraph is available on Windows, Mac OS and Linux systems. igraph has extensive, robust and easily accessible functionality giving the user access to most classic and many very recent and innovative graph theoretic algorithms.	Kininmonth *et al.* (2010) Andrello *et al.* (2013)
Marine Geospatial Ecology Tools (MGET): Open source geoprocessing for marine research and conservation URL: http://mgel.env.duke.edu/ mget (accessed 25 May 2017) **References:** Roberts *et al.* (2010) Treml *et al.* (2012)	(MGET) is a free, open-source geoprocessing toolbox for ArcGIS that can help solve a wide variety of marine research, conservation and spatial planning problems. The 'Connectivity Analysis' Tool simulates marine larval transport based on biological parameters (*e.g.*, reproductive output, competency characteristics, behaviour, mortality) and various hydrodynamic data products. The tool facilitates habitat data development and hydrodynamic data download and running the dispersal simulations.	Schill *et al.* (2015) Mora *et al.* (2012)
NetLogo: a multiagent programmable modelling environment. URL: https://ccl.northwestern .edu/netlogo/ (accessed 25 May 2017) **Reference:** Wilensky (1999) NetLogo. Center for Connected Learning and Computer-Based Modelling, Northwestern University, Evanston, IL	A free and open-source cross-platform stand-alone programming environment requiring Java and runs on Windows, Mac OS, and Linux-based systems. This individual-based modelling environment is quite flexible. The network extension allows for performing network analysis within the NetLogo environment, including path distances, centrality and clustering algorithms.	Hovel & Regan (2008)
PATHMATRIX: A GIS tool to compute effective distances among samples. URL: http://cmpg.unibe.ch/ software/pathmatrix/ (accessed 25 May 2017) **Reference:** Ray (2005)	Pathmatrix is an extension to ESRI's ARCVIEW 3.x software. This tool computes functional connectivity distances based on a least-cost path algorithms connecting habitat patches. Habitat data and a cost surface are required to calculate three distance measures: least-cost distance, length of the least-cost path and Euclidean distance.	Fontaine *et al.* (2007)

into three different structural levels of analysis: node and neighbourhood-level metrics, components or subgraphs and network-wide analyses. A constructed graph may be enough for visualisation purposes and some level of spatial planning, yet to quantify more nuanced or emergent patterns in connectivity or to inform management decisions, network-based measures of population connectivity are needed. For an in-depth review of network metrics refer to Rayfield *et al.* (2011).

Clearly articulating the connectivity-related question, the scale of interest and the node / edge formulation is the critical first step in approaching the analysis of a marine population connectivity network. Areas of investigation might include:

- Studying the strength and persistence of downstream connectivity from focal source patches (*e.g.,* to what extent do individual source sites contribute to downstream site persistence?)
- Identifying the proportion of individuals originating at a habitat patch that return to their natal patch (*e.g.,* local retention – to what degree are individual sites self-sustaining?)
- Discovering key pathways and connections within the network (*e.g.,* which habitat patches or populations contribute to the overall cohesion of the network in terms of their likelihood of serving as critical stepping stone sites?)
- Investigating interactions between network components (*e.g.,* how does the population connectivity network naturally cluster into groups of strongly connected sites?)

Below, we describe common network-based approaches for addressing these problems from node-level analyses to network-wide analyses. Following this, we briefly discuss several 'classic' network topologies and highlight key behaviours and emergent properties that may be useful in understanding marine population connectivity networks and developing testable hypotheses.

10.4.1   Node and Neighbourhood-level Metrics

Node-level metrics describe local properties related to the individual connectedness of a node. These metrics, however, do not focus on elements independently; instead, they describe the role of a node in relation to its neighbours (Rayfield *et al.* 2011). Centrality measures, for example, are considered node-level metrics as they are calculated for each particular node or edge, but they are used to visualise and quantify network-wide patterns of connectivity and topology. Betweenness centrality is the proportion of shortest paths that pass through a particular node or link (Freeman 1978; Urban *et al.* 2009) and is a measure of the role each node or link plays in network traffic. A path is a route through a graph from one node to another; if two nodes are not nearest neighbours, the path between them will contain one or more intermediary nodes. Mapping betweenness scores on a graph is an effective way of identifying important dispersal pathways and highlighting nodes that, due to their position in the network, are key for population connectivity. For example, Treml *et al.* (2008) used betweenness scores to reveal the most used dispersal routes of coral larvae across different seasons in the Tropical Pacific. Another measure of centrality is closeness, which is the inverse of the average path length from a node to each of its neighbours (Urban & Keitt 2001). A node's degree is the number of neighbours for each node: out-degree is the number of connections linking to downstream neighbours and in-degree is the number of connections linking to upstream neighbours. A hub is a node that is connected to many other nodes (*i.e.*, a high-degree node). A node's neighbourhood is the set of neighbours immediately upstream and downstream, or more generally, the set of nodes that are within a given number of connections from a primary node. Consequently, neighbourhood level metrics can compare attributes of a focal node to those of the entire neighbourhood to give insights into the degree to which a node acts as a local source or sink (Rayfield *et al.* 2011). Another way of quantifying neighbourhood topology is through the identification of motifs (Proulx *et al.* 2005). Motifs are small, repeated patterns that occur significantly more often than expected from random networks. Network motifs are common across different networks (*e.g.*, food webs, gene regulatory networks, transcription networks; Milo *et al.* 2002), but they are yet to be identified in habitat networks. Rayfield *et al.* (2011) propose that the presence and number of source-sink motifs could indicate ongoing route-specific flux from sources to neighbouring sinks.

10.4.2 Components, Subgraphs and Clusters

Component and cluster measures are the next structural level in network analysis. A graph component is a connected subgraph (Urban *et al.* 2009), or group of interconnected nodes separated from the rest of the seascape. Movement can occur between any two nodes in a component, but cannot occur between nodes in different components (Minor & Urban 2008). Clustering refers to the probability that two nearest neighbours of the same node are also mutual neighbours (*i.e.*, one's friends tend to be friends with each other) and can be assessed with the clustering coefficient. When a connected graph (*i.e.*, a graph where paths exist between every pair of nodes) is broken into subgraphs or components by the removal of a key node, the graph has a cut-node at that point. Component and cluster level metrics focus, for example, on the number of cut-nodes or total number of nodes in a component (*i.e.*, component order).

Detecting and characterizing network community structure is an important area of research in network theory (Clauset *et al.* 2004; Palla *et al.* 2005). Methods for detecting groups in networks can be divided into two main areas: one is known as 'graph partitioning' and is mainly used in computer science and the other is 'hierarchical clustering' or 'community structure detection', developed by sociologists and, more recently, biologists (Newman 2006). Researchers in different fields have proposed multiple community detection algorithms (Fortunato 2010), but a standard benchmark in the field is the method proposed by Girvan & Newman (2002), which is based on the idea of using centrality measures to find community boundaries. The algorithm simply (i) calculates betweenness for all edges in the network, (ii) removes edges with highest betweenness, (iii) recalculates betweenness affected by the removal and (iv) repeats from (ii) until no edges remain. More recently, a method that has given good results is based on optimised function called modularity (Q = number of edges within communities – expected number of such edges placed at random; Newman 2006). A positive modularity implies possible presence of community structure and its value can be optimised by a range of algorithms, such as the leading eigenvector algorithm implemented in igraph (Csardi & Nepusz 2006).

10.4.3 Graph-level Metrics

Graph-level metrics assess connectivity properties of the entire graph. The total number of nodes is the graph's order and quite often remains constant. Yet, the total number of connections (*i.e.*, graph's size) and the strength and distribution of those links generally show spatial and temporal variations. Identifying these variations is crucial for conservation planning. A fundamental algorithm in graph theory is the Dijkstra (1959) solution to find the shortest route between any two nodes. This is used to identify the longest of the shortest paths between any two nodes, or the graph's diameter. Characteristic path length is the average shortest path length between all pairs of nodes in the network. The graph diameter is indicative of the trafficking speed through a network, whereas the characteristic path length describes the density of the network. The degree distribution of a network is the probability distribution of the number of edges a node can have across the entire network (West 2001), a skewed distribution and several large hubs suggest robustness to node removal and fast spread across the network (Minor & Urban 2008; Proulx *et al.* 2005). Furthermore, relationships between species networks and spatial networks (*i.e.*, where nodes have a fixed geographic location) can be modelled using meta-networks. For example, Devoto *et al.* (2014) used meta-networks to study the spatial and temporal use of floral resources by bumble bees. They defined meta-networks as the set of local plant communities interlinked by bumble bees foraging at different times of the year and used these data to model colony survival under different combinations of patch size and bumble bee flight distance.

10.4.4 Insights from Classic Networks

Graph theory has been recently incorporated in the field of landscape ecology (Urban and Keitt 2001) and there is a solid foundation originating from computer and social sciences that can be used to analyse connectivity in ecological networks (Barabasi & Albert 1999; Watts & Strogatz 1998). All networks – from friendship groups to connected habitat patches – have emergent properties. Network topology is one such emergent property and is particularly interesting because it affects network characteristics such as connectivity, resilience to disturbance and the spread of information or diseases (Minor & Urban 2008). In most habitat networks, an intermediate level of connectivity is preferred; if a network is too disconnected, patches will be isolated from each other and, if it is highly connected, a disease or invasive species may spread quickly. Resilience, or network robustness, is defined in the network literature as the number of nodes that can be removed from a network without altering its connectivity. Robustness is strongly correlated to the node-degree distribution, which is a property of the network's topology and networks with a significant variance in node connectivity will be more robust to random removal of nodes (Albert *et al.* 2000).

Any given network can have the following nonexclusive general topologies: regular, planar, random or complex, which includes small-world and scale-free (Minor & Urban 2008; Newman 2003). In regular networks, each node has exactly the same number of links (*i.e.*, same node degree) and each node connects to all of its nearest neighbours. This topology is rarely observed in nature and, thus, will not be discussed further. It is not clear what constitutes the topology of a 'typical' habitat network and a single landscape may have different topologies depending on what focal species are examined. Nonetheless, using this framework to investigate seascape connectivity, independent of a particular context, makes it possible to make comparisons across a broad spectrum of applications and to identify consistently emerging properties and patterns. Understanding the rules that these networks follow will in turn lead to more informed conservation efforts (Minor & Urban 2008).

10.4.5   Planar Networks

Planar networks are two-dimensional arrangements where edges do not cross each other and, thus, nodes are only connected to geographic neighbours (*i.e.*, adjacent nodes) and are linked to more distant ones through stepping-stone nodes. Several important characteristic of planar networks include slow movement across the network due to the lack of shortcuts and a high local clustering coefficient. An urban street network, for example, where nodes are intersections linked by streets or edges, is a real-world planar network. In this kind of network, to move between two nonadjacent intersections, a pedestrian must walk across all intervening nodes. In contrast, an airline network is not planar; a passenger that boards a plane can arrive at any particular destination without passing through every city in between. In this vein, connectivity derived from flying animals moving between habitats can be considered a nonplanar network. However, a dispersing bird searching for new territory may sample all patches in between before moving permanently away from its original habitat. Therefore, in many situations planar networks may be suitable null models for seascape connectivity due to the restrictions of geography on functional connections (Urban *et al.* 2009).

10.4.6   Random Networks

Random networks are constructed by placing links between any two nodes based on a simple probability (Strogatz 2001). In these networks, the node-degree distribution is unimodal and most nodes have approximately the same number of edges (*i.e.*, there are no dominant hubs). Accordingly, this network topology generally does not display strong clustering and has a lower tolerance to random perturbations (Albert *et al.* 2000). In addition, movement or spread across this type of network will be quicker than a planar network but not as efficient as scale-free or small-world networks (Newman 2003). Until recently, all complex networks were treated as random networks following the models introduced by Erdős and Rényi (1959). Scientists have now recognised that most self-forming networks have more complex features or emergent properties, such as scale-free or small-world properties (Albert & Barabasi 2002; Watts & Strogatz 1998). Consequently, random networks will rarely represent features and behaviours observed in habitat networks.

10.4.7   Scale-free Networks

A scale-free network is characterised by the preferential attachment of links to certain nodes in a rich-gets-richer process (Barabasi 2009), also known as the 'Matthew Effect' (Merton 1968). This means that highly connected nodes acquire more links than those that are less connected, leading to the natural emergence of a few highly connected hubs and many low-degree nodes (Minor & Urban 2008). Therefore, in scale-free networks node degree distribution follows a continuously decreasing function or power law distribution. A typical example of a scale-free network is the World Wide Web, within which a few highly connected sites have millions of connections and are responsible for holding the network together (*e.g.*, Google). An important characteristic of scale-free networks is that they are resilient to random node removal, but vulnerable to targeted node failure of highly connected hubs (Barabasi 2009). In a scale-free habitat network, if a disease outbreak started at a random, low-degree node, it would have a low spread rate since these nodes have few connections. However, if the disease infected a major hub, it would quickly spread throughout the entire network. It has been suggested that from a conservation point of view, a scale-free network may be an ideal topology for a habitat network (Minor & Urban 2008) where conservation efforts could focus on protecting large hub patches from removal (*e.g.*, clearing and development). These strong hubs could be managed and monitored for the health of the desirable species, as well as to prevent the spread of invasive species, disease, or other disturbances. In addition, an appropriate distribution of these highly connected hubs throughout the landscape could both help isolate disturbances while maintaining population connectivity.

10.4.8 Small-world Networks

Small-world networks are graphs characterised by shortcuts that allow rapid and direct flow between distant nodes and thus these networks have a small diameter relative to the number of nodes (Watts and Strogatz 1998). In the context of seascapes, shortcuts may arise from strong and rare natural events, such as storms or atypical weather events, or from human-mediated transport (*e.g.*, ballast water). In this way, propagules may travel much farther than they could under normal or natural conditions. Small-world networks tend to have a high clustering coefficient compared with random graphs (Watts & Strogatz 1998) which may be a desired property in habitat networks, enhancing the stability of populations. Although small-world properties have been described for the Great Barrier Reef (Kininmonth *et al.* 2010), few other marine habitat examples exist.

10.5 Case Study in Marine Connectivity: Hawaiian Islands

In this case study, we investigated the spatial patterns in marine population connectivity for corals in the main Hawaiian islands to inform the regional conservation management strategy. We developed a representative coral connectivity network for the islands and used network analysis to identify conservation priorities. Here, our interest was in maintaining the short-term persistence of key populations, ensuring the long-term cohesion of the coral metapopulation, and to discover the emergent structure of coral connectivity across the region. The key goals were to: (i) quantify the capacity for coral populations to self-sustain by retaining locally produced offspring (*i.e.*, estimate local retention); (ii) estimate the value of each population in contributing larvae to downstream destination sites (*i.e.*, quantify source strength); (iii) map the core multigenerational dispersal pathways through the coral network by identifying important stepping-stone populations; and (iv) identify suites of subpopulations that are highly interconnected through dispersal, highlighting the emergent clustered structure within the coral network. Below, we briefly describe the biophysical modelling approach, the construction of the network model and the analysis utilised to achieve these goals. Finally, the key results and conservation recommendations are summarised.

The main Hawaiian islands span nearly 900 km, extending from the largest and youngest island of Hawai'i in the southeast to Nihoa in the northwest (Figure 10.2). Most of the nearshore environment surrounding the islands is made up of coral reefs that support a wide diversity of fishes and invertebrates. For this research, we focused on a generic coral profile, which broadly represents four common coral species in the region (*Porites lobata*, *P. compressa*, *Pocillopora meandrina* and *Montipora capita*).

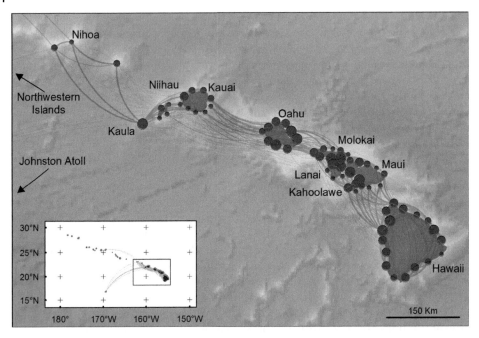

Figure 10.2 The Hawaiian island seascape where the coral population connectivity was modelled and subsequent network analysis used to identify potential conservation priorities. Coral connectivity was modelled for all reef habitat along the Hawaiian archipelago and Johnston Atoll (see text for biophysical parameters). Reef patches are shown as red nodes (sizes scaled to availability of suitable habitat). The dispersal-based coral population network is drawn with blue links where the directionality is implied by following the arcs in a clockwise direction. True north is aligned vertically and data presented in the Mercator projection (165°W central meridian).

Our modelled coral lives on reefs at depths less than 100 m where the benthic habitat has been mapped and classified using high-resolution satellite imagery and diver surveys (Franklin *et al.* 2013). Individual reefs were identified and unique identifiers assigned, each represented by a node in the network where node size reflects the abundance of corals present (Figure 10.2). Johnston Atoll and the northwest Hawaiian islands were included as unique patches in the model to account for any substantial influx of larvae from these areas into the main islands. A biophysical larval dispersal model (Roberts *et al.* 2010; Treml *et al.* 2012) was used to simulate the spawning, transport and settlement of coral larvae from all individual reefs in the system. Spawning occurred following all full moons during the summer (June through September) for all years when hydrodynamic data were available (2009–2013). Larvae were allowed to be passively transported by the currents (4 km, daily resolution) for up to 100 days and allowed to settle in suitable habitat from seven days after spawning. Mortality during the larval phase was modelled using a Weibull distribution (Connolly & Baird 2010). After all individual spawning events were simulated, the entire ensemble was aggregated into a single migration matrix representing the likelihood that individuals settling to a destination site (j, column) came from each potential source site (i, row) in any modelled year. To restrict our conservation analysis to only strong demographically significant linkages, values in the migration matrix that were less than 0.05

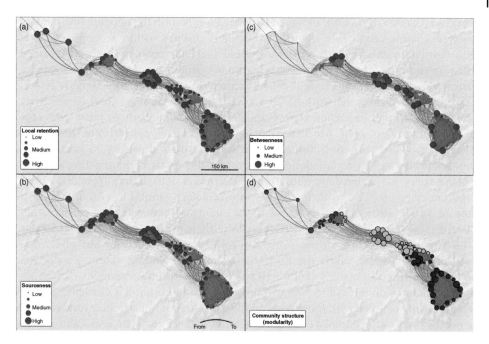

Figure 10.3 Results of the network analysis completed for the system of reefs and coral dispersal dynamics along the Hawaiian islands. The strong demographically significant linkages are shown as a network of blue arcs where the directionality is implied by following them in a clockwise direction. Nodes in all networks represent unique reef habitat patches and are scaled and / or coloured to represent results. (a) The relative rate of local retention. (b) The sourceness of each patch, quantified by calculating the total flow of larvae from the focal patch to all downstream nearest neighbours. (c) Betweenness centrality reflects each site's contribution to the overall cohesion of the coral network and identifies important stepping-stone sites. (d) A network modularity algorithm was used to partition the network into densely clustered communities. Each unique community is shown in a unique colour (node size is shown relative to reef area).

were removed. All subsequent analyses were performed using this simplified coral connectivity network (blue arcs in Figure 10.2).

The relative levels of local retention were extracted from the probability matrix for each site prior to calculating the migration matrix. This simple proxy generally reveals lower likelihoods along exposed sites and greater local retention in more sheltered areas and within larger reef patches (Figure 10.3a). The source strength of each patch was quantified by calculating the total flow of larvae from every source patch to all of the downstream nearest neighbours. This neighbourhood metric combines the size of the source patch with the weight, direction and number of outgoing linkages. This source-ness measure (Figure 10.3b) identifies all reefs patches that substantially contribute to the local rescue effect at downstream sites. To quantify the degree to which each site contributes to the overall cohesion of the coral network, we calculated the between-ness centrality (Borgatti 2005; Freeman 1979) on a log-transformed migration matrix, $\log([\mathbf{M}]^{-1})$, to transform weights to a more appropriate distance measure. This centrality measure identifies potentially important stepping-stone sites within the network that consistently fall along the shortest paths between populations (Figure 10.3c). Finally, the emergent clustering of the coral network was revealed using a network modularity algorithm (Clauset *et al.* 2004) designed to optimally partition the network into densely clustered communities. This algorithm identified clusters within a simplified coral network whereby directionality was removed and dual links between sites were replaced by the maximum edge weight (greatest migration). The resultant five relatively dense clusters were mapped using a unique colour for each community of nodes (Figure 10.3d).

Taken together, these network-based analyses on the Hawaiian coral connectivity network reveal some important conservation considerations. Perhaps most striking is the role of the O'ahu reefs. Most reef patches around O'ahu appear to be adequately self-seeding, while also contributing significantly to their downstream neighbours. These reefs also appear to fall along an important dispersal corridor, acting as a key stepping-stone for movement both to the northwest into Kaua'i (primarily from O'ahu's north shore) and to the southeast into Moloka'i. Similar to O'ahu, the north shore of Kaua'i also appears to be relatively important across all the conservation objectives. Western Moloka'i and west Maui and their nearshore islands, also appear to be important stepping stones along the coral dispersal corridor out of the big island of Hawai'i (particularly the Kona coast along the western shore). Finally, the community structure analysis (Figure 10.3d) revealed that the emergent clusters of coral populations do not fall strictly within island communities. For example, O'ahu's reefs are tightly coupled with those reefs on the east coast of Kaua'i and several sites on the northwest coast of Moloka'i. Similarly, several reefs along Maui and Kaho'olawe's south coast appear to be strongly linked to those along the north shore of Hawai'i. Coral reef conservation efforts within the main Hawaiian islands may be well served in considering the upstream / downstream island ecological neighbours (*sensu* Treml & Halpin 2012) and recognizing the key connectivity functions of O'ahu and Kaua'i and the dispersal corridor through the southern islands.

10.6 Conclusions and Future Research Priorities

Network models of marine connectivity offer a simple, yet robust and flexible framework for exploring questions related to the flow of individuals among habitat patches or populations. Connectivity fundamentally involves understanding dependencies within populations, linkages among populations, groups of populations within communities and the biophysical drivers that impact them. Key considerations include understanding which of the linkages and habitat patches are of greatest value (and when), how the networked system is organised and how this may impact system dynamics across the seascape. This knowledge brings us closer to understanding the subtle and emergent properties of marine population networks, allowing us to make more spatially and ecologically informed decisions about how and where to focus effort to achieve desired outcomes; for example, ensuring a minimum degree of connectivity across the system, or determining how to most effectively avoid the spread of undesirable species.

Scientists and researchers now have the ability to develop marine population network models and investigate connectivity and the key environmental drivers in ways that were impossible even five to ten years ago. New remote sensing technologies have increased the extent, spatial and temporal resolution and diversity of data for monitoring marine systems. For example, advances in multibeam sonar are capable of delivering spatial measurements of geophysical features and habitat at a 10 m resolution or better and new public satellite platforms are being developed that are capable of generating images with high spatial (10 m, Drusch *et al.* 2012) and temporal (10 minutes, Bessho *et al.* 2016) resolution. Inexpensive drones can be fitted with camera systems for monitoring shallow areas and autonomous underwater vehicles (AUVs) are an increasingly affordable and available option for acquiring data in deep regions. Tagging strategies are rapidly evolving, providing a more detailed picture of species in transit and are identifying how small changes in an organism's behaviour can generate significant differences in movement patterns over large distances. In tandem with this, computing power has dramatically increased as a result of improvements in processing and data input-output technologies, as well as the development of software libraries capable of splitting large problems into smaller, coordinated and parallel tasks. Large and complex network structures can now be processed, characterised and visualised in milliseconds, and emerging machine-learning algorithms can be used to traverse extensive collections of data to discover new and unexpected patterns and relationships.

All of these developments point to unprecedented opportunities to use network approaches to solve previously intractable research problems in marine ecology and demonstrate how innovations across a broad spectrum of research can collectively lead to advances in new areas and different disciplines. Connectivity can be a deep and complex subject, but there is also a certain intuition to it and even general audiences are able to appreciate the importance of interdependencies between individuals, populations and ecosystems. By cultivating a better understanding of how natural systems operate and improved knowledge of how impacts may cascade through the system, we can be more effective and efficient with conservation and management efforts. As a result, we can improve the likelihood that marine systems and the species that inhabit them will persist and thrive, preserving a rich array of ecosystem services and providing a broad range of benefits to humanity in general.

10.7   Acknowledgements

We appreciate the assistance and insights from the Marine Connectivity Working Group at the University of Melbourne, particularly Emily Fobert, Valeriya Kormyakova, Jack O'Connor, Francisca Samsing and Michael Sievers, and guidance from Simon Pittman. This chapter is published with the permission of the CEO, Geoscience Australia.

References

Albert R, Barabasi AL (2002) Statistical mechanics of complex networks. Reviews of Modern Physics 74: 47–97.

Albert R, Jeong H, Barabasi AL (2000) Error and attack tolerance of complex networks. Nature 406: 378–382.

Almany GR, Berumen ML, Thorrold SR, Planes S, Jones GP (2007) Local replenishment of coral reef fish populations in a marine reserve. Science 316: 742–744.

Anadón JD, D'Agrosa C, Gondor A, Gerber LR (2011) Quantifying the spatial ecology of wide-ranging marine species in the Gulf of California: implications for marine conservation planning. PloS One 6(12): e28400.

Andrello M, Mouillot D, Beuvier J, Albouy C, Thuiller W, Manel S (2013) Low connectivity between Mediterranean marine protected areas: a biophysical modelling approach for the dusky grouper *Epinephelus marginatus*. PloS One 8(7): e68564.

Barabasi AL (2009) Scale-free networks: A decade and beyond. Science 325: 412–413.

Barabasi AL, Albert R (1999) Emergence of scaling in random networks. Science 286: 509–512.

Beger M, McGowan J, Treml EA, Green AL, White AT, Wolff NH, Klein CJ, Mumby PJ, Possingham HP (2015) Integrating regional conservation priorities for multiple objectives into national policy. Nature Communications 6: article no. 8208.

Bélisle M (2005) Measuring landscape connectivity: The challenge of behavioural landscape ecology. Ecology 86: 1988–1995.

Bender DJ, Tischendorf L, Fahrig L (2003) Using patch isolation metrics to predict animal movement in binary landscapes. Landscape Ecology 18: 17–39.

Bessho K, Date K, Hayashi M, Ikeda A, Imai T, Inoue H, Kumagai Y, Miyakawa T, Murata H, Ohno T, Okuyama A (2016) An introduction to Himawari-8/9 – Japan's new-generation geostationary meteorological satellites. Journal of the Meteorological Society of Japan Ser II 94(2): 151–183.

Borgatti SP (2005) Centrality and network flow. Social Networks 27: 55–71.

Botsford LW, White JW, Coffroth MA, Paris CB, Planes S, Shearer TL, Thorrold SR, Jones GP (2009) Connectivity and resilience of coral reef metapopulations in marine protected areas: matching empirical efforts to predictive needs. Coral Reefs 28: 327–337.

Bradbury IR, Hubert S, Higgins B, Bowman S, Borza T, Paterson IG, Snelgrove PV, Morris CJ, Gregory RS, Hardie D, Hutchings JA (2013) Genomic islands of divergence and their consequences for the resolution of spatial structure in an exploited marine fish. Evolutionary Applications 6(3): 450–461.

Burgess SC, Treml EA, Marshall DJ (2012) How do dispersal costs and habitat selection influence realized population connectivity? Ecology 93: 1378–1387.

Calabrese JM, Fagan WF (2004) A comparison-shopper's guide to connectivity metrics. Frontiers in Ecology and the Environment 2: 529–536.

Clauset A, Newman MEJ, Moore C (2004) Finding community structure in very large networks. Physical Review E 70(6): 066111.

Connolly SR, Baird AH (2010) Estimating dispersal potential for marine larvae: dynamic models applied to scleractinian corals. Ecology 91: 3572–3583.

Cowen RK, Gawarkiewicz G, Pineda J, Thorrold SR, Werner FE (2007) Population connectivity in marine systems: An overview. Oceanography: Special issue on marine population connectivity 20: 14–21.

Cowen RK, Sponaugle S (2009) Larval dispersal and marine population connectivity. Annual Review of Marine Science 1: 443–466.

Csardi G, Nepusz T (2006) Igraph software package for complex network research. Complex Systems 1695.

DeAngelis DL, Grimm V (2014) Individual-based models in ecology after four decades. F1000Prime Reports 6(39): 6.

Devoto M, Bailey S, Memmott J (2014) Ecological meta-networks integrate spatial and temporal dynamics of plant-bumble bee interactions. Oikos 123: 714–720.

Dijkstra EW (1959) A note on two problems in connection with graphs. Numerische Mathematik 1(1): 269–271.

Drusch M, Del Bello U, Carlier S, Colin O, Fernandez V, Gascon F, Hoersch B, Isola C, Laberinti P, Martimort P, Meygret A (2012) Sentinel-2: ESA's optical high-resolution mission for GMES operational services. Remote Sensing of Environment 120: 25–36.

Erdős P, Rényi A (1959) On random graphs. Publicationes Mathematicae 6: 290–297.

Espinoza M, Lédée EJ, Simpfendorfer CA, Tobin AJ, Heupel MR (2015) Contrasting movements and connectivity of reef-associated sharks using acoustic telemetry: implications for management. Ecological Applications 25(8): 2101–2118.

Fontaine MC, Baird SJ, Piry S, Ray N, Tolley KA, Duke S, Birkun A, Ferreira M, Jauniaux T, Llavona A, Öztürk B (2007) Rise of oceanographic barriers in continuous populations of a cetacean: the genetic structure of harbour porpoises in Old World waters. BMC Biology 5(1): 1.

Fortunato S (2010) Community detection in graphs. Physics Reports 486: 75–174.

Franklin EC, Jokiel PL, Donahue MJ (2013) Predictive modelling of coral distribution and abundance in the Hawaiian Islands. Marine Ecology Progress Series 481: 121–132.

Freeman LC (1979) Centrality in social networks conceptual clarification. Social Networks 1(3): 215–239.

Funk WC, McKay JK, Hohenlohe PA, Allendorf FW (2012) Harnessing genomics for delineating conservation units. Trends in Ecology and Evolution 27: 489–496.

Gagnaire P-A, Broquet T, Aurelle D, Viard F, Souissi A, Bonhomme F, Arnaud-Haond S, Bierne N (2015) Using neutral, selected, and hitchhiker loci to assess connectivity of marine populations in the genomic era. Using neutral, selected, and hitchhiker loci to assess connectivity of marine populations in the genomic era. Evolutionary Applications 8: 769–786.

Galpern P, Manseau M, Fall A (2011) Patch-based graphs of landscape connectivity: A guide to construction, analysis and application for conservation. Biological Conservation 144: 44–55.

Galuardi B, Royer F, Golet W, Logan J, Neilson J, Lutcavage M (2010) Complex migration routes of Atlantic bluefin tuna (*Thunnus thynnus*) question current population structure paradigm. Canadian Journal of Fisheries and Aquatic Sciences 67: 966–976.

Girvan M, Newman MEJ (2002) Community structure in social and biologial networks. Proceedings of the National Academy of Sciences 99(12)7821–7826.

Grimm V, Railsback S (2005) Individual-based modelling and ecology. Princeton University Press Princeton, NJ.

Harary F (1969) Graph Theory. Addison-Wesley Publishing Co, Reading, MA.

Hazen EL, Maxwell SM, Bailey H, Bograd SJ, Hamann M, Gaspar P, Godley BJ, Shillinger GL (2012) Ontogeny in marine tagging and tracking science: technologies and data gaps. Marine Ecology Progress Series 457: 221–240.

Hovel KA, Regan HM (2008) Using an individual-based model to examine the roles of habitat fragmentation and behaviour on predator-prey relationships in seagrass landscapes. Landscape Ecology 23(1): 75–89.

Irisson J-O, Paris C, Guigand C, Planes S (2010) Vertical distribution and ontogenetic 'migration' in coral reef fish larvae. Limnology and Oceanography 55: 909–919.

Irisson J-O, Paris C, Leis JM, Yerman MN (2015) With a little help from my friends: Group orientation by larvae of a coral reef fish. PLoS One 10: e0144060.

Jackson AM, Semmens BX, Sadovy de Mitcheson Y, Nemeth RS, Heppell SA, Bush PG, Aguilar-Perera A, Claydon JA, Calosso MC, Sealey KS, Schärer MT (2014) Population structure and phylogeography in Nassau grouper (*Epinephelus striatus*), a mass-aggregating marine fish. PLoS One 9: e97508.

Jacoby DMP, Freeman R (2016) Emerging network-based tools in movement ecology. Trends in Ecology Evolution 31: 301–314.

Jones GP, Milicich MJ, Emslie MJ, Lunow C (1999) Self-recruitment in a coral reef fish population. Nature 402: 802–804.

Jones GP, Planes S, Thorrold SR (2005) Coral reef fish larvae settle close to home. Current Biology 15: 1314–1318.

Jönsson BF, Watson JR (2016) The timescales of global surface-ocean connectivity. Nature Communications 7: 11239.

Kininmonth SJ, De'ath G, Possingham HP (2010) Graph theoretic topology of the great but small barrier reef world. Theoretical Ecology 3: 75–88.

Kool JT, Moilanen A, Treml EA (2013) Population connectivity: recent advances and new perspectives. Landscape Ecology 28: 165–185.

Kool JT, Paris CB, Barber PH, Cowen RK (2011) Connectivity and the development of population genetic structure in Indo-West Pacific coral reef communities. Global Ecology and Biogeography 20: 695–706.

Kough AS, Paris CB, Butler MJ, IV (2013) Larval connectivity and the international management of fisheries. PLoS One 8(6): e64970.

Lett C, Ayata S-D, Huret M, Irisson J-O (2010) Biophysical modelling to investigate the effects of climate change on marine population dispersal and connectivity. Progress in Oceanography 87: 106–113.

Liggins L, Treml EA, Riginos C (2013) Taking the plunge: An introduction to undertaking seascape genetic studies and using biophysical models. Geography Compass 7: 173–196.

Lowe WH, Allendorf FW (2010) What can genetics tell us about population connectivity? Molecular Ecology 19: 3038–3051.

Magris RA, Treml EA, Pressey RL, Weeks R (2016) Integrating multiple species connectivity and habitat quality into conservation planning for coral reefs. Ecography 39: 649–664.

McGarigal K, Cushman SA, Neel MC, Ene E (2002) FRAGSTATS: spatial pattern analysis program for categorical maps. University of Massachusetts, Amherst, MA.

McGarigal K, Marks BJ (1994) FRAGSTATS: Spatial pattern analysis program for categorical maps. University of Massachusetts, Amherst, MA.

McRae BH (2006) Isolation by resistance. Evolution 60: 1551–1561.

Merton RK (1968) The Matthew effect in science. The reward and communication systems of science are considered. Science 159(3810): 56–63.

Meynecke JO, Lee SY, Duke NC, Warnken J (2007) Relationships between estuarine habitats and coastal fisheries in Queensland, Australia. Bulletin of Marine Science 80(3): 773–793.

Milo R, Shen-Orr S, Itzkovitz S, Kashtan N, Chklovskii D, Alon U (2002) Network motifs: Simple building blocks of complex networks. Science 298: 824–827.

Minor ES, Urban DL (2008) A graph-theory framework for evaluating landscape connectivity and conservation planning. Conservation Biology 22(2): 297–307.

Mora C, Treml EA, Roberts J, Crosby K, Roy D, Tittensor DP (2012) High connectivity among habitats precludes the relationship between dispersal and range size in tropical reef fishes. Ecography 35(1): 89–96.

Murphy HM, Jenkins GP, Hamer PA, Swearer SE (2011) Diel vertical migration related to foraging success in snapper *Chrysophrys auratus* larvae. Marine Ecology Progress Series 433: 185–194.

Neubauer P, Shima JS, Swearer SE (2010) Scale-dependent variability in Forsterygion lapillum hatchling otolith chemistry: implications and solutions for studies of population connectivity. Marine Ecology Progress Series 415: 263–274.

Newman MEJ (2003) The structure and function of complex networks. SIAM Review 45: 167–256.

Newman MEJ (2006) Modularity and community structure in networks. Proceedings National Academy Sciences USA 103: 8577–8582.

North EW, Gallego A, Petitgas P (2009) Manual of recommended practices for modelling physical – biological interactions during fish early life. ICES Cooperative Research Report 1(295): 1–112.

Palla G, Derenyi I, Farkas I, Vicsek T (2005) Uncovering the overlapping community structure of complex networks in nature and society. Nature 435: 814–818.

Paris CB, Helgers J, Van Sebille E, Srinivasan A (2013) Connectivity Modelling System: A probabilistic modelling tool for the multi-scale tracking of biotic and abiotic variability in the ocean. Environmental Modelling and Software 42: 47–54.

Pittman SJ, Monaco ME, Friedlander AM, Legare B, Nemeth RS, Kendall MS, Poti M, Clark RD, Wedding LM, Caldow C (2014) Fish with chips: tracking reef fish movements to evaluate size and connectivity of Caribbean marine protected areas. PLoS One 9: e96028.

Planes S, Jones GP, Thorrold SR (2009) Larval dispersal connects fish populations in a network of marine protected areas. Proceedings of the National Academy of Sciences 106: 5693–5697.

Pringle C (2003) What is hydrologic connectivity and why is it ecologically important? Hydrological Processes 17: 2685–2689.

Proulx SR, Promislow DEL, Phillips PC (2005) Network thinking in ecology and evolution. Trends in Ecology and Evolution 20: 345–353.

Ray N (2005) PATHMATRIX: a geographical information system tool to compute effective distances among samples. Molecular Ecology Notes 5(1): 177–180.

Rayfield B, Fortin MJ, Fall A (2011) Connectivity for conservation: a framework to classify network measures. Ecology 92: 847–858.

Riginos C, Crandall ED, Liggins L, Bongaerts P, Treml EA (2016) Navigating the currents of seascape genomics: how spatial analyses can augment population genomic studies. Current Zoology 62(6): 581–601.

Roberts JJ, Best BD, Dunn DC, Treml EA, Halpin PN (2010) Marine Geospatial Ecology Tools: An integrated framework for ecological geoprocessing with ArcGIS, Python, R, MATLAB, and C plus. Environmental Modelling and Software 25: 1197–1207.

Rogers LA, Olsen EM, Knutsen H, Stenseth NC (2014) Habitat effects on population connectivity in a coastal seascape. Marine Ecology Progress Series 511: 153–163.

Saura S, Torné J (2009) Conefor Sensinode 2.2: a software package for quantifying the importance of habitat patches for landscape connectivity. Environmental Modelling and Software 24(1): 135–139.

Selkoe KA, D'Aloia CC, Crandall ED, Iacchei M, Liggins L, Puritz JB, von der Heyden S, Toonen RJ (2016) A decade of seascape genetics: contributions to basic and applied marine connectivity. Marine Ecology Progress Series 554: 1–19.

Selkoe KA, Henzler CM, Gaines SD (2008) Seascape genetics and the spatial ecology of marine populations. Fish and Fisheries 9: 363–377.

Schill SR, Raber GT, Roberts JJ, Treml EA, Brenner J, Halpin PN (2015) No reef is an island: Integrating coral reef connectivity data into the design of regional-scale marine protected area networks. PloS One 10(12): e0144199.

Shima JS, Swearer SE (2009) Larval quality is shaped by matrix effects: implications for connectivity in a marine metapopulation. Ecology 90: 1255–1267.

Siegel DA, Kinlan BP, Gaylord B, Gaines SD (2003) Lagrangian descriptions of marine larval dispersion. Marine Ecology-Progress Series 260: 83–96.

Strogatz SH (2001) Exploring complex networks. Nature 410: 268–276.

Swearer SE, Caselle JE, Lea DW, Warner RR (1999) Larval retention and recruitment in an island population of a coral-reef fish. Nature 402: 799–802.

Tischendorf L, Fahrig L (2000) On the usage and measurement of landscape connectivity. Oikos 90: 7–19.

Treml E, Ford JR, Black KP, Swearer SE (2015) Identifying the key biophysical drivers, connectivity outcomes, and metapopulation consequences of larval dispersal in the sea. Movement Ecology 3: 1–16.

Treml E, Halpin PN (2012) Marine population connectivity identifies ecological neighbours for conservation planning in the Coral Triangle. Conservation Letters 5: 441–449.

Treml E, Halpin P, Urban D, Pratson L (2008) Modelling population connectivity by ocean currents: a graph-theoretic approach for marine conservation. Landscape Ecology 23: 19–36.

Treml E, Roberts JJ, Chao Y, Halpin PN, Possingham HP, Riginos C (2012) Reproductive output and duration of the pelagic larval stage determine seascape-wide connectivity of marine populations. Integrative and Comparative Biology 52: 525–537.

Turner MG, Gardner RH, O'Neill RV (2001) Landscape ecology in theory and practice. Springer-Verlag, New York, NY.

Urban D, Keitt T (2001) Landscape connectivity: A graph-theoretic perspective. Ecology 82: 1205–1218.

Urban D, Minor ES, Treml EA, Schick RS (2009) Graph models of habitat mosaics. Ecology Letters 12: 260–273.

Watts DJ, Strogatz SH (1998) Collective dynamics of 'small-world' networks. Nature 393: 440–442.

West DB (2001) Introduction to Graph Theory (2nd edtion). Prentice-Hall, Upper Saddle River, NJ.

Wilensky U (1999) NetLogo (and NetLogo User Manual), Center for Connected Learning and Computer- Based Modeling, Northwestern University, Evanston, Illinois. http://ccl.northwestern.edu/netlogo/ (accessed 14 June 2017).

Wood S, Paris CB, Ridgwell A, Hendy EJ (2014) Modelling dispersal and connectivity of broadcast spawning corals at the global scale. Global Ecology and Biogeography 23(1): 1–1.

11

Linking Landscape and Seascape Conditions: Science, Tools and Management

Kirsten L. L. Oleson, Kim A. Falinski, Donna-marie Audas, Samantha Coccia-Schillo, Paul Groves, Lida Teneva and Simon J. Pittman

11.1   Introduction

The regions of the earth where the land and sea interact form a dynamic and interconnected boundary, or ecotone, referred to as the coastal zone. Over geological time, sea-level changes have meant that much of the landscape at the coastal zone that we see today has been at various stages of submergence and emergence. The geomorphology of the coastline itself reflects the exposure to the sea at a range of spatial and temporal scales. In a profound way, too, the biophysical characteristics of the sea are inextricably linked to the land because the very mineral composition of the sea, particularly the salinity, is in part derived from riverborne materials and runoff from land as a result of millions of years of land erosion. In effect, the water cycle, combined with riverine erosive forces on land, drives many biogeochemical cycles in the ocean through land-sourced nutrients. As such, the connectivity of land to sea has been important for maintaining the productivity of marine and estuarine habitats and supporting the evolution of human civilization.

More recently in our history, coastal waters globally have experienced a substantial net loss of biogenic ecosystem structure – *e.g.*, seagrasses (Waycott *et al.* 2009), coral reefs (Fabricius 2005, 2011), salt marshes (Deegan *et al.* 2012) – which is connected to large localized increases in terrestrial material entering coastal waters. These changes have occurred primarily through suspended organic and inorganic material flowing to the sea through river discharge, nutrients and other contaminants delivered through groundwater, surface water flow (*i.e.*, runoff) across the land and engineered point discharges such as sewage pipes and storm water drainage. At broader spatial and temporal scales, the changing climate has increased rainfall in some areas, which has contributed to greater suspended sediment load in riverine discharges and in other regions it has reduced rainfall, decreasing sediment load in places like the Yellow River in China (Walling & Fang 2003).

Landscape structure is one of the most important land factors influencing nutrient and organic matter runoff. The biogeochemical condition of coastal marine waters can be influenced dramatically by the spatial patterning of vegetation, soils, geology, precipitation and human activity on land, even processes occurring hundreds of kilometres inland from the coast (Bartley *et al.* 2014b). For instance, coral reefs off

Seascape Ecology, First Edition. Edited by Simon J. Pittman.
© 2018 John Wiley & Sons Ltd. Published 2018 by John Wiley & Sons Ltd.

the coast of central Kenya have been exposed to an increase in suspended sediment load from Kenya's second largest river (390 km long Athi-Galana-Sabaki River), which drains a land basin area of 70 000 km^2. In Kenya, changing agricultural land use patterns since the 1900s and increasing population density in the river catchment together with deforestation and severe droughts have caused increasing sediment loads recorded in coral cores offshore (Fleitmann *et al.* 2007). Similarly, in North America, 200 years of farming-related nutrient loading of the Mississippi River has had a significant negative impact on river water quality leading to extreme nutrification of coastal waters in the Gulf of Mexico (Turner & Rabalais 2003). Sediment and coral cores have revealed that the Great Barrier Reef off the coast of Queensland, Australia has experienced increased sediment loads and nutrients since the 1900s associated with agricultural land use practices (Kroon *et al.* 2016). In many areas, changing land use and land cover (*e.g.*, intensive agriculture, deforestation, ranching, development) drive sedimentation by exacerbating erosion processes (Walling 2005; Houser *et al.* 2006; Porto *et al.* 2013; Wilkinson *et al.* 2013; Bartley *et al.* 2014a) and alter the landscape's ability to regulate water and retain sediment (Karr & Schlosser 1978; Hupp *et al.* 2009).

Watershed impairment has been correlated with notable declines in coastal and marine ecosystem health (Cabaço *et al.* 2008; Gorman *et al.* 2009; Burke *et al.* 2011). Increased fluxes of nutrients and sediment to coastal environments (Syvitski *et al.* 2005; Walling 2006; Alvarez-Cobelas *et al.* 2008) are deleterious to marine ecosystems (Airoldi 2003; Fabricius 2005; Halpern *et al.* 2008; Knee *et al.* 2008; Street *et al.* 2008) (Figure 11.1). Increased nutrients such as nitrogen and phosphorus entering coastal waters around the world from runoff and groundwater (Jones *et al.* 2001; Foley *et al.* 2005; Chen & Hong 2012; Howarth *et al.* 2012) have led to eutrophication through enhanced algal productivity and subsequent drawdown of oxygen in coastal water columns, creating dead zones around the world (Dodds 2006; Diaz & Rosenberg 2008). In addition, organic pollutants, such as dichlorodiphenyltrichloroethane (DDT), polychlorinated biphenyl (PCB) (Klumpp *et al.* 2002; Pait *et al.* 2007; Dachs & Méjanelle 2010) and heavy metals (Arifin *et al.* 2012; El-Serehy *et al.* 2012) are pervasive anthropogenic contributions to coastal waters. These compounds bioaccumulate and have the potential to affect the physiology of marine life and may affect human health through consumption of seafood.

For example, in coral reefs specifically, excess nutrients fuel macroalgae growth, which can then lead to smothering of coral and a net reduction in coral cover (McCook 1999; Fabricius 2005). Broad-scale correlations suggest that nutrient loading can accelerate the coral bleaching response (Carilli *et al.* 2009; Wooldridge 2009) and increase disease prevalence and severity (Vega Thurber *et al.* 2014). Although sediment is often loaded with nutrients and other contaminants, the presence of sediment in the water column also has impacts on coral reefs and coral physiology, including reduced coral growth (Erftemeijer *et al.* 2012), recruitment decline (Perez *et al.* 2014), slower photosynthesis in coral symbionts (Fabricius 2011) and various disturbed physiological and ecological processes in reef fish (DeMartini *et al.* 2013; Johansen & Jones 2013; Wenger *et al.* 2014). Knowledge about how stressors interact remains a key research question (Hughes & Connell 1999; Darling & Côté 2008; Dunne 2010). In systems where multiple stressors occur, lack of understanding of interactions can hamper management efforts (Gurney *et al.* 2013). Declines in reef health and structure can affect the delivery of ecosystem

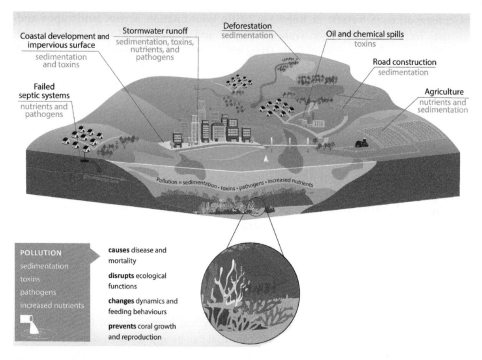

Figure 11.1 Schematic diagram of pathways through which land use affects nearshore coral reef environment. *Source:* Adapted by S. Pittman with permission from National Oceanic and Atmospheric Administration.

goods and services critical to the food security, livelihoods, economies and culture of coastal populations (Moberg & Folke 1999).

In this chapter, we explore how landscape ecology concepts and techniques can improve our understanding of the potential effects of landscape patterning on seascape condition. We focus on spatial patterns at spatial scales that are operationally relevant to management in order to help bridge the gap between ecological science and decision making in landscape planning. Because of the sensitivity of shallow coral reef ecosystems to land-based sources of pollution and the rapid changes to landscape structure taking place within watersheds in the tropics, we focus our review and case studies on land-sea connectivity in the tropics including Hawai'i, the US Caribbean and Australia. We concentrate on the application of spatial data and modelling tools that have been applied to predict and map runoff and we use these studies to illustrate knowledge gaps, highlight suitable spatially explicit modelling tools and techniques and the implications for land-sea management. Priority areas for future research are presented to help guide the next steps in this rapidly emerging area of applied interdisciplinary seascape ecology.

Various management frameworks have emerged in an attempt to integrate management-relevant knowledge across landscapes and seascapes (*e.g.*, Integrated Coastal Management, Ecosystem-Based Management, Integrated Island Management, *etc.*) (Leslie & McLeod 2007; Jupiter *et al.* 2014; Reuter *et al.* 2016). Although an integrated approach has been favoured for more than a decade (Cicin-Sain & Belfiore 2005; Stoms *et al.* 2005) and may be necessary for successful resource management and conservation (Silverstri & Kershaw 2010), coordination across landscapes and seascapes has been comprehensively developed and implemented only recently and on a rather limited geographical scale (Beger *et al.* 2010; Ruttenberg *et al.* 2011; Lebel 2012). Integrated management faces a number of hurdles, including basic knowledge gaps in how land and sea are functionally connected and the relationship between ecological functions and delivery of ecosystem goods and services, which hinder the ability to set explicit conservation objectives for these cross-realm processes (Tallis *et al.* 2008; Álvarez-Romero *et al.* 2011). Moreover, the planning process usually considers only terrestrial or marine ecosystems, rather than accounting for social, political and ecological linkages between the systems (Pressey *et al.* 2007; Beger *et al.* 2010; Klein *et al.* 2012). Resulting governance arrangements rarely fit the system well, leading to environmental degradation and increased conflict (Crowder *et al.* 2006; Cumming *et al.* 2013). In short, landscape ecology techniques linking land and sea could directly inform new management paradigms.

11.2 Landscape Ecology as a Guiding Framework for Integrated Land-Sea Management

Landscape ecology provides a conceptual and analytical framework for understanding the causes and consequences of spatial patterns that can be applied across spatially complex and interconnected terrestrial and aquatic systems (Boström *et al.* 2011). In the 1970s, ecologists Gene Likens and Herbert Bormann called for a more holistic approach in addressing problems associated with runoff with an explicit need to consider landscape mosaics. The authors wrote: 'legislation which ignores the biospheric perspective, or the complexity of the landscape mosaic, is ultimately naïve' (Likens & Bormann 1974).

Using a landscape ecology perspective, a landscape can be examined as a system of discrete, internally homogeneous habitat patches (*i.e.*, patch-mosaic model), or as a continuously varying surface such as a terrain (*i.e.*, gradient model) (Turner 1989; Dunning *et al.* 1992; McGarigal *et al.* 2009). The ecological and physical dynamics between habitat patches are governed by patch quality, structure and spatial arrangement and can be affected by changes at various spatial and temporal scales (Turner 1989; Dunning *et al.* 1992; Boström *et al.* 2011). Terrain function is influenced by attributes such as slope, topographic complexity and other morphological variables (Pike *et al.* 2009). Wiens (2002), writing on the application of landscape ecology to riverine systems, proposed a focus on six central themes: (i) patches differ in quality; (ii) patch boundaries affect flows; (iii) patch context matters; (iv) connectivity is critical; (v) organisms are important; and (vi) scale is important. Landscape ecology provides many spatially explicit tools and techniques to describe landscape patterning. This includes spatial pattern metrics ranging from simple patch attributes such as cover type, area, number of patches, shape and orientation, to more complex indices that measure the degree of fragmentation or connectivity of plant and soil patterns (McGarigal 2006).

Changes to patch boundaries, composition and connectivity can alter the landscape's delivery of ecosystem goods and services (Lotze *et al.* 2011). A prime terrestrial example of the importance of measuring landscape change has been highlighted by the impact of land cover on erosion processes, which has received much attention since the Landsat satellite data program first revealed the global extent of land transformation in the 1970s (Foley *et al.* 2005). Vegetated areas modify land surface characteristics and water flow through evapotranspiration, interception, infiltration, percolation and absorption (Bosch & Hewlett 1982; LeBlanc *et al.* 1997; Ludwig *et al.* 2005) and therefore can affect important ecosystem services including water yield, sediment and nutrient retention, flood control and groundwater recharge (Szilassi *et al.* 2006; Van Rompaey *et al.* 2007; Van Dessel *et al.* 2008). Other studies have highlighted the role of connectivity in physical, biological and community ecological processes at various scales, with important implications for ecosystem services such as pollination, aquatic habitat, water yield and fisheries (Hovel & Lipcius 2001; Wiens 2002; Ricketts *et al.* 2008; Bennett *et al.* 2009).

Several spatial pattern metrics have been developed specifically to quantify landscape patterning relative to water flow. In semi-arid catchments in Spain, Mayor *et al.* (2008) quantified the average length of runoff pathways across three catchments with patchy vegetation, classifying bare soil as runoff sources and vegetation as runoff sinks. The connectivity metric (in this case flowlength) was positively related to runoff and sediment yields and correctly ranked the three catchments according to total runoff. In Niger, central Africa, spatial pattern metrics quantified the change in fragmentation of patchy shrub landscapes from 1960 to 1992 finding that fragmentation of vegetation reduced water retention and significantly increased surface flow (Wu *et al.* 2000). Ludwig *et al.* (2007) developed a set of pattern metrics using algorithms that reflect the way in which spatial configuration of vegetation cover and terrain affect soil loss. A metric called the leakiness index (Figure 11.2) indicates the potential for landscapes to leak or retain soil

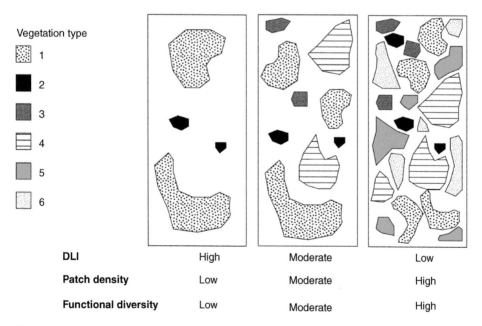

Figure 11.2 The Directional Leakiness Index (DLI) is a metric of the spatial arrangement and size of bare, 'interpatch' areas. Studies have found DLI is a good indicator of runoff and sediment. Here, conceptual landscapes are arranged from highest to lowest leakiness. *Source:* Adapted by Cecilia Leviol.

sediments. When applying the directional leakiness index to three Australian savanna landscapes, Ludwig *et al.* (2002) were able to effectively classify the landscapes along a function-dysfunction continuum, where the most dysfunctional landscapes were less able to retain resources than the more functional landscapes. Similarly, Bautista *et al.* (2007) found that while vegetated patch density was inversely related to runoff, the grain size and connectivity of the bare, interpatch pattern, measured as the leakiness index, had the best correlation with runoff and sediment. These metrics tend to be used as cost-effective indicators of landscape condition and are a simple alternative to complex modelling approaches. They are not intended to model the complex hydrological processes used to predict actual amounts of runoff although many other software programs are available to do so (see 3.0 Evaluating the connections between land and sea).

11.3   Modelling and Evaluating the Connections between Land and Sea

The connection between land and sea and particularly assessment of land-based anthropogenic threat to the condition of marine coastal waters, has been approached a number of ways: (i) measuring threat exposure; (ii) spatial modelling of land-sea processes; and (iii) decision analysis and support.

11.3.1   Measuring Threat Exposure from Land-based Sources

Threat exposure can be measured using coastal water quality monitoring, coral cores (Lewis *et al.* 2007), ground-based LiDAR (James *et al.* 2007), or remote sensing (Chérubin *et al.* 2008; Devlin *et al.* 2012; Evans *et al.* 2012; Álvarez-Romero *et al.* 2013). Direct measurement has the advantage of estimating the export of land-based sources of pollution over time and can establish a baseline condition for the watershed above the sampling point. Identifying source sediments is often a challenge that requires extensive geomorphological surveys, chemical fingerprinting investigations, or synchronized *in situ* monitoring (Walling 2005). However, monitoring programs require significant effort in data collection to support causal analyses and may only find weak correlation between sediment loads at the coast and their source within the watershed (Hamel *et al.* 2015). Even point data collection using georeferenced samples, such that the location of sampling is known, is limited in spatial extent relative to runoff events. Furthermore, most measurements are unspecific in terms of identifying the source of the stressor, which limits their usefulness in designing management controls. One way to overcome the challenge of identifying causation is to use fingerprinting techniques such as radio-isotopic analysis, which can help pinpoint the origin of certain pollutants in time and space (Phillips & Gregg 2003; Walling 2005). Monitoring efforts are constrained by the time period they are measuring and often sampling efforts miss intermittent storm events that may contribute a bulk of the land-based pollutant export (Sahoo *et al.* 2006). Another limitation of directly measuring sediment and nutrient concentrations in the near shore is that it does not evaluate or predict threats under alternative management or environmental conditions, which is why many turn to modelling.

Models can be used to represent complex spatially dynamic processes from terrestrial watersheds to seascapes. Modelling tools differ in their sensitivity to landscape pattern and, depending upon their level of integration across the land-sea interface, may be more or less useful for improving our understanding of how landscape patterns and seascape patterns are related. Moreover, some of the tools identify simple and inexpensive landscape metrics that can be used as cost effective proxies to guide management, obviating the need for more complex modelling in cases where simple will suffice. Their power lies in being able to simulate varying conditions and predict threat levels where field measurements are scarce or even unavailable.

11.3.2.1   Spatial Proxies

The simplest modelling uses readily available spatial data on land cover to create indirect diagnostic proxies using a geographical information system (GIS) software. Examples include proxies of watershed health (Rodgers *et al.* 2012) and landscape development intensity indices (LDI) (Oliver *et al.* 2011). Indirect proxies can easily reflect changes in stressors due to land use changes, but these generally are not calibrated to observations. These simple spatial proxy models are cost-effective options in places where data or technical capacity is scarce and where information to support decision making is required rapidly.

11.3.2.2   Hydrological Models

Increasing in modelling sophistication, deterministic biophysical models simplify the system to the key biophysical processes producing sediment and nutrients and can be calibrated to specific sites using observational data. Common approaches use empirical, physical, or statistical models to estimate potential pollution loads at the shore. At the core of each of these models is the digital elevation model (DEM), which is used to estimate where and how water and sediment move downhill and daily or annual rainfall to predict when runoff happens. Significant effort has been put into the development of sediment and nutrient export models for average annual loads and storm loads (Merritt *et al.* 2003; Vigerstol & Aukema 2011) and the models have been applied in many locations globally to explore land-based sources of pollution (Raclot & Albergel 2006).

Most of these models are based on the Universal Soil Loss Equation (USLE), or the revised or modified versions of the USLE (Renard *et al.* 1991), which estimates annual soil loss caused by sheet and rill erosion, using factors of rainfall energy, soil resistance to dislodging and transport, topography and management. Watershed-scale sediment and nutrient export models can be divided into two categories – those that predict sediment and nutrient fluxes by first predicting water discharge on a daily basis and those that estimate soil erosion and nutrient loading rates and then use a delivery ratio approach to estimate loads at the coast (Borah & Bera 2003). Models in the first category include SWAT (Arnold *et al.* 1998), HSPF, WATEM-SEDEM (Verstraeten *et al.* 2002) and AnnAGNPS (Polyakov *et al.* 2007). Models in the second category include custom-based GIS methods (Mishra *et al.* 2006; Ricker *et al.* 2008; Chou 2010; Jain & Das 2010), Nonpoint Source Pollution and Erosion

Comparison Tool's (N-SPECT) annual sediment estimates and the Integrated Valuation of Ecosystem Services and Tradeoffs (InVEST) Sediment Delivery Ratio model (Hamel *et al.* 2015; Sharp *et al.* 2015). An entirely different class of hydrologic models are process- or physics-based, such as WEPP (Mahmoodabadi *et al.* 2014), GSSHA (Downer & Ogden 2002), DHSVM or DWSM (Borah *et al.* 2002) and MIKE-SHE (Refshaard *et al.* 1995), using the Richards equation or other differential equations to estimate surface and subsurface flow. All of these models require some degree of spatial input and calibration data, such as DEMs, land cover, soil properties and precipitation.

11.3.2.3 Nearshore Dynamics

A major limitation of several of the hydrological models mentioned above is that the predicted loadings for runoff stop at the land-sea interface. Rarely are landscape models integrated with hydrodynamic models capable of predicting dispersion of materials after entering the sea. Seawater hydrodynamics and seafloor bathymetry and regional weather patterns create currents and waves that disperse and advect sediment plumes from land. Sediment modelling in the coastal zone therefore requires reliable bathymetry data. In part, the progress in seamless modelling from land to sea has been hampered by the lack of integrated terrain models that extend from watersheds to the ocean floor. Where bathymetry is available, a nearshore gap in the bathymetry often exists due to depth restrictions for survey ships. Hogrefe *et al.* (2008) addressed this data gap by filling the void with depth data derived from satellite imagery (pseudo-bathymetry) to create a seamless coastal terrain model for Tutuila Island (American Samoa). The authors proposed that topographically defined units (referred to as marine-terrestrial units – MTUs) that span the land-sea interface should enable quantitative analyses of material and energy exchange that will help to identify the impact of terrestrial inputs to near shore marine environments. Some good examples have now emerged, for instance, integrating laser altimetry (LiDAR) across land and shallow seas (even in turbid waters), together with multibeam for deeper waters, where vertical datum have been aligned (Gesch & Wilson 2001). The technical challenges such as the need for sufficient overlap between the terrestrial and very shallow marine datasets are now being overcome with the latest advances in multisensor imaging systems for coastline mapping (*e.g.*, The Coastal Zone Mapping and Imaging Lidar system (CZMIL)). Lidar can characterize reef geomorphology (Zawada & Brock 2009), which affects dispersion of terrestrial materials including residence time and deposition by affecting currents and sediment resuspension (Storlazzi *et al.* 2009). Even in places where bathymetry data are sufficient, the lack of detailed nearshore current mapping hampers efforts to map sediment and nutrient residence times. Storlazzi *et al.* (2006) provides an example of the level of detail needed to characterize currents and water column properties. Notably, linking land-based models that have poor or even no calibration data to nearshore models describing complex nearshore processes can compound errors and uncertainties.

A few examples extend land-based pollution into the nearshore environment through dispersion functions. For example, Burke *et al.* (2011), Klein *et al.* (2014) and Orlando & Yee (2017) use a sediment delivery model linked to a simple coastal plume model to assess relative exposure of reefs to pollution. For the Meso-American reef, Paris & Cherubin (2008) integrated watershed modelling using the N-SPECT tool together with ocean colour from satellites and a 3D regional ocean circulation model (ROMS) to predict sediment and nutrient discharge and map plumes of buoyant matter effecting coral reefs. The ultimate goal was to identify the origin of the buoyant matter arriving onto specific reefs by establishing connectivity between individual watersheds and coral reefs on a seasonal and annual basis. In another recent study, Tulloch *et al.* (2016) connected N-SPECT to a coastal transport model to prioritize marine management.

Dispersion and advection of sediment pollution is complicated, however, by the fact that pollutants chemically react with seawater and have different physical properties like grain size and colour. The added complexity of three dimensions means that in many cases, numerical simulations are still the norm for single discharge locations. While few watershed sediment export models explicitly consider grain size, grain size becomes critical to transport in the coastal zone. Just as land-based models that are available to predict sediment export first need to constrain the hydrological system, coastal models that predict sediment transport first estimate hydrodynamics, including wave dynamics and longshore and cross-shore currents. Coastal sediment transport models use bathymetry and wave height as the basic physical inputs. Models that have been applied to tropical environments include XBEACH, Delft 3D and ROMS, amongst many others (Papanicolaou *et al.* 2008). Understanding nutrient advection and dispersion is currently only modelled numerically, because of the many chemical reactions between dissolved and particulate nutrients that occur when nutrients reach seawater with different pH and temperature (Jickells 1998). Further complicating the nutrient story is that nutrients can enter the coastal zone through both groundwater and surface flow and groundwater discharge can be a significant source of nitrogen and other nutrients (Windom *et al.* 2006; Swarzenski *et al.* 2013). Effectively modelling dispersion, advection, settlement and resuspension of terrestrial inputs to the coastal environment is one of the highest priority data needs for understanding the spatial distribution of threats to coastal marine habitat and will likely remain an active area of study.

11.3.2.4 Ecological Response and Social-Ecological Systems Models

The logical next step is to assess the ecological response to the stressor reaching the nearshore environment. Connecting all the way to an ecological model of the reef to predict ecological response to pollution requires expertise, data and time that most managers do not have, so alternative techniques have been used to capture the relationship between land and sea. For example, some researchers have shown correlation between coarse proxies of the state of watershed modification and the condition of adjacent coral reefs (Oliver *et al.* 2011; Rodgers *et al.* 2012). Others used statistical analysis to evaluate impacts of stressors on reef state (Jouffray *et al.* 2015). Orlando and Yee (2016) went a step further, building statistical relationships between sediment load and the nearshore ecosystem's capacity to deliver multiple ecosystem goods and services. Yet another approach used expert opinion to assess vulnerability of coastal habitats to combinations of stressors globally (Halpern *et al.* 2008). Furthermore, a recent paper used a Bayesian Belief Network model to show that reducing sediment inputs from catchments was one of the most effective investments to improve the condition of coral reefs in Moreton Bay, Australia (Gilby *et al.* 2016).

While the deterministic, process-based biophysical models can be a powerful tool in understanding unidimensional relationships between land and sea, models in support of land-sea ecosystem-based management need to encompass a more thorough understanding of the system to be able to predict the response of the ecological system to myriad disturbances or management actions. The models need to have humans as an integral component of the system, both as drivers of anthropogenic threats, as well as users of ecosystem goods and services (Levin *et al.* 2009). These socio-ecological systems models need to capture interactions between system components and management sectors, cumulative / synergistic impacts of disturbances, human behaviour driving threats and human dependencies on resources (Weijerman *et al.* 2015). That said, the complexity of the social-ecological model can range from simple (sometimes qualitative) to complex (Verburg *et al.* 2015; Weijerman *et al.* 2015) and focused on key ecosystem components / processes tailored to the question, data and expertise at hand (Tallis *et al.* 2008).

One example of a simple, qualitative modelling approach is conceptual ecosystem models (CEMs), a method of diagramming social-ecological system components and relationships. CEMs illustrate the best understanding of system dynamics, key processes and connections between ecosystem components, while highlighting human influence and social values (Gross 2003; Kelble *et al.* 2013). When done in a participatory manner, they can help merge existing scientific and community knowledge (Kelble *et al.* 2013) and collect information and observations from diverse sources, which is an especially important function when data are scarce for any part of the system (Hohenthal *et al.* 2015). CEMs can moreover help reveal gaps in knowledge (Manley *et al.* 2000), which can guide future scientific investigation, *i.e.*, the need to assess how ecosystem goods and services delivery will vary under different water quality standards (Yee *et al.* 2015). CEMs have been used to describe land-sea system interactions in Florida as part of the National Oceanic and Atmospheric Administration's (NOAA's) Integrated Ecosystem Assessment process for the Florida Keys and Dry Tortugas marine ecosystem (Kelble *et al.* 2013) and US Environmental Protection Agency's efforts to set water quality criteria in the US Virgin Islands and Puerto Rico and watershed management in Guanica Bay Puerto Rico (Yee *et al.* 2015). They can also be used as an initial framework to develop quantitative models, such as Bayesian Belief Networks that can be used to simulate management outcomes (Nyberg *et al.* 2006; Fletcher *et al.* 2014).

Simple, dynamic, quantitative models that capture system feedbacks with a few mathematical equations have been used to reveal nonlinear interactions between coral reef system variables (see review in Weijerman *et al.* 2015). These models pare down the system to a handful of processes thought to be driving the system's dynamics and can be used to understand a system's response to perturbations and identify the most sensitive variables and important interactions, helping managers avoid unwanted surprises, including tipping points (Verburg *et al.* 2015; Weijerman *et al.* 2015). These models may have limited use in guiding management in situations where multiple stressors interact, however, as they are typically designed for single stressors.

Intermediate models, defined by their broader system perspective that encompass key physical and anthropogenic drivers of ecological change, are useful for understanding the link between environmental forcing and biological processes (Weijerman *et al.* 2015). These models build out parts of the system to fully capture dynamics, while simplifying other components. While it is relatively rare for intermediate models to include broader human dimensions, one recent body of work links discrete biophysical models with socio-economic models, including agent-based models describing human behaviour related to land use, to run dynamic scenario analysis (Bolte & Vaché 2010). An issue with intermediate models is that they are often cobbled together by linking individual models, which complicates interpretation and compounds uncertainties (Pascual *et al.* 1997; Lorek & Sonnenschein 1999).

Complex models, which capture many more of the processes and two-way interactions across the entire system, can be informative, but require significant effort, data and are therefore costlier to develop. For integrated land-sea systems, these models need to include biogeochemical, microbial and additional ecosystem processes to capture synergistic effects of a wide range of disturbances, including nutrients, sediment and fishing (Weijerman *et al.* 2015). They must incorporate feedbacks and interactions or risk missing tipping points where abrupt system change occurs (Verburg *et al.* 2015). Atlantis is an example of an end-to-end system dynamics model that was developed for ecosystem-based fisheries management strategy evaluation (Fulton *et al.* 2011). It has recently been applied to Pacific reef systems (Weijerman *et al.* 2014) and has the capacity to analyse impacts on reef fish of fishing and land-based pollution. Most system models do not include broader human dimensions, such as societal interactions and impacts and anthropogenic drivers (Verburg *et al.* 2015). However, some innovations exist, for example, using Bayesian models to simulate human behaviour (Sebastian & McClanahan 2013); altering threat level through fisheries management (Ainsworth *et al.* 2008); or by integrating an agent-based model (Yñiguez *et al.* 2008) (see Chapter 13 in this book). Recent advances in systems models that explicitly couple human and natural dynamics (such as MIMES – Boumans *et al.* 2015) hold promise in their ability to capture both how humans alter the biophysical environment and benefit from it, but require deep understanding of system processes and feedbacks to build and parameterize (Boumans *et al.* 2015) and have yet to be developed for land-sea systems. A major drawback of complex models is their uncertainty and error; as the model becomes more complex, uncertainty in assumptions about the model structure and parameterization can reduce the reliability of the simulated projections and limit management salience and uptake (Draper 1995; Weijerman *et al.* 2015).

11.3.3 Decision Analysis and Support

The third category of tools focuses on decision analysis and support. Integrating terrestrial and marine management can have socially beneficial outcomes, as the potential is greater to optimally allocate resource use and management actions across land and marine stressors (Tallis *et al.* 2008; Klein *et al.* 2012, 2014; Álvarez-Romero *et al.* 2015). The integrated scope of coastal management requires decision makers from multiple stakeholder groups and agencies to coordinate efforts, collaboratively defining the problem, setting common goals / outcomes, generating a set of possible alternative measures that attack the problem from all angles, predicting outcomes, choosing a course of action to implement that balances objectives and carefully monitoring eventual response to enable adaptive response. Solutions should draw from management actions on land and in the sea to maximize benefits and minimize costs (Álvarez-Romero *et al.* 2011). Various studies have shown how decision science can lead to more optimal solutions than other prioritization approaches (White *et al.* 2012a; Makino *et al.* 2013; Klein *et al.* 2014; Tulloch *et al.* 2016). Decision analysis can facilitate the connection between best available science and management decisions, help highlight and prioritize goals and facilitate communication with stakeholders about coastal issues, tradeoffs and management alternatives (Maguire 2004). It can also highlight where additional information would be more / less useful (Yokota & Thompson 2004). For management and conservation purposes, spatially explicit decision support systems need to accurately depict the complex processes and feedbacks connecting land- and seascapes in a given geography and include transparent techniques for setting spatial conservation, restoration and management priorities (Van Kouwen *et al.* 2007; Tallis *et al.* 2008; Beger *et al.* 2010; Álvarez-Romero *et al.* 2013).

Simulation models like those described above that predict outcomes of management actions enable tradeoff analysis and decision optimization, which can provide directly actionable advice for managers. The general idea is to identify a combination of decisions that provide the best results in terms of the objectives that the decision maker seeks to achieve. An optimization problem requires specification of alternative actions, an objective function that describes the fundamental objectives of the decision makers and a model that functionally connects decisions, relevant variables and outcomes. Various approaches exist to solving decision problems, ranging from simple (decision trees and influence diagrams) to complex (quantitative optimization models). Finding optimal solutions in complex, multiobjective contexts requires adaptive stochastic dynamic programming, which incorporates uncertainty in the objective and system dynamics as well as structural uncertainty (Conroy & Peterson 2013). Often these decision algorithms are paired with ecological models, *e.g.*, Ecosim with Ecopath and socio-economic production functions to optimize across multiple social and ecological outcomes (White *et al.* 2012a). The utility of using optimization algorithms, such as via the planning software MARXAN, in land-sea planning has been demonstrated numerous times via modelling catchments and coral reefs in the data-poor region of Vanua Levu, Fiji (Klein *et al.* 2012; Makino *et al.* 2013; Klein *et al.* 2014); the data-rich region of Queensland, Australia and the Great Barrier Reef (Beger *et al.* 2010), as well as areas of California and Indonesia (Watts *et al.* 2009). Relatively simple models paired with optimization algorithms have provided insights to seemingly intractable multiobjective decision problems in support of marine spatial planning in the United States (White *et al.* 2012b; Lester *et al.* 2013) and elsewhere.

11.4   Case Studies

Management of the land-sea interface can be informed by landscape patterns at different scales. For instance, patterns at the island scale can help managers select watersheds to restore, while patterns at a finer subwatershed scale can guide placement of specific actions within watersheds. Below are four case studies – one from Hawai'i, two from the Caribbean and one from Australia – that demonstrate how understanding patterns can inform management decision making. In each case, land cover plays a crucial role in determining runoff into the nearshore environment. The modelling detail needed in each case is defined by the type of decision being made.

11.4.1   Hawai'i

Land-based sources of sediments and nutrients directly impact the water quality of coastal Hawai'i and constitute one of the principal threats to the health of shallow coral reefs ecosystems in the main Hawaiian islands (Dugan 1977; Wolanski *et al.* 2003; Fabricius 2005; Friedlander *et al.* 2005; Erftemeijer *et al.* 2012). Land use has undergone massive transformation in the past two centuries, from landscapes dominated by native forests, grasslands and subsistence agriculture, to cattle ranches and plantation agriculture that started in the early to mid-1800s and ended in 1990s–2000s, to current urban and suburban development (MacLennan 2014). To most effectively address threats from land-based sources of pollution and enhance coral reef resilience to climate change, a paradigm of 'ridge-to-reef' management is increasingly common throughout the Pacific (Richmond *et al.* 2007). In Hawai'i this return to integrated management represents a modern version of traditional ridge-to-reef (*ahupua'a*) management (Jokiel *et al.* 2011) and has been codified in the Hawai'i Coral Reef Strategy (Gombos *et al.* 2010) and the recent iteration of the US National Ocean Policy (Executive Order 13547 2010). Rodgers *et al.* (2012) corroborated the validity of the ridge-to-reef relationship for the main Hawaiian islands by showing a strong correlation between a qualitative watershed health index and reef health index.

11.4.1.1 Estimating Spatial Patterns of Erosion from Land Cover Change and Exposure of Reefs in Maui

Land-based sources of pollution actively threaten the coral reef ecosystem of West Maui, a state and US federal government priority site (Sustainable Resources Group International 2012). In accordance with the Hawai'i Coral Reef Strategy, broad management efforts to improve reef health span the five coastal watersheds (totalling 97.37 km^2) and the land-sea interface. While both sediment and nutrients are principal pollutants causing significant impacts on the reef (Sustainable Resources Group International 2012), management effort to date has focused on reducing sediment, in part because nutrients are delivered largely via submarine groundwater discharge, which is far harder to manage and the principal source of nutrients is well known (wastewater) (Soicher & Peterson 1997; Glenn *et al.* 2013). Given the diversity of threats to coral reef health, there is a critical need to evaluate tradeoffs associated with the multiple potential resource management strategies and actions.

The objectives of the project were to quantify the relative levels of land-based sources of pollution, map hillslope erosion sources and develop an understanding of the consequences of land-based pollution for the near-shore environment and its ecosystem services. The project used a spatial hydrological model to estimate and map the intensity of land-based stressors reaching the near shore and approximated reef exposure using simple dispersion functions. Many process-based deterministic models, ranging from simple to data-hungry and complex, exist to predict sediment loads from watersheds. The extreme gradients in topography and precipitation and heterogeneity in geology presented by Hawaiian systems typically fall outside the range of most hydrological models. Various models have been parameterized and used in Hawai'i (Downer & Ogden 2002; Sahoo *et al.* 2006; Polyakov *et al.* 2007; Apple 2008; Verger *et al.* 2008; Gaut 2009; Goldstein *et al.* 2012) (Table 11.1), but most require extensive system-specific calibration data. The InVEST SDR (sediment delivery ratio) model, by contrast, is a relatively simple, customizable version of the Universal Soil Loss Equation that is paired with a sediment delivery model (Tallis *et al.* 2008; Hamel *et al.* 2015). It is a comparatively parsimonious model that can simulate changes in annual sediment as a result of changes to land cover and land use. It maps erosion on a pixel basis, although results

Table 11.1 Studies using modelling approaches to examine patterns of runoff in the Hawaiian islands.

LBSP	Why?	Model	Watershed	Citation
Sediment	Pollutant loads	ALAWAT	Ala Wai, Oʻahu	(Freeman & Fox 1995)
Flow/ Sediment	Assessment	HEC-1 + mass balance	Nawiliwili, Kauaʻi	(El-Kadi *et al.* 2003)
Sediment	Calibration	AnnAGNPs	Field experiment	(Bingner & Theurer 2001)
Sediment	Assessment	GSSHA	Makua, Oʻahu; Wailea Point, Maui	(Harrelson & Zakikhani 2006)
Sediment	Land use decisions	N-SPECT	Hilo, Hawaiʻi	(Okano 2009)
Sediment	Assessment	MikeSHE	Manoa, Oʻahu	(Sahoo *et al.* 2006)
Sediment	Rainfall changes	AnnAGNPs	Hanalei, Kauaʻi	(Polyakov *et al.* 2007)
Sediment	Restoration planning	N-SPECT	Kailua, Oʻahu	
Sediment	Assessment	DHSVM	Hamakua, Hawaiʻi	(Verger *et al.* 2008)
Sediment	Assessment	HSPF	Kaneohe, Oʻahu Waiʻulaʻula, Hawaiʻi	(Apple 2008)
Sediment	Best management practices	N-SPECT	Waiʻulaʻula, Hawaiʻi	(Gaut 2009)
Sediment	Assessment	N-SPECT, AnnAGNPs	Halawa, Oʻahu	(Cheng 2007)
Sediment	Land-use scenarios	N-SPECT	Pelekane, Hawaiʻi	(Group 70 International 2015)
Nitrogen	Land use scenarios	InVEST	North Shore, Oʻahu	(Goldstein *et al.* 2012)
Sediment	Feral ungulates	N-SPECT	Honolua, Maui	(TerrAqua Environmental Consulting 2013)
Sediment/ Nutrients	Best management practices	N-SPECT	Kahana, Honokahua and Honolua, Maui	(Group 70 International 2015)
Sediment	Model analysis	GSSHA	Halawa, Oʻahu	(Nielson & Francis in press)
Sediment	Model analysis	InVEST SDR	Statewide	(Falinski 2016)
Sediment	Road rehabilitation	InVEST SDR	West Maui	(Oleson *et al.* 2017)

are more accurate at the subwatershed scale. Falinski (2016) customized the InVEST model to Hawaiʻi, specifically incorporating information on available sediment, which was derived from high-resolution spatial land use imagery, to account for volcanic soils.

Insights from landscape ecology are inherent in the InVEST SDR. The InVEST SDR model runs on a pixel basis, the size of which is dependent on the digital elevation data. The sediment delivered to the stream from any given pixel is a function of erosion within that pixel, as calculated by USLE, as well as features of the upslope area draining / transporting sediment into that pixel and sediment retention along the downslope path as it travels to the stream. As the cover factor and slope play a key role, the model could be sensitive to landscape features, such as vegetated patches or bare interpatches within the landscape, if input data's scale is fine enough. The model results in a number of indicators (*i.e.*, erosion rates, total erosion per watershed, total load at pour points)

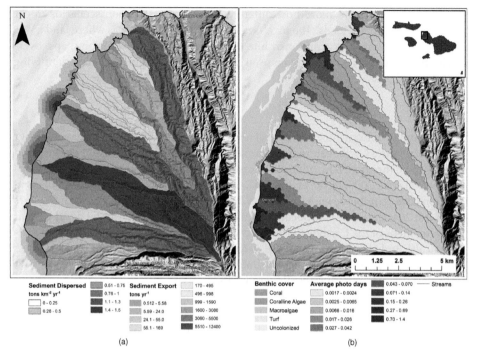

Sediment Dispersed tons km⁻² yr⁻¹		Sediment Export tons yr⁻¹		Benthic cover		Average photo days		
	0.51 - 0.75		170 - 495	Coral		0.0017 - 0.0024		0.043 - 0.070 — Streams
0 - 0.25	0.76 - 1		496 - 998	Coralline Algae		0.0025 - 0.0065		0.071 - 0.14
0.26 - 0.5	1.1 - 1.3	0.512 - 5.58	999 - 1590	Macroalgae		0.0066 - 0.016		0.15 - 0.26
	1.4 - 1.5	5.59 - 24.0	1600 - 3080	Turf		0.017 - 0.026		0.27 - 0.69
		24.1 - 55.0	3090 - 5500	Uncolonized		0.027 - 0.042		0.70 - 1.4
		55.1 - 169	5510 - 12400					

(a)   (b)

Figure 11.3 Erosion, sediment exposure of coral reefs and recreation in West Maui. (a) Sediment export (tons/year) estimates by subwatershed calculated using InVEST SDR 3.2, coupled with a sediment dispersion model to estimate spatial distribution of sediment dispersed (ton/km²/year) in the coastal zone. (b) Coral cover based on NOAA benthic data and recreation intensity based on InVEST Recreation model.

and can calculate a sediment retention index, which is the avoided sediment loss due to land cover. The effect on these indicators of different landscape configurations could be helpful for understanding the optimal spatial arrangement of management.

To understand better the consequences of sediment runoff for the near-shore marine environment, the Focal Statistics function in Spatial Analyst of ArcGIS was used to map the extent to which the sediment might disperse from 19 pour points into the near shore environment. The resulting dispersion (Figure 11.3) is a simplistic representation of the potential footprint and intensity of sediment plumes, but ignores any near-shore processes (see discussion above). Nonetheless, it can serve to help identify priority areas for sediment management, particularly as this is an area where coral reef health is a priority and many tourists visit this coast and would prefer clear water. A map of coral cover was added to highlight where pollution was likely to impact reefs and their goods and services, which include habitat, snorkeling and fishing. To illustrate the potential impact on tourism and recreational services, the sediment map was overlaid with a map of tourism visitation (Figure 11.3) (using InVEST recreation model, which uses photo uploads to estimate recreation intensity (Wood *et al.* 2013; Keeler *et al.* 2015)).

The results of the sediment model provided managers with information about the role of land cover on erosion. Because the SDR approach captures spatial patterns of land cover, it can be used to explore how different landscape restoration scenarios might impact coastal sediment loads. The pixel-scale maps highlighted areas where land cover and physical characteristics combined into erosion hotspots. The erosion maps, over-laid with the coral reef and tourism maps, were used in workshops aimed at identifying priority watersheds.

11.4.2 Caribbean

The health and structural integrity of coral reefs in the US Caribbean, as with many other locations across the Caribbean region, has severely declined over the past century and continues to be threatened by multiple stressors including land-based sources of pollu-tion (Pandolfi *et al.* 2005; Burke *et al.* 2011). Urbanization of watersheds and changes in vegetation cover, combined with hurricanes, has led to increased runoff to coastal waters over the last 20 years in the Caribbean region (Jackson *et al.* 2014). Coastal development in the watersheds of the Caribbean islands is widely considered to be negatively impact-ing the health and viability of nearshore coral reefs (Larsen & Webb 2009; Warne *et al.* 2005). Coral reefs are sensitive to high sedimentation rates, which can negatively impact coral reef health and growth rates (Rogers 1990).

11.4.2.1 Summit to Sea Runoff Modelling for St John, US Virgin Islands

In the US Virgin Islands (USVI), coastal development, pollution and runoff are con-sidered an increasing threat to coral reef and human health (Rogers & Beets 2001; Rothenberger *et al.* 2008). Observations of coral reef condition before and after urban development have shown increased coral stress at reef sites exposed to the highest rates of sedimentation and greater susceptibility to bleaching during a thermal stress event (Nemeth & Nowlis 2001). More broadly, a stress gradient resulting from runoff and impacting coral reef health has been identified from near-shore to offshore waters across the insular shelf of the USVI (Smith *et al.* 2008). Due to the ecological and economic importance of maintaining safe bathing water and good coral reef health, an understanding of the spatial and temporal patterns of runoff is an ongoing priority for land use planning in the USVI.

The geology, topography and vegetation of the USVI are key factors influencing coastal hydrology. Although the volcanic islands have no perennial streams and near-shore waters have relatively minor influence from ground water, the watersheds do have steep slopes and increasing amounts of impervious surfaces, which can create high velocity runoff and soil erosion. More than 80% of the island of St John has slopes exceeding 35%. The bulk of water flow occurs via large gullies (guts) that dissect the steep slopes resulting in material entering the bay as point sources where the gullies join the coastal waters (Brooks *et al.* 2007). On St John, upslope sediment produc-tion has been linked to increased sediment in downstream marine environments, particularly where the removal of vegetation and construction of unpaved roads have greatly intensified erosion rates (Ramos Scharrón & MacDonald 2005). Sedimentation from unpaved roads can be 300–900% higher than that experienced in undisturbed watersheds (Rogers 2006). Sediment delivery to coastal waters in the early 1990s was estimated at three to four times the long-term historic rate due largely to unpaved roads (Macdonald *et al.* 1997). Sediment cores collected in Coral Bay, the largest bay on St John with highest recent rates of urbanization on the island revealed that marine surface sediments exhibited a dramatic increase in land derived sediment input since the 1950s (Brooks *et al.* 2007). Mitigation of runoff has occurred in some heavily impacted watersheds through funding to support communities in control of runoff and development of watershed management plans across the region (Whitall *et al.* 2014).

Although several detailed studies have been conducted for individual watersheds and bays on St John this case study demonstrates the use of spatially explicit data applied to construct a synoptic island-wide model of predicted runoff and to identify exposure to coral reefs. The approach builds on an earlier US Caribbean project called 'Summit to Sea', a collaboration between the US National Oceanic and Atmospheric Administration (NOAA) and the World Resources Institute, to develop spatial models of sediment load and coral reef exposure (World Resources Institute & NOAA 2005). The approach used spatial and statistical techniques to characterize watersheds across the USVI with regard to relative erosion rates and the threat of land-based sources of sediment and pollutant delivery to coastal waters. A simplified version of the Revised Universal Soil Loss Equation (RUSLE) using slope, land cover, precipitation and soil characteristics was applied, as well as indicators of road density and erosivity by watershed. This case study was developed by NOAA Biogeography Branch to update the methods of the earlier study by using newer and higher resolution data.

The spatial analysis of land-based sources of threat to coral reefs has several main components:

Delineation of watersheds (basins) based on a hydrologically corrected digital elevation (DEM) data set (30 m resolution). That is, the locations of known waterways were added to the elevation model. These basins reflect all land areas discharging to a single coastal location (pour point).

Analysis of relative vulnerability of land to erosion based on slope, precipitation (peak rainfall month) and the soil characteristics (within each 30 m resolution grid cell).

Analysis of the relative erosivity of land given current land cover within each 30 m resolution grid cell. In addition, the relative sediment delivery for each pour point is estimated (Figure 11.4a). Land cover categories were reclassified to relative erosion rates, ranging from 15 (for forest) to 220 for barren land. The erosion factor (K-factor) was obtained from the Soil Survey Geographic database (SSURGO) of the US Department of Agriculture. Since not all erosion makes its way to the sea, sediment delivery ratios (based on watershed size) were applied in order to estimate relative sediment delivery at the pour point. Relative erosion rates and sediment delivery are used as a proxy for both sediment and pollution delivery. Two indicators of erosion within the watershed were calculated for each basin: (i) average erosion rates for the basin and (ii) total relative erosion within the basin. An indicator of relative sediment delivery at the river mouth was estimated by multiplying total relative erosion in the basin by the sediment delivery ratio (SDR) for the basin, which is a function of watershed size. The SDR reflects the percentage of erosion within the basin that reaches the pour point.

In the absence of a hydrodynamic model, a simple diffusion model was applied with a maximum extent of 1 km from the pour point using spatial interpolation. To help evaluate the models, 1 km satellite derived ocean colour data from the MERIS sensor (Medium Resolution Imaging Spectrometer) was processed to map turbidity concentrations across the shelf. A multiyear (2003–2011) synthesis of the turbidity was mapped around the US Virgin Islands by computing the 90% quantile value (*i.e.*, the mean of the highest values) for all the MERIS images in a year, across all years (Figure 11.4c).

In addition, the Landscape Development Intensity (LDI) Index was calculated following the methods of Oliver *et al.* (2011) for comparison with the Summit to Sea hydrological models to evaluate the utility of LDI as a rapid assessment approach that estimates cumulative threats based almost entirely on patch characteristics of a single dataset (*i.e.*, land cover / land use). The LDI was calculated by quantifying the area and percentage of each land cover class within each watershed (Figure 11.4b). The LDI coefficient was multiplied by the proportion of each land cover type for individual watersheds and then all coefficients were summed together to calculate a single value, which represents an index value for each watershed. Both models used the same land cover data classified from aerial imagery acquired in 2012.

Comparison of the two models (hydrological model versus land use as a proxy) with satellite derived data on the distribution of turbidity suggests that overall the hydrological model (summit to sea) provides a more detailed representation of coral reef exposure to runoff than the LDI model (Figure 11.4). In addition, the hydrological model enables land managers to identify areas with high runoff potential. The LDI model is likely to give a useful island-wide perspective where pockets of high urbanization exist regardless of the terrain, however, on islands with steep-sided vegetated watersheds there is a benefit of integrating slope and vegetation together with spatially accurate entry points at the land-sea interface. The LDI provides an initial cost-effective and rapid depiction of the spread of urbanization and comparison of the extent of human conversion of land among watersheds. This can be easily tracked over time and quantified in a cost-effective way using landscape metrics. Both models agree that runoff is greatest in the bays adjacent to watersheds with highest LDI values (*i.e.*, most urbanized), specifically on the south coast and least from the highly vegetated watersheds on the north coast. This is corroborated by actual water colour reflectance from the MERIS satellite data (mean values 2003–2011). The weakness in both approaches relates to an absence of nearshore circulation models that are required to predict the dispersion of runoff. Dispersal models are an important and often missing component of land-sea models due to the difficulty in deriving reliable models of current patterns in structurally complex nearshore waters. Furthermore, rarely do studies explicitly link the spatial structure of the seascape to that of the neighbouring landscape, yet with new data rapidly emerging these types of studies are increasingly becoming feasible.

11.4.2.2 Land-Sea Decision Support Modelling for the Northeast Marine Corridor, Puerto Rico

A collaboration between NatureServe and NOAA sought to support management of coastal ecosystems by modelling land-based threats to coral reef ecosystems in northeast Puerto Rico. This particular region of northeast Puerto Rico is home to the Río Fajardo watershed, which has been contributing large amounts of sediment to nearshore waters and has experienced increased urbanization and soil erosion rates over the last century (Ramos Scharrón & MacDonald 2005). The project integrated multiple spatial models to achieve the goal of modelling the exposure of coral reefs during multiple runoff scenarios and the identification of watersheds contributing elevated levels of sediment and pollution to coastal waters. NOAA's Nonpoint Source Pollution and Erosion

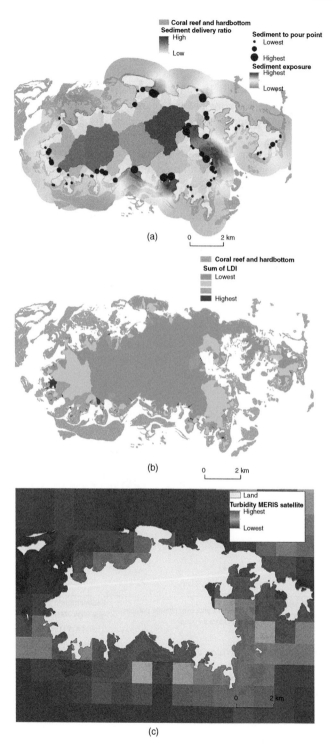

Figure 11.4 Models of runoff for the Caribbean island of St John (US Virgin Islands). (a) Summit-to-sea hydrological model of runoff showing pour points and predicted exposure of coral reefs to sediment. (b) Landscape development intensity index (LDI) based on land use classified from aerial imagery. (c) Satellite-derived water colour from the MERIS sensor (2003–2011).

Tool (OpenNSPECT https://nspect.codeplex.com/, accessed 25 May 2017) was applied to model and map cumulative exposure of nearshore waters from land-based sources of pollution (Step 1). Integrated within NatureServe's Vista decision support tool (Vista), multiple exposure scenarios were examined (Step 2). This coupled software then fed data into a spatial prioritization algorithm (Step 3) using the Marxan software (Ball & Possingham 2000) to prioritize land and sea zones where management actions could be focused to mitigate runoff. Here we describe the process and discuss some of the benefits of this multimodel approach.

In Step 1, OpenNSPECT was applied to model erosion and the accumulation of sediment, nutrients and other pollutants using hydrological modelling based on land cover, land use, soil type, precipitation, digital elevation and watershed boundaries. Model outputs provided information on the amount of sediment, nitrogen and phosphorus levels generated per pixel in the watershed, as well as the locations and amounts where they accumulate. Figure 11.5 (left panel) shows sedimentation plumes and the distribution of coral reefs in a subregion of the study area. The model output from OpenNSPECT shows high sediment values inside the Río Fajardo watershed and along the river. To identify and quantify places in the watershed where effective mitigation strategies could

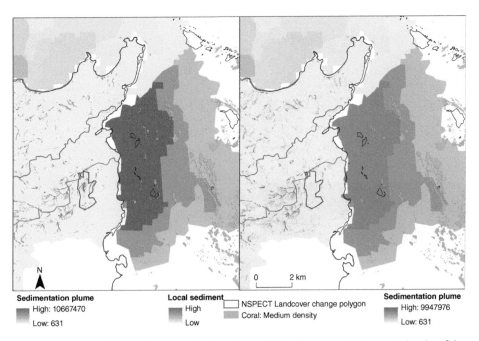

Figure 11.5 Local sediment effects ranging from low to high values across the terrestrial realm of the Northeast Corridor of Puerto Rico, as well as sedimentation plumes in the marine realm. The Río Fajardo is show in blue, as well as locations of coral reefs in the marine realm. Decreases in local sediment effects (right-hand map) are observed after land cover changes were made in NSPECT to the polygon outlined in red for our mitigation scenario. The resulting marine plume's accumulated sediment values also decreased.

take place, OpenNSPECT was again used to produce new hydrological results based on alternative land cover scenarios. For example, a $1320\,km^2$ area of high sediment generation was observed near the Fajardo river that is currently occupied by agricultural land cover. To simulate a mitigation scenario, the agricultural land cover at this site was converted to forest cover and sedimentation and pollutant load outputs were reanalysed through OpenNSPECT (Figure 11.5 right panel). Decreases in local sediment effects were observed, as well as accumulated sediment effects, which resulted in decreased values to the sedimentation plumes in the marine environment.

In Step 2, the results from OpenNSPECT were integrated into Vista in a more holistic scenario that incorporates terrestrial and marine stressors and conservation uses. An element response model was constructed in Vista to model how the condition of each element (*e.g.*, coral types) changes under different runoff exposure scenarios. Low element condition values resulted in low coral viability when a condition threshold was applied.

In Step 3, the Marxan optimization tool was used to demonstrate efficient areas for conservation focus based on the lowest 'cost' per area. Several Marxan analyses were computed within Vista. This novel application integrated terrestrial and marine biodiversity elements with locations of high exposure to land-based sources of pollution as targets within Marxan. The area surrounding the Río Fajardo river was identified as an ecologically important area due to its Marxan irreplaceability values, as well as the distribution of marine elements in the area. These include observations of manatee, sea turtle and richer coral species diversity, as well as the Río Fajardo's designation as a priority protection area (Ramos Scharrón & MacDonald 2005). To examine conflicts occurring in this region, the Vista Site Explorer tool was used for site level analysis of the marine elements. Figure 11.6 shows the output, where high sedimentation plumes were observed, originating from the Río Fajardo watershed. The analyses suggested that areas of medium density coral were considered incompatible with the level of exposure to stressors. Direct stressors from a nearby marina location and indirect stressors from sedimentation plumes are both currently affecting this site.

This decision support toolkit approach supports modelling potential and actual conflicts in both the terrestrial and marine realms through scenarios and for calculating the benefits of conservation action on both direct and indirect threats to achieve fully integrated land-sea conservation. Additional scenarios could be computed to examine the runoff patterns when different sized patches of land are changed in composition and spatial configuration. Does fragmentation of the vegetation or creation or removal of riparian corridors of vegetation or continuous belts of coastal wetlands influence coral reef condition? Where are the most cost effective places for watershed interventions that will reduce the highest runoff loading for the least economic cost?

11.4.3   Australia

In 2009, the Great Barrier Reef Marine Park Authority (the Authority) released a Great Barrier Reef Outlook Report that highlighted water quality and coastal development as two major threats to the health of the coral reef ecosystem, a conclusion reiterated in the recent 2014 report. Managing the health of the reef has often been defined by issues

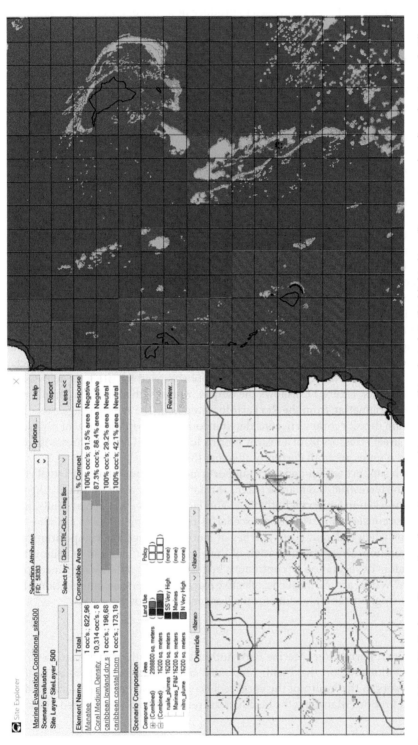

Figure 11.6 Vista Site Explorer. Two marine elements are noted to occur in this selected site (highlighted in red): manatee and medium density corals. All areas of medium density coral and manatee occurring on this site are incompatible and display a negative response to the scenario components listed below, which include a marina, as well as very high nitrogen and sediment plumes.

such as water quality, salinity and acid sulphate soils. Programs and projects have been and continue to be, initiated to address these issues; however, they are often in response to a single issue and, as a result, are very targeted and focused. Over the last seven years, the Authority has been working to change this single issue focus with the introduction of programs that seek to consolidate and understand the values, functions and services provided by the catchment and how these physical, biological and biogeochemical processes support the health of the reef.

In response to the 2009 Outlook Report, the Authority began analysing existing terrestrial spatial data to understand the changes that have taken place in the adjacent Great Barrier Reef catchment. The catchment is approximately $424\,000\,km^2$ and has been heavily modified since European settlement, especially south of Cooktown. The catchment continues to experience pressure from agriculture, industry, urban expansion and mining development and its associated infrastructure. While Queensland's vegetation mapping indicated a high percentage (64%) of intact regional ecosystem vegetation, more detailed analysis revealed that the estimate was driven by one large, remote area. Further analysis showed that although large areas of the catchment are intact with almost 10% within protected areas, open grazing occurs throughout most of these ecosystems as 74% of the catchment is cattle grazed, with most of the remaining areas of the catchment under other developed land uses.

The floodplains in the catchment naturally provide an important role of slowing the velocity of water and the associated sediment, nutrients and pollutants that run off the land. However, modifications to floodplains have changed this natural hydrological function, largely in response to the need to redirect or remove water from developed areas. As a result, more water is typically moving faster out of the catchments and therefore scouring banks. These high velocity flows have altered processes such as the recharge of groundwater in the catchment and increased the delivery of pollutants to the reef. Major changes and modifications have occurred in the coastal zone, with land use activities such as cropping and grazing extending into marginal areas due to pressures to extend production.

At the local scale, pressure from coastal development has led to highly modified coastal ecosystems. Estuarine ecosystems have been highly modified to allow for expansion of extensive cattle grazing and intensive agriculture such as sugar cane; for example, through bunding (retaining walls) to prevent tidal flow. These areas can then retain freshwater but can be poorly drained and choked with weeds resulting in poor water quality. These changes reduce the feeding areas of many aquatic species, which in turn may impact on commercial and recreational fisheries. Importantly, around 33% of saltmarshes, which are important fish and crustacean foraging areas, have been bunded to increase the extent of grazing land and further areas have been cleared for sugarcane.

Based on concerns about the consequences of land use change for sediment, nutrient and pesticide runoff onto the reef and the effects on habitats and species, the Authority has also been developing a suite of tools that include 'Bluemaps' and an Ecological Process Calculator. The Bluemaps map the frequency of hydrological connections in the pre-European state of the catchment and are based on existing data such as floodplain extent, highest astronomical tide, storm surge, Queensland wetlands and a wet vegetation signature (used as a pseudo for groundwater dependent ecosystems). The Ecological Processes Calculator is a tool for eliciting expert opinion to assess the changes to ecological processes provided by catchment ecosystems that support the health and resilience of the reef. Experts engage and populate the Calculator by participating at workshops on a regional scale; this provides an opportunity to tailor the information at a basin scale. The Calculator compares the capacity of pre-European (pre-clear) coastal ecosystem ecological processes to those of a present day catchment made up of natural and modified ecosystems using a relative scoring system. This information is then used with marine resource information to identify where ecological function and process has been modified or lost and management actions can then be targeted to highlight areas for system repair. The Calculator can also be used to estimate the impacts of improved practices (current best practice) on the ecological processes provided at a broad functional scale. These tools have been designed to be used together to assist in identifying areas in the catchment for protection, restoration or management of coastal ecosystems and their functions.

11.4.3.1 Edgecumbe Bay Receiving Waters (Gregory and Eden Lassie Creek Sub-basins)

Edgecumbe Bay is a north facing bay that forms part of the Proserpine Basin in the Mackay Whitsunday region. The Whitsunday region is Ngaro people's country and the area contains some of the oldest archaeological sites discovered in the Great Barrier Reef World Heritage Area that date occupation back over 9000 years ago. Of the original 4723 hectares of estuaries in the Edgecumbe Bay Gregory and Eden Lassie creek sub-basins (Figure 11.7 panel a), 90% currently remain (Figure 11.7 panel b). The most intact areas, however, are only protected under the Queensland *Fisheries Act* 1994 and around 10% are grazed. Forested floodplains in this area declined from 12 hectares to 10 hectares with 90% of the remaining used for grazing and 10% sited within forestry reserves. Of the original 54 582 hectares of forests, around 61% of remain with 15% protected and 77% used for grazing. Around 48% of pre-European woodlands are remaining, with 97% of the remaining woodlands grazed. Around 95% of rainforest areas remain with about 52% in protected areas and 44% allocated as forestry reserves. Nearly all heath and shrublands remain intact as part of forestry reserves on high terrain. The dominant land use in the Edgecumbe Bay sub catchments is now pasture grazing occurring on former forest and woodland areas and there has also been an increase in the extent of sugar cane between 1999 and 2009 (Figure 11.8 panel a; Figure 11.8 panel b).

Ecosystem and land use mapping can be combined with maps of the hydrological connections to better understand the areas of strongest connectivity to the reef. The Bluemaps identify where and how frequently the landscape is wet. Mapping the hydrologically connection (*i.e.*, wet frequency) can provide an important added dimension for management. For instance, in the areas most frequently connected to the Bay (Very Frequently Connected: Figure 11.9) pasture grazing, grazing of ponded pastures and aquaculture (prawn farm) were identified as the dominant uses. The areas most connected to the reef likely have the highest potential to impact the reef and these specific land uses may be producing sediment, nutrients and other pollution. The combination of maps can identify places in the catchment where the restoration of ecological processes will be most effective for reducing impacts of runoff to the marine environment.

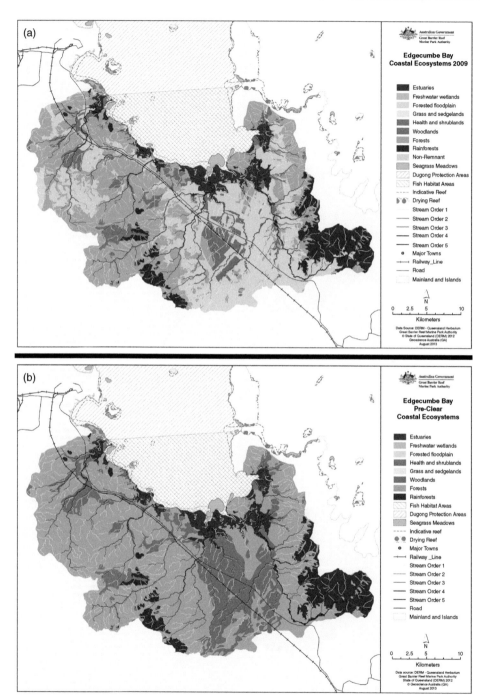

Figure 11.7 (a) Postclearance (2009) map of Edgecumbe Bay remaining coastal ecosystems showing where vegetation has been modified or lost. (b) Preclearance (pre-European) coastal ecosystem map of the Edgecumbe Bay sub-basins.

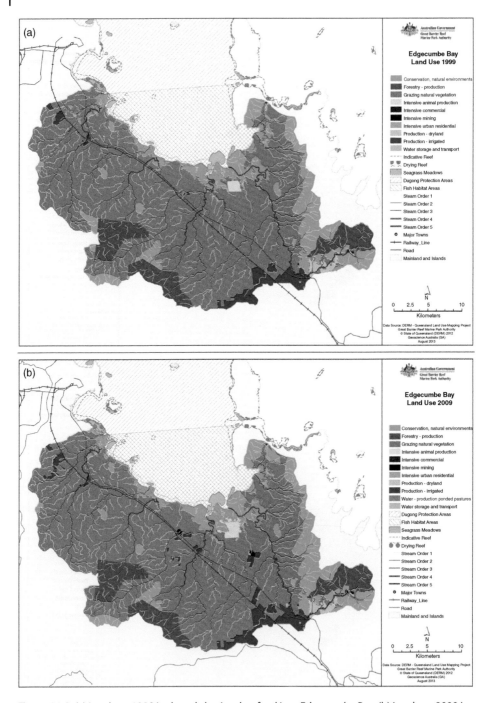

Figure 11.8 (a) Land use 1999 in the sub-basins that feed into Edgecumbe Bay. (b) Land use 2009 in the sub-basins that feed into Edgecumbe Bay.

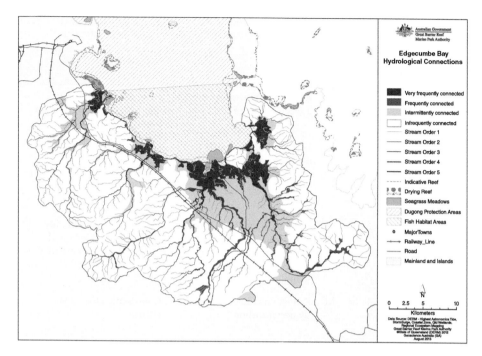

Figure 11.9 Hydrological connections mapping, 'Bluemaps', for the subcatchments.

Turning to the Ecological Processes Calculator, the systems values matrix (Table 11.2) shows the level of change to ecological processes within the Edgecumbe Bay subcatchments using the Bluemaps to define areas most affected by changes in capacity. The greatest loss in capacity for most ecological processes is evident in the Bluemaps *infrequently* and *intermittently connected* areas. Summarized results (Table 11.3) from the assessment provide a good basis for making better informed management decisions. Having stakeholder engagement in the process maximizes the opportunity for uptake from the community to improve ecosystem services derived from functional ecological processes.

The future for system repair across the reef's catchment is exciting and ground breaking with the recognition that we need to forge and strengthen partnerships between agencies and stakeholders and work collaboratively to develop a strong understanding of the catchments in which we live. As we move into a new era of catchment and marine management, the Authority's coastal ecosystems program has highlighted the need for developing an agreed ecosystem health with our partners and stakeholders. This must be based on a holistic adaptive management approach, supported by consistent whole of system understanding and guided by an agreed toolbox of data and information. This type of understanding will allow for targeted on ground works that will benefit the long term heath of the reef.

Table 11.2 Changes to ecological processes in very frequently connected (VFC), frequently connected (FC), intermittently connected (INT) and infrequently connected (INF) areas of the catchment.

Ecological process	VFC	FC	INT	INF	Comment	Implications for ecosystem services
Landscape processes	M	M	P	M	Reduction in overland flows as a result of Peter Faust Dam; Increase in runoff as a result of land clearing and cattle grazing; Over extraction of groundwater	Increase in event water quantity, decline in ambient water quantity due to extraction. Lowering of groundwater table with possible implications for remnant MNES vegetation, reduced fish habitat due to increased ephemerality
Physical processes	G	G	M	M	Loss of groundcover, riparian vegetation, poor land management and gully erosion resulting in increased fine sediments in run-off	Increase inshore turbidity resulting in less light and subsequent loss in extent and density of seagrass.
Chemical processes	G	M	M	M	Reduction in the capacity to sequester carbon	Contribution to climate change
Biogeochemical processes	G	G	M	M	Reduction in capacity of the area to cycle nutrients before they reach the bay due to fast flows	Declining water quality in the bay and beyond
Biological processes	G	M	G	VG	Some loss of important fish habitat; barriers to fish passage, loss of riparian	Reduction in fish habitat, foraging and nursery areas due to reduced connectivity and changed hydrology.

Note: MNES = Matters of national and state environmental significance.

Table 11.3 Ecosystem services and actions that can improve the capacity of the Edgecumbe Bay catchment to deliver them. Note: Very frequently connected (VFC), frequently connected (FC), intermittently connected (INT) and infrequently connected (INF).

Hazard reduction	Prioritize best management practices on sloping land that are present in the upper catchment Prioritize best management practices on grazing of natural areas in the upland (INF) areas Assess road culverts for erosion risk through a fluvial geomorphological assessment
Water quality protection	Improve uptake of best management practices in grazing and irrigated cropping
Climate regulation	Build resilience of coastal ecosystems through other identified actions
Biodiversity	Promote a systems understanding of the value of the Gregory River for its biodiversity values to subcatchment landholders Install in-stream habitats in priority waterways, including the Gregory River

(Continued)

Table 11.3 (Continued)

	Assess and reconnect waterways in the VFC and FC areas
	Install fish passage devices in VFC and FC areas in the Gregory and Eden Lassie Rivers
	Reinstate riparian areas around in-river refugia areas and areas of groundwater expression
Erosion protection	Promote reduced boating speeds in estuaries
	Reinstate riparian vegetation as identified by fluvial geomorphology assessment
Provision of food	Improve quality and quantity of fish habitat in the Gregory River
Nutrient cycling	Ensure no further degradation of water quality in this subcatchment

11.5 Towards Applying Landscape Ecology to Land-Sea Modelling and Management

A large and growing body of evidence suggests that landscape patterns matter for terrestrial processes and ecosystem services, some of which, like sediment retention and nutrient processing, are critical to coastal health. Landscape ecology has played a significant role in identifying the importance of habitat patterns and fragmentation on terrestrial ecological processes (*e.g.,* ecosystem connectivity, surface water flow, *etc.*) (Hobbs 1994; Wilcove *et al.* 1998; Fausch *et al.* 2002; Torrubia *et al.* 2014) and the impacts for soil erosion rates (Saunders *et al.* 1991; Dynesius & Nilsson 1994; Sahin & Hall 1996; Turner *et al.* 2001; Costa *et al.* 2003; Ziegler *et al.* 2007). A comparably smaller literature reveals similar relationships between seascape patterns and marine ecosystem processes and ecosystem services. For instance, the influence of seagrass patterns on scallop survival (Irlandi *et al.* 1995), the role of large-scale habitat patterns on species connectivity (Grober-Dunsmore *et al.* 2009) and the influence of structure on reef fish assemblages (Grober-Dunsmore *et al.* 2007). Because patterns are important for ecosystem function, they affect the landscape's ability to provide ecosystem goods and services in both terrestrial and marine realms. For example, research on Thai coastal habitats highlighted the importance of habitat size for wave attenuation (Barbier *et al.* 2008) and work in the Amazon showed how deforestation affects myriad services, including carbon storage and modulation of infectious disease (Foley *et al.* 2007).

Despite this impressive body of research, major gaps exist in our understanding of the connections between terrestrial and coastal patterns across an integrated land-seascape, including how disturbed patterns and geometries across the land-sea interface affect present and future seascape ecological processes (Pittman 2013) and ecosystem service delivery. The interaction between landscapes and seascapes is, of course, not a one-way process (Polis *et al.* 2004). Each system affects the other through the exchange of energy and materials mediated by waves, wind and temperature, chemical and geological constituents like water, sediment and nutrients and organisms and their bodies' materials. Processes linking land and sea include the protection of shorelines by coastal habitats such as reefs and mangroves, inputs from rivers and groundwater, coastal wetlands processing pollution, turtles nesting, seabirds enriching soils by depositing guano on land and the movement of euryhaline species between land and sea, such as salmon migrating up river to spawn (Álvarez-Romero *et al.* 2011). Given the clear linkages across the land-sea continuum, it is surprising that few landscape ecology-informed studies have focused on the dynamic boundary between land and sea (Wedding *et al.* 2011). Characteristics of the land-sea interface could greatly influence how (even if) patterns translate across realms, for instance by altering the spatial and temporal patterns of pollution. Although there is an emerging field of research in land-sea connectivity, our ability to understand interactions between terrestrial and coastal biophysical processes across various spatial and temporal scales is still limited. All of this suggests a critical need for basic science.

Emerging science could then be used to derive spatial pattern metrics that can parsimoniously and cost-effectively represent patterns affecting key cross-realm processes. In line with terrestrial metrics such as DLI, researchers in the marine realm have borrowed and developed new metrics meaningful for seascapes, such as a fragmentation index (Sleeman *et al.* 2005), patch isolation, contagion and habitat richness (Wedding *et al.* 2011). That said, the marine realm offers additional challenges related to the dynamics at land-sea boundary and the three-dimensionality of the marine environment and few metrics have been developed to deal with these (Wedding *et al.* 2011).

Numerous modelling tools work within the limitations posed by the lack of basic science describing key processes and interactions within land-seascape systems. Based on their complexity, modelling tools can provide more or less specificity and the level of specificity needed depends on the decision context. Useful spatial models exist that generate basic information with relatively little effort by capitalizing on existing tools and data. Modifications to the LDI should continue to be investigated to increase its capacity to rapidly map potential for runoff and integrate other spatial data to refine the model. For example, when the LDI has been combined with spatial pattern metrics, specifically patch area, the model has better predicted catchments delivering greater nutrient loads to coastal waters (Carey *et al.* 2011). Similarly, spatial pattern metrics could also boost deterministic spatial models, such as N-SPECT or InVEST, where the aim is to predict sediment and nutrient loads. More nuanced information may be needed if management seeks to understand the absolute levels of stressors, socioeconomic drivers of land use change, ecological response to multiple stressors or management action, or socio-economic impacts of ecological change; in these cases, a more complex, dynamic, systems model can target key processes or the entire system. Many of these tools have capabilities to deepen our understanding of the emergence of patterns, as well as their consequences, even if they have rarely been applied to this end.

More acute even than our knowledge gaps in biophysical relationships across the land-sea interface are the gaps in our understanding of how these patterns interact with human socioeconomic systems; for instance, how human systems drive biophysical change and how they alter the seascape's ability to deliver socially valuable ecosystem goods and services. Human activities increasingly disturb and adapt coastal landscapes and have led to a significant degradation of salt marshes (50%, United States), mangroves (>35% globally degraded), coral reefs (>30% globally degraded, 60% threatened) and seagrass meadows (>25% globally degraded) (Kennish 2001; Valiela *et al.* 2001; Orth *et al.* 2006; Burke *et al.* 2011). Human activities can both amplify and erode land-sea exchanges, for example, resource extraction alters food webs and impacts habitats; anthropogenic effluents from land introduce nutrients; aquaculture facilities modify coastlines and introduce disease; dams interrupt hydrological regimes and migration pathways; introduced species change ecosystem dynamics; and climate change has broad impacts ranging from weather, phenology and the local persistence of species (Álvarez-Romero *et al.* 2011). Degradation of coastal systems undermines their ability to provide ecosystem goods and services that underpin human wellbeing for millions of people globally (Burke *et al.* 2011; Pandolfi *et al.* 2011).

A greater integration of multidisciplinary knowledge is required to understand the complex land-sea social-ecological system, for example, integrated studies using data and expertise on the coupled hydrological-vegetation system of watersheds; dispersal modelling and biochemical analyses to determine the fate and biological consequences of materials when entering the coastal waters and marine life; ecological models to assess reef system response; ecological economics to assess wellbeing consequences; and engineering and planning to develop mitigation strategies. Such studies can be organized around frameworks such as the DPSIR (Driver-Pressure-State-Impact-Response) and could operationalize this quantitatively with spatial tools and landscape ecology techniques to understand the linkages between drivers of spatial patterning in watersheds and consequences for seascape condition (Lewison *et al.* 2016).

A fundamental goal of holistic, ecosystem-based management is to balance multiple, often competing objectives across the land-seascape. Land-sea connectivity has been recognized as a strong factor in the delineation of marine conservation planning priorities in many locations (Lagabrielle *et al.* 2009; Beger *et al.* 2010; Klein *et al.* 2012; Makino *et al.* 2013), but integrated land-sea planning needs to incorporate objectives for the processes that connect land and sea, as well as the threats in one that affect the other, at a relevant scale (Leslie & McLeod 2007). Solutions should draw from management actions on land and in the sea to maximize benefits and minimize costs (Álvarez-Romero *et al.* 2011). Various studies have shown how decision science can lead to far more optimal solutions than other prioritization approaches (White *et al.* 2012a; Makino *et al.* 2013; Klein *et al.* 2014). The spatial nature of the systems is fundamental to optimizing solutions – the relative costs and benefits of different spatial arrangements of management determine their relative priority. The optimization algorithm needs to consider how patterns affect either the cost or benefit of a given solution set. For instance, if certain habitat configurations are better at providing ecosystem services, then this should be reflected in their assigned benefit. Similarly, certain land use patterns may be far costlier to fix than others (*e.g.*, because they are dispersed over rugged terrain). Optimizing management with an eye towards patterns could have significant social benefits. The useful role of seascape ecology as counterpart and complementary to landscape ecology can help effectively draw interdisciplinary land-sea ecological and social knowledge into cross-jurisdictional policy making and management plan implementation process. Not accounting for landscape-seascape connectivity may ultimately prevent successful resource management and conservation (Silverstri & Kershaw 2010).

This chapter set out to explore how landscape ecology concepts and techniques can build our understanding of the land-sea system and support cross-realm management. It seems there is a rich future for expanding basic science of social-ecological system patterns, using the tools at hand and perhaps developing yet more to explore patterns and integrating these into spatial management across the land-sea interface. Few examples exist where spatially explicit and quantitative measurements of landscape patterning have been linked to spatially explicit measures of seascape patterning. In particular, future integration of landscape ecology into a holistic framework for investigating land-sea connectivity could inform pathways of more effective integrated management by delineating more explicitly the connectivity between patterns and processes and the consequences for natural capital and its socio-economic benefit stream.

References

Ainsworth CH, Varkey DA, Pitcher TJ (2008) Ecosystem simulation models of Raja Ampat, Indonesia, in support of ecosystem based fisheries management. In Bailey M, Pitcher TJ (eds) Ecological and Economic Analyses of Marine Ecosystems in the Bird's Head Seascape, Papua, Indonesia: II. Fisheries Centre Research Reports 16(1). Fisheries Centre, University of British Columbia, Canada, pp. 3–124.

Airoldi L (2003) The effects of sedimentation on rocky coast assemblages. In Gibson R, Atkinson R (eds) Oceanography and Marine Biology: An Annual Review. CRC Press, Boca Raton, FL.

Alvarez-Cobelas M, Angeler DG, Sánchez-Carrillo S (2008) Export of nitrogen from catchments: A worldwide analysis. Environmental Pollution 156: 261–269.

Álvarez-Romero JG, Adams VM, Pressey RL, Douglas M, Dale AP, Augé AA, Ball D, Childs J, Digby M, Dobbs R, Gobius N, Hinchley D, Lancaster I, Maughan M, Perdrisat I (2015) Integrated cross-realm planning: A decision-makers' perspective. Biological Conservation 191: 799–808.

Álvarez-Romero JG, Devlin M, Teixeira da Silva E, Petus C, Ban NC, Pressey RL, Kool J, Roberts JJ, Cerdeira-Estrada S, Wenger AS, Brodie J (2013) A novel approach to model exposure of coastal-marine ecosystems to riverine flood plumes based on remote sensing techniques. Journal of Environmental Management 119: 194–207.

Álvarez-Romero JG, Pressey RL, Ban NC, Vance-Borland K, Willer C, Klein CJ, Gaines SD (2011) Integrated land-sea conservation planning: The missing links. Annual Review of Ecology, Evolution, and Systematics 42: 381–409.

Apple M (2008) Applicability of the Hydrological Simulation Program-Fortran (HSPF) for Modelling Runoff and Sediment in Hawaii Watersheds. PhD thesis. University of Hawaii at Manoa, Honolulu, HI.

Arifin Z, Puspitasari R, Miyazaki N (2012) Heavy metal contamination in Indonesian coastal marine ecosystems: A historical perspective. Coastal Marine Science 35: 227–233.

Arnold JG, Srinivasan R, Muttiah RS, Williams JR (1998) Large area hydrologic modelling and assessment – Part 1: Model development. Journal of the American Water Resources Association 34: 73–89.

Ball IR, Possingham HP (2000) MARXAN (V1.8.2) Marine Reserve Design using Spatially Explicit Annealing. A manual prepared for the Great Barrier Reef Marine Park Authority. University of Adelaide, Adelaide.

Barbier EB, Koch EW, Silliman BR, Hacker SD, Wolanski E, Primavera J, Granek EF, Polasky S, Aswani S, Cramer LA, Stoms DM, Kennedy CJ, Bael D, Kappel CV, Perillo GME, Reed DJ (2008) Coastal ecosystem-based management with nonlinear ecological functions and values. Science 319: 321–323.

Bartley R, Bainbridge ZT, Lewis SE, Kroon FJ, Wilkinson SN, Brodie JE, Silburn DM (2014a) Relating sediment impacts on coral reefs to watershed sources, processes and management: A review. Science of the Total Environment 468–469: 1138–1153.

Bartley R, Corfield JP, Hawdon AA, Kinsey-Henderson AE, Abbott BN, Wilkinson SN, Keen RJ (2014b) Can changes to pasture management reduce runoff and sediment loss to the Great Barrier Reef? The results of a 10-year study in the Burdekin catchment, Australia. The Rangeland Journal 36: 67–84.

Bautista S, Mayor AG, Bourakhouadar J, Bellot J (2007) Plant spatial pattern predicts hillslope runoff and erosion in a semiarid Mediterranean landscape. Ecosystems 10: 987–998.

Beger M, Linke S, Watts M, Game E, Treml E, Ball I, Possingham HP (2010) Incorporating asymmetric connectivity into spatial decision making for conservation. Conservation Letters 3: 359–368.

Bennett EM, Peterson GD, Gordon LJ (2009) Understanding relationships among multiple ecosystem services. Ecology Letters 12: 1394–1404.

Bingner R, Theurer F (2001) Topographic factors for RUSLE in the continuous-simulation, watershed model for predicting agricultural, non-point source pollutants (AnnAGNPS). Proceedings of the American Society of Agricultural Engineers. Agricultural Research Service.

Bolte J, Vaché K (2010) Envisioning Puget Sound Alternative Futures: PSNERP Final Report. Oregon State University, Corvallis, OR.

Borah DK, Bera M (2003) Watershed-scale hydrologic and nonpoint-source pollution models: Review of applications. Transactions of the ASAE 47: 789–803.

Borah DK, Xia R, Bera M (2002) DWSM-A dynamic watershed simulation model. In Singh VP & Frevert DK (eds) Mathematical Models of Small Watershed Hydrology and Applications. Water Resources Publications, Highlands Ranch, CO, pp. 113–166.

Bosch JM, Hewlett JD (1982) A review of catchment experiments to determine the effect of vegetation changes on water yield and evapotranspiration. Journal of Hydrology 55: 3–23.

Boström C, Pittman SJ, Simenstad C, Kneib R (2011) Seascape ecology of coastal biogenic habitats: advances, gaps, and challenges. Marine Ecology Progress Series 427: 191–217.

Boumans R, Roman J, Altman I, Kaufman L (2015) The Multiscale Integrated Model of Ecosystem Services (MIMES): Simulating the interactions of coupled human and natural systems. Ecosystem Services 12: 30–41.

Brooks GR, Devine B, Larson RA, Rood BP (2007) Sedimentary development of Coral Bay, St John, USVI: a shift from natural to anthropogenic influences. Caribbean Journal of Science 43: 226–243.

Burke L, Reytar K, Spalding M, Perry A (2011) Reefs at risk revisited. World Resources Institute, Washington, DC.

Cabaço S, Santos R, Duarte CM (2008) The impact of sediment burial and erosion on seagrasses: A review. Estuarine, Coastal and Shelf Science 79: 354–366.

Carey RO, Migliaccio KW, Li Y, Schaffer B, Kiker GA, Brown MT (2011) Land use disturbance indicators and water quality variability in the Biscayne Bay Watershed, Florida. Ecological Indicators 11: 1093–1104.

Carilli JE, Norris RD, Black BA, Walsh SM, McField M (2009) Local stressors reduce coral resilience to bleaching. PloS One 4: e6324.

Chen N, Hong H (2012) Integrated management of nutrients from the watershed to coast in the subtropical region. Current Opinion in Environmental Sustainability 4: 233–242.

Cheng CL (2007) Evaluating the performance of AnnAGNPS and N-SPECT for tropical conditions. MS, University of Hawaii at Manoa, Honolulu, HI.

Chérubin LM, Kuchinke CP, Paris CB (2008) Ocean circulation and terrestrial runoff dynamics in the Mesoamerican region from spectral optimization of SeaWiFS data and a high resolution simulation. Coral Reefs 27: 503–519.

Chou W-C (2010) Modelling watershed scale soil loss prediction and sediment yield estimation. Water Resources Management 24: 2075–2090.

Cicin-Sain B, Belfiore S (2005) Linking marine protected areas to integrated coastal and ocean management: A review of theory and practice. Ocean and Coastal Management 48: 847–868.

Conroy MJ, Peterson JT (2013) Decision Making in Natural Resource Management: A Structured, Adaptive Approach. John Wiley & Sons, Inc., Hoboken, NJ.

Costa MH, Botta A, Cardille JA (2003) Effects of large-scale changes in land cover on the discharge of the Tocantins River, Southeastern Amazonia. Journal of Hydrology 283: 206–217.

Crowder LB, Osherenko G, Young OR, Airamé S, Norse EA, Baron N, Day JC, Douvere F, Ehler CN, Halpern BS, Langdon SJ, McLeod KL, Ogden JC, Peach RE, Rosenberg AA, Wilson JA (2006) Resolving mismatches in US ocean governance. Science 313: 617.

Cumming G, Olsson P, Chapin F, Holling C (2013) Resilience, experimentation, and scale mismatches in social-ecological landscapes. Landscape Ecology 28: 1139–1150.

Dachs J, Méjanelle L (2010) Organic pollutants in coastal waters, sediments, and biota: A relevant driver for ecosystems during the Anthropocene? Journal of the Coastal and Estuarine Research Federation 33: 1–14.

Darling ES, Côté IM (2008) Quantifying the evidence for ecological synergies. Ecology Letters 11: 1278–1286.

Deegan LA, Johnson DS, Warren RS, Peterson BJ, Fleeger JW, Fagherazzi S, Wollheim WM (2012) Coastal eutrophication as a driver of salt marsh loss. Nature 490: 388–392.

DeMartini E, Jokiel P, Beets J, Stender Y, Storlazzi C, Minton D, Conklin E (2013) Terrigenous sediment impact on coral recruitment and growth affects the use of coral habitat by recruit parrotfishes (F. Scaridae). Planning and Management 17: 417–429.

Devlin MJ, McKinna LW, Álvarez-Romero JG, Petus C, Abott B, Harkness P, Brodie J (2012) Mapping the pollutants in surface riverine flood plume waters in the Great Barrier Reef, Australia. Marine Pollution Bulletin 65: 224–235.

Diaz RJ, Rosenberg R (2008) Spreading dead zones and consequences for marine ecosystems. Science 321: 926–929.

Dodds WK (2006) Nutrients and the 'dead zone': the link between nutrient ratios and dissolved oxygen in the northern Gulf of Mexico. Frontiers in Ecology and the Environment 4: 211–217.

Downer C, Ogden F (2002) GSSHA User's Manual, Gridded Surface Subsurface Hydrologic Analysis Version 1.43 for WMS 6.1. Engineer Research and Development Center Technical Report. Vicksburg, MS.

Draper D (1995) Assessment and propagation of model uncertainty. Journal of the Royal Statistical Society Series B (Methodological) 57: 45–97.

Dugan GL (1977) Water Quality of Normal and Storm-Induced Surface Water Runoff: Kaneohe Bay Watershed, Oahu, Hawaii. Water Resources Research Center, University of Hawaii at Manoa, Honolulu, HI.

Dunne RP (2010) Synergy or antagonism – interactions between stressors on coral reefs. Coral Reefs 29: 145–152.

Dunning JB, Danielson BJ, Pulliam HR (1992) Ecological processes that affect populations in complex landscapes. Oikos 65: 169–175.

Dynesius M, Nilsson C (1994) Regulation of river systems in the northern third of the world. Science 266.

El-Kadi AI, Fujioka RS, Liu CCK, Yoshida K, Vithanage G, Pan Y, Farmer J (2003) Assessment and Protection Plan for the Nawiliwili Watershed: Phase 2 – Assessment of Contaminant Levels. Water Resources Research Center, University of Hawai'i at Manoa, Honolulu, Hawai'i.

El-Serehy HA, Aboulela H, Al-Misned F, Kaiser M, Al-Rasheid K, El-Din HE (2012) Heavy metals contamination of a Mediterranean coastal ecosystem, eastern Nile Delta, Egypt. Turkis Journal of Fisheries and Aquatic Sciences 12: 751–760.

Erftemeijer PLA, Riegl B, Hoeksema BW, Todd PA (2012) Environmental impacts of dredging and other sediment disturbances on corals: A review. Marine Pollution Bulletin 64: 1737–1765.

Evans RD, Murray KL, Field SN, Moore JAY, Shedrawi G, Huntley BG, Fearns P, Broomhall M, McKinna LIW, Marrable D (2012) Digitise this! A quick and easy remote sensing method to monitor the daily extent of dredge plumes. PloS One 7: e51668.

Executive Order 13547 (2010) Stewardship of the ocean, our coasts, and the Great Lakes. Federal Register 75(140): 43023–43027.

Fabricius KE (2005) Effects of terrestrial runoff on the ecology of corals and coral reefs: review and synthesis. Marine Pollution Bulletin 50: 125–146.

Fabricius KE (2011) Factors determining the resilience of coral reefs to eutrophication: A review and conceptual model. In Dubinsky Z, Stambler N (eds) Coral Reefs: An Ecosystem in Transition. Springer Netherlands, Dordrecht.

Falinski KA (2016) Predicting Sediment Export into Tropical Coastal Ecosystems to Support Ridge to Reef Management. PhD thesis. University of Hawai'i Mānoa, Honolulu, HI.

Fausch KD, Torgersen CE, Baxter CV, Li HW (2002) Landscapes to riverscapes: bridging the gap between research and conservation of stream fishes. BioScience 52: 483.

Fleitmann D, Dunbar RB, McCulloch M, Mudelsee M, Vuille M, McClanahan TR, Cole JE, Eggins S (2007) East African soil erosion recorded in a 300 year old coral colony from Kenya. Geophysical Research Letters 34(4): L04401.

Fletcher PJ, Kelble CR, Nuttle WK, Kiker GA (2014) Using the integrated ecosystem assessment framework to build consensus and transfer information to managers. Ecological Indicators 44: 11–25.

Foley JA, Asner GP, Costa MH, Coe MT, Defries R, Gibbs HK, Howard EA, Olson S, Patz J, Ramankutty N, Snyder P (2007) Amazonia revealed: Forest degradation and loss of ecosystem goods and services in the Amazon Basin. Frontiers in Ecology and the Environment 5: 25–32.

Foley JA, Defries R, Asner GP, Barford C, Bonan G, Carpenter SR, Chapin FS, Coe MT, Daily GC, Gibbs HK, Helkowski JH, Holloway T, Howard EA, Kucharik CJ, Monfreda C, Patz.JA, Prentice IC, Ramankutty N, Snyder PK (2005) Global consequences of land use. Science 309: 570.

Freeman W, Fox J (1995) ALAWAT: A spatially allocated watershed model for approximating stream, sediment, and pollutant flows in Hawaii, USA. An International Journal for Decision Makers, Scientists and Environmental Auditors 19: 567–577.

Friedlander AM, Aeby G, Brown E, Clark A, Coles S, Dollar S, Hunter C, Jokiel P, Smith J, Walsh B, Williams I, Wiltse W (2005) The state of coral reef ecosystems of the Main Hawaiian Islands. In Waddell J (ed.) The state of coral reef ecosystems of the United States and Pacific Freely Associated States: 2005. NOAA Technical Memorandum NOS NCCOS 11. NOAA/NCCOS Center for Coastal Monitoring and Assessment's Biogeography Team, Silver Spring, MD, pp. 222–269.

Fulton EA, Link JS, Kaplan IC, Savina-Rolland M, Johnson P, Ainsworth C, Horne P, Gorton R, Gamble RJ, Smith ADM, Smith DC (2011) Lessons in modelling and management of marine ecosystems: the Atlantis experience. Fish and Fisheries 12: 171–188.

Gaut K (2009) Application of the nonpoint source pollution and erosion comparison tool (N-SPECT) model for resource management in the Waiulaula watershed, island of Hawai'i. MS. University of Hawai'i at Hilo, Hilo, Hawai'i.

Gesch D, Wilson R (2001) Development of a seamless multisource topographic/bathymetric elevation model of Tampa Bay. Marine Technology Society Journal 35: 58–64.

Gilby BL, Olds AD, Connolly RM, Stevens T, Henderson CJ, Maxwell PS, Tibbetts IR, Schoeman DS, Rissik D, Schlacher TA (2016) Optimising land-sea management for inshore coral reefs. PloS One 11: e0164934.

Glenn CR, Whittier RB, Dailer ML, Dulaiova H, El-Kadi AI, Fackrell J, Kelly JL, Waters CA, Sevadjian J (2013) Lahaina groundwater tracer study – Lahaina, Maui, Hawai'i, Final Report. Prepared for the State of Hawai'i Department of Health, the US Environmental Protection Agency, and the US Army Engineer Research and Development Center. School of Ocean and Earth Science and Technology, Department of Geology and Geophysics, University of Hawai'i at Manoa Honolulu, Hawai'i .

Goldstein JH, Caldarone G, Duarte TK, Ennaanay D, Hannahs N, Mendoza G, Polasky S, Wolny S, Daily GC (2012) Integrating ecosystem-service tradeoffs into land-use decisions. Proceedings of the National Academy of Sciences of the United States 109: 7565.

Gombos M, Komoto J, Lowry K, MacGowan P (2010) Hawai'i Coral Reef Strategy: Priorities for Management in the Main Hawaiian Islands 2010–2020. The State of Hawai'i, Honolulu, HI.

Gorman D, Russell BD, Connell SD (2009) Land-to-sea connectivity: linking human-derived terrestrial subsidies to subtidal habitat change on open rocky coasts. Ecological Applications 19: 1114–1126.

Grober-Dunsmore R, Frazer TK, Beets JP, Lindberg WJ, Zwick P, Funicelli NA (2007) Influence of landscape structure on reef fish assemblages. Landscape Ecology 23: 37–53.

Grober-Dunsmore R, Pittman SJ, Caldow C, Kendall MS, Frazer TK (2009) A landscape ecology approach for the study of ecological connectivity across tropical marine seascapes. In Nagelkerken I (ed.) Ecological Connectivity Among Tropical Coastal Ecosystems. Springer, Dordrecht.

Gross JE (2003) Developing Conceptual Models for Monitoring Programs. National Park Service, Ft. Collins, CO.

Group 70 International (2015) Kahana, Honokahua and Honolua Watersheds Characterization Report. Honolulu, HI.

Gurney GG, Melbourne-Thomas J, Geronimo RC, Aliño PM, Johnson CR, Ferse SCA (2013) Modelling coral reef futures to inform management: Can reducing local-scale stressors conserve reefs under climate change? PloS One 8(11): e80137.

Halpern BS, Walbridge S, Selkoe KA, Kappel CV, Micheli F, D'Agrosa C, Bruno JF, Casey KS, Ebert C, Fox HE, Fujita R, Heinemann D, Lenihan HS, Madin EMP, Perry MT, Selig ER, Spalding M, Steneck R, Watson R (2008) A global map of human impact on marine ecosystems. Science 319: 948.

Hamel P, Chaplin-Kramer R, Sim S, Mueller C (2015) A new approach to modelling the sediment retention service (InVEST 3.0): Case study of the Cape Fear catchment, North Carolina, USA. Science of the Total Environment 524–525: 166–177.

Harrelson DW, Zakikhani M (2006) Environmental Assessment of Makua Military Reservation in Hawaii. Engineer Research and Development Center, Vicksburg, MS.

Hobbs RJ (1994) Landscape ecology and conservation: Moving from description to application. Pacific Conservation Biology 1: 170–176.

Hogrefe KR, Wright DJ, Hochberg EJ (2008) Derivation and integration of shallow-water bathymetry: implications for coastal terrain modelling and subsequent analyses. Marine Geodesy 31: 299–317.

Hohenthal J, Owidi E, Minoia P, Pellikka P (2015) Local assessment of changes in water-related ecosystem services and their management: DPASER conceptual model and its application in Taita Hills, Kenya. International Journal of Biodiversity Science, Ecosystem Services and Management 11: 225–238.

Houser JN, Mulholland PJ, Maloney KO (2006) Upland disturbance affects headwater stream nutrients and suspended sediments during baseflow and stormflow. Journal of Environmental Quality 35: 352–365.

Hovel KA, Lipcius RN (2001) Habitat fragmentation in a seagrass landscape: patch size and complexity control blue crab survival. Ecology 82: 1814–1829.

Howarth R, Swaney D, Billen G, Garnier J, Hong B, Humborg C, Johnes P, Mörth C-M, Marino R (2012) Nitrogen fluxes from the landscape are controlled by net anthropogenic nitrogen inputs and by climate. Frontiers in Ecology and the Environment 10: 37–43.

Hughes TP, Connell JH (1999) Multiple stressors on coral reefs: a long-term perspective. Limnology and Oceanography 44: 932–940.

Hupp CR, Pierce AR, Noe GB (2009) Floodplain geomorphic processes and environmental impacts of human alteration along coastal plain rivers, USA. Wetlands 29: 413–429.

Irlandi EA, Ambrose Jr WG, Orlando BA (1995) Landscape ecology and the marine environment: how spatial configuration of seagrass habitat influences growth and survival of the bay scallop. Oikos 72(3): 307–313.

Jackson J, Donovan M, Cramer K, Lam V (2014) Status and Trends of Caribbean Coral Reefs: 1970–2012. Global Coral Reef Monitoring Network, Washington, DC.

Jain MK, Das D (2010) Estimation of sediment yield and areas of soil erosion and deposition for watershed prioritization using GIS and remote sensing. Water Resources Management 24: 2091–2112.

James LA, Watson DG, Hansen WF (2007) Using LiDAR data to map gullies and headwater streams under forest canopy: South Carolina, USA. Catena 71: 132–144.

Jickells TD (1998) Nutrient biogeochemistry of the coastal zone. Science 281: 217–222.

Johansen JL, Jones GP (2013) Sediment-induced turbidity impairs foraging performance and prey choice of planktivorous coral reef fishes. Ecological Applications 23: 1504–1517.

Jokiel PL, Rodgers KS, Walsh WJ, Polhemus DA, Wilhelm TA (2011) Marine resource management in the Hawaiian Archipelago: The traditional Hawaiian system in relation to the Western approach. Journal of Marine Biology 2011: 1–16.

Jones KB, Neale AC, Nash MS, Van Remortel RD, Wickham JD, Riitters KH, O'Neill RV (2001) Predicting nutrient and sediment loadings to streams from landscape metrics: a multiple watershed study from the United States Mid-Atlantic Region. Landscape Ecology 16: 301–312.

Jouffray J-B, Nyström M, Norström AV, Williams ID, Wedding LM, Kittinger JN, Williams GJ (2015) Identifying multiple coral reef regimes and their drivers across the Hawaiian archipelago. Philosophical Transactions of the Royal Society of London B: Biological Sciences 370: 20130268.

Jupiter SD, Jenkins AP, Long WJL, Maxwell SL, Carruthers TJB, Hodge KB, Govan H, Tamelander J, Watson JEM (2014) Principles for integrated island management in the tropical Pacific. Pacific Conservation Biology 20: 193–205.

Karr JR, Schlosser IJ (1978) Water resources and the land-water interface. Science 201: 229–234.

Keeler BL, Wood SA, Polasky S, Kling C, Filstrup CT, Downing JA (2015) Recreational demand for clean water: evidence from geotagged photographs by visitors to lakes. Frontiers in Ecology and the Environment 13: 76–81.

Kelble CR, Loomis DK, Lovelace S, Nuttle WK, Ortner PB, Fletcher P, Cook GS, Lorenz JJ, Boyer JN (2013) The EBM-DPSER conceptual model: integrating ecosystem services into the DPSIR framework. PLoS One 8: e70766.

Kennish MJ (2001) Coastal salt marsh systems in the US: A review of anthropogenic impacts. Journal of Coastal Research 17: 731–748.

Klein CJ, Jupiter SD, Selig ER, Watts ME, Halpern BS, Kamal M, Roelfsema C, Possingham HP (2012) Forest conservation delivers highly variable coral reef conservation outcomes. Ecological Applications 22: 1246–1256.

Klein CJ, Jupiter SD, Watts M, Possingham HP (2014) Evaluating the influence of candidate terrestrial protected areas on coral reef condition in Fiji. Marine Policy 44: 360.

Klumpp DW, Huasheng H, Humphrey C, Xinhong W, Codi S (2002) Toxic contaminants and their biological effects in coastal waters of Xiamen, China: I. Organic pollutants in mussel and fish tissues. Marine Pollution Bulletin 44: 752–760.

Knee KL, Layton BA, Street JH, Boehm AB, Paytan A (2008) Sources of nutrients and fecal indicator bacteria to nearshore waters on the north shore of Kauai (Hawaii, USA). Estuaries and Coasts 31: 607–622.

Kroon FJ, Thorburn P, Schaffelke B, Whitten S (2016) Towards protecting the Great Barrier Reef from land-based pollution. Global Change Biology 22(6): 1985–2002.

Lagabrielle E, Rouget M, Payet K, Wistebaar N, Durieux L, Baret S, Lombard A, Strasberg D (2009) Identifying and mapping biodiversity processes for conservation planning in islands: A case study in Réunion Island (western Indian Ocean). Biological Conservation 142: 1523–1535.

Larsen MC, Webb RMT (2009) Potential effects of runoff, fluvial sediment, and nutrient discharges on the coral reefs of Puerto Rico. Journal of Coastal Research 25(1): 189–208.

Lebel L (2012) Governance and coastal boundaries in the tropics. Current Opinion in Environmental Sustainability 4: 243–251.

LeBlanc RT, Brown RD, FitzGibbon JE (1997) Modelling the effects of land use change on the water temperature in unregulated urban streams. Journal of Environmental Management 49: 445–469.

Leslie HM, McLeod KL (2007) Confronting the challenges of implementing marine ecosystem-based management. Frontiers in Ecology and the Environment 5: 540–548.

Lester SE, Costello C, Halpern BS, Gaines SD, White C (2013) Evaluating tradeoffs among ecosystem services to inform marine spatial planning. Marine Policy 38: 80–89.

Levin PS, Fogarty MJ, Murawski SA, Fluharty D (2009) Integrated ecosystem assessments: developing the scientific basis for ecosystem-based management of the ocean. PLoS Biol 7: e1000014.

Lewis SE, Shields GA, Kamber BS, Lough JM (2007) A multi-trace element coral record of land-use changes in the Burdekin River catchment, NE Australia. Palaeogeography, Palaeoclimatology, Palaeoecology 246: 471–487.

Lewison RL, Rudd MA, Al-Hayek W, Baldwin C, Beger M, Lieske SN, Jones C, Satumanatpan S, Junchompoo C, Hines E (2016) How the DPSIR framework can be used for structuring problems and facilitating empirical research in coastal systems. Environmental Science and Policy 56: 110–119.

Likens GE, Bormann FH (1974) Linkages between terrestrial and aquatic ecosystems. BioScience 24(8): 447–456.

Lorek H, Sonnenschein M (1999) Modelling and simulation software to support individual-based ecological modelling. Ecological Modelling 115: 199–216.

Lotze HK, Coll M, Magera AM, Ward-Paige C, Airoldi L (2011) Recovery of marine animal populations and ecosystems. Trends in Ecology and Evolution 26: 595–605.

Ludwig JA, Bastin GN, Chewings VH, Eager RW, Liedloff AC (2007) Leakiness: a new index for monitoring the health of arid and semiarid landscapes using remotely sensed vegetation cover and elevation data. Ecological Indicators 7(2): 442–454.

Ludwig JA, Eager RW, Bastin GN, Chewings VH, Liedloff AC (2002) A leakiness index for assessing landscape function using remote sensing. Landscape Ecology 17(2): 157–171.

Ludwig J, Tongway D (2002) Clearing savannas for use as rangelands in Queensland: altered landscapes and water-erosion processes. The Rangeland Journal 24: 83–95.

Ludwig J, Wilcox BP, Breshears DD, Tongway DJ, Imeson AC (2005) Vegetation patches and runoff-erosion as interacting ecohydrological processes in semiarid landscapes. Ecology 86: 288–297.

Macdonald LH, Anderson DM, Dietrich WE (1997) Paradise threatened: land use and erosion on St John, US Virgin Islands. Environmental Management 21: 851–863.

MacLennan CA (2014) Sovereign Sugar: Industry and Environment in Hawai'i. University of Hawaii Press, Honolulu, HI.

Maguire LA (2004) What can decision analysis do for invasive species management? Risk Analysis 24: 859–868.

Mahmoodabadi M, Ghadiri H, Rose C, Yu B, Rafahi H, Rouhipour H (2014) Evaluation of GUEST and WEPP with a new approach for the determination of sediment transport capacity. Journal of Hydrology 513: 413–421.

Makino A, Beger M, Klein CJ, Jupiter SD, Possingham HP (2013) Integrated planning for land–sea ecosystem connectivity to protect coral reefs. Biological Conservation 165: 35–42.

Manley PN, Zielinski WJ, Stuart CM, Keane JJ, Lind AJ, Brown C, Plymale BL, Napper CO (2000) Monitoring ecosystems in the Sierra Nevada: the conceptual model foundation. Environmental Monitoring and Assessment 64: 139–152.

Mayor ÁG, Bautista S, Small EE, Dixon M, Bellot J (2008) Measurement of the connectivity of runoff source areas as determined by vegetation pattern and topography: a tool for assessing potential water and soil losses in drylands. Water Resources Research 44(10): W10423.

McCook LJ (1999) Macroalgae, nutrients and phase shifts on coral reefs: scientific issues and management consequences for the Great Barrier Reef. Journal of the International Society for Reef Studies 18: 357–367.

McGarigal K (2006) Landscape Pattern Metrics. John Wiley & Sons, Ltd, Chichester.

McGarigal K, Tagil S, Cushman SA (2009) Surface metrics: an alternative to patch metrics for the quantification of landscape structure. Landscape Ecology 24: 433–450.

Merritt WS, Letcher RA, Jakeman AJ (2003) A review of erosion and sediment transport models. Environmental Modelling & Software 18: 761–799.

Mishra SK, Tyagi JV, Singh VP, Singh R (2006) SCS-CN-based modelling of sediment yield. Journal of Hydrology 324: 301–322.

Moberg F, Folke C (1999) Ecological goods and services of coral reef ecosystems. Ecological Economics 29: 215–233.

Nemeth RS, Nowlis JS (2001) Monitoring the effects of land development on the near-shore reef environment of St Thomas, USVI. Bulletin of Marine Science 69: 759–775.

Nielson J, Francis O (in press) An evaluation of GSSHA and comparison of the Engelund-Hansen and Kilinc-Richardson sediment transport equations in steep Hawaiian terrain. Hydrology.

Nyberg JB, Marcot BG, Sulyma R (2006) Using Bayesian belief networks in adaptive management. Canadian Journal of Forest Research 36: 3104–3116.

Okano DMJ (2009) Community watershed management: does the use of a watershed model affect the process? PhD dissertation. University of Hawaiʻi at Mānoa, Honolulu.

Oleson KLL, Falinski K, Lecky J, Rowe C, Kappel C, Selkoe K, White C (2017) Upstream solutions to coral reef conservation: the payoffs of smart and cooperative decision-making. Journal of Environmental Management 191: 8–18.

Oliver LM, Lehrter JC, Fisher WS (2011) Relating landscape development intensity to coral reef condition in the watersheds of St Croix, US Virgin Islands. Marine Ecology Progress Series 427: 293–302.

Orlando J, Yee S (2017) Linking terrifenous sediment delivery to declines in coral reef ecosystem services. Estuaries and Coasts 40(2): 359–375.

Orth RJ, Carruthers TJB, Dennison WC, Duarte CM, Fourqurean JW, Heck KL, Hughes AR, Kendrick GA, Kenworthy WJ, Olyarnik S, Short FT, Waycott M, Williams SL (2006) A global crisis for seagrass ecosystems. BioScience 56: 987–996.

Pait AS, Whitall DR, Jeffrey CFG, Caldow C, Mason AL, Christensen JD, Monaco ME, Ramirez J (2007) An Assessment of Chemical Contaminants in the Marine Sediments of Southwest Puerto Rico. NOAA Technical Memoranda. NOAA, Silver Spring, MD.

Pandolfi JM, Connolly SR, Marshall DJ, Cohen AL (2011) Projecting coral reef futures under global warming and ocean acidification. Science 333: 418.

Pandolfi JM, Jackson JBC, Baron N, Bradbury RH, others (2005) Are US coral reefs on the slippery slope to slime? Science 307: 1725.

Papanicolaou AN, Elhakeem M, Krallis G, Prakash S, Edinger J (2008) Sediment transport modelling review – current and future developments. Journal of Hydraulic Engineering 134: 1–14.

Paris C, Chérubin L (2008) River-reef connectivity in the Meso-American Region. Journal of the International Society for Reef Studies 27: 773–781.

Pascual MA, Kareiva P, Hilborn R (1997) The Influence of model structure on conclusions about the viability and harvesting of Serengeti wildebeest. Conservation Biology 11: 966–976.

Perez K, Rodgers KS, Jokiel PL, Lager CV, Lager DJ, Hay M (2014) Effects of terrigenous sediment on settlement and survival of the reef coral *Pocillopora damicornis*. PeerJ 2: e387.

Phillips DL, Gregg JW (2003) Source partitioning using stable isotopes: coping with too many sources. Oecologia 136: 261–269.

Pike RJ, Evans IS, Hengl T (2009) Geomorphometry: a brief guide. Geomorphometry: concepts, software, applications 33: 3–30.

Pittman SJ (2013) Seascape ecology: A new science for the spatial information age. Marine Scientist 44: 20–23.

Polis GA, Power ME, Huxel GR (2004) Food Webs at the Landscape Level. University of Chicago Press, Chicago, IL.

Polyakov V, Fares A, Kubo D, Jacobi J, Smith C (2007) Evaluation of a non-point source pollution model, AnnAGNPS, in a tropical watershed. Environmental Modelling and Software 22: 1617–1627.

Porto P, Walling DE, Callegari G (2013) Using 137Cs and 210Pbex measurements to investigate the sediment budget of a small forested catchment in southern Italy. Hydrological Processes 27: 795.

Pressey RL, Cabeza M, Watts ME, Cowling RM, Wilson KA (2007) Conservation planning in a changing world. Trends in Ecology and Evolution 22: 583–592.

Raclot D, Albergel J (2006) Runoff and water erosion modelling using WEPP on a Mediterranean cultivated catchment. Physics and Chemistry of the Earth 31: 1038–1047.

Ramos Scharrón CE, MacDonald LH (2005) Measurement and prediction of sediment production from unpaved roads, St John, US Virgin Islands. Earth Surface Processes and Landforms 30: 1283–1304.

Refshaard JC, Storm B, Singh VP (1995) MIKE SHE. In Singh VP (ed.) Computer Models of Watershed Hydrology. Water Resources Publications, Denver, CO, pp. 809–846.

Renard KG, Foster GR, Weesies GA, Porter JP (1991) RUSLE: Revised universal soil loss equation. Journal of Soil and Water Conservation 46: 30–33.

Reuter K, Juhn D, Grantham HS (2016) Integrated land-sea management: recommendations for planning, implementation and management. Environmental Conservation 43(02): 1–18.

Richmond RH, Rongo T, Golbuu Y, Victor S, Idechong N, Davis G, Kostka W, Neth L, Hamnett M, Wolanski E (2007) Watersheds and coral reefs: conservation science, policy, and implementation. BioScience 57: 598–607.

Ricker MC, Odhiambo BK, Church JM (2008) Spatial analysis of soil erosion and sediment fluxes: a paired watershed study of two Rappahannock River tributaries, Stafford County, Virginia. Environmental Management 41: 766–778.

Ricketts TH, Regetz J, Steffan-Dewenter I, Cunningham SA, Kremen C, Bogdanski A, Gemmill-Herren B, Greenleaf SS, Klein AM, Mayfield MM, others (2008) Landscape effects on crop pollination services: are there general patterns? Ecology Letters 11: 499–515.

Rodgers Ku, Kido M, Jokiel P, Edmonds T, Brown E (2012) Use of integrated landscape indicators to evaluate the health of linked watersheds and coral reef environments in the Hawaiian Islands. Environmental Management 50: 21–30.

Rogers CS (1990) Responses of coral reefs and reef organisms to sedimentation. Marine Ecology Progress Series 62(1): 185–202.

Rogers CS (2006) Threats to coral reefs in the US Virgin Islands. Sixteenth Meeting of the US Coral Reef Task Force, Status of USVI Coral Reefs Workshop, 24 to 28 October, St Thomas, United States Virgin Islands.

Rogers CS, Beets J (2001) Degradation of marine ecosystems and decline of fishery resources in marine protected areas in the US Virgin Islands. Environmental Conservation 28: 312–322.

Rothenberger P, Blondeau J, Cox C, Curtis S, Fisher WS, Hills-Starr Z, Jeffrey CFG, Kadison E, Lundgren I, Miller WJ, Muller E, Nemeth R, Paterson S (2008) The state of coral reef ecosystems of the US Virgin Islands. In Waddell JE, Clarke AM (eds) The State of Coral Reef Ecosystems in the United States and Pacific Freely Associated States. NOAA/NCCOS Center for Coastal Monitoring and Assessment's Biogeography Team, Silver Spring, MD.

Ruttenberg BI, Hamilton SL, Walsh SM, Donovan MK, Friedlander A, Demartini E, Sala E, Sandin SA (2011) Predator-induced demographic shifts in coral reef fish assemblages. PloS One 6: e21062.

Sahin V, Hall MJ (1996) The effects of afforestation and deforestation on water yields. Journal of Hydrology 178: 293–309.

Sahoo GB, Ray C, De Carlo EH (2006) Calibration and validation of a physically distributed hydrological model, MIKE SHE, to predict streamflow at high frequency in a flashy mountainous Hawaii stream. Journal of Hydrology 327: 94–109.

Saunders DA, Hobbs RJ, Margules CR (1991) Biological consequences of ecosystem fragmentation: A review. Conservation Biology 5: 18–32.

Sebastian CR, McClanahan TR (2013) Description and validation of production processes in the coral reef ecosystem model CAFFEE (Coral-Algae-Fish-Fisheries Ecosystem Energetics) with a fisheries closure and climatic disturbance. Ecological Modelling 263: 326.

Sharp R, Tallis HT, Ricketts T, Guerry AD, Wood SA, Chaplin-Kramer R, Nelson E, Ennaanay D, Wolny S, Olwero N, Vigerstol K, Pennington D, Mendoza G, Aukema J, Foster J, Forrest J, Cameron D, Arkema K, Lonsdorf E, Kennedy C, Verutes G, Kim CK, Guannel G, Papenfus M, Toft J, Marsik M, Bernhardt J, Griffin R, Glowinski K, Chaumont N, Perelman A, Lacayo M, Mandle L, Hamel P, Vogl AL, Rogers L, Bierbower W (2015) InVEST 3.2 User's Guide. Stanford University, University of Minnesota, The Nature Conservancy, and World Wildlife Fund, Stanford, CA.

Silverstri S, Kershaw F (2010) Framing the Flow: Innovative Approaches to Understand, Protect, and Value Ecosystem Services across Linked Habitat. UNEP World Conservation Monitoring Centre, Cambridge.

Smith TB, Nemeth RS, Blondeau J, Calnan JM, Kadison E, Herzlieb S (2008) Assessing coral reef health across onshore to offshore stress gradients in the US Virgin Islands. Marine Pollution Bulletin 56: 1983–1991.

Soicher AJ, Peterson FL (1997) Terrestrial nutrient and sediment fluxes to the coastal waters of west Maui, Hawai'i. Oceanographic Literature Review 12: 1562.

Stoms DM, Davis FW, Andelman SJ, Carr MH, Gaines SD, Halpern BS, Hoenicke R, Leibowitz SG, Leydecker A, Madin EMP, Tallis H (2005) Integrated coastal reserve planning: making the land-sea connection. Frontiers in Ecology and the Environment 3: 429–436.

Storlazzi CD, Field ME, Bothner MH, Presto MK, Draut AE (2009) Sedimentation processes in a coral reef embayment: Hanalei Bay, Kauai. Marine Geology 264: 140–151.

Storlazzi CD, McManus MA, Logan JB, McLaughlin BE (2006) Cross-shore velocity shear, eddies and heterogeneity in water column properties over fringing coral reefs: West Maui, Hawaii. Continental Shelf Research 26: 401–421.

Street JH, Knee KL, Grossman EE, Paytan A (2008) Submarine groundwater discharge and nutrient addition to the coastal zone and coral reefs of leeward Hawai'i. Marine Chemistry 109: 355–376.

Sustainable Resources Group International (2012) Wahikuli-Honokowai watershed management plan: Volume 1, Watershed characterization. West Maui Ridge 2 Reef Initiative and NOAA Coral Reef Conservation Program, Silver Spring, MD.

Swarzenski P, Dulaiova H, Dailer M, Glenn C, Smith C (2013) A geochemical and geophysical assessment of coastal groundwater discharge at select sites in Maui and O'ahu, Hawai'i. In Wetzelhuetter C (ed.) Groundwater in the coastal zones of Asia-Pacific. Springer, New York, NY.

Syvitski JPM, Vorosmarty CJ, Kettner AJ, Green P (2005) Impact of humans on the flux of terrestrial sediment to the global coastal ocean. Science 308: 376.

Szilassi P, Jordan G, van Rompaey A, Csillag G (2006) Impacts of historical land use changes on erosion and agricultural soil properties in the Kali Basin at Lake Balaton, Hungary. Catena 68: 96–108.

Tallis H, Ferdana Z, Gray E (2008) Linking terrestrial and marine conservation planning and threats analysis. Conservation Biology 22: 120.

TerrAqua Environmental Consulting (2013) Modelling feral ungulate impacts in the Honolua Watershed. West Maui Mountain Watershed Partnership.

Torrubia S, McRae BH, Lawler JJ, Hall SA, Halabisky M, Langdon J, Case M (2014) Getting the most connectivity per conservation dollar. Frontiers in Ecology and the Environment 12: 491–497.

Tulloch VJD, Brown CJ, Possingham HP, Jupiter SD, Maina JM, Klein C (2016) Improving conservation outcomes for coral reefs affected by future oil palm development in Papua New Guinea. Biological Conservation 203: 43–54.

Turner M (1989) Landscape ecology: The effect of pattern On process. Annual Review of Ecology and Systematics 20: 171–197.

Turner M, Gardner R, O'Neill R (2001) Landscape ecology in theory and practice: Pattern and process. Springer New York, New York, NY.

Turner RE, Rabalais NN (2003) Linking landscape and water quality in the Mississippi River Basin for 200 years. BioScience 53: 563–572.

Valiela I, Bowen JL, York JK (2001) Mangrove forests: One of the world's threatened major tropical environments. BioScience 51: 807–815.

Van Dessel W, Van Rompaey A, Poelmans L, Szilassi P (2008) Predicting land cover changes and their impact on the sediment influx in the Lake Balaton catchment. Landscape Ecology 23: 645–656.

Van Kouwen F, Dieperink C, Schot P, Wassen M (2007) Applicability of decision support systems for integrated coastal zone management. Coastal Management 36: 19–34.

Van Rompaey A, Krasa J, Dostal T (2007) Modelling the impact of land cover changes in the Czech Republic on sediment delivery. Land Use Policy 24: 576–583.

Vega Thurber RL, Burkepile DE, Fuchs C, Shantz AA, McMinds R, Zaneveld JR (2014) Chronic nutrient enrichment increases prevalence and severity of coral disease and bleaching. Global Change Biology 20(2): 544–554.

Verburg PH, Dearing JA, Dyke JG, van der Leeuw S, Seitzinger S, Steffen W, Syvitski J (2015) Methods and approaches to modelling the Anthropocene. Global Environmental Change 39: 328–340.

Verger RP, Augustijn DCM, Booij MJ, Fares A, Erdbrink CD, Os vAG (2008) Modelling of a vanishing Hawaiian stream with DHSVM.

Verger RP, Augustijn DCM, Booij MJ, Fares A, Erdbrink CD, Os vAG (2008) Modelling of a vanishing Hawaiian stream with DHSVM. Proceedings of the NCR-days 2008. NCR publication 33-2008. Netherlands Centre for River Studies, Enschede.

Verstraeten G, Oost K, Rompaey A, Poesen J, Govers G (2002) Evaluating an integrated approach to catchment management to reduce soil loss and sediment pollution through modelling. Soil Use and Management 18: 386–394.

Vigerstol KL, Aukema JE (2011) A comparison of tools for modelling freshwater ecosystem services. Journal of Environmental Management 92: 2403–2409.

Walling DE (2005) Tracing suspended sediment sources in catchments and river systems. Science of the Total Environment 344: 159–184.

Walling DE (2006) Human impact on land–ocean sediment transfer by the world's rivers. Geomorphology 79: 192–216.

Walling DE, Fang D (2003) Recent trends in the suspended sediment loads of the world's rivers. Global and Planetary Change 39: 111–126.

Warne AG, Webb RM, Larsen MC (2005) Water, sediment, and nutrient discharge characteristics of rivers in Puerto Rico, and their potential influence on coral reefs. Scientific Investigations Report 2005-5206. US Department of the Interior, US Geological Survey, Denver, CO.

Watts ME, Ball IR, Stewart RS, Klein CJ, Wilson K, Steinback C, Lourival R, Kircher L, Possingham HP (2009) Marxan with zones: Software for optimal conservation based land- and sea-use zoning. Environmental Modelling and Software 24: 1513–1521.

Waycott M, Duarte CM, Carruthers TJB, Orth RJ, Dennison WC, Olyarnik S, Calladine A, Fourqurean JW, Kenneth L Heck, Jr., Hughes AR, Kendrick GA, Kenworthy WJ, Short FT, Williams SL (2009) Accelerating loss of seagrasses across the globe threatens coastal ecosystems. Proceedings of the National Academy of Sciences of the United States of America 106: 12377–12381.

Wedding LM, Lepczyk CA, Pittman SJ, Friedlander AM, Jorgensen S (2011) Quantifying seascape structure: extending terrestrial spatial pattern metrics to the marine realm. Marine Ecology Progress Series 427: 219–232.

Weijerman M, Fulton EA, Janssen ABG, Kuiper JJ, Leemans R, Robson BJ, van de Leemput IA, Mooij WM (2015) How models can support ecosystem-based management of coral reefs. Progress in Oceanography 138: 559–570.

Weijerman M, Kaplan I, Fulton E, Gordon B, Grafeld S, Brainard R (2014) Design and Parameterization of a Coral Reef Ecosystem Model for Guam. NOAA Technical Memorandum. NOAA, Honolulu, HI.

Wenger AS, McCormick MI, Endo GGK, McLeod IM, Kroon FJ, Jones GP (2014) Suspended sediment prolongs larval development in a coral reef fish. The Journal of Experimental Biology 217: 1122.

Whitall D, Menza C, Hill R (2014) A baseline assessment of Coral and Fish Bays (St John, USVI) in support of ARRA watershed restoration activities. NOAA Technical Memorandum 178. National Ocean Service, National Centers for Coastal Ocean Science, Silver Spring, MD.

White C, Costello C, Kendall BE, Brown CJ (2012a) The value of coordinated management of interacting ecosystem services. Ecology Letters 15: 509–519.

White C, Halpern BS, Kappel CV (2012b) Ecosystem service tradeoff analysis reveals the value of marine spatial planning for multiple ocean uses. Proceedings of the National Academy of Sciences of the United States 109: 4696.

Wiens JA (2002) Riverine landscapes: taking landscape ecology into the water. Freshwater Biology 47: 501–515.

Wilcove DS, Rothstein D, Dubow J, Phillips A, Losos E (1998) Quantifying threats to imperiled species in the United States. BioScience 48: 607–615.

Wilkinson SN, Hancock GJ, Bartley R, Hawdon AA, Keen RJ (2013) Using sediment tracing to assess processes and spatial patterns of erosion in grazed rangelands, Burdekin River basin, Australia. Agriculture, Ecosystems and Environment 180: 90–102.

Windom HL, Moore WS, Niencheski LFH, Jahnke RA (2006) Submarine groundwater discharge: A large, previously unrecognized source of dissolved iron to the South Atlantic Ocean. Marine Chemistry 102: 252–266.

Wolanski E, Richmond R, McCook L, Sweatman H, others (2003) Mud, marine snow and coral reefs: The survival of coral reefs requires integrated watershed-based management activities and marine conservation. American Scientist 91: 44–51.

Wood SA, Guerry AD, Silver JM, Lacayo M (2013) Using social media to quantify nature-based tourism and recreation. Scientific reports 3: 2976.

Wooldridge SA (2009) Water quality and coral bleaching thresholds: Formalising the linkage for the inshore reefs of the Great Barrier Reef, Australia. Marine Pollution Bulletin 58: 745–751.

World Resources Institute, NOAA (2005) Land-based sources of threat to coral reefs in the US Virgin Islands. Washington, DC.

Wu XB, Thurow TL, Whisenant SG (2000) Fragmentation and changes in hydrologic function of tiger bush landscapes, south-west Niger. Journal of Ecology 88: 790–800.

Yee SH, Carriger JF, Bradley P, Fisher WS, Dyson B (2015) Developing scientific information to support decisions for sustainable coral reef ecosystem services. Ecological Economics 115: 39.

Yñiguez AT, McManus JW, DeAngelis DL (2008) Allowing macroalgae growth forms to emerge: Use of an agent-based model to understand the growth and spread of macroalgae in Florida coral reefs, with emphasis on Halimeda tuna. Ecological Modelling 216: 60–74.

Yokota F, Thompson KM (2004) Value of information analysis in environmental health risk management decisions: past, present, and future. Risk Analysis 24: 635–650.

Zawada DG, Brock JC (2009) A multiscale analysis of coral reef topographic complexity using Lidar-derived bathymetry. Journal of Coastal Research 10053: 6–15.

Ziegler AD, Giambelluca TW, Plondke D, Leisz S, Tran LT, Fox J, Nullet MA, Vogler JB, Troung DM, Vien TD (2007) Hydrological consequences of landscape fragmentation in mountainous northern Vietnam: buffering of Hortonian overland flow. Journal of Hydrology 337: 52–67.

PART IV

People and Seascapes

12

Advancing a Holistic Systems Approach in Applied Seascape Ecology

Simon J. Pittman, Chris A. Lepczyk, Lisa M. Wedding and Camille Parrain

12.1 Introduction

Coastal ecosystems provide vital benefits and services to communities worldwide (Beaumont *et al.* 2007; Barbier *et al.* 2011) and are often exploited with insufficient knowledge of their ecological interdependencies, including linkages with people in the system. We are increasingly aware of the diverse range of services associated with marine ecosystems, as well as the vulnerability of these services to exploitation, pollution, global climate change and other environmental changes (Millennium Ecosystem Assessment 2005; Rocha *et al.* 2015). As responsible marine stewards, a major societal priority is to find ways to harness marine resources while safeguarding, or even enhancing, ecosystem biodiversity, functional integrity and resilience to disturbance to ensure long-term sustainability. Scientific support for marine stewardship is vital to informed decision making, particularly for actions such as the development of effective place-based conservation measures, understanding and mitigating threats to ecosystem health and informing adaptation strategies to global climate change (Lubchenco *et al.* 2003; Crowder & Norse 2008; Foley *et al.* 2010). This scientific challenge is a formidable one because managing for sustainable development in a spatially heterogeneous and dynamic environment, with multiple competing societal interests, requires the synthesis of vast amounts of data if we are to develop more comprehensive models about the key relationships between people and marine ecosystem structure, function and change (Folke *et al.* 2005). To better understand and manage the marine environment for ecological and social outcomes a conceptual and operational framework is urgently needed that will integrate appropriate ecological realism within a pragmatic transdisciplinary and multiscale systems approach (Holling *et al.* 1978). While this goal once appeared insurmountable due to data limitations, the data needs for representing marine patterns and processes are fast becoming less of a hindrance than the ability to represent adequately the dynamic linkages among the complex multidimensional attributes of the system. Since the late 1990s there have been rapid technological advances in environmental sensors and considerable investments in long-term monitoring, including remote sensing systems for acquisition of marine environmental data, as well as increases in the diversity and geographical scope of mapped socio-economic data (Caldow *et al.* 2015). Alongside advances in data acquisition, we have also seen considerable progress in the sophistication of analytical tools, including spatial modelling allowing us to better visualise, predict and explain pattern-process relationships across complex ecosystems. For some exceptionally data-rich areas of the earth we are moving ever closer to the goal of representing eco-logically meaningful realism in models at operationally relevant scales that adequately support the needs of marine ecosystem-based management.

In addition to technological advances, the theoretical and operational advancements in studies of SES (social-ecological systems) (Berkes & Folke 1998; Anderies *et al.* 2004; Olsson *et al.* 2004; Folke *et al.* 2005; Ostrom 2009; Leslie *et al.* 2015) have supported a more comprehensive understanding of complex marine spaces driven by a need to support management for sustainable systems. Yet, current tools used to integrate SES frameworks and characterise complex systems interactions fail to do so in a spatially explicit way, resulting in suboptimal decision making, particularly for prioritising actions, assessing risk and anticipating the consequences of change. Harnessing both the technological innovations involved in the data revolution and building from the theoretical advances in the social sciences to include a spatial component is critical to the advancement of a holistic approach to seascape ecology. In order to guide successful decision making in ocean governance, we now urgently need to develop a greater holistic understanding of the complex biophysical and social-ecological spatiotemporal dynamics of seascapes.

Although it has been long acknowledged that a diverse range of spatial patterns and ecological processes operate to control marine ecosystem dynamics (Stommel 1963; Levin 1992), marine science had, until recently, rarely integrated people and society into systems models. In the past decade, however, systems modelling has gained new traction via the rise of a coupled human-nature systems perspective and the emergence of ecosystem-based management and its extension into marine spatial planning, which have highlighted the importance of incorporating human activities, values and socio-economic variables (Martin & Hall-Arber 2008; Caldow *et al.* 2015; Kittinger *et al.* 2014; Cornu *et al.* 2014). Rarer still, are holistic systems models capable of explaining relationships between the spatial patterning in the marine environment and the consequences for social and ecological variables across multiple scales. The absence of a holistic framework and SES models in seascape ecology represent an unbridged chasm hindering the scientific potential for seascape ecology to support marine ecosystem-based management presenting a priority for ecologists to address. If a methodological framework does not exist to incorporate cross-scale interactions between human dimensions and biophysical spatial patterning, then those social-ecological structures and interactions cannot adequately be represented in the models and ultimately cannot be incorporated into management decisions. Key actors, relationships and pieces of the seascape will be missing in management actions. Not only will ignoring the characteristics of complex adaptive systems distort our picture of how these systems work, but it will reduce the effectiveness of marine policies and management actions aimed at maintaining or restoring environmental resilience (Levin *et al.* 2013). As we transition towards a seascape ecology approach that fully integrates the biophysical system, social system and marine governance, we need to embrace the complexity of human relationships within ecosystems, including their social, cultural, political, economic and environmental dimensions, in order to develop and implement viable management strategies (Fulton *et al.* 2011). Making good use of the wide range of available information, which is both spatially explicit and multiscale, requires a suitable conceptual and analytical framework capable of incorporating and synthesising complex information in an operationally useful way.

12.1.1   What can Landscape Ecology Offer?

Landscape ecology provides a way of thinking about the evolution and dynamics of spatially heterogeneous spaces that explicitly focuses on linkages between spatial structure, ecological function and change. The discipline has a strong link between theory and application supported by a wide range of tools and analytical techniques for quantifying and analysing the geometry of patterning across spatial scales and for investigating scale effects and spatial processes. Of special relevance to a holistic approach, landscape ecology has a history of integrating multiple disciplines to address complex cross-scale questions, with roots reaching back to the eighteenth century biogeographer and explorer Alexander von Humboldt (1769–1859) who attempted to unify the sciences under a holistic perception of the universe. In 1939, the German plant geographer and cartographer Carl Troll coined the term landscape ecology, 'landschaftsökologie', based upon his work with aerial photographs for mapping patchiness in the spatial distribution of plant communities to understand ecological processes (Troll 1939). Some 30 years later, Troll defined landscape ecology as 'the study of the main complex causal relationships between the life communities and their environment in a given section of a landscape' (Troll 1968, 1971). Through its origin in biogeography, landscape ecology has since evolved as an interdisciplinary field that uses a systems science perspective to integrate biophysical and social components (Naveh 2000; Wu 2006). Nevertheless, geographical differences in the rise of landscape ecology exist.

Traditionally, the dominant European approach has been based on a holistic society-centred perspective (*i.e.*, human centred perspective) focusing on solving societal issues, whereas the dominant North American approach has been focused on an ecological-centred spatial perspective, characterised by a question driven scientific framework and application towards natural resource management (Wiens 1999; Bastian 2001; Wu 2006). The ecological science view has been criticised for being too reductionist and analytical (mechanistic and hypothetico-deductive), making it inadequate for addressing pressing and complex socio-ecological environmental management problems (Gallopín *et al.* 2001; Liu *et al.* 2007). More recently, however, a shift has occurred in North American landscape ecology towards a more supportive role for conservation and management. The development of a holistic systems perspective, which integrates SES, where societal (human) and ecological (biophysical) subsystems function in mutual interaction, gathered pace in the 1990s, as it became apparent that important emergent traits of system behaviour such as vulnerability, resilience and adaptive capacity could only be understood through a holistic system approach (Gallopín 2006). Specifically, in 1998, a call for a broadening of the scope of landscape ecology to include 'biophysical and societal causes and consequences of landscape heterogeneity' and 'the mutual interactions between landscapes and humans, as affecting the cultural human perception of landscapes and their sociological, spiritual and psychotherapeutic effects' was made at the Holistic Landscape Ecology symposium held at the Fifth World Congress of the International Association of Landscape Ecology

(Palang *et al.* 2000). This holistic landscape ecology perspective, although not central to all landscape ecology thinking, was considered as a core paradigm in the rise of landscape ecology (Naveh 1982, 2000; Zonneveld 1982). Naveh (1990) believed that a holistic integrative landscape ecology perspective had great potential in bridging the gap between the biological sciences and human ecology. Coupling the reductionist question driven science with the holistic human-centred perspective has the potential to be very effective for investigating SES. Furthermore, what permeates both European and North American schools of thought and unifies landscape ecology as a discipline is the focus on spatial heterogeneity (*i.e.,* pattern) and the consequences for ecological processes, with an explicit consideration of scale (Forman 1995; Wu and Hobbs 2002; Wu 2006).

12.1.2   A Shift towards a more Holistic Systems Approach for Marine Stewardship

In order to help understand and guide decision making in marine governance there is an urgent need for a holistic systems approach in marine science, similar to that which exists in a terrestrial context. In fact, a global movement has been under way with a shift towards a more holistic SES approach sometimes referred to as marine ecosystem-based management (EBM) (Arkema *et al.* 2006; Leslie & McLeod 2007; Crowder & Norse 2008). Ecosystem-based management has been defined as 'an integrated approach to management that considers the entire ecosystem, including humans' and has a goal of maintaining ecosystems that are productive, healthy and resilient in order to meet the needs of society (Rosenberg & McLeod 2005). With EBM, more emphasis is placed on the interconnectedness and interdependent nature of ecosystem components, including land-sea interactions and the importance of linkages between structure and function (Leslie & McLeod 2007; Curtin & Prellezo 2010; Agardy *et al.* 2011). Furthermore, using EBM within a spatially explicit context allows for systems to have spatially unique attributes that can interact with one another.

To emphasise the applied utility of landscape ecology to EBM, Wu (2006) presented a view of landscape ecology that outlined six ways that landscape ecology can form a unifying framework to support the implementation of EBM: (i) scaling landscapes based on spatial scales relevant to people; (ii) providing a hierarchical and integrative ecological basis for studying systems at multiple scales; (iii) providing holistic approaches to socioecological systems; (iv) providing theory and methods for studying spatial heterogeneity and linking patterns to processes; (v) providing methods and metrics for quantifying sustainability and (vi) providing theory and methods for scaling and uncertainty analysis. While ecosystem structure and functions determine the natural capital stock and affect the flow and the quality of ecosystem goods and services to human societies, it remains challenging to operationally and quantitatively model these linkages. These challenges are particularly important when considering sustainability issues. For instance, understanding how spatial heterogeneity may promote ecosystem resilience and affect the flow of ecosystem services is of paramount importance for both understanding and managing terrestrial landscapes (Turner 2013). As a means to begin integrating a landscape framework into a management paradigm Wu (2013) adapted the Millennium Ecosystem Assessment framework and placed landscape composition and configuration at the core because landscape pattern creates, mediates and impedes ecosystem service provisioning. As a result, landscape-specific ecosystem services will occur whereby different composition and spatial configuration results in different kinds of services, each of which may have different management needs. Similarly, for seascapes we can expect a dynamic relationship between the spatial patterning in the marine environment and the patterning of ecosystem services and human wellbeing (Figure 12.1).

Seascape ecology, the marine counterpart of landscape ecology, has made significant progress recently in the quantitative multiscale analyses of spatial patterning in marine animal ecology through a fusion of traditional marine ecology and geosciences within a predominantly reductionist scientific perspective. Similar to landscape ecology, seascape ecology can be defined as the study of spatial patterning and its causes and ecological consequences (Pittman *et al.* 2011). However, unlike landscape ecology, the holistic perspective that has evolved in terrestrial systems has not yet been coupled with the nascent and largely biophysically based seascape ecology perspective. This lack of integration is reflected in the fact that unlike land management, few concepts and techniques from landscape ecology have been applied to the design and management of spaces at sea. Such a knowledge gap constrains our understanding of how human-environment interactions drive ecological outcomes in coastal and ocean ecosystems and limits the approaches practitioners can take to achieve better social and ecological outcomes (Arkema *et al.* 2015). Human relationships with ocean environments are diverse and include social, cultural, political and economic dimensions (McCay 1978; Cinner & David 2011; Kittinger *et al.* 2012). Furthermore, the application of landscape ecology to seascapes has been relatively narrow in focus and reductionist in methodology, with most applications examining the influence of shallow water patch mosaics on mobile marine fauna, such as fish and crustaceans (see Chapter 5 in this book). Few examples exist that have integrated the spatial patterning of human activities, values, perceptions, or socio-economic variables into seascape ecology studies. However, it is critical for effective marine planning that managers understand the dynamic spatial patterning of human uses of the sea and assess the impacts to ecosystem function, such as how ecosystem services can be influenced by changes in seascape patterning (see Chapter 13 in this book). As management transitions towards ecosystem-based approaches that fully integrate the biophysical and social systems and their respective institutions, we need to embrace the complexity of human relationships with ecosystems, including their social, cultural, political and economic dimensions, in order to develop and implement viable management strategies (Fulton *et al.* 2011). To achieve such an understanding, an expanded conceptual framework is needed for seascape ecology whereby we broaden the research focus towards a systems-based approach that includes methodological pluralism, working across disciplines and considers multiple scales. To support a movement towards such an understanding we propose broadening the existing seascape ecology framework towards a more holistic systems approach. Specifically, we advocate for a seascape ecology framework that is more inclusive of people and our complex and dynamic interactions within the seascape. In particular, we consider that people directly and indirectly create, modify and respond to spatial heterogeneity within the social-ecological system.

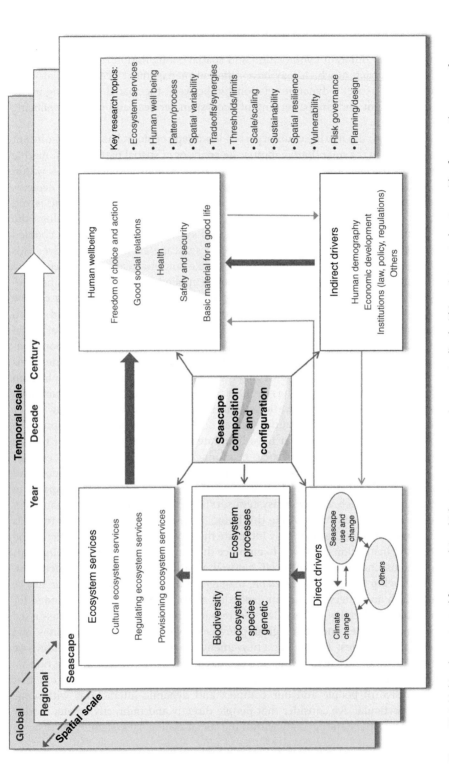

Figure 12.1 Multiscale conceptual framework in which seascape structure is conceptualised within an integrated system with a focus on the nexus of ecosystem services and human wellbeing. Key components, interactions, drivers of change and relevant research topics are listed. *Source:* Adapted from Wu 2013 and Millennium Ecosystem Assessment 2005.

12.2   People as Part of the Seascape

People and their activities, values, perceptions and interactions with the seascape are integral to a holistic systems approach to seascape ecology. As such, we suggest that human characteristics (*e.g.,* values, perception, culture) and activities (*e.g.,* resource use, behaviour, population distribution, physiology, psychology, economics) be treated in the ocean the same way they are on land by systems ecologists. The rationale here is simple, humans evolved through natural selection as an integral part of nature and therefore whatever people do is natural and part of a natural system. While we acknowledge that some human activities are socially undesirable, we do not consider them separate from nature. That said, it is necessary to recognise that humans are a dominant species capable of changing system stability by altering environmental constraints, process rates and biotic structures (O'Neill 2001).

Our integrated ecosystem perspective, which deviates from a human-centred perspective, also has implications for semantics. As such, we avoid the dichotomy in language that was traditionally used in ecology, but is becoming less of an issue, whereby humans and nature are communicated as separate entities in the same sentence. For example, we avoid referring to social and ecological systems as different entities and avoid a binary classification of threats and stressors as 'human and natural processes' or 'anthropogenic and natural disturbances' or 'human and natural drivers of change'. Similarly, we avoid distinguishing between 'animals and humans' since we too are animals. The focus on semantics is intentional and not intended to be pedantic as we wish to encourage integration of multiple disciplines within a logical science-based holistic ecosystems approach. As such, we encourage exploration, integration and hybridisation of concepts and techniques from multiple disciplines to transcend some of the cultural limitations in developing a more holistic understanding of seascape ecology. Essentially, we propose that bringing together multiple perspectives can be more effective at addressing complex problems in a spatiotemporal context than any single perspective. There is a clear need for innovation in our seascape ecology approach, which utilises both objective (*e.g.,* quantitative mapping of human uses) and subjective approaches (*e.g.,* employing qualitative data) to spatially characterise activities, values and perceptions of the ocean. Integrating these socio-ecological attributes are important for providing a suite of cultural ecosystem services that are often overlooked in current marine mapping efforts, but which are most valued by communities and critical to address and integrate into successful marine governance strategies. Without including such human dimensions, seascape management can fail, just as many terrestrial management actions have in which the human dimension was neglected. Like many species, we are capable of modifying ecosystem structure and function and in places have disrupted and destabilised the system's capacity for productivity and maintenance of biodiversity and have been responsible for state changes to systems. For example, over half of the world's oceans are strongly impacted by multiple interacting stressors driven by human activity (Halpern *et al.* 2008). Consequently, many ecosystem services provided by coastal and marine ecosystems are in decline (Millennium Ecosystem Assessment 2005).

Within landscape ecology humans are integrated in different ways, ranging from humans as an ecological component creating and responding to spatial heterogeneity to considering the landscape as a total human ecosystem (Naveh 2000; Wu 2006). Such a holistic and transdisciplinary paradigm shift is beginning to change the science and practice of adaptive resources management in the ocean through approaches such as EBM (Leslie & McLeod 2007; Berkes 2012). In Europe, the seascape is beginning to be considered with such a human centered perspective whereby the seascape is recognised as a key element of individual and social wellbeing (Musard *et al*. 2014). For example, The UK government's advisor for the natural environment, Natural England, communicates the term seascape with a distinctive human perspective as follows: 'seascape, like landscape, is about the relationship between people and place' and is defined specifically as 'an area of sea, coastline and land, as perceived by people, whose character results from the actions and interactions of land with sea, by natural and / or human factors' (Natural England 2012). This seascape description was adapted from the definition of landscape put forward by the European Landscape Convention 2000. Natural England's society-centred holistic view of seascape highlights the diversity of variables that can be attributed to a seascape when embracing an inclusive approach (Figure 12.2).

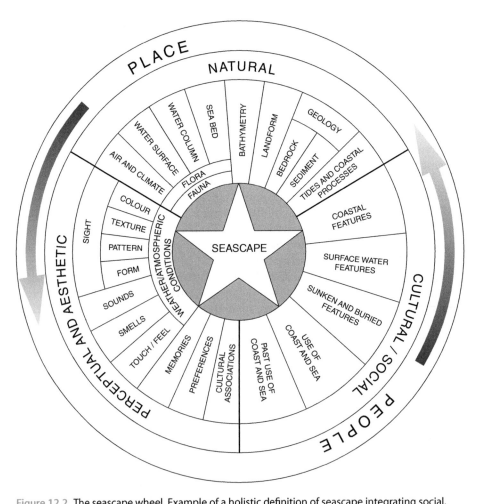

Figure 12.2 The seascape wheel. Example of a holistic definition of seascape integrating social, cultural, psychological and environmental attributes while retaining a distinct societal-centred approach. *Source:* Natural England (2012).

The framework has been applied to guide spatial characterisations of coastal zones in England to inform marine spatial planning (Sarlöv Herlin 2016) as directed under the Marine and Coastal Access Act 2009, but was not intended to provide an operational systems model for ecological study of seascapes. In another example, Conservation International (CI) defines seascapes scaled to management units as: 'a network of marine protected areas, typically large, multiple-use marine areas, where governments, private organisations and other key stakeholders work together to conserve the diversity and abundance of marine life and promote human well-being.' The seascape approach developed by CI aims to improve ecological and socioeconomic outcomes through community-based management focused on restoration of priority species and habitats, sustainable use practices and benefits to human wellbeing through an EBM approach. Notably, neither Natural England nor CI specifically focus on incorporating landscape ecology concepts or techniques into their frameworks. Hence, the linkages between spatial patterning of the seascape and ecological processes are not evident.

12.3 How Holistic Systems Science can Help Seascape Ecology

Coastal and marine ecosystems, like all ecological systems, are characterised by spatially complex and dynamic biophysical patterns. Complexity often results from nonlinear interactions among system components, which frequently lead to emergent properties and unexpected dynamics sometimes leading to surprising and unpredictable outcomes (Levin 1992; Wu & David 2002). Coupled systems exhibit a range of complex behaviours, including nonlinear dynamics, feedback loops and thresholds, which are not evident in single-discipline studies (Liu *et al.* 2007). Sometimes region-specific and nonintuitive outcomes can result in failed management goals and therefore interactions between ecosystem components and emergent properties cannot be ignored (Levin 2006). Although the dominant reductionist scientific approach has greatly informed society, it is clear that oversimplifying a complex system to a small number of parts for the purpose of meeting the criteria for experimental science (*i.e.*, falsification, parsimony) can increase uncertainty in system behaviour. In contrast, holism in science is an approach that emphasises the study of complex systems. Three key characteristics of a holistic framework are that they: (i) consider and use a multidisciplinary, interdisciplinary, or transdisciplinary approach; (ii) a focus on behaviour of complex systems; and (iii) recognise feedback within systems as a crucial element for understanding system behaviour.

The systems approach has developed to capture and model complex systems by providing a holistic conceptual and operational framework to examine connectedness, relationships and context whereby a complex system is both composed of parts and exhibits emergent properties that are more than the sum of the parts. Systems thinking focuses on the basic principles of organisation and sets out to identify and understanding the key causal linkages between different variables at different scales and how these components interact to determine the dynamics of the system, the generation of emergent properties, adaptive behaviour (*e.g.*, resilience to disturbance) and transformation (*e.g.*, phase shifts, regime shifts and tipping points) (Levin & Lubchenco 2008; Liu *et al.* 2015). The systems approach has been highly effective in modelling resilience to disturbance in complex adaptive ecosystems (Holling 1986; Walters 1986; Folke 2006; Liu *et al.* 2015).

12.3.1 Properties of an Ecological Systems Approach

Ecosystems are open systems that have emergent holistic system properties and can be organised with hierarchical spatial structure connected by cross-scale interactions (cycling, feedback loops, energy flows) (Patten 1975; Odum 1983). A feedback process or loop is a two-way interaction that can be either reinforcing (positive) or dampening (negative) on the trajectory of a system (Chapin *et al.* 2009). Seascapes are complex systems and, as such, will exhibit a number of characteristics typical of all ecosystems including nonlinearity, emergent properties, self-organisation, multiscale patterns and processes with cross-scale coupling and irreducible uncertainty due to complex unpredictable behaviour. These cross-scale interactions can result in nonlinear dynamics and produce thresholds with pronounced implications for systems behaviour (Peters *et al.* 2007; Soranno *et al.* 2014). A major challenge in ecology is to understand the processes behind these interactions, especially for forecasting future dynamics.

In the early stages of any ecological system study it is valuable to draw a diagram of the system with all the components in order to represent the portion of the world of interest to the researchers (Kitching 1983). The diagram will define the system's boundaries and connections inside and outside the system, where relevant linkages are made to external components and neighbouring systems. Once completed the system diagram can then be used to define the appropriate spatial and temporal scales (domains / ecological neighbourhoods) of each component pattern and process and of the entire system.

Even a systems approach will inevitably be a form of simplification of true system complexity, but through attention to comprehensive inclusivity it is more likely to capture key components and interactions that can help reduce uncertainty and increase predictability. Often a multimodel solution is required to represent an integrated ecosystem or to couple subsystems. For example, to understand and manage change in coral reef ecosystems in Mexico, Melbourne-Thomas *et al.* (2011) integrated two separate decision support models incorporating interacting biophysical and socioeconomic processes. The biophysical model is multiscale, using dynamic equations to capture local-scale ecological processes on individual reefs, with reefs connected at regional scales by the ocean transport of larval propagules. The agent-based socioeconomic model simulates changes in tourism, fisheries and urbanisation (Figure 12.3). Models such as these are usually developed by a transdisciplinary team of experts who identify the drivers, key interactions, operational scales and the data required. The models enable us to identify and quantify the interactions between components and can be used to formulate mathematical simulations to predict potential impacts and to examine and evaluate tradeoffs and monitor change. The system can also be simulated within a spatial framework as a map-based decision support tool to model management scenarios enabling users with different interests to vary scenarios in order to examine tradeoffs and evaluate impacts as a consequence of environmental change (*e.g.,* SimReef and ReefGame, Cleland *et al.* 2010). However, not all interactions were modelled and the authors acknowledge the difficulty in modelling the feedback between reef condition

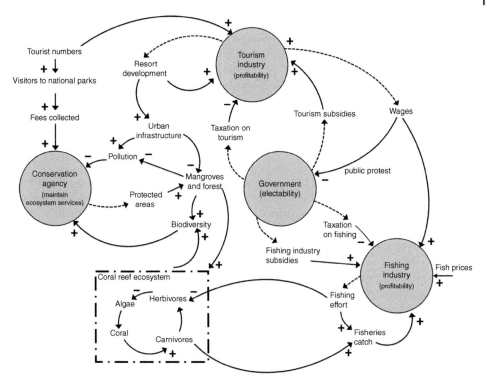

Figure 12.3 An example of a causal network of a coral reef-human ecosystem in Mexico, which incorporates connections between ecosystem structure, fishing, conservation, economics and governance. Grey circles represent major socioeconomic drivers. Broken lines represent decisions made by a group of stakeholders. Positive (+) and negative (−) feedback relationships are marked alongside arrows. *Source:* Adapted from Cleland *et al.* (2010).

and tourism and the lack of information available on reproductive dynamics of species and of socioeconomic dynamics, which can influence predictions of reef futures (Melbourne-Thomas *et al.* 2011). In general, one of the greatest challenges in modelling tightly coupled integrated ecosystems is still to find ways of incorporating sufficient complexity to reduce uncertainty to operationally useful levels (Schlueter *et al.* 2012).

12.3.2   The Rise of Whole-of-System Modelling

The shift towards EBM has resulted in the development of a wide range of new whole-of-system analytical tools to model and simulate complex systems using spatial frameworks (Kaplan & Marshall 2016; Planque 2016). So called 'end-to-end' models that combine reductionist and holistic approaches have been used to examine the influence of large numbers of processes (top-down and bottom-up controls operating simultaneously) on ecosystem dynamics, to examine various marine scenarios in support of tradeoff analyses and to predict marine futures under different environmental drivers such as climate change (Rose *et al.* 2010). It is becoming apparent that multiple models of different size, complexity and scope when combined (*i.e.,* ensemble approaches) provide the required flexibility to address the needs of environmental management (Fulton *et al.* 2015). For example, modelling ecosystem dynamics for the North Atlantic required integration of multiple models including ocean dynamics, biogeochemical cycling, biology of phytoplankton and zooplankton, larval distributions, animal behaviour of exploited species, food webs, habitat, human effects and bioeconomics to guide governance and policy (Holt *et al.* 2014). This example represents the shift towards modelling SES in marine systems and can be used as a basis to expand future modelling efforts to include both objective and subjective (*e.g.*, human values, perception) data sets that will be critical to the advancement of a holistic approach to seascape.

In Australia, Fulton *et al.* (2011) describes a flexible modular whole-of-system agent-based model (InVitro), which consists of five main submodels all of which incorporate uncertainty: (i) biophysics, which defines the natural environment, (ii) socio-economics, which defines human behaviour, (iii) industry, which defines large-scale economic drivers and institutions, (iv) management, involving decision making and (v) monitoring and assessment. This model uses a grid for bathymetry, sediments, nutrients, some habitat types and benthic biomass pools; patches for other kinds of coastal habitats (*e.g.*, mangroves); polygonal plumes for contaminants; regional subpopulations for many mid-trophic groups; school-sized aggregations for commercial invertebrates; individual or small groups of large-bodied or high trophic level vertebrates (*e.g.*, sharks or whale sharks); and nodes for roads, ports and accommodation networks (for tourist use) (reproduced from Fulton 2010). The inclusion of human behaviour, management and economic drivers and institution represent a step towards integrating a distinctive human perspective into seascape modelling that represents a more holistic relationship between people and place at sea.

Spatially explicit agent-based models provide another useful tool for linking the spatial dynamics of ecosystem service flows to spatial patterning on the landscape (Johnson *et al.* 2012; Bagstad *et al.* 2013). Service Path Attribution Networks (SPANs), a spatial tool for mapping and quantifying ecosystem service flows was applied to the Puget Sound region of Oregon where more than 80% of tidal wetlands have been lost and watershed have been dramatically modified by removal of old growth forests during the past 50 years with consequences for ecosystem service flows. The multimodel methodology uses artificial intelligence assisted modelling to select the most suitable models (including mixing data-driven and hypothesis-driven models) based on context specific data availability and an understanding of ecosystem services. The resultant maps identify the spatial distribution of services including major sources and sinks, as well as regions critical to maintaining the supply and flow of benefits for specific beneficiary groups. The impacts on human wellbeing for specific beneficiary groups from a proposed landscape alteration can be evaluated (Bagstad *et al.* 2014). Although the technique has only been applied to terrestrial landscapes, if the relevant spatial data were available, such an approach could be implemented to investigate the seascape drivers and pathways of ecosystem service flows across the seascape and to model ecosystem service flows across the land-sea interface. The ecosystem service lens in this study allows for the modelling approach to address the complexity of human relationships within ecosystems, including cultural and socio-economic factors that result in increased human health, livelihoods and wellbeing when sustainably managed.

12.4   Connecting Seascape Patterns to Human Health, Livelihoods and Wellbeing

Just as patterns on the landscape effect processes, the spatial patterning of the seascape influences a wide range of processes. However, within seascapes the linkages are rarely examined along the causal loop from the driver of change to seascape structure (*i.e.*, pattern) through modifications to ecological processes, including socio-economic and health implications for people. For example, what are the consequences of mangrove fragmentation and loss on coastal systems, including the health and wellbeing of coastal communities?

Just as in terrestrial systems, the spatial patterning we perceive as seascape structure and the dynamic processes of interest determine how investigators frame a seascape study, the research questions asked and the spatiotemporal scale(s) at which measurements are made (Allen & Hoekstra 1991). For instance, an oceanographer will typically ask different questions focused at different scales than a physiologist. A seascape ecologist will have a predisposition to perceive spatial patterning of patches, patch-mosaics and seafloor terrain, which is thought to influence ecological processes. For example, a seascape ecologist may wish to investigate the effect of spatial configuration of different patch types across the seascape on functional connectivity (Figure 12.4), or the influence of seafloor terrain complexity on patterns of biodiversity, or the consequences of fragmentation of seagrass beds on predator-prey dynamics. Collaboration with a physiologist would likely deepen the focus on physiological consequences to an organism as a function of seascape structure, an economist would shed light on the consequence for ecosystem service value, such as fisheries or ecotourism, while a historian's perspective of historical change would broaden the temporal scope of the study.

The temporal dimension has often been neglected in marine ecology, but will be particularly insightful when understanding the trajectory of an ecosystem in terms of historical factors that shape communities. Fisheries scientists have sometimes been criticised for working with time series that are too short to determine realistic baselines, or relying on anecdotal evidence from fishers' recollections from which to measure trends, evaluate economic losses and set restoration targets (Pauly 1995). To address some questions focused on understanding linkages between seascape structure and implications for cultural and economic dimensions it may be necessary to scale the seascape to people's experience of seascapes. This scaling could be done through the ecological neighbourhood approach (Addicott *et al.* 1987) whereby a key process, such as movements of fish, defines the spatial and temporal domain. For example, for fishers, vessel monitoring system data combined with knowledge of any gear depth information can define an ecological space use pattern in three dimensions (x, y, & z [depth]) over time. The study can then set this human defined domain within an appropriate spatial hierarchy, which may include within-domain patterns and processes, as well as the broader spatial seascape context surrounding the focal domain. Unlike some landscapes, or the object view in geography, the seascape unit, particularly pelagic seascape spaces, may not be easily defined using distinct structural boundaries. As a result, a scaling protocol will need to be developed guided by ecologically meaningful rationale wherever possible, or by some operationally relevant boundaries representing a jurisdictional space for management. For both landscapes and seascapes, it is relevant to distinguish whether we view the world as continuous (a field view) or discrete (object view) as that will guide

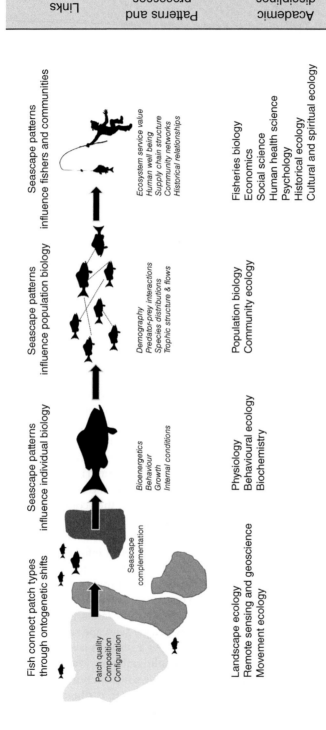

Figure 12.4 A heuristic model linking nearshore benthic seascape patterning (mangroves, seagrasses, coral reefs) to fishing communities showing how seascape structure influences functional connectivity, which in turn influences individual biology and population structure with multiple consequences for fishing communities. The figure provides some examples of relevant patterns and processes and the specialist disciplines that would each offer a different perspective and focal scale(s) through which to study the phenomena.

the specific manner by which we observe, represent and evaluate the spatial nature of the system.

With the multiple perspectives and scales that can all be relevant for an operational approach to seascape ecology, it is important to avoid scale-ambiguous terminology and instead select quantitative estimates (tens of centimetres; 100s of kilometres). As such, we avoid the terms 'seascape scale' and 'seascape level' because they are uninformative regarding spatial scale and organisational level and completely open to interpretation and can be misleading (much as 'landscape scale' has caused challenges on land) (Allen 1998). We also avoid the use of unsubstantiated relative adjectives such as 'large', 'big' and 'small' scale. In cartography, for example, a large-scale map is more detailed (finer resolution) than a small-scale map.

A major challenge in marine spatial planning and seascape ecology is incorporating key ecological dynamics in the planning process because spatial data are often described on conventional maps, which tend to represent dynamic components as static patterns and multidimensional structure as two-dimensional maps. Essentially this 2D approach represents the planning area as a static flatscape. While still useful for many applications, in reality many key patterns of interest change daily, seasonally and over longer timescales. Short-term studies (*e.g.*, 1–2 years) should be combined with longer term studies (multiyear, interdecadal) because of temporal scale dependency defined by the duration of observation (Liu *et al.* 2015). For instance, short-term studies may capture immediate changes in system behaviour, but longer term studies will be able to account for processes occurring over a different time scale, which can aid in interpreting temporal patterning such as time lags, cumulative effects, legacy effects and rare events that may not be measurable over shorter periods of observation. Incorporating patterning relevant to operational scales of human activities is important, for example, spatial management decisions such as the placement of shipping lanes, wind farms, marine protected areas typically are fixed in place for many years.

Advances in spatial analyses tools such as geographical information systems are being developed to work with multidimensional ecological data. The oceans will eventually need to be modelled in four dimensions (4D). The concept of a 4D ocean is not new, the concept of ocean space from a holistic socio-ecological system science approach, which includes human activities, was developed in the 1960s and 1970s (Stel 2016). However, the term was most often used in research on maritime law, policy and resource use, yet was not defined explicitly as a 4D construct. Progress in the visualisation of 4D voxel-based models is underway in several fields such as cartography, ocean circulation modelling and studies of animal movements (see Chapter 7 in this book).

12.5 Conclusions and Future Research Priorities

Our dependence on reliable information on the ecological and social linkages between seascape structure, function and change is increasing as society becomes more closely interconnected with the sea. To better understand and manage the coasts and oceans for ecological and social outcomes, seascape ecology studies should progress in a manner that uses elements of a holistic systems science approach that incorporates human dimensions. Several major conceptual considerations are important to keep in mind for the advancement and future development of a holistic transdisciplinary framework approach for seascape ecology. Specifically, we must move beyond the few marine studies that have adopted a seascape ecology approach to model the ecological consequences of spatial patterning in the seascape on individual marine organisms, populations and communities to investigate the cascading social effects through human health, livelihoods and wellbeing. This advancement will require a typology of distal drivers of seascape condition and methodological approaches to map culture, values and perception across the seascape.

A holistic systems science approach offers an appropriate scientific framework to understand SES structure and function and anticipate the impacts of environmental change. However, transitions require system innovations together with cultural shifts towards a greater integration of multiple perspectives and hybridisation of techniques. Technological progress and cultural shifts will need to be accompanied by a focus on increasing our understanding of the causes and consequences of spatial patterning in the sea if we are to ensure that the conceptual core of seascape ecology occupies a central position in the framework. A focus on predicting how changing spatial patterns leads to ecosystem tipping points, threshold effects and regime shifts will be high priorities in applications to resilience-based management and broader societal-centred EBM. Furthermore, by using this focus on changing spatial patterns we can address fundamental questions, such as how does seascape patterning influence spatial social-ecological resilience and which drivers contribute most to system vulnerability and robustness?

After nearly three decades of valuable research that led to real world conservation and management outcomes, a review (Turner *et al.* 2013) noted where the field stands and what the most important questions are for the future in terms of applying landscape ecology to ecosystem-based management with a focus on understanding ecosystem services. Using these questions as a starting point for needed research in seascape ecology, the following translate important terrestrial questions into marine questions together with several additional research priorities to advance a holistic systems science approach in seascape ecology.

- What types and levels of spatial heterogeneity contribute to sustained production of coastal ecosystem services and what types and levels do not? Seascapes are dynamic, all seascapes are unique and there is no optimal seascape mosaic that will increase all ecosystem services. However, some seascape configurations will be optimal for some species and the ability to classify seascape types in a way that better incorporates human dimensions could support efforts to prioritise ocean spaces in management strategies.
- Where in the coastal environment do suites of ecosystem services respond similarly, or in opposite directions, to anticipated changes in the environment? and what are the mechanisms behind such synergies and tradeoffs? Understanding the kind, amount, distribution and patterning of multiple ecosystem services across heterogeneous seascapes is critical for evaluating synergies and tradeoffs among ecosystem services in the development of management priorities.
- What are the implications for resilience and vulnerability of coastal ecosystem services of anticipated trajectories of seascape change? Anticipating seascape changes and how the benefits people derive from a region will be affected by such changes are difficult, but methods from landscape ecology can contribute to addressing this challenge. The integration of distal drivers (*e.g.*, governance, economics, values, perception, culture) through a holistic systems approach in seascape ecology will support

that needed shift beyond understanding the relationship between primary proximate drivers (*e.g.,* climate change, resource extraction, land-based pollution) and resilient marine systems.

- To what degree can seascape pattern be purposefully managed to enhance the resilience of coastal ecosystem services in the face of changing proximate and distal drivers? Understanding the mechanisms behind synergies and tradeoffs among ecosystem services can help identify ecological leverage points where small management investments can yield substantial benefits.
- How well will understanding of past seascape dynamics and coastal ecosystem services inform the future?

Beyond these five questions derived from the landscape ecology literature, there are several important research directions that we need to address to advance a holistic systems approach in seascape ecology:

- A greater understanding of the causal chain of linkages from seascape patterning to social conditions is required to advance a holistic systems approach in seascape ecology together with a broadening of the human dimensions to include consequences to both objective and subjective factors constituting human health, wellbeing and quality of life, which are rarely integrated in marine SES models. Landscape ecology provides both theoretical background and the methodological approach for addressing the issues of scaling that are fundamental to nature-society interactions (Wu 2006) and the advancement of SES models in seascape ecology.
- In addition to the potential benefits and challenges to incorporating spatial patterning into systems models, many other challenges lay ahead related to measuring uncertainty in forecasting spatial dynamics, scale effects and the development of best practices in applications to management (Fulton 2010; Rose *et al.* 2010; Fulton *et al.* 2015; Kaplan & Marshall 2016). Where spatially explicit model outputs are produced for management planning the research team will need to provide sufficient quantitative validation of performance to allow users to both measure map accuracy, identify the source of errors and to visualise the spatial patterns of error across the region of interest. Exploring where the relationships between components appear to breakdown in their explanation of observations, for example, by using replicate models and varying the linkages (Thiele & Grimm 2015).
- Do ocean tipping points exist in the spatial configuration of seascape structure beyond which we can expect to see accelerated decline in community wellbeing based on proximate or distal drivers of ecosystem state?
- Experimental approaches are still important for addressing questions on the consequences of changes in broad scale patterning. We will need to take advantage of modifications to seascape structure where intentional change, or change due to non-human processes (*e.g.,* storms), have influenced seascape patterning to investigate the consequences for ecosystem services and human wellbeing. This can also be done through comparative studies of seascape with different patterning where the methods are guided by landscape ecology concepts and techniques.

Finally, like landscape ecology in the mid-1980s, seascape ecology is in its infancy as a discipline. We have already learned a great deal about marine systems from some initial integrations from landscape and systems ecology, but much remains to be done. However, given maturity of other ecological fields, seascape ecology may be able to advance a much more rapid rate than its terrestrial counterpart by using existing knowledge and experience. As such, seascape ecology holds great utility and promise for advancing both science and management of the world's oceans in the twenty-first century.

References

Addicott JF, Aho JM, Antolin MF, Padilla DK, Richardson JS, Soluk DA (1987) Ecological neighbourhoods: scaling environmental patterns. Oikos 1: 340–346.

Agardy T, Di Sciara GN, Christie P (2011) Mind the gap: addressing the shortcomings of marine protected areas through large scale marine spatial planning. Marine Policy 35(2): 226–232.

Allen TF (1998) The landscape 'level' is dead; persuading the family to take it off the respirator. In Peterson DL, Parker VT (eds) Ecological Systems: Theory and Applications. Columbia University Press, New York, NY, pp. 35–54.

Allen TF, Hoekstra TW (1991) Role of heterogeneity in scaling of ecological systems under analysis. In Ecological Heterogeneity. Springer, New York, NY, pp. 47–68.

Anderies JM, Janssen MA, Ostrom E (2004) A framework to analyze the robustness of social-ecological systems from an institutional perspective. Ecology and Society 9(1): 18.

Arias-Gonzalez J, Johnson C, Seymour RM, Perez P, Alino P (2011) Scaling up models of the dynamics of coral reef ecosystems: an approach for science-based management of global change. In Z Dubinsky Z, Stamler, N (eds) Coral Reefs: An Ecosystem in Transition. Springer, Berlin, pp. 373–388.

Arkema KK, Abramson SC, and Dewbury BM (2006) Marine ecosystem-based management: from characterization to implementation. Frontiers in Ecology and Environment 4: 525–532.

Arkema KK, Verutes GM, Wood SA, Clarke-Samuels C, Rosado S, Canto M, Rosenthal A, Ruckelshaus M, Guannel G, Toft J, Faries J, Silver JM, Griffin R, Guerry AD (2015) Embedding ecosystem services in coastal planning leads to better outcomes for people and nature. Proceedings of the National Academy of Sciences 112(24): 7390–7395.

Bagstad KJ, Semmens DJ, Waage S, Winthrop R (2013) A comparative assessment of decision-support tools for ecosystem services quantification and valuation. Ecosystem Services 5: 27–39.

Bagstad KJ, F Villa, D Batker, J Harrison-Cox, B Voigt, GW Johnson (2014) From theoretical to actual ecosystem services: mapping beneficiaries and spatial flows in ecosystem service assessments. Ecology and Society 19(2): 64.

Barbier EB, Hacker SD, Kennedy C, Koch EW, Stier AC, Silliman BR (2011) The value of estuarine and coastal ecosystem services. Ecological Monographs 81(2): 169–193.

Bastian O (2001) Landscape Ecology: Towards a unified discipline? Landscape Ecology 16: 757–766.

Beaumont NJ, Austen MC, Atkins JP, Burdon D, Degraer S, Dentinho TP, Derous S, Holm P, Horton T, Van Ierland E, Marboe AH (2007) Identification, definition and quantification of goods and services provided by marine biodiversity: implications for the ecosystem approach. Marine Pollution Bulletin 54(3): 253–265.

Berkes F (2012) Implementing ecosystem-based management: evolution or revolution? Fish and Fisheries 13: 465–476.

Berkes F, Folke C (eds) (1998) Linking Social and Ecological Systems: Management Practices and Social Mechanisms for Building Resilience. Cambridge University Press, Cambridge.

Caldow C, Monaco ME, Pittman SJ, Kendall MS, Goedeke TL, Menza C, Kinlan BP, Costa BM (2015) Biogeographic assessments: a framework for information synthesis in marine spatial planning. Marine Policy 51: 423–432.

Chapin III FS, Folke C, Kofinas GP (2009) A framework for understanding change. In Chapin FS, III, Kofinas GP, Folke, C (eds) Principles of Ecosystem Stewardship: Resilience-based Natural Resource Management in a Changing World. Springer New York, NY, pp. 3–23.

Cinner JE, David G (2011) The human dimensions of coastal and marine ecosystems in the western Indian Ocean. Coastal Management 39(4): 351–357.

Cleland D, Dray A, Perez P and Geronimo R (2010) SimReef and ReefGame: gaming for integrated reef research and management. In D. Cleland, J. Melbourne-Thomas, M. King & G. Sheehan (eds), Building Capacity in Coral Reef Science: An Anthology of CRTR Scholars' Research 2010. St Lucia: University of Queensland, pp. 123–129.

Cornu EL, Kittinger JN, Koehn JZ, Finkbeiner EM, Crowder LB (2014) Current practice and future prospects for social data in coastal and ocean planning. Conservation Biology 28: 902–911.

Crowder L, Norse E (2008) Essential ecological insights for marine ecosystem-based management and marine spatial planning. Marine Policy 32(5): 772–778.

Curtin R, Prellezo R (2010) Understanding marine ecosystem based management: a literature review. Marine Policy 34(5): 821–830.

Foley MM, Halpern BS, Micheli F, Armsby MH, Caldwell MR, Crain CM, Prahler E, Rohr N, Sivas D, Beck MW, Carr MH (2010) Guiding ecological principles for marine spatial planning. Marine Policy 34(5): 955–966.

Folke C (2006) Resilience: the emergence of a perspective for social-ecological systems analyses. Global Environmental Change 16: 253–267.

Folke C, Hahn T, Olsson P and Norberg J (2005) Adaptive governance of social-ecological systems. Annual Review Environmental Resource 30: 441–473.

Forman RT (1995) Some general principles of landscape and regional ecology. Landscape Ecology 10(3): 133–142.

Fulton EA (2010) Approaches to end-to-end ecosystem models. Journal of Marine Systems 81(1): 171–183.

Fulton EA, Boschetti F, Sporcic M, Jones T, Little LR, Dambacher JM, Gray R, Scott R, Gorton R (2015) A multi-model approach to engaging stakeholder and modellers in complex environmental problems. Environmental Science and Policy 48: 44–56.

Fulton EA, Smith ADM, Smith DC, van Putten IE (2011) Human behaviour: the key source of uncertainty in fisheries management. Fish and Fisheries 12: 2–17.

Gallopín GC (1991) Human dimensions of global change: linking the global and the local processes. International Social Science Journal 130: 707–718.

Gallopín GC (2006) Linkages between vulnerability, resilience, and adaptive capacity. Global Environmental Change 16(3): 293–303.

Gallopín GC, Funtowicz S, O'Connor M, Ravetz J (2001) Science for the twenty-first century: from social contract to the science core. International Social Science Journal 53(2): 219–229.

Halpern BS, Walbridge S, Selkoe KA, Kappel CV, Micheli F, D'Agrosa C, Bruno JF, Casey KS, Ebert C, Fox HE, Fujita R (2008) A global map of human impact on marine ecosystems. Science 319(5865): 948–952.

Holling CS (1978) Adaptive Environmental Assessment and Management. Volume 3. International Series on Applied Systems Analysis. John Wiley & Sons, Inc., New York, NY.

Holling CS (1986) The resilience of terrestrial ecosystems: local surprise and global change. In Clark WC, Munn RE (eds), Sustainable Development of the Biosphere. Cambridge University Press, London, pp. 292–317.

Holt J, Allen JI, Anderson TR, Brewin R, Butenschön M, Harle J, Huse G, Lehodey P, Lindemann C, Memery L, Salihoglu B (2014) Challenges in integrative approaches to modelling the marine ecosystems of the North Atlantic: physics to fish and coasts to ocean. Progress in Oceanography 129: 285–313.

Johnson GW, Bagstad KJ, Snapp RR, Villa F (2012) Service path attribution networks (SPANs): a network flow approach to ecosystem service assessment. International Journal of Agricultural and Environmental Information Systems 3(2): 54–71.

Kaplan IC, Marshall KN (2016) A guinea pig's tale: learning to review end-to-end marine ecosystem models for management applications. ICES Journal of Marine Science doi: 10.1093/icesjms/fsw047.

Kitching RL (1983) Systems ecology: An introduction to ecological modelling. University of Queensland Press, Australia.

Kittinger JN, Koehn JZ, Le Cornu E, Ban NC, Gopnik M, Armsby M, Brooks C, Carr MH, Cinner JE, Cravens A, D'Iorio M, Erickson A, Finkbeiner EM, Foley MM, Fujita R, Gelcich S, Martin KS, Prahler E, Reineman DR, Shackeroff J, White C, Caldwell MR, Crowder LB (2014) A practical approach for putting people in ecosystem-based ocean planning. Frontiers in Ecology and the Environment 12(8): 448–456.

Leslie HM, Basurto X, Nenadovic M, Sievanen L, Cavanaugh KC, Cota-Nieto JJ, Erisman BE, Finkbeiner E, Hinojosa-Arango G, Moreno-Báez M, Nagavarapu S (2015) Operationalizing the social-ecological systems framework to assess sustainability. Proceedings of the National Academy of Sciences 112(19): 5979–5984.

Leslie HM, McLeod KL (2007) Confronting the challenges of implementing marine ecosystem-based management. Frontiers in Ecology and the Environment 5(10): 540–548.

Levin SA (1992) The problem of pattern and scale in ecology: The Robert H. MacArthur Award Lecture. Ecology 73(6): 1943–1967.

Levin SA (2006) Learning to live in a global commons: socioeconomic challenges for a sustainable environment. Ecological Research 21(3): 328–333.

Levin SA, Lubchenco J (2008) Resilience, robustness, and marine ecosystem-based management Bioscience 58(1): 27–32.

Levin S, Xepapadeas T, Crépin AS, Norberg J, de Zeeuw A, Folke C, Hughes T, Arrow K, Barrett S, Daily G, Ehrlich P (2013). Social-ecological systems as complex adaptive systems: modelling and policy implications. Environment and Development Economics 18(02): 111–132.

Liu J, Dietz T, Carpenter SR, Alberti M, Folke C, Moran E, Pell AN, Deadman P, Kratz T, Lubchenco J, Ostrom E (2007) Complexity of coupled human and natural systems. Science 317(5844): 1513–1516.

Liu J, Mooney H, Hull V, Davis SJ, Gaskell J, Hertel T, Lubchenco J, Seto KC, Gleick P, Kremen C, Li S (2015) Systems integration for global sustainability. Science 347(6225): 1258832.

Lubchenco J, Palumbi SR, Gaines SD, Andelman S (2003) Plugging a hole in the ocean: the emerging science of marine reserves. Ecological Applications 13(1): S3–7.

Martin KS, Hall-Arber M (2008) The missing layer: geo-technologies, communities, and implications for marine spatial planning. Marine Policy 32(5): 779–786.

McCay BJ (1978) Systems ecology, people ecology, and the anthropology of fishing communities. Human Ecology 6(4): 397–422.

Melbourne-Thomas J, Johnson CR, Perez P, Eustache J, Fulton EA, Cleland D (2011) Coupling biophysical and socioeconomic models for coral reef systems in Quintana Roo, Mexican Caribbean. Ecology and Society 16(3): 23.

Millennium Ecosystem Assessment (2005) Ecosystems and human well-being: biodiversity Synthesis. World Resources Institute, Washington, DC.

Musard O, Le Dû-Blayo L, Parrain C, Clément C (2014) Landscape emerging: a developing object of study. In Musard O, Le Dû-Blayo L, Francour P, Beurier J-P, Feunteun E, Talassinos L (eds.) Underwater Seascapes. From Geographical to Ecological Perspectives. Springer International Publishing, Cham, pp. 27–39.

Natural England (2012) An approach to seascape character assessment. Natural England Commissioned Report NECR105. Natural England, Bristol.

Naveh Z (1982) Landscape ecology as an emerging branch of human ecosystem science. Advances in Ecology Research 12: 189–237.

Naveh Z (1990) Landscape ecology as a bridge between bioecology and human ecology. In Svobodova H (ed.) Cultural Aspects of Landscape. PUDOC, Wageningen.

Naveh Z (2000) What is holistic landscape ecology? A conceptual introduction. Landscape and Urban Planning 50(1–3): 7–26.

Odum HT (1983) Systems Ecology: An Introduction. John Wiley & Sons, New York, NY.

Olsson P, Folke C, Berkes F (2004) Adaptive co-management for building resilience in social-ecological systems. Environmental Management 34(1): 75–90.

O'Neill RV (2001) Is it time to bury the ecosystem concept? (with full military honors, of course!). Ecology 82(12): 3275–3284.

Ostrom E (2009) A general framework for analyzing sustainability of social-ecological systems. Science. 325(5939): 419–422.

Palang H, Mander Ü, Naveh Z (2000) Holistic landscape ecology in action. Landscape and Urban Planning 50(1): 1–6.

Patten BC (1975) Systems Analysis and Simulation in Ecology. Volume III. Elsevier, New York, NY.

Pauly D (1995) Anecdotes and the shifting baseline syndrome of fisheries. Trends in Ecology and Evolution 10(10): 430.

Peters DPC, Bestelmeyer BT and Turner MG (2007) Cross-scale interactions and changing pattern–process relationships: consequences for system dynamics. Ecosystems 10: 790–796.

Pittman SJ, Kneib RT, Simenstad CA (2011) Practicing coastal seascape ecology. Marine Ecology Progress Series 427: 187–190.

Planque B (2016) Projecting the future state of marine ecosystems, 'la grande illusion'? ICES Journal of Marine Science 73(2): 204–208.

Rocha J, Yletyinen J, Biggs R, Blenckner T, Peterson G (2015) Marine regime shifts: drivers and impacts on ecosystems services. Philosophical Transactions of the Royal Society B 370(1659): 20130273.

Rose KA, Allen JI, Artioli Y, Barange M, Blackford J, Carlotti F, Cropp R, Daewel U, Edwards K, Flynn K, Hill SL (2010) End-to-end models for the analysis of marine ecosystems: challenges, issues, and next steps. Marine and Coastal Fisheries 2(1): 115–130.

Rosenberg AA, McLeod KL (2005) Implementing ecosystem-based approaches to management for the conservation of ecosystem services. Marine Ecology Progress Series 300: 270–274.

Sarlöv Herlin I (2016) Exploring the national contexts and cultural ideas that preceded the Landscape Character Assessment method in England. Landscape Research 41(2): 175–85.

Schlueter M, McAllister RR, Arlinghaus R, Bunnefeld N, Eisenack K, Hoelker F, Milner-Gulland EJ, Müller B, Nicholson E, Quaas M, Stöven M (2012) New horizons for managing the environment: a review of coupled social-ecological systems modelling. Natural Resource Modelling 25(1): 219–272.

Soranno PA, Cheruvelil KS, Bissell EG, Bremigan MT, Downing JA, Fergus CE, Filstrup CT, Henry EN, Lottig NH, Stanley EH, Stow CA, Tan PN, Wagner T, Webster KE (2014) Cross-scale interactions: quantifying multi-scaled cause-effect relationships in macrosystems. Frontiers in Ecology and Environment 12: 65–73.

Stel JH (2016) Ocean space and sustainability. In Heinrichs H, Martens P, Michelsen G, Wiek A (eds) Sustainability Science. Springer, Dordrecht, pp. 193–205.

Stommel H (1963) Varieties of oceanographic experience. Science 139(3555): 572–576.

Thiele JC, Grimm V (2015) Replicating and breaking models: good for you and good for ecology. Oikos 124(6): 691–696.

Troll C (1939) Luftbildplan und ökologische Bodenforschung. In Zeitschrift der Gesellschaft für Erdkunde zu Berlin (7/8): 241–298.

Troll C (1968) Landschaftsökologie. In Tuxen B (ed.) Pflanzensoziologie und Landschaftsökologie, Berichte das 1963 internalen symposiums der internationalen vereinigung fur vegetationskunde. Junk, The Hague, pp 1–21.

Troll C (1971) Landscape ecology (geoecology) and biogeocenology – a terminology study. Geoforum 8(71): 43–46.

Turner MG, Donato DC, Romme WH (2013) Consequences of spatial heterogeneity for ecosystem services in changing forest landscapes: priorities for future research. Landscape Ecology 28(6): 1081–1097.

Walters CJ (1986) Adaptive Management of Renewable Resources. Macmillan, New York, NY.

Wiens JA (1999) Toward a unified landscape ecology. In Wiens JA, Moss MR (eds) Issues in Landscape Ecology. 5th IALE-World Congress, 29 July–3 August 1999, Snowmass, CO, pp. 148–151.

Wu J (2006) Cross-disciplinarity, landscape ecology, and sustainability science. Landscape Ecology 21: 1–4.

Wu J (2013) Landscape sustainability science: ecosystem services and human well-being in changing landscapes. Landscape Ecology 28(6): 999–1023.

Wu J, David JL (2002) A spatially explicit hierarchical approach to modelling complex ecological systems: theory and applications. Ecological Modelling 153(1): 7–26.

Wu J, Hobbs R (2002) Key issues and research priorities in landscape ecology: An idiosyncratic synthesis. Landscape Ecology 17: 355–365.

Zonneveld IS (1982) Land(scape) ecology, a science or a state of mind. In Tjallingii SP, de Veer AA (eds) Perspectives in Landscape Ecology. Wageningen, Netherlands, pp. 9–16.

13

Human Ecology at Sea: Modelling and Mapping Human-Seascape Interactions

Steven Saul and Simon J. Pittman

13.1   Introduction

All life forms, including humans, modify the environment in which they live in order to accommodate their needs and desires. A recent archaeological study suggests that there are no longer any places on earth unaffected by human activity and that this has been the case for thousands of years (Boivin *et al.* 2016). Yet, the true size and scope of our ecological footprint remains largely unknown (White 1967). Studies that have mapped the global footprint of humans on the ocean suggest that most of the global ocean space is now affected by cumulative impacts from human activities generated from both land and sea-based activities (Figure 13.1; Halpern *et al.* 2008). All too often, insufficient emphasis and funding is placed on understanding the human dimensions of marine ecosystems (Branch *et al.* 2006; Fulton *et al.* 2011). This is an ongoing challenge in marine spatial planning where 'the social landscape of the marine environment is undocumented and remains a missing layer in decision-making' (Martin & Hall-Arber 2008). Insufficient attention to the socioeconomic drivers of systems structure and function has sometimes led to either the failure of, or unintended consequences of, management measures meant to balance human benefit with resource sustainability (Hanna 2001; Acheson 2006; Clark 2006; Milner-Gulland 2012).

Where human-seascape information is available it is typically accompanied by high uncertainty due to data-collection methods and patchy coverage in time and space. When managing natural resources, the costs of uncertainty are derived from how well managers can predict human response to regulations under consideration. The challenges of working with uncertainty are exemplified by the debate over the amount of warming degrees the earth may experience in response to anthropogenic atmospheric carbon emissions. For instance, the change in warming degrees is estimated by the Intergovernmental Panel on Climate Change to be between 1.5 and 4.5 °C over the next century (IPCC 2014). As a result of uncertainties, there is great deliberation over the economic and social effects such warming may have on the costs and benefits humans receive from nature (Wagner & Weitzman 2015). Costs can manifest through overpreparing for potential environmental changes that never occur (*e.g.*, reinforcing buildings to withstand stronger storms, which never occur in an area), or by underpreparing and suffering damages that result from changing climatic conditions (*e.g.*, flooding and coastal erosion). Predictability is complicated by adaptive system feedback

Seascape Ecology, First Edition. Edited by Simon J. Pittman.
© 2018 John Wiley & Sons Ltd. Published 2018 by John Wiley & Sons Ltd.

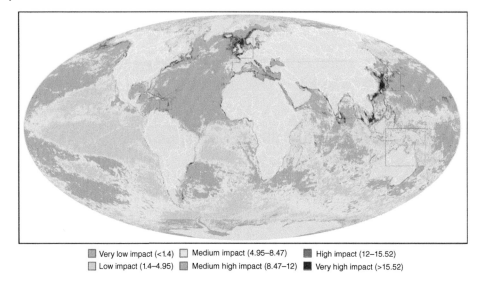

Very low impact (<1.4)   Medium impact (4.95–8.47)   High impact (12–15.52)
Low impact (1.4–4.95)   Medium high impact (8.47–12)   Very high impact (>15.52)

Figure 13.1 Map from Halpern *et al*. (2008) illustrating cumulative human impact across 20 ocean ecosystem types. *Source:* Reproduced with permission from the American Association for Advancement of Science.

interactions including human behavioural responses and how behaviour affects the environment (Bath 1998; Wagner & Weitzman 2015). As a result, management of seascapes is largely concerned with evaluating risks to ecosystem structure and function with a focus on directly managing human activities rather than the resources. In fisheries-focused EBM, human behaviour is often considered a major driver of change and a source of uncertainty facing fisheries managers (McCay 1978; Fulton *et al*. 2011; Berkes 2012).

Understanding what people do to impact seascapes and how they respond to seascape structure, function and change is integral to a holistic approach to seascape ecology (see Chapter 12 in this book). The well-known microbiologist, Robert L. Starkey, argued that whatever people do, whether considered to be good or bad for our environment, happens in the course of natural events (Starkey 1976). Starkey argued that humans evolved as part of nature such that all human activity can be considered natural, even if we sometimes don't like or agree with the actions or their ecosystem effects. This inclusive perspective, which will later be defined as a key principle in a holistic systems-based approach, recognizes humans as animal members of the system, interacting with the biotic and abiotic features of their ecosystem (see Chapter 12 in this book). In marine ecosystem-based management (EBM), people are considered a part of nature, with human activities recognized as important and influential ecological processes within a complex adaptive and coupled social-ecological system (Berkes & Folke 1998; Liu *et al*. 2007; see also Chapter 12 in this book).

In this chapter, we discuss tools and concepts from spatial ecology that can be used to investigate geographic patterns of human activity and influence across the seascape. The premise of this discussion will focus around the fact that humans are a part of the ecosystem rather than an external force operating on it. This chapter focuses on providing the reader with quantitative spatial techniques, including some familiar to landscape ecologists, which can be used to measure human use across the seascape to advance a holistic understanding of marine systems and to support effective marine stewardship. In doing so, the dynamic coupling that exists between people and the surrounding environment is highlighted and integrated approaches that involve interdisciplinary teams (ecology, economics, sociology, anthropology) are advocated. Emphasis is placed on techniques that help researchers and managers understand the spatial patterning of seascapes and the distributions of human activities at sea. As such, many of the techniques discussed in this chapter will be applicable to quantifying both human activity, as well as other ecosystem processes. This chapter will first discuss the importance of evaluating spatial patterns and considering spatial scale when conducting natural resource evaluations. Second, the chapter introduces several quantitative approaches that can be applied across the seascape to evaluate the spatial and temporal patterns of human use. Each approach is accompanied by examples to demonstrate how the technique was applied in the context of marine spatial management.

13.2 Seascape Ecology, Spatial Patterns and Scale

The concepts and methods in this chapter are discussed within the context of seascape ecology, which refers to the application of the theoretical and analytical frameworks of landscape ecology to the marine environment (Pittman *et al.* 2011). Landscape ecology, now a mature discipline, has been defined in different ways throughout its evolution (Wu 2013; Turner 2015). A common thread, as the discipline evolved, has been the integration and quantification of human use patterns. Naveh & Lieberman (1994) defined landscape ecology as a 'branch of modern ecology that deals with the interrelationship between man and his open and built-up landscapes' (Naveh & Lieberman 1994). In search of principles to integrate human cultural aspects into landscape ecology, Nassauer (1995) proposed that 'human landscape perception, cognition and values directly affect the landscape and are affected by the landscape'. On land, much work has focused on integrating landscape ecology into urban planning because of its emphasis on the interrelationship between landscape patterns and socio-ecological processes at different scales, while encouraging place-based research that integrates ecology with planning, design and other social sciences (Leitão & Ahern 2002; Wu 2008; Pickett *et al.* 2016). These holistic socio-ecological perspective in landscape ecology, with a focus on understanding how landscape patterns influence wellbeing and sustainability, also hold promise for application to the marine environment.

Applying the principles of landscape ecology to human use of the seascape involves studying the spatial patterns, interactions and bidirectional feedbacks that occur between people and the marine ecosystems they use and affect. This extension of landscape ecology also acknowledges that human use patterns are influenced by people's knowledge of the spatial and temporal distribution of marine resources and, conversely, in many locations, human use patterns directly affect the spatial and temporal distributions of marine resources. Given the complexities of these processes and their interactions, the techniques outlined in this chapter will revolve around applying a systems-based approach to seascape ecology. Application of a systems-based approach has long been advocated for and applied in the terrestrial ecology realm (Naveh 2000; Oreszczyn 2000; Atkins *et al.* 2011), but has more recently found favour with the marine community under the banner of the 'ecosystem approach' to assessment and management. In theory, the ecosystem approach is meant to 'address the various natural and anthropogenic pressures faced by the key components of marine systems simultaneously' (Link & Browman 2014).

Over the years, the majority of ecosystem-based research has disproportionally focused on studying and modelling the nonhuman components of the ecosystem, with less emphasis placed on understanding and modelling human behaviour. Many of the modelling efforts that have been developed as part of the ecosystem modelling paradigm have not quantitatively represented human behaviour and its nuances in a spatially explicit and ecologically meaningful way. For example, ecosystem-based fisheries models have gone to great lengths to develop biological realism, such as the interaction of predator-prey relationships, together with planktonic and detrital dynamics and although complex spatial dynamics in fisher behaviour has long been recognized (Schaefer 1957; McCay 1978), conventional fisheries models have typically represented fishing effects very simply, often just by applying a uniform mortality value across spatially and temporally heterogeneous populations. This is sometimes due to high uncertainty in the reliability of the fisheries data and an unfamiliarity with the importance of spatial and temporal heterogeneity in fishing effort. Though the emphasis towards replicating relevant complex biological dynamics in the ocean is critically important, such models are increasingly being sought after to provide resource management advice. As a result, equal attention toward representing the complexities of human behaviour, as that given to the other biotic and abiotic components is becoming increasingly important. A number of innovative approaches have recently been explored to improve our understanding and representation of the human component within ecological models such as in developing more realistic bioeconomic fisheries fleet models and 'whole of system' models integrating social-ecological systems that incorporate aspects of human behaviour and implications for human wellbeing (*e.g.,* Fulton 2010; Griffith *et al.* 2012; Fulton *et al.* 2015; Kroetz & Sanchiro 2015; see also Chapter 12 in this book). In spatially and temporally heterogeneous systems, which experience fishing, the spatial dispersal of the harvesting sector is just as important to model as other dynamic ecosystem processes such as biological dispersal processes.

A movement to integrate social and ecological systems is underway in ecology, but rarely do these models examine the consequences of spatially explicit seascape patterns for human behaviour at sea. An unusual early exception is that of Sanchirico & Wilen (1999), whereby a spatially explicit bioeconomic model was developed that incorporated complex spatial patch dynamics. To address questions such as: How does the spatial pattern of effort depend upon the type of biological system the industry is exploiting? the authors characterized a resource population structure using concepts from metapopulation biology with a focus on patchiness, heterogeneity and interconnections among and between patches. In this three-patch model system, a patch was defined as a location in space that contains, or has the potential to contain, an aggregation of biomass. Patches are connected by dispersal processes and are affected by the spatial distribution of harvesting effort. Dispersal or diffusion mechanisms linking space and time are a critical component of spatial-dynamic systems (Smith *et al.* 2009). This modelling approach allowed the model to examine edge effects where patches existed in a linear arrangement and the flow of biomass between patches at the edges and interior could be examined, as well as implications for economic parameters including the cost of exploitation. By quantifying patchiness and connectivity, the model revealed that the spatial pattern of effort for a mobile and economically responsive fishing fleet is driven fundamentally by patch-specific cost-price ratios. Of additional interest is that the model recognizes gradient structures existing between patches in both economic (*i.e.*, movements of fishing effort) and biological variables (*i.e.*, movements of marine biomass) created by density. Sanchirico & Wilen (2005) compared spatial (patchy) and nonspatial (spatially uniform) solutions finding significant differences suggesting that a spatially differentiated approach allowing interaction between spatial gradients in economic and biological variables will reduce errors in estimates of fishing effort, harvesting and biomass levels across space. In addition to consideration of spatial heterogeneity in human behaviour, studies are now also beginning to consider heterogeneity in a wide variety of linked variables such as income, prices and ecological impacts. Working on the red sea urchin fishery in California, Wilen *et al.* (2002) demonstrated an economics-based model of fisher behaviour (participation and spatial choice) linked to a biological model of metapopulation dynamics to show that modelling spatial behaviour strongly affects the predicted outcomes of management policies. To this end, more work is needed to improve the representation and quantification of human dynamics and the linkages to other structural patterning of the seascape in more realistic spatially complex environment, both as part of already built ecosystem models and as standalone analyses in support of management across the seascape.

13.2.1   Scale and Scaling

Scale refers to the spatial, temporal, quantitative, or analytical dimensions used to measure and study any phenomenon (Gibson *et al.* 2000). Scale effects and scaling is an equally important consideration in the study of human-focused spatial patterns as it is in any other topic in seascape ecology (see Chapter 4 of this book). Scale mismatches between dynamic ecological systems and social scales informing human behaviour and decision making in resource management are widespread and can result in suboptimal performance in management objectives (Gibson *et al.* 2000; Cumming *et al.* 2006; Wilson 2006). The spatial distribution of a natural resource typically determines how humans are distributed geographically (Luck 2006; Boivin *et al.* 2016), although the correlation between the spatial distribution of human uses and the spatial distribution of the targeted resource is contingent on how much information resource users have about the spatial distribution of the resource (Alessa *et al.* 2008). Some landscape ecologists focusing on human dimensions have recommended that environmental patterning be scaled by human processes (Nassauer 1995). This can be achieved by understanding the space-use patterns over time (daily home range, monthly, seasonal, annual *etc.*) and can be measured and analysed with telemetry and home range estimators much the same as we would do for any other animal in movement ecology (see also Chapter 12 in this book). In this way we can map the boundaries of an ecological neighbourhood for a person or group of people. Metcalfe *et al.* (2016) collaborated with fishers in the Republic of Congo to map and characterize human space-use behaviour (distance offshore and water depth) using Global Positioning System tracking devices equipped with motion detectors.

It is important to quantify and report the scale(s) selected (geographical / temporal extent and spatial / temporal resolution) to anchor the study in a specific space and time domain, although this is rarely done. This will avoid ambiguity and facilitate comparative analyses and to help investigate scale dependency in ecological relationships (Schneider 2001; see also Chapter 4 in this book). Ultimately, scale of a study should not be selected arbitrarily but instead must be guided by a process of interest, the research question(s) and information need. Additional important fundamental considerations related to scale and scaling are: (i) how does scale, extent and resolution affect the identification of patterns? (Wu 2004); (ii) how does scale affect the explanation of social phenomena? (iii) can theories and interpretations of system function derived from observations at one scale be generalized to another scale?

Matching the spatial scales at which sampling, assessment and management of a resource occurs is critical for producing appropriate regulatory actions that best fit both the managed resource and the users of that resource (Sale 1998). This is especially relevant, for example, when establishing spatial regulations such as marine protected areas and marine spatial planning, yet rarely considering in a meaningful, quantitative way (Claudet *et al.* 2010). Thus, there is a great need to carefully identify and define the appropriate spatial and temporal scales relevant to a specific process or research question as an initial step in any seascape ecology study.

13.3 Human Use Data Types and Geographical Information Systems

There are generally two types of data used when considering human behaviour and decision making: stated preference and revealed preference. Stated preference data refers to information that was collected from individuals under hypothetical scenarios. This could be data collected by implementing a questionnaire, or a choice experiment where respondents must choose what they prefer from a series of hypothetical options. Revealed preference data is information collected by observing people's actual decisions in the real world and the conditions under which they make these decisions. Experiments can be conducted to collect this kind of information, but must occur in real life settings (*i.e.*, real time and space). Here we discuss some techniques for gathering information on human use patterns and the benefits and pitfalls of some of these methodologies. Tools designed to collect information for other purposes can, at times, also be used to quantify human use patterns across the seascape. Finally, in many cases, existing data can be leveraged in creative ways to understand human use. The benefits and challenges to working with existing data (often referred to as 'data of opportunity') are discussed.

The first question one needs to consider is how does one know that they have the 'right' data? This is largely dependent on the research and management questions under consideration and the time and spatial scales that they encompass. The data used to answer a research question or develop management strategies should reflect the important components of the system (both human and natural) that are affected and addressed by the research question or management action under consideration. Attempting to represent and account for every single process within a system doesn't make sense and is typically limited by the available data.

13.3.1   Mapping Human Behaviour across the Seascape

Often, the best predictors of future human behaviour can be understood by studying the past decisions made under a similar set of circumstances (Aarts *et al.* 2006; Ouellette & Wood 1998). Thus, forecasting human behaviour, such as how people may respond to a proposed regulation, can be understood and modelled from past observations of human decision making made under a variety of conditions. Later sections of this chapter will discuss how decision-making behaviour can be modelled. Panel datasets, those with repeated observations of people's decisions made under different environmental conditions, are often used to develop models of human behaviour. Often, panel datasets (often referred to as revealed preference data) may not be directly available, but can be developed by combining and rearranging several different existing datasets. For example, as mentioned above, the fishing industry is legally required to submit their logbook data to the government, which contains daily information on when and where they fished, what and how much they caught, as well as the gear they used, how much and how long they spent fishing with that gear (revealed preference data). Although the purpose for collecting this information as legally mandated is to calculate trends in catch and population abundance, the data can also be used to address three important socioeconomic questions: when people fish, where they fish and when they return to port. In order to forecast decision making under different sets of conditions, one must compute the probability that such decisions were made in the past, across a range of different conditions. In order to accomplish this, a panel dataset must be constructed by combining the logbook data with information on weather conditions, economic conditions (such as fuel and fish prices), regulatory measures and their spatial knowledge of relative resource distribution. The information from these different data sources should be merged together based on common spatial and temporal criteria (Saul & Die 2016). Revealed preference data (that representing observed human decision making) is often not available, or does not exist. In this case, stated preference data can be collected to help quantify human behaviour. Although conducting fieldwork in coastal communities to gather this kind of information is typically informative and fun, like any other field campaign, it poses a number of challenges. First, such work needs to be funded, which can be costly, especially if your study site is in a remote location, or far from your home. Second, any work with human subjects, even if it is just asking them to complete a survey, must be approved by an institutional review board to be sure it complies with ethical standards. Third, it can be difficult to recruit participants to your study, as they may be reluctant to share their personal information, even if they are ensured anonymity. This is especially true if mistrust exists between resource users and management agencies and the information requested in your study is perceived by resource users as being used against them in some way. It can also be difficult to find mutually available times and connect with resource users, especially if their work requires them to be at sea for long periods. Fourth, data provided through questionnaires or interviews carry inherent limitations such as recall bias, when respondents, who are asked to reflect back on their past behaviour, tend to provide a more optimistic view of their past behaviours and decisions (Trumbo *et al.* 2011; Loomis 2014; see Podsakoff *et al.* 2003 for review of questionnaire limitations). Respondents also tend to recall events in the distant past, as if they had occurred more recently (*e.g.,* telescope bias, Dex 1995).

Despite these limitations, in many cases, gathering field data is sometimes the only way to understand human use patterns across the seascape. Two instruments that can be used to collect quantitative data on human behaviour include questionnaires and stated preference experiments. Questionnaires are advantageous in that they are practical and cost effective to implement, allow the researcher to collect a large volume of quantitative information in a relatively short period of time and can be used as an exploratory tool to create new theories or develop hypotheses. Stated preference experiments involve using game theory to present an individual with choices in an experimental setting (typically using a computer simulation), under different sets of simulated environmental and economic conditions and recording the decisions they make under those conditions. This allows the researcher to understand how a respondent's choice may change under different state conditions, or different attribute characteristics and is a way of developing a *de facto* panel dataset when one does not exist (Train 2009). Discrete choice experiments tend to be more expensive (compensation to incentivize participation and pay someone to develop the software) and are more time consumptive of participants (usually taking hours to complete), in comparison to questionnaires. However, the data collected during discrete choice experiments are typically considered more robust then that collected during a questionnaire because the responses are thought to be less biased by social desirability (Fifer *et al*. 2014).

13.3.1.1 Remote Sensing

In recent years, remote sensing from satellites and aircraft have made it possible to acquire large amounts of spatially explicit data in a reasonably fast and cost-effective process. This is a particularly useful tool for remote locations (Douvere 2008). Satellite data has been used creatively to map human use patterns and interactions with the environment (Geoghegan *et al*. 1998). The most common use of the technology with respect to mapping human resource use has been to observe spatial change detection over time (for review articles see Singh 1989; Coppin *et al*. 2004; Tewkesbury *et al*. 2015; Willis 2015). Spatial change detection refers to the comparison of two, or more, georectified images of the same location to determine how the distribution and use of the habitat in that location has changed overtime. While chance detection studies and techniques are useful towards understanding how human activity has affected the landscape or seascape over time, they typically do not measure or observe human activity directly and cannot unravel the mechanisms that drive human behaviour (Fox *et al*. 2003). Instead, remote sensing data can measure the spatial context of social phenomena, by cataloguing the magnitude and direction of changes caused by a human activity (Patino & Duque 2013). In order to understand the mechanisms behind these changes, some studies have linked remote sensing data with other social science research techniques such as surveys and interviews in order to elucidate the anthropogenic mechanisms driving observed changes (Dennis *et al*. 2005; Taubenböck *et al*. 2009; Herrmann *et al*. 2014). The coupling of remotely sensed information with socioeconomic data has also been used to track the spread of diseases such as malaria (Ohemeng & Mukherjee 2015) and determine the risks of natural disasters on property values, such as from a flood (Taubenböck *et al*. 2011), fire (Dennis *et al*. 2005), or erosion (Leh *et al*. 2013).

Remotely sensing human behaviour patterns from orbit in real time (not just how human use patterns have shaped the landscape) is more challenging. This is because the spatial resolution of most spaceborne sensors available for public use, the most sensitive of which can be resolved down to about a meter, too coarse to track individual people. Despite this, several marine-focused studies have used remote sensing technology creatively to map human use patterns. For example, Davies *et al.* (2016) used satellite imagery to map night-time artificial lights from coastal development, offshore infrastructure, shipping and fishing boats to determine exposure of marine protected areas to light pollution. Rowlands *et al.* (2012) used night time light patterns detected by spaceborne satellite to spatially characterize fishing activity. Other spectrally based remote sensing technologies such as hyperspectral imaging from aircraft, which collects at a finer resolution and can be further enhanced using spectral unmixing techniques, could be very effective at measuring human activity however it has not conventionally been used for this purpose. One reason for this may be because deploying sensors from aircraft are costly and gathering repeated observations over time can become very expensive. In addition, aerial photography can also be a useful source of information, especially if a high-resolution camera is used. With the increasing advancement of drone technology and its availability to civilians at competitive prices, such airborne sensors may become much easier and more cost effective to deploy. It is also noteworthy that the military regularly uses drone technology to observe human behaviour for national security purposes. Thermal infrared sensors may also be a useful tool for sensing human use of the seascape. While difficult to use over land since emissivity varies as land type changes, over the oceans, emissivity is known and nearly constant. As a result, any additional heat signatures from people or ships can be detected.

A number of other remote sensors observe and quantify human use in different ways such as micro electromechanical sensors, image sensors, radio frequencies identifiers and pressure sensors (Amato *et al.* 2013). For example, some fisheries are required to use video cameras on commercial fishing vessels in order to measure the catch and monitor for illegal fishing activity. Mandatory satellite vessel monitoring systems (VMS) are becoming more commonplace on ships and have improved the ability to monitor marine protected area compliance and determine human use patterns across the seascape (Pedersen *et al.* 2009). VMS units report a vessel's location back to a satellite at regular time intervals and these data are now becoming available to scientists and resource managers for use mapping fishing patterns across the seascape (Witt & Godley 2007). Another similar initiative is a project called Global Fishing Watch operated by SkyTruth, a not-for-profit organization that uses data transmitted to a satellite from a ship's Automatic Identification System (AIS), an electronic unit used by ships to avoid collisions at sea during reduced visibility. Similar to VMS data, AIS has been used to enforce spatial closures and for marine spatial planning in Kiribati (Figure 13.2; McCauley *et al.* 2016). Here, six months of post-closure monitoring revealed only one case of fishing activity in the Phoenix Island Protected Area (PIPA), and this vessel was fined by regulatory authorities in Kiribati. Vessel characteristic information is typically transmitted together with location information for both AIS and VMS. Organizations (such as SkyTruth and NOAA Fisheries) are using this information, together with their location, to develop algorithms that determine when a vessel is engaged in fishing activity with its gear deployed. Such spatially explicit information can be used to map marine corridors for vessel activity, identify and measure edge effects, such as where fishing intensity is concentrated along the boundary of no-take marine reserves (a spatial phenomenon known as fishing the line) and evaluate the displacement of fishing effort (Stelzenmüller *et al.* 2008).

Aerial surveys have been used to help quantify the magnitude and spatial distribution of recreational human use across the seascape. In south Florida, aerial surveys are

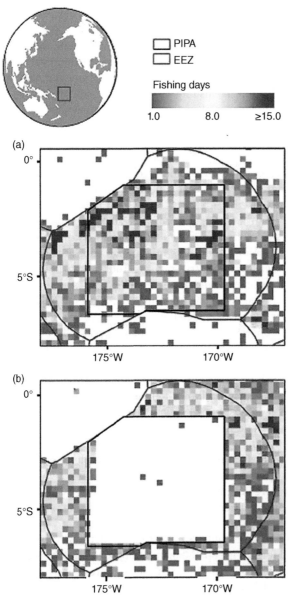

Figure 13.2 Automatic Identification System (AIS) data showing fishing intensity (purse seine and long line) patterns before (a) and after (b) establishing a marine reserve Phoenix Island Protected Area (PIPA) in Kiribati. *Source:* McCauley *et al.* 2016.

regularly used to collect data on seascape use by private and commercial boats. Use categories included fishing, diving and snorkeling, sightseeing, picnicking and sunbathing, among others (Gorzelany 2005a, 2005b, 2009; Behringer & Swett 2011). Information collected by these surveys in Florida helps to guide the development of management measures and focus their emphasis on the areas that receive the most use (for example, installation of mooring buoys in areas with most use). In addition, when overlaid with habitat maps in a geographic information system (to be discussed below), such information on human use patterns can provide valuable insight into the types of habitats that are visited and their relative use levels. For example, aerial surveys determined that vessels were often found to cluster around benthic features that contained reef hard bottom (Behringer & Swett 2011). Similarly, another study that mapped the spatial location of fishing boats correlated their spatial occurrence with a higher incidence of marine debris (Bauer *et al.* 2008). Debris on the seafloor was associated with the presence of ledges because the more complex seafloor relief attracted recreational fishers and scuba divers, therefore linking benthic seascape structure to human space-use patterns, behaviour at sea and consequently to measurable and undesirable impacts.

13.3.1.2 Participatory Mapping and Spatial Analysis

Spatially explicit data is increasingly required to support spatial decision making in marine planning (Purkis & Klemas 2011). Geographical Information Systems (GIS) are useful ways to compile and visualize the data available, understand its spatial extent and where it overlaps with other datasets and determine the appropriate analytical approach for addressing research. GIS have statistical tools built in to conduct modelling and analysis and characterize spatial uncertainty. Some spatial decision support tools have been developed exclusively for marine spatial planning. For example, SeaSketch, an interactive web-based application, allows users to participate in the ocean planning process (McClintock 2013; McClintock *et al.* 2016). Both spatial and nonspatial data are uploaded to the software from any participating user for inclusion in the project. Using simple sketching tools, stakeholders can develop spatial plans and share their plans with the group. The strength of this tool is that it allows a diverse group of stakeholders to contribute their local knowledge and management advice for consideration. Contingent on the kinds of data included in the project, reports can be easily generated for each proposed management scenario that provide information on the types of habitats protected, social and economic costs and benefits and other metrics. One of its more notable applications has been the Safe Passage Project, in southern California, which has worked to develop marine spatial planning solutions to balance commercial shipping needs with conservation priorities, such as endangered whales vulnerable to ship strikes (McClintock *et al.* 2016).

The SeaSketch framework is an example of participatory GIS (pGIS) which seeks to involve stakeholders, community members, managers and scientists alike in the production and use of geographical information (Dunn 2007; Radil & Junfeng 2016). Examples of successful implementation of pGIS across the seascape include: determining the placement of wind farms (Mekonnen & Gorsevski 2015); resolving gear conflicts between small-scale and large-scale fishing boats in Vietnam (Thinh *et al.* 2016); designing a marine protected area (Aswani & Lauer 2006); and resolving conflicts among different uses of a coral reef in the coastal zone (Levine & Feinholz 2015). In the US Virgin Islands (eastern Caribbean), Loerzel *et al.* (2017) developed a web-based mapping tool to harness local expert knowledge from the occupational SCUBA diving community on coral reef characteristics, health, human uses, threat and perceived resilience to change. The pGIS process enabled the project to comprehensively identify coral reefs used for commercial, recreational and scientific purposes providing valuable information, which contributed to a spatial prioritization for conservation actions.

13.3.1.3 Social Sensing

A variant on pGIS, called social sensing, acknowledges that the spectral reflectance data provided by most remote sensing platforms cannot be used to infer socioeconomic information about the features that they sense. For example, although spectral sensors can detect landscape and infrastructure features, such as roads, they lack the capacity to extract the socioeconomic attributes and human dynamics about these features, such as the movement patterns and daily activities of people when they use these roads (Liu *et al.* 2015). As a result, the approach advocates combining typical remote sensing information (such as satellite imagery) with unconventionally considered, remotely sensed data that contains social information. Such data sources have included taxi trajectories, mobile phone records, social media or social networking data, or the activity on 'smart cards' scanned to board public transportation (Liu *et al.* 2015). Studies of human mobility using cell phone locations suggested that, in general, our movement patterns are relatively simple and predictable with a high degree of temporal and spatial regularity characterized by return visits to a few highly frequented locations (González *et al.* 2008). In the nearshore marine environment, cell phone activity could be used to map human use across the seascape. In Estonia, a study was conducted using cell phone data to assess the coastal locations favoured by tourists (Ahas *et al.* 2007), while cellphone-based reporting has been considered to gather self-report data on recreational fishing activity to improve fisheries management (Baker & Oeschger 2009).

13.3.1.4 Mapping Ecosystem Services

Economists and policy makers have recently embraced a construct of natural capital accounting called ecosystem services, a mechanism to define and quantify the direct and indirect benefits that wildlife, biodiversity, and ecosystems provide to people for human wellbeing (Millennium Ecosystem Assessment 2005; Boyd & Banzhaf 2007). These ecosystem services are often spatially mapped and categorized using an accounting system as part of a valuation exercise in order to assist decision makers in evaluating the tradeoffs between alterative management measures (Millennium Ecosystem Assessment 2005). Quantifying and mapping ecosystem services helps managers identify the ecosystem components and their locations that provide substantial services and merit either protection or sustainable management (Posner *et al.* 2016). In addition, conducting these evaluations at regular time intervals can enable managers and business interests to identify and prepare for changes to ecosystem service provisions (Petter *et al.* 2012). Spatially mapping ecosystem services contributes to a systems-based approach toward evaluating management strategies because it demonstrates to stakeholders and managers alike the full suite of ways that components of the ecosystem can contribute to human wellbeing. When considering the management question to be addressed, ecosystem services should be evaluated and mapped at the appropriate spatial and temporal scales and can be analysed at multiple scales (Grêt-Regamey *et al.* 2014).

The literature is replete with references to ecosystem services and varying approaches on how to characterize and map such services (*i.e.*, Ott & Staub 2009; Carpenter *et al.* 2009; Haines-Young & Potschin 2011; Staub *et al.* 2011; Martínez-Harms & Balvanera 2012). A review of various different decision-support tools that exist for quantifying and valuing ecosystem services is provided by Bagstad *et al.* (2013). Due to the number of disparate tools that exist, there is a need for the ecosystem services community to develop and embrace a more standardized process to map and model ecosystem services (Boyd & Banzhaf 2007; Crossman *et al.* 2013). Rarely are insights from landscape ecology considered such as the influence of spatial configuration of the seascape on the type, quality and value of services. Barbier (in Chapter 15 in this book) discusses the importance of connectivity between patches of mangroves, seagrasses and coral reefs on ecosystem service values.

One popular tool for mapping and categorizing ecosystem services is the Integrated Valuation of Ecosystem Services and Tradeoffs (InVEST) modelling toolbox. InVEST is a suite of models and algorithms used to map, model and value multiple ecosystem services, as well as project the status of these ecosystem services under different future scenarios. The tool can be used across various environments (terrestrial, freshwater and marine) and contains a number of marine specific models for understanding how ecosystem benefits can be realized under different human use scenarios across the seascape (Guerry *et al.* 2012). For example, the tool can be used to examine the effect of altering land use on the spatial patterns of terrestrial runoff to marine waters (see Chapter 11 in this book). In Colombia, South America, InVEST was applied to evaluate the risk to coastal property due to inundation from storm surge if mangrove areas were converted to development, and valuing the service that the mangroves provided to local subsistence fishers. Spatial data on land elevation, bathymetry and coastal habitat types, together with historical storm information on wind speeds, directions and wave heights were used to model inundation scenarios. Model results showed that areas currently occupied by mangrove forest would experience flooding during a typical storm, however water would not infiltrate beyond the mangrove forest. A 20% reduction in mangrove cover predicted minimal inundation on lands adjacent to the mangrove forest, with land loss valued at 2.4 million US dollars (Figure 13.3). The 40% and 60% reduction in mangrove coverage scenarios, however, forecast a substantial portion of land in the northeast part of the property to be at least partially flooded.

13.4   Modelling Human-Seascape Interactions with a Systems Approach

The use of quantitative and spatially explicit modelling tools to understand and quantify the dynamics within a system and how components interact with one another constitutes a systems-based approach to ecology. The systems-based philosophy emphasizes how ecosystem function can be influenced by human intervention and use and strongly encourages interdisciplinary collaboration between biologists, ecologists, economists, anthropologists and all relevant stakeholder groups (Berkes & Folke 1998; Glaeser *et al.* 2009; see also Chapter 12 in this book). This section will introduce the reader to different types of modelling frameworks, connect the data sources discussed in previous sections to their use and application in these modelling frameworks and provide some examples.

For purposes of discussing seascape ecology and human behaviour, the term 'model' (and by extension modelling) will be used to refer to a computer-based program that imitates or represents the operations of real-world processes within a system. What is meant by the term 'system' is largely dependent on the objectives and location of a study. In the context of seascape ecology, a system usually refers to the collection of entities and

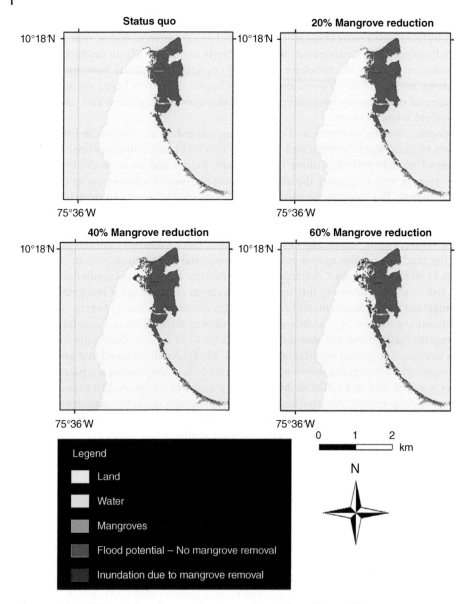

Status quo
20% Mangrove reduction
40% Mangrove reduction
60% Mangrove reduction
Legend
Land
Water
Mangroves
Flood potential – No mangrove removal
Inundation due to mangrove removal

Figure 13.3 InVEST model results mapping the predicted effect of three different mangrove removal scenarios on coastal inundation during storm events. Mangroves are located in all flood potential areas, coloured in dark blue, under the status quo scenario of no mangrove removal.

processes (*i.e.*, fish, currents, seagrass, fishing vessels, oil rigs, *etc.*) that interact with one another (Law 2007). The state of a system is defined as a set of variables that describe the characteristics of a system at a given point in time (Law 2007). For example, on a given day these can represent the weather conditions such as wind speed, oceanic conditions (*e.g.*, wave height, sea temperature), economic conditions (*e.g.*, unemployment rate or annual income), or regulatory conditions.

Systems modelling is an important analytical approach for scientists and resource managers to understand and use, as it is one of the most effective ways to communicate system complexity and interdependencies to a broad audience. Scenario modelling is useful for addressing different management hypotheses, quantifying the costs and benefits of each management alternative and understanding how they may affect the management of ecosystem services. For example, systems modelling can inform and quantify the degree that ecosystem restoration may re-establish the functions of direct value to humans and the ability of such ecological systems to cope with future disturbance (Moberg & Rönnbäck 2003; Weijerman *et al.* 2015).

13.4.1 Custom-built Statistical Models

Ecological resource assessments typically use custom-built statistical modelling approaches to represent the ecological and human components of a system. The strategy behind employing this class of model for resource assessment and management is to estimate the parameters that describe the natural and human components of the system and their interactions, by finding a numerical solution to the suite of equations governing these processes. The collections of equations, which represent the ecological and human processes (*i.e.,* animal growth, animal reproduction, resource harvest), are often linked together using a likelihood function. The parameters in the model are numerically adjusted by the computer in an iterative process (maximum likelihood), until the likelihood value cannot be further maximized, at which point an estimated solution is reached. Data providing observations of the natural system are incorporated into the model and the computer algorithm adjusts the parameter values in order to most closely replicate (*i.e.,* fit) the patterns seen in the data. Once a solution is reached, the resulting estimated parameters are assumed to represent the current state of that system. Using these parameter estimates, the possible effect of various management scenarios can be forecast into the future under different assumed state conditions (Fournier *et al.* 2011).

Many fisheries stock assessment models are custom-built statistical models. An example is the tool, Stock Synthesis, most frequently used by NOAA Fisheries to evaluate the status of fish populations. This tool includes a population simulation model to represent abundance and mortality, an observation model that relates the simulated population to the observable data (landings, fish size and abundance trends) by simulating the fishing process and fitting the simulated fishery observations to the real data and a statistical model that adjusts the parameters to achieve the best fit (Methot & Wetzel 2012). The final set of estimated parameters is then used to forecast the fish harvest levels and population trends one could expect under different proposed management measures. The book published by Quinn & Deriso (1999) is a comprehensive reference on this subject.

Though the utility of sophisticated stock assessment tools such as Stock Synthesis have proven useful in successfully managing fisheries resources around the world, they often poorly represent the seascape ecology discipline in several important ways. Firstly, most stock assessment models only represent the dynamics of a single species without considering its predator-prey interactions. Secondly, most stock assessments do not consider the spatial dimensions of either the fish or the fishers, though Stock Synthesis can handle large-scale spatial partitions. Predator-prey dynamics and spatial considerations in fisheries are typically reserved for ecosystem-based models (such as Atlantis, Ecopath with

Ecosim) and not built into regular single species assessments. This is largely due to the lack of available data to parameterize such processes and the time that it takes to put together such elaborate ecosystem models. Furthermore, ecosystem-based fisheries models are still in their infancy. Their growth and development has largely been made possible in recent years through the availability of inexpensive computing power. As such, they often take years to parameterize and can be difficult and unstable during optimization due to the sheer magnitude of parameters to estimate and the processes they represent.

Thirdly and most importantly, mainstream fisheries modelling platforms have over-simplified their representation of human behaviour. Research on the behaviour and decision-making processes of fishers has been given considerably less attention in comparison to advances in our understanding of marine ecology (Branch *et al.* 2006). In most cases, the intricacies of the fishing process are reduced down to a function representing the size of fish a particular fishing gear can capture (selectivity), the probability of capture by the gear (catchability) and a fishing mortality parameter. In reality however, we know that fishers are heterogeneous and don't necessarily operate as rational profit maximizers (Eggert & Kahui 2013; Nguyen & Leung 2013). Gaps in understanding of fisher behaviour have sometimes led to unintended and surprising responses to management interventions and have been identified as a key impediment to progress in ecosystem-based fishery management (Fulton *et al.* 2011).

13.4.2   Predefined Statistical Routines

The second category of models to discuss relative to human use of the seascape is pre-defined statistical models. This category is very broad and encompasses models that may or may not be spatial, may fit to time series or spatial data, or both and can range in complexity from large likelihood maximization exercises, to simple t-tests, ANOVA, or linear regression. Predefined statistical models assume that the observed data you are analysing (*i.e.*, the size of 100 snapper collected from the Atlantic Ocean) was collected from and is representative of a greater population (*i.e.*, the size distribution of all the snapper fish in the Atlantic Ocean). This greater population is meant to represent all of the items in the universe from which you have a small sample (McCullagh 2002). The assumptions behind predefined statistical models are usually quite strict and assume that your data (and the variability contained in your data) sufficiently approximates a particular probability distribution, which is the same distribution as that found in the greater population. Statistical inference is defined as the process of fitting scientific data to a variety of statistical models and, comparing the fit of these models to select the one that best approximates the process you are trying to represent (Burnham and Anderson 1998). This can be used to compare different parameterizations of both pre-defined statistical routine models, as well as custom-built models as described above. Akaike's Information Criterion (AIC) has emerged as the current standard for comparing competing statistical models. The text by Burnham and Anderson (1998) provides an excellent guide for making inferences from scientific data using statistical modelling.

There are numerous software tools available for statistical analysis and each have their strengths and weaknesses. Further, many of these tools have spatial components that can be used to evaluate spatially explicit data collected across the seascape (*i.e.*, ESRI's ArcMap, The R-Project, MATLAB, SPSS, Stata, *etc.*). A seascape specific, statistically driven decision support tool for marine spatial planning is MARXAN, which stands for Spatially Explicit Annealing. MARXAN is standalone software, which uses statistical optimization routines to generate different possible systems of marine reserves. Its strength lies in its ability to achieve user specified conservation and socioeconomic goals by objectively determining the size, shape and geographic placement of protected areas. The statistical algorithm accomplishes this by minimizing the costs and maximizing the benefits across many potential planning options (Esfandeh *et al.* 2015). MARXAN and it's built in statistical optimizer, has been used in a variety of settings to determine the effective size and placement of marine protected areas, given the ecosystem services provided though human use of the seascape (Fraschetti *et al.* 2009).

The field of spatial statistics is an area where seascape ecology has greatly benefited from its land-based counterpart. Much of the methodologies that make up the current suite of spatial statistics arose out of necessity to answer natural resource related questions across different terrestrial disciplines. These included mining engineering and soil sciences, which facilitated the development of geostatistics and spatial prediction to identify the location of mineral resources and agriculture and forestry, which facilitated the development of randomization and block designs (Gelfand *et al.* 2010). An increased interest in spatial and space-time problems, likely ushered in through the availability of inexpensive computers, has led to the growth and application of spatial statistics to marine spatial planning. Where point data are patchy and a continuous surface is more desirable, spatial prediction techniques can rapidly provide spatial information on patterns of human use across the seascape using techniques such as inverse distance weighting, variogram modelling and kriging, generalized linear modelling-based regression and regression kriging (Cressie 1993; Dale & Fortin 2014).

13.4.3 Discrete Choice Models

A special group of predefined statistical models used to quantify and predict human behaviour are discrete choice models (also referred to as random utility models). These econometric models use binomial and multinomial logit and probit functions to represent and describe the decision-making process of people. Such models rely on either revealed or stated preference data in the form of a panel dataset for estimation as discussed earlier in this chapter. Like all statistical models, the factors included in discrete choice models to describe the state variables that affect an individual's decision making typically don't include all of the factors that go into the decision-making process. Humans are complex creatures and the factors that motivate our behaviours and decisions, are not always apparent or observable. As a result, the unexplained factors that go into an individual's decision making are grouped into an error term, which is not observed. Thus, the probability that an individual making a decision will select a particular outcome from the set of all possible outcomes is calculated as a probability. Information detailing how to structure and estimate these types of models is best provided by Train (2009).

Specific to our human relationship with the seascape, these models have been used in a variety of ways including to assess the nonmarket benefits of seagrass restoration (Börger & Piwowarczyk 2016), value recreational beach use (Lew & Larson 2008), manage tourism use of marine resources (Mejía & Brandt 2015), understand spatial fishing patterns (Abbott & Wilen 2011; Alvarez *et al.* 2014; Davies *et al.* 2014; Saul &

Die 2016) and study how multiple seascape uses can best be managed (Tidd *et al.* 2015). Recently, Saul & Die (2016) used discrete choice models to quantify the drivers behind the decision-making behaviours of commercial fishers who participate in the reef fish fishery on the West Florida Shelf. Three main decisions were modelled: participation, site choice and trip termination.

A panel dataset was constructed by merging daily logbook observations with other datasets containing state information for a given day or time period. Vessel logbook data contained information on when a vessel started a fishing trip, the location they fished, what they caught and when they returned to port. Data on vessel characteristics were incorporated from NOAA's vessel operating unit data. Landing sites and port locations for each trip were obtained from corresponding NOAA dealer records. Spatially explicit daily wind speed came from the NOAA National Data Buoy Center. Real weekly diesel fuel price came from the Energy Information Administration and was corrected by the Consumer Price Index (CPI). The daily price of fish and expected revenue were calculated from landings data for the five reef fish species that affect markets and corrected by the CPI. Expected revenue at each potential fishing location choice was calculated for these five fish species as the product CPI adjusted fish price, monthly variation in abundance and each vessel's relative fishing power. Regulatory history was obtained from the most recent NOAA Fisheries Southeast Data Assessment and Review process stock assessments for commercially important reef fish species and was incorporated into the decision models (Saul & Die 2016).

Results show that some factors identified as statistically significant influence the spatial and temporal distribution of fisher effort (*i.e.,* weather such as wind speed, price changes, fuel costs, expected revenue). For example, the top two panels in Figure 13.4 depict the partial probabilities that different wind speeds contribute to the participation decision and the partial probability that filling up the fish hold contributes to a decision to return to port. The bottom two panels in Figure 13.4 show the effect of fuel price and wind speed on fishing location choice, with red circles depicting the locations that were statistically significant for each effect. Estimated model results were used to parameterize human behaviour in an agent-based simulation model (further discussed below). Behavioural studies such as this are important to help inform scientists and managers about the human use patterns that occur across the seascape in response to new regulations or changing environmental conditions. Awareness of the magnitude and direction of such human behaviours can help reduce management uncertainty and avoid surprising and unexpected responses to newly implemented fishing regulations (Branch *et al.* 2006; Fulton *et al.* 2011; Saul & Die 2016).

13.4.4   Simulation Modelling

In many cases, the system being studied may be highly complex and, as such, analytical or numerical solutions may not be attainable. When this is the case, the system must then be studied using simulation modelling. In simulation modelling, the model is iteratively run with different sets of inputs to study how these different input combinations affect a priori specified output performance measures. Like statistical models, simulation models can be broken down into numerical (mathematic or statistical) simulation models and agent-based simulation models. Simulation modelling is an effective tool in operations research as it allows the user to recreate a problem and explore alternative

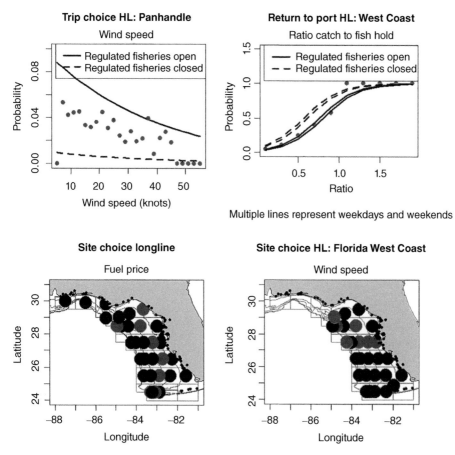

Figure 13.4 Discrete choice model fit results showing some of the factors that influenced various modelled fisher decisions off the west coast of Florida. The top two panels show the partial probabilities that different wind speeds and catches relative to fish hold capacity affect the decisions to take a fishing trip and return to port respectively. Red points represent the observed values, while black lines are model fits. Bottom panels spatially depict the fishing locations for which parameters for fuel price and wind speed were significant using red circles. *Source:* Saul & Die (2016).

ways that the system may respond to different possible interventions. In many cases, it is not physically possible, moral, legal, or cost effective to conduct such experiments in the real world (*i.e.,* test what would happen if all the sharks were removed from a system, test whether adding an extension to a factory would increase productivity). As a result, simulation modelling is increasingly recognized as an important decision support tool. Due to their complexity, simulation modelling lends itself well to representing the dynamics of coupled human-natural systems across the seascape.

The book by Law (2007) is one of the most comprehensive guides devoted to all aspects of simulation modelling. Law (2007) recognizes that developing a simulation model is not a linear process and in many cases, the model development process itself is iterative, where the developer may need to go back a previous step before progressing. In this spirit, Law (2007) outlines ten steps practitioners follow when developing a simulation model (Figure 13.5). This workflow, when used to develop models to test

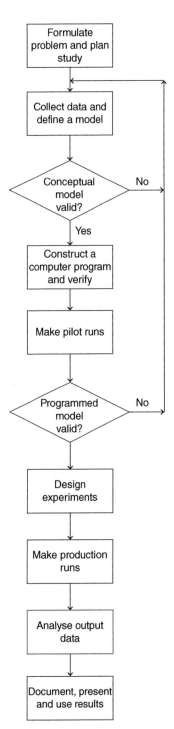

Figure 13.5 Steps to conduct a sound simulation study. *Source:* Adapted from Law (2007).

potential marine management measures, is sometimes referred to as management strategy evaluation (Butterworth & Punt 1999; Sainsbury *et al.* 2000). Note that the two main points of revision occur in validating the model assumptions and validating the model itself. A clear list of model assumptions should be produced when the available data reflecting the processes being modelled has been reviewed and when the simulation model structure has been defined. This should take place before programming begins to avoid the need to reprogram later. Validation of the model itself is another critical step and one that is often inquired about in great detail by those the results of the model may affect (*i.e.*, policy makers and other stakeholders). Model validation simply means comparing the patterns that the model produces and forecasts, with actual human patterns across the seascape.

There are various ways to validate a simulation model. If data are available from the real world that represents aspects of the system you are trying to model, then you could compare performance measures and patterns produced by the simulation with those of the actual system. When validating a model in this way, the data from the system being compared to the simulation model should not have been used in any way to parameterize the simulation model. A model's robustness can also be evaluated by conducting sensitivity analyses, which explores how a range of input values for each parameter affects the model's performance. Those factors found during sensitivity analysis to significantly impact performance measures, need to then be considered very carefully and where possible, verified against data for accuracy. Finally, the analyst together with stakeholders and subject matter experts should review the model carefully for correctness. Involvement of stakeholders early and often throughout model development, validation and implementation is critically important to achieving the goals behind developing the model in the first place.

13.4.5 Agent-based Models

Agent-based modelling (also sometimes referred to as individual-based modelling or IBMs) is an approach to simulating systems (see also Chapter 8 in this book). These models comprise autonomous, interacting individuals, with characteristics and behaviours, parameterized at the individual level. Applications of agent-based modelling (ABM) have been used in a variety of disciplines (from economics and social science to biology and ecology) and commercial applications (military, transportation, *etc.*) to model complex systems and study the emergent properties that result from individuals' interactions and behaviours (Bankes 2002; Bonabeau 2002). With ABMs, an individual's behaviours and interrelationships are defined by rule-based expressions. Agents that are similar may be grouped into the same class, with which they share a collection of behaviours. The state of a particular agent is defined by discrete values contained within attribute variables. Though an agent may belong to a particular class, each agent possesses a unique set of attribute values. Patterns emerge from system dynamics as agents interact with one another and their surrounding environment at discrete time intervals (Grimm *et al.* 2005).

This modelling approach is well suited to studying complex ecological systems both in the terrestrial and aquatic realms, where overall system dynamics are highly complex, nonlinear and contain many parameters to estimate (Uchmanski & Grimm 1996; Grimm 1999). Such systems can be very difficult to represent using equation-based, numerical / statistical simulation models due to their nonlinearity. This may cause the optimizer to find a local maxima or minima and think a solution has been reached, when in reality, a better solution actually exists but was not found by the algorithm. The agent-based modelling approach is not meant to replace equation-based modelling, but is complementary and can incorporate fitted parameters from equation-based models to represent individual agent behaviours and ecosystem function (Fahse *et al*. 1998; Parunak *et al*. 1998; Wilson 1998). Key challenges to address relate to the type of information used to parametrize decision making, which can be a set of rule-based decisions based on theory or on observations of real decision making. Whether the parametrization is entirely statistical versus using adaptive behavioural rules or some hybrid will affect model outcomes (Filatova *et al*. 2013).

13.4.6 Pattern-oriented Modelling

Validation of agent-based models is often disproportionately criticized compared to their numerical counterparts, because agent-based models do not fit to data or optimize a likelihood function. Therefore, parameter fitting and model rigor cannot be evaluated using typical fitting metrics such as likelihood ratio tests or AIC. Despite this, however, Grimm *et al*. (2005) developed a framework for validating such models called pattern-oriented modelling. Although developed specifically for agent-based models, this technique can and should be used to validate all models. In addition, the pattern-oriented modelling approach is well suited to the discussion in this chapter, which has considered human use spatial patterns across the seascape. The pattern-oriented approach involves comparing alternative models of the same system and process using what Grimm (2005) refers to as 'inference'. The process involves implementing alternative theories of an agent's decisions using various characteristic patterns at both the individual and higher levels and then testing how well these models reproduce the patterns seen in the real world. Models that fail to reproduce the characteristic patterns observed in nature are rejected and improved upon, while models that replicate the human and natural patterns observed in the system are considered to be robust to the rules and assumptions used (Grimm *et al*. 2005, 2006).

Comparison of the patterns emergent from the simulation model, with those observed in the system for which there is data, can be done in various different ways depending on the data that is available. Traditional practitioners who are new to agent-based modelling may feel more comfortable with a statistical comparison. For example, an agent-based model developed to represent the migration of reef fish in the Gulf of Mexico was validated using a tagging study conducted in the same location. Firstly, a potential model of fish location was developed and, in the simulation, a simulated tagging study was conducted analogous to that conducted in the real world for which data was available. Secondly, results from the simulation were statistically evaluated against the tagging data by using Wilcoxon rank sum or Student's *t* tests to compare the following emergent patterns represented as probability distributions: linear-movement speed between the tagging and recapture locations, the distance moved between the tagging and recapture locations, the distribution of fish lengths between

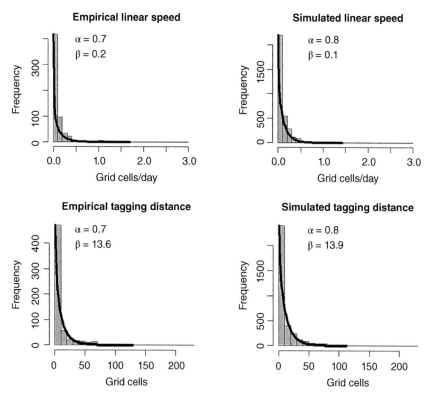

Figure 13.6 Statistical comparison of population level patterns emerging from an agent-based model, with population level patterns from field collected empirical data. Simulated linear fish speeds (Wilcoxon rank-sum test: $p = 0.2933$) and tagging distance (Wilcoxon rank-sum test: $p = 0.780$) emergent from the simulation model for red grouper are compared to the same metrics calculated from a field tagging experiment. Alpha and beta are the parameters of the fitted gamma distributions used to approximate the simulated and empirical probability density functions depicted in the graphs. *Source:* Saul *et al.* (2012).

tagging and recapture and the spatial distribution of recaptured fish (Figure 13.6; Saul *et al.* 2012).

If data comparing simulated output with real world observations is not available or accessible to the investigator, a more visual approach may be used. For example, the agent-based simulation model of reef fish in the Gulf of Mexico that was described above, also modelled the two primary fishing fleets that capture reef fish. As alluded to earlier in this chapter, individual vessel behaviour in the simulation model was parameterized using fitted discrete choice models. As the model was developed and refined, spatial results mapping fisher use of the seascape in the simulation were visually compared with plots of vessel monitoring system (VMS) data using pattern-oriented modelling (Figure 13.7). Statistical comparison was not able to be conducted because we were not permitted access to the VMS data itself due to confidentiality of the data. Consequently, note that as a result, VMS data was not used to parameterize the simulation in any way. Visual comparison of these two maps and their spatial patterns show that the underlying processes programmed and parameterized that

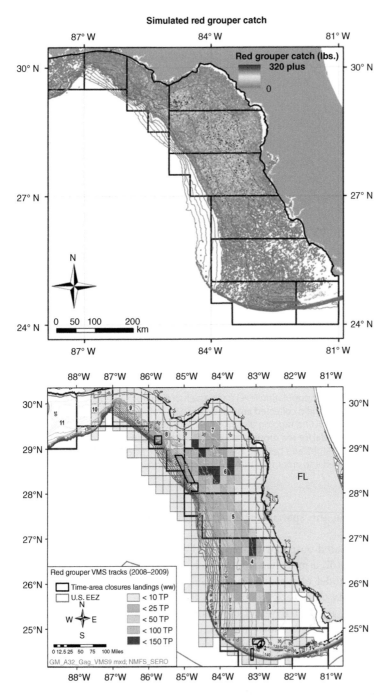

Figure 13.7 Visual comparison of simulated emergent spatial catch patterns of red grouper predicted by an agent-based model, with actual spatial catch observations of red grouper provided by vessel monitoring system (VMS) data. *Source:* Saul (2012).

govern individual fish and vessel behaviours and decisions, when they interact, recreate the spatial patterns observed in the real system. Thus, one can conclude that the individual level behaviours are reflective of what may occur on an individual level in the real system (Saul 2012).

There are many other examples from fishery science in which agent-based modelling was the approach of choice. One such model is a spatially explicit ABM for modelling vessel movements and alternative effort allocation patterns of Danish fishing vessel activities (Bastardie *et al.* 2010, 2014). The effects of effort reallocation on total fuel consumption and energy efficiency in relation to the quantity and value of landings were investigated for alternative scenarios of fuel-saving behaviour. Another ABM was used to model the effects of an individual transferable quota system in a multispecies, multisector fishery and the model was applied to the Coral Reef Fin Fish Fishery in Queensland, Australia (Little *et al.* 2009). Other examples include (i) a model of a fishery targeting different species in different areas, developed to analyse the implications of taking into account the response of fishing fleets to regulatory controls (Soulie & Thebaud, 2006); (ii) an interactive model designed to better understand the interactions between regional and local drivers strongly influencing the health of coral reefs in the Yucatan peninsula, Mexico (Perez *et al.* 2009); and (iii) an agent-based fishery management model of Hawaii's longline fishery (Yu *et al.* 2009).

13.5   Conclusions and Future Research Priorities

The knowledge gap on human use patterns and decision-making behaviour across the seascapes and the ecological and economic consequences of these human-seascape interactions presents both a problem hindering effective spatial management and an exciting frontier for new interdisciplinary research. Clearly, more work is needed to better operationalize the human dimension in seascape ecology (Samhouri *et al.* 2013; Link & Browman 2014; Dolan *et al.* 2016). Unprecedented opportunity now exists for the study of human-seascape relationship through theoretical advances in modelling socio-ecological systems together with advances in geospatial technologies for spatial data acquisition and sophisticated spatial modelling algorithms. These improvements are critical steps toward advancing the field of seascape ecology and progressing ecosystem-based management, which is dependent on the inclusion of human use pattern information.

Special attention must be focused on understanding and communicating uncertainty. As discussed in the introduction to the chapter, uncertainty is expensive, particularly when it leads to management failures. The cost of uncertainty can manifest itself either in foregone use of ecosystem services provided by the seascape, which could have been more aggressively exploited, or conversely in the unsustainable overuse of resources and their depletion. Cressie (1993) begins his book on spatial statistics by referring to statistics as the 'science of uncertainty', which 'attempts to model order in disorder' (Cressie 1993). We can extend this mantra to include the suite of quantitative tools discussed in this chapter. It is important to be aware that increasing the complexity of a model may not necessarily reduce uncertainty because uncertainty in one component can propagate and multiply as it cascades through the whole system. As pointed out by Oreskes (2003) 'a complex model may be more realistic, yet more uncertain'.

In some cases, a modular approach may be appropriate for building systems models where greater attention is given to the development of specialist submodels, which are then integrated to create the whole system. Some suggestions for how this can be done and associated discussion of advantages and disadvantages are provided by Voinov & Shugart (2013).

Managers and scientists often have a disproportionate understanding of the human use patterns that occur within marine systems, in comparison to our understanding of the biological population dynamics and oceanographic dynamics. As a result, the complexity of human use across the seascape is often overly simplified down to one or several simple parameters. Implementing management action without a comprehensive understanding of human use patterns and how they interact with natural system dynamics may occur for several reasons: in response to a governmental mandate or law, pressure from lobby groups or lawsuits, or because local scientists and managers lack the technical expertise or funding to conduct such analyses (Turner *et al.* 2016). This sometimes results in the implementation of management measures that fall short of achieving their desired conservation or management objectives. This can result in surprising, unconsidered outcomes as resource users adapt and respond to the newly implemented regulation in unanticipated ways (Hilborn 1985; Lane 1988; Hilborn & Walters 1992; Branch *et al.* 2006; Fulton *et al.* 2011). Thus, an important recommendation is to develop capacity-building initiatives for resource scientists and managers, to train them on the use of quantitative techniques for evaluating human use patterns across the seascape and using the results to develop management.

The quantitative tool one selects for resource and policy evaluation must match the data available and the spatial and temporal scale of the study. In data limited locations, analysts may only be able to view several layers of coarsely defined spatial data using GIS software, with spatial planning based simply on where the spatial layers overlap and therefore where competing interests intersect. In locations where sufficient information exists, an estimation or simulation model is advised, even if it is simple, to help scientists and managers quantify uncertainty and test the implementation of different hypothesized management measures. Further, even a highly stylized simulation model, meaning one that is mostly theoretical due to the absence of sufficient data on the system, can go a long way towards helping managers and resource users better comprehend the complexities of a system and the direction that their actions may perturb system dynamics (Carpenter & Gunderson 2001).

Data collection campaigns and data mining initiatives should be structured to address management questions, with model development and parameterization as the end goal if possible. This will help identify gaps where additional information may need to be gathered, to which data collection efforts can be refocused. Future research priorities should focus on developing improved data collection programs that capture information on the behaviours and decision making of human seascape use patterns across different sectors (*i.e.*, fisheries, mining, shipping, tourism and recreation). This can include socioeconomic surveys, as well as theoretical and applied experiments (*i.e.*, game theory). This is important as models and statistical analyses and the predictions they provide are only as good as the data used to parameterize or fit them (*i.e.*, garbage in, garbage out mantra).

Collecting field data, whether biological or socioeconomic, is difficult and expensive. All too often, data from field work ends up not being analysed, or fully exploited.

This is typically due to a lack of time, personnel, computational resources, desire and / or capacity (*i.e.,* statistical knowledge). In a sense, the next generation of seascape ecologists needs to be prepared to be a sort of 'analytical jack of all trades and master of some' to be successful. The ability to evaluate existing data, identify gaps, collect data to fill missing information and synthesize diverse information using statistical analyses, GIS systems, or spatially explicit models. Furthermore, seascape ecologists need to have a basic interdisciplinary knowledge of (and respect for) the terminology and methodology used by other experts of a range of complimentary disciplines such as ecologists, economists, sociologists, geographers, psychologists and others. The best way to achieve this synergy is by establishing interdisciplinary teams of individuals from appropriate disciplines.

Finally, where seascape ecology is being applied to management (*e.g.,* marine spatial planning) a transdisciplinary approach will be required whereby managers and policy makers and possibly a range of stakeholders are included from the very beginning, during problem conceptualization all the way through to the implementation of regulations (Caldow *et al.* 2015). Many of the tools that this chapter discussed are spatial, from GIS datasets to spatially explicit agent-based models. Spatial tools and the maps they produce are excellent ways to engage stakeholders in the mapping and modelling process. For example, individuals watching a spatial simulation that shows fishing vessels leaving a port and operating in different locations can relate to what they are seeing on the computer screen and provide valuable feedback on the parameterization of the processes in the model and the realism it may or may not represent. Finally, stakeholders who have participated in the process often become important allies and help to educate their resource using colleagues on the importance of the conservation objectives under consideration.

Here we provide several priority areas for future research that represent new areas of research with potential to advance the integration of human spatial dynamics into seascape ecology and bridge the information gap from seascape ecology to marine stewardship strategies such as marine spatial planning:

- As highlighted in our chapter, continued effort is required to improve the detailed representation of spatial heterogeneity in socio-economic systems models to boost predictive performance of human use patterns and the relationships between human decision making and biophysical conditions.
- As model complexity increases so models become increasingly hard to analyse, understand and interpret (Voinov & Shughart 2013). Consequently, research is required to discover the optimal spatial complexity and spatial scales necessary to develop reliable model systems.
- An important and largely unexplored focus of investigation in human-seascape relationships is the application of spatial pattern metrics to quantify human use patterns across the seascape. Landscape ecology offers a plethora of algorithms that can be applied to maps regardless of what they represent including socioeconomic data (*i.e.,* corridors and connecting, edge effects, patchiness, patch geometry). Human distributions and activities have quantifiable patterning that can be linked to patterning in the surrounding seascape. Taking a landscape ecology approach to investigating human ecology will provide fresh insights into the human dimensions of seascape ecology, its structure, function and change.

References

Aarts H, Verplanken B, Knippenberg A (2006) Predicting behaviour from actions in the past: repeated decision making or a matter of habit? Journal of Applied Social Psychology 28: 1355–1374.

Abbott JK, Wilen JE (2011) Dissecting the tragedy: a spatial model of behaviour in the commons. Journal of Environmental Economics and Management 62: 386–401.

Acheson JM (2006) Institutional failure in resource management. Annual Review of Anthropology 35: 117–134.

Ahas R, Aasa A, Mark U, Pae T, Kull A (2007) Seasonal tourism spaces in Estonia: case study with mobile positioning data. Tourism Management 28: 898–910.

Alessa L, Kliskey A, Brown G (2008) Social-ecological hotspots mapping: a spatial approach for identifying coupled social-ecological space. Landscape and Urban Planning 85: 27–39.

Alvarez S, Larkin S, Whitehead J, Haab T (2014) A revealed preference approach to valuing non-market recreational fishing losses from the deepwater horizon oil spill. Journal of Environmental Management 145: 199–209.

Amato A, Di Lecce V, Piuri V (2013) Sensors for human behaviour analysis. In Amato A, Di Lecce V, Piuri V (eds) Semantic Analysis and Understanding of Human Behaviour in Video Streaming. Springer Science and Business Media, New York, NY.

Aswani S, Lauer M (2006) Incorporating fishermen's local Knowledge and behaviour into geographical information systems (GIS) for designing marine protected areas in Oceania. Human Organization 65: 81–102.

Atkins JP, Burdon D, Elliott M, Gregory AJ (2011) Management of the marine environment: integrating ecosystem services and societal benefits with the DPSIR framework in a systems approach. Marine Pollution Bulletin 62: 215–226.

Bagstad KJ, Semmens DJ, Waage S, Winthrop R (2013) A comparative assessment of decision-support tools for ecosystem services quantification and valuation. Ecosystem Services 5: e27–e39.

Baker MS, Oeschger I (2009) Description and initial evaluation of a text message based reporting method for marine recreational anglers. Marine and Coastal Fisheries: Dynamics, Management and Ecosystem Science 1: 143–154.

Bankes SC (2002) Agent-based modelling: a revolution? Proceedings of the National Academies of Science USA 99: 7199–7200.

Bastardie F, Nielsen JR, Andersen BS, Eigaard OR (2010) Effects of fishing effort allocation scenarios on energy efficiency and profitability: an individual-based model applied to Danish fisheries. Fisheries Research 106: 501–516.

Bastardie F, Nielsen JR, Miethe T. (2014) DISPLACE: a dynamic, individual-based model for spatial fishing planning and effort displacement-integrating underlying fish population models. Canadian Journal of Fisheries and Aquatic Sciences 71: 366–386.

Bath AJ (1998) The role of human dimensions in wildlife resource research in wildlife management. Ursus 10: 349–355.

Bauer LJ, Kendall MS, Jeffrey CFG (2008) Incidence of marine debris and its relationships with benthic features in Gray's Reef National Marine Sanctuary, southeast USA. Marine Pollution Bulletin 56: 402–413.

Behringer DC, Swett RA (2011) Determining Vessel Use Patterns in the Southeast Florida Region. Southeast Florida Coral Reef Initiative, Florida Department of Environmental Protection, Miami Beach, FL, pp 1–87.

Berkes F (2012) Implementing ecosystem-based management: evolution or revolution? Fish and Fisheries 13: 465–476.

Berkes F, Folke C (1998) Linking Social and Ecological Systems: Management Practices and Social Mechanisms for Building Resilience. Cambridge University Press, Cambridge.

Boivin NL, Zeder MA, Fuller DQ, Crowther A, Larson G, Erlandson JM, Denham T, Petraglia MD (2016) Ecological consequences of human niche construction: examining long-term anthropogenic shaping of global species distributions. Proceedings of the National Academies of Science USA 113: 6388–6396.

Bonabeau, E (2002) Agent-based modelling: methods and techniques for simulating human systems. Proceedings of the National Academies of Science USA 99: 7280–7287.

Börger T, Piwowarczyk J (2016) Assessing non-market benefits of seagrass restoration in the Gulf of Gdańsk. Journal of Ocean and Coastal Economics 3(1): 1–20.

Boyd J, Banzhaf S (2007) What are ecosystem services? The need for standardized environmental accounting units. Ecological Economics 63: 616–626.

Branch TA, Hilborn R, Haynie AC, Fay G, Flynn L, Griffiths J, Marshall KN, Randall JK, Scheuerell JM, Ward EJ, Young M (2006) Fleet dynamics and fishermen behaviour: lessons for fisheries managers. Canadian Journal of Fisheries and Aquatic Sciences 63: 1647–1668.

Burnham KP, Anderson DR (1998) Model Selection and Multimodel Inference, A Practical Information-Theoretic Approach (2nd edition). Springer Science and Business Media, New York, NY.

Butterworth DS, Punt AE (1999) Experiences in the evaluation and implementation of management procedures. ICES Journal of Marine Science 56: 985–998.

Caldow C, Monaco ME, Pittman SJ (2015) Biogeographic assessments: a framework for information synthesis in marine spatial planning. Marine Policy 51: 423–432.

Carpenter SR, Gunderson LH (2001) Coping with collapse: ecological and social dynamics in ecosystem management. BioScience 51: 451–457.

Carpenter SR, Mooney HA, Agard J, Caplstrano D, DeFries RS, Diaz S, Dietz T, Duralappah AK, Oteng-Yeboah A, Pereira HM, Perrings C, Reid WV Sarukhan J, Scholes RJ, Whyte A (2009) Science for managing ecosystem services: beyond the Millennium Ecosystem Assessment. Proceedings of the National Academies of Sciences 106: 1305–1312.

Clark CW (2006) The Worldwide Crisis in Fisheries: Economic Models and Human Behaviour. Cambridge University Press, New York, NY.

Claudet J, García-Charton JA, Lenfant P (2010) Combined effects of levels of protection and environmental variables at different spatial resolutions on fish assemblages in a marine protected area. Conservation Biology 25: 105–114.

Coppin PR, Jonckheere I, Nackaerts K, Muys B, Lambin E (2004) Review article digital change detection methods in ecosystem monitoring: a review. International Journal of Remote Sensing 25: 1565–1596.

Cressie NAC (1993) Statistics for Spatial Data. John Wiley & Sons, New York, NY.

Crossman ND, Burkhard B, Nedkov S, Willemen L, Petz K, Palomo I, Drakou EG, Martín-Lopez B, McPhearson T, Boyanova K, Alkemade T, Egoh B, Dunbar MB, Maes J (2013) A blueprint for mapping and modelling ecosystem services. Ecosystem Services 4: 4–14.

Cumming GS, Cumming DH, Redman CL (2006) Scale mismatches in social-ecological systems: causes, consequences, and solutions. Ecology and Society 11(1): 14.

Dale MRT, Fortin MJ (2014) Spatial Analysis: A Guide for Ecologists. Cambridge University Press, Cambridge.

Davies TW, Duffy JP, Bennie J, Gaston KJ (2016) Stemming the tide of light pollution encroaching into marine protected areas. Conservation Letters 9: 164–171.

Davies TK, Mees CC, Milner-Gulland EJ (2014) Modelling the spatial behaviour of a tropical tuna purse seine fleet. PLOS One 9: e114037.

Dennis RA, Mayer J, Applegate G, Chokkalingam U, Colfer CJP, Kurniawan I, Lachowski H, Maus P, Permana RP, Ruchiat Y, Stolle F, Suyanto, Tomich TP (2005) Fire, people and pixels: linking social science and remote sensing to understand underlying causes and impacts of fires in Indonesia. Human Ecology 33: 465–504.

Dex S (1995) The reliability of recall data: a literature review. Bulletin of Sociological Methodology 49: 58–89.

Dolan TE, Patrick WS Link JS (2016) Delineating the continuum of marine ecosystem-based management: a US fisheries reference point perspective. ICES Journal of Marine Science 73: 1042–1050.

Douvere F (2008) The importance of marine spatial planning in advancing ecosystem-based sea use management. Marine Policy 32: 762–771.

Dunn CE (2007) Participatory GIS – a people's GIS. Progress in Human Geography 31: 616–637.

Eggert H, Kahui V (2013) Reference-dependent behaviour of paua (abalone) divers in New Zealand. Applied Economics 45: 1571–1582.

Esfandeh S, Kaboli M, Eslami-Andargoli L (2015) A Chronological review on application of MARXAN tool for systematic conservation planning in landscape. International Journal of Engineering and Applied Sciences 2: 6–17.

Fahse L, Wissel C, Grimm V (1998) Reconciling classical and individual-based approaches in theoretical population ecology: a protocol for extracting population parameters from individual-based models. The American Naturalist 152(6): 838–852.

Fifer S, Rose J, Greaves S (2014) Hypothetical bias in stated choice experiments: is it a problem? And if so, how do we deal with it? Transportation Research Part A 61: 164–177.

Filatova T, Verburg PH, Parker DC, Stannard CA (2013) Spatial agent-based models for socio-ecological systems: challenges and prospects. Environmental Modelling and Software 45: 1–7.

Fournier DA, Skaug HJ, Ancheta J, Ianelli J, Magnusson A, Maunder MN, Nielsen A, Sibert J (2011) AD Model Builder: using automatic differentiation for statistical inference of highly parameterized complex nonlinear models. Optimization Methods and Software 27: 233–249.

Fox J, Mishra V, Rindfuss R, Walsh S (eds) (2003) People and the Environment: Approaches for Linking Household and Community Surveys to Remote Sensing and GIS. Kluwer Academic Publishers, Boston, MA.

Fraschetti S, d'Ambrosio P, Micheli F, Bussotti S, Pizzolante F, Terlizzi A (2009) Planning marine protected areas in a human-dominated seascape. Marine Ecology Progress Series 375: 13–24.

Fulton EA (2010) Approaches to end-to-end ecosystem models. Journal of Marine Systems 81(1): 171–183.

Fulton EA, Boschetti F, Sporcic M, Jones T, Little LR, Dambacher JM, Gray R, Scott R, Gorton R (2015) A multi-model approach to engaging stakeholder and modellers in complex environmental problems. Environmental Science and Policy 48: 44–56.

Fulton EA, Smith ADM, Smith DC, van Putten IE (2011) Human behaviour: the key source of uncertainty in fisheries management. Fish and Fisheries 12: 2–17.

Gelfand AE, Diggle PJ, Fuentes M, Guttorp P (eds) (2010) Handbook of Spatial Statistics. CRC Press, Boca Raton, FL.

Geoghegan J, Prichard L, Ogneva-Himmelberger Y, Chowdhury RR, Sanderson S, Turner BL (1998) 'Socializing the pixel' and 'pixelizing the social' in land-use and land-cover change. In Liverman DM, Moran E, Rindfuss R, Stern P (eds) People and Pixels: Linking Remote Sensing and Social Science. Committee on the Human Dimensions of Global Environmental Change. National Research Council, National Academies Press, Washington, DC.

Gibson CC, Ostrom E, Ahn T (2000) The concept of scale and the human dimensions of global change: a survey. Ecological Economics 32 (2): 217–239.

Glaeser B, Bruckmeier K, Glaser M, Krause G (2009) Social-ecological Systems analysis in coastal and marine areas: A path toward integration of interdisciplinary knowledge. Current Trends in Human Ecology 183(203): 183–203.

González MC, Hidalgo CA, Barabasi AL (2008) Understanding individual human mobility patterns. Nature 453(7196): 779–782.

Gorzelany, JF (2005a) Recreational Boat Traffic Surveys of Broward County, Florida. Mote Marine Lab, Mote Technical Report No. 1017. Sarasota, FL.

Gorzelany, JF (2005b) Assessment of Boat Traffic Patterns and Vessel Wake Effects at Blowing Rocks Marina, Martin County, Florida, Final report. Mote Marine Lab, Mote Technical Report No. 1058, Sarasota, FL.

Gorzelany, JF (2009) Recreational Boating Activity in Miami-Dade County. Mote Marine Lab, Mote Technical Report No. 1357, Sarasota, FL.

Grêt-Regamey A, Weibel B, Bagstad KJ, Ferrari M, Geneletti D, Klug H, Schirpke U, Tappeiner U (2014) On the effects of scale for ecosystem services mapping. PloS One 9: e112601.

Griffith GP, Fulton EA, Gorton R, Richardson AJ (2012) Predicting interactions among fishing, ocean warming, and ocean acidification in a marine system with whole-ecosystem models. Conservation Biology 26(6): 1145–1152.

Grimm V (1999) Ten years of individual-based modelling in ecology: what have we learned and what could we learn in the future? Ecological Modelling 115: 129–148.

Grimm V, Berger U, Bastiansen F, Eliassen S, Ginot V, Giske J, Goss-Custard J, Grand T, Heinz SF, Huse G, Huth A, Jepsen JU, Jørgensen C, Mooij WM, Muller B, Pe'er G, Piou C, Railsback SF, Robbins AM, Robbins MM, Rossmanith E, Ruger N, Strand E, Souissi S, Stillman RA, Vabø R, Visser U, DeAngelis DL (2006) A standard protocol for describing individual-based and agent-based models. Ecological Modelling 198: 115–126.

Grimm V, Revilla E, Berger U, Jeltsch F, Mooij WM, Railsback SF, Thulke H, Weiner J, Wiegand T, DL DeAngelis (2005) Pattern-oriented modelling of agent-based complex systems: lessons from ecology. Science 310: 987–991.

Guerry AD, Ruckelshaus MH, Arkema KK, Bernhardt JR, Guannel G, Kim C, Marsik M, Papenfus M, Toft JE, Verutes G, Wood SA, Beck M, Chan F, Chan KMA, Gelfenbaum G, Gold BD, Halpern BS, Labiosa WB, Lester SE, Levin PS, McField M, Pinsky ML, Plummer M, Polasky S, Ruggiero P, Sutherland DA, Tallis H, Day A, Spencer J (2012)

Modelling benefits from nature: using ecosystem services to inform coastal and marine spatial planning. International Journal of Biodiversity Science, Ecosystem Services, and Management 8: 107–121.

Haines-Young R, Potschin M (2011) Common International Classification of Ecosystem Services (CICES) 2011 Update. Paper prepared for discussion at the expert meeting on ecosystem accounts organized by the UNSD, the EEA and the World Bank, London, December 2011. EEA/BSS/07/007. European Environment Agency, Nottingham.

Halpern BS, Walbridge S, Selkoe KA, Kappel CV, Micheli F, D'Agrosa C, Bruno JF, Casey KS, Ebert C, Fox HE, Fujita R, Heinemann D, Lenihan HS, Madin EMP, Perry MT, Selig ER, Spalding M, Steneck R, Watson R (2008) A global map of human impact on marine ecosystems. Science 319: 948–952.

Hanna S (2001) Managing the human-ecological interface: marine resources as example and laboratory. Ecosystems 4: 736–741.

Herrmann SM, Sall I, Sy O (2014) People and pixels in the Sahel: A study linking coarse-resolution remote sensing observations to land users' perceptions of their changing environment in Senegal. Ecology and Society 19: 29.

Hilborn R (1985) Fleet dynamics and individual variation: why some people catch more fish than others. Canadian Journal of Fisheries and Aquatic Sciences 42: 2–13.

Hilborn R, Walters C (1992) Quantitative Fisheries Stock Assessment. Choice, Dynamics and Uncertainty. Chapman & Hall, New York, NY.

IPCC (2014) Climate Change 2014 Synthesis Report. Contribution of Working Groups I, II and III to the Fifth Assessment Report of the Intergovernmental Panel on Climate Change (Core Writing Team, R.K. Pachauri and L.A. Meyer (eds)). IPCC, Geneva.

Kroetz K, Sanchirico J (2015) The bioeconomics of spatial-dynamical systems in natural resource management. Annual Review of Resource Economics 7: 16.1–16.19.

Lane DE (1988) Investment decision-making by fishermen. Canadian Journal of Fisheries and Aquatic Sciences 45: 782–796.

Law AM (2007) Simulation Modelling and Analysis. McGraw-Hill, New York, NY.

Leh M, Bajwa S, Chaubey I (2013) Impact of land use change on erosion risk: an integrated remote sensing, geographic information system, and modelling methodology. Land Degradation and Development 24: 409–421.

Leitão AB, Ahern J (2002) Applying landscape ecological concepts and metrics in sustainable landscape planning. Landscape and Urban Planning 59(2): 65–93.

Levine AS, Feinholz CL (2015) Participatory GIS to inform coral reef ecosystem management: mapping human coastal and ocean uses in Hawaii. Applied Geography 59: 60–69.

Lew DK, Larson DM (2008) Valuing a beach day with a repeated nested logit model of participation, site choice and stochastic time value. Marine Resource Economics 23: 233–252.

Link JS, Browman HI (2014) Integrating what? Levels of marine ecosystem-based assessment and management. ICES Journal of Marine Science 71: 1170–1173.

Little LR, Punt AE, Mapstone BD, Begg GA, Goldman B, Williams AJ (2009) An agent-based model for simulating trading of multi-species fisheries quota. Ecological Modelling 220: 3404–3412.

Liu J, Dietz T, Carpenter SR, Alberti M, Folke C, Moran E, Pell AN, Deadman P, Kratz T, Lubchenco J, Ostrom E (2007) Complexity of coupled human and natural systems. Science 317(5844): 1513–1516.

Liu Y, Liu X, Gao S, Gong L, Kang C, Zhi Y, Chi G, Shi L (2015) Social sensing: a new approach to understanding our socioeconomic environments. Annals of the Association of American Geographers 105: 512–530.

Loerzel J, Goedeke TL, Dillard MK, Brown G (2017) SCUBA divers above the waterline: Using participatory mapping of coral reefs conditions to inform reef management. Marine Policy 76: 79–89.

Loomis JB (2014) Strategies for overcoming hypothetical bias in stated preference surveys. Journal of Agriculture and Resource Economics 39: 34–46.

Luck GW (2006) The relationships between net primary productivity, human population density and species conservation. Journal of Biogeography 34: 201–212.

Martin KS, Hall-Arber, M (2008) The missing layer: geo-technologies, communities, and implications for marine spatial planning. Marine Policy 32: 779–786.

Martínez-Harms M, Balvanera P (2012) Methods for mapping ecosystem service supply: a review. International Journal of Biodiversity Science, Ecosystem Services, and Management 8: 17–25.

McCay BJ (1978) Systems ecology, people ecology, and the anthropology of fishing communities. Human Ecology 6(4): 397–422.

McCauley DJ, Woods P, Sullivan B, Bergman B, Jablonicky C, Roan A, Hirshfield M, Boerder K, Worm B (2016) Ending hide and seek at sea: new technologies could revolutionize ocean observation. Science 351: 1148–1150.

McClintock W (2013) GeoDesign: optimizing stakeholder-driven marine spatial planning. Proceedings of the Marine Safety and Security Council, The Coast Guard Journal of Safety and Security at Sea, Fall 2013: 63–67.

McClintock W, Paul E, Burt C, Bryan T (2016) McClintock Lab: SeaSketch. Designing for Our Oceans, http://www.seasketch.org (accessed 15 June 2017).

McCullagh P (2002) What is a statistical model? The Annals of Statistics 30: 1225–1310.

Mejía CV, Brandt S (2015) Managing tourism in the Galapagos Islands through price incentives: a choice experiment approach. Ecological Economics 117: 1–11.

Mekonnen AD, Gorsevski PV (2015) A web-based participatory GIS (PGIS) for offshore wind farm suitability within Lake Erie, Ohio. Renewable and Sustainable Energy Reviews 41: 162–177.

Metcalfe K, Collins T, Abernethy KE, Boumba R, Dengui J-C, Miyalou R, Parnell RJ, Plummer KE, Russell DJF, Safou GK, Tilley D, Turner RA, VanLeeuwe H, Witt MJ, Godley BJ (2016) Addressing uncertainty in marine resource management; combining community engagement and tracking technology to characterise human behaviour. Conservation Letters 10.1111/conl.12293.

Methot RD, Wetzel CR (2012) Stock synthesis: a biological and statistical framework for fish stock assessment and fishery management. Fisheries Research 142: 86–99.

Millennium Ecosystem Assessment. (2005) Ecosystems and Human Well-being: Synthesis. Island Press, Washington, DC.

Milner-Gulland EJ (2012) Interactions between human behaviour and ecological systems. Philosophical Transactions of the Royal Society B: Biological Sciences 367: 270–278.

Moberg F, Rönnbäck P (2003) Ecosystem services of the tropical seascape: interactions, substitutions, and restoration. Ocean and Coastal Management 46: 27–46.

Nassauer JI (1995) Culture and changing landscape structure. Landscape Ecology 10: 229–237.

Naveh Z (2000) What is holistic landscape ecology? A conceptual introduction. Landscape and Urban Planning 50: 7–26.

Naveh Z, Lieberman AS (1994) Landscape ecology: theory and application, 2nd edition. Springer, New York, NY.

Nguyen Q, Leung P (2013) Revenue targeting in fisheries. Environment and Development Economics 18: 559–575.

Ohemeng FD, Mukherjee F (2015) Modelling the spatial distribution of the Anopheles mosquito for malaria risk zoning using remote sensing and GIS: a case study in the Zambezi Basin, Zimbabwe. International Journal of Applied Geospatial Research 6: 7–20.

Oreskes N (2003) The role of quantitative models in science. In Canham *et al.* (eds) Models in ecosystem science. Princeton University Press, Princeton, NJ.

Oreszczyn S (2000) A systems approach to the research of people's relationships with English hedgerows. Landscape and Urban Planning 50: 107–117.

Ott W, Staub C (2009) Welfare-significant environmental indicators. A feasibility study on providing a statistical basis for the resources policy. Summary. Environmental Studies No. 0913. Federal Office for the Environment, Bern, Switzerland.

Ouellette JA, Wood W (1998) Habit and intention in everyday life: the multiple processes by which past behaviour predicts future behaviour. Psychological Bulletin 124: 54–74.

Parunak HVD, Savit R, Riolo RL (1998) Agent-based modelling vs. equation-based modelling: a case study and users guide. Proceedings of the First International Workshop on Multi-agent Systems and Agent-based Simulation. Lecture Notes in Computer Science 1534: 10–25.

Patino JE, Duque JC (2013) A review of regional science applications of satellite remote sensing in urban settings. Computers, Environment and Urban Systems 37: 1–17.

Pedersen SA, Fock H, Krause J, Pusch C, Sell AL, Böttcher U, Rogers SI, Sköld M, Skov H, Podolska M, Piet GJ (2009) Natura 2000 sites and fisheries in German offshore waters. ICES Journal of Marine Science 66: 155–169.

Perez P, Dray A, Cleland D, Arias-González JE (2009) An agent-based model to address coastal management issues in the Yucatan Peninsula, Mexico. Proceedings of the World IMACS/MODSIM Congress 18: 72–79.

Petter M, Mooney S, Maynard SM, Davidson A, Cox M, Horosak I (2012) A methodology to map ecosystem functions to support ecosystem services assessments. Ecology and Society 18(1): 31.

Pickett STA, Cadenasso ML, Childers DL, McDonnell MJ, Zhou W (2016) Evolution and future of urban ecological science: ecology in, of, and for the city. Ecosystem Health and Sustainability 2(7): e01229.

Pittman SJ, Kneib R, Simenstad C, Nagelkerken I (2011) Seascape ecology: application of landscape ecology to the marine environment. Marine Ecology Progress Series 427: 187–190.

Podsakoff PM, MacKenzie SB, Lee J, Podsakoff NP (2003) Common method biases in behavioural research: A critical review of the literature and recommended remedies. Journal of Applied Psychology 88: 879–903.

Posner SM, McKenzie E, Ricketts TH (2016) Policy impacts of ecosystem services knowledge. Proceedings of the National Academies of Science of the USA 113: 1760–1765.

Purkis S, Klemas V (2011) Remote Sensing and Global Environmental Change. John Wiley & Sons, Ltd., Hoboken, NJ .

Quinn TJ, Deriso RB (1999) Quantitative Fish Dynamics. Oxford University Press, New York, NY.

Radil SM, Junfeng, J (2016) Public participatory GIS and the geography of inclusion. The Professional Geographer 68: 202–210.

Rowlands G, Purkis S, Riegl B, Metsamaa L, Bruckner A, Renaud P (2012) Satellite imaging coral reef resilience at regional scale. A case-study from Saudi Arabia. Marine Pollution Bulletin 64: 1222–1237.

Sainsbury KJ, Punt AE, Smith ADM (2000) Design of operational management strategies for achieving fishery ecosystem objectives. ICES Journal of Marine Science 57: 731–741.

Sale PF (1998) Appropriate spatial scales for studies of reef-fish ecology. Australian Journal of Ecology 23: 202–208.

Samhouri JF, Haupt AJ, Levin PS, Link JS, Shuford R (2013) Lessons learned from developing integrated ecosystem assessments to inform marine ecosystem-based management in the USA. ICES Journal of Marine Science 71(5): 1205–1215.

Sanchirico JN, Wilen JE (1999) Bioeconomics of spatial exploitation in a patchy environment. Journal of Environmental Economics and Management 37(2): 129–150.

Sanchirico JN, Wilen JE (2005) Optimal spatial management of renewable resources: matching policy scope to ecosystem scale. Journal of Environmental Economics and Management 50(1): 23–46.

Saul S (2012) An Individual-based Model to Evaluate the Effect of Fisher Behaviour on Reef Fish Catch per Unit Effort. PhD Dissertation. University of Miami, Miami, Florida, FL.

Saul S, Die D (2016) Modelling the decision making behaviour of fishers in the reef fish fishery on the west coast of Florida. Human Dimensions of Wildlife 21: 567–586.

Saul S, Die D, Brooks EN, Burns K (2012) An individual-based model of ontogenetic migration in reef fish using a biased random walk. Transactions of the American Fisheries Society 141: 1439–1452.

Schaefer MB (1957) Some considerations of population dynamics and economics in relation to the management of the commercial marine fisheries. Journal of the Fisheries Research Board of Canada 14(5): 669–681.

Schneider DC (2001) The rise of the concept of scale in ecology The concept of scale is evolving from verbal expression to quantitative expression. BioScience 51(7): 545–553.

Singh A (1989). Review article: digital change detection techniques using remotely-sensed data. International Journal of Remote Sensing 10: 989–1003.

Smith MD, Sanchirico J, Wilen J (2009) The economics of spatial-dynamic processes. Applications to renewable resources. Journal of Environmental Economics and Management 57(1): 104–121.

Soulie J, Thebaud O (2006) Modelling fleet response in regulated fisheries: an agent-based approach. Mathematical and Computer Modelling 44: 553–564.

Starkey RL (1976) In the Course of Natural Events. Proceedings of the Third International Biodegradation Symposium. Sharpley JM, Kaplan AM (eds), Applied Science Publishers Ltd, London.

Staub C, Ott W, Heusi F, Klingler G, Jenny A, Hacki M, Hauser A (2011) Indicators for Ecosystem Goods and Services: Framework, Methodology and Recommendations for a

Welfare-related Environmental Reporting. Environmental Studies No. 1102, Federal Office for the Environment, Bern, Switzerland.

Stelzenmüller V, Maynou F, Bernard G, Cadiou G, Camilleri M, Crec'hriou R, Criquet G, Dimech M, Esparza O, Higgins R, Lenfant P (2008) Spatial assessment of fishing effort around European marine reserves: implications for successful fisheries management. Marine Pollution Bulletin 56(12): 2018–2026.

Taubenböck H, Wurm M, Netzband M, Hendrik Z, Roth A, Rahman A, Dech S (2011) Flood risks in urbanized areas – multi-sensoral approaches using remotely sensed data for risk assessment. Natural Hazards and Earth System Sciences 11: 431–444.

Taubenböck H, Wurm M, Setiadi N, Gebert N, Roth A, Strunz G, Birkmann J, Dech S (2009) Integrating remote sensing and social science. Joint Urban Remote Sensing Event 2009: 1–7.

Tewkesbury AP, Comber AJ, Tate NJ, Lamb A, Fisher PF (2015) A critical synthesis of remotely sensed optical image change detection techniques. Remote Sensing of Environment 160: 1–14.

Thinh NA, Huan NC, Thanh NV, Tuyen LT, Ly TTP, Nam NM (2016) Spatial conflict and priority for small-scale fisheries in near-shore seascapes of the central coast Vietnam. Journal of Geography and Regional Planning 9: 28–35.

Tidd AN, Vermard Y, Marchal P, Pinnegar J, Blanchard JL, Milner-Gulland EJ (2015) Fishing for space: fine-scale multi-sector maritime activities influence fisher location choice. Public Library of Science One 10: e0116335.

Train KE (2009) Discrete Choice Methods with Simulation. Cambridge University Press, New York, NY.

Trumbo C, Lueck M, Marlatt H, Peek L (2011) The effect of proximity to hurricanes Katrina and Rita on subsequent hurricane outlook and optimistic bias. Risk Analysis 31: 1907–1918.

Turner BL, Esler KJ, Bridgewater P, Tewksbury J, Sitaws N, Abrahams B, Chapman FS, Chowdhury RR, Christie P, Diaz S, Firth P, Knapp CN, Kramer J, Leemans R, Palmer M, Pietri D, Pittman J, Sarukhán J, Shackleton R, Seidler R, van Wilgen B, Mooney H (2016) Socio-environmental systems (SES) research: what have we learned and how can we use this information in future programs. Current Opinion in Environmental Sustainability 19: 160–168.

Turner MG (2015) Twenty-five years of United States landscape ecology: looking back and forging ahead. In Barrett GW, Barrett TL, Wu J (eds) History of Landscape Ecology in the United States. Springer, New York, NY.

Uchmanski J, V Grimm (1996) Individual-based modelling in ecology: what makes the difference? Trends in Ecology and Evolution 11: 437–441.

Voinov A, Shugart HH (2013) 'Integronsters', integral and integrated modelling. Journal of Environmental Modelling and Software 39: 149–158.

Wagner G, Weitzman ML (2015) Climate Shock: The Economic Consequences of a Hotter Planet. Princeton University Press, Princeton, NJ.

Weijerman M, Fulton EA, Janssen AB, Kuiper JJ, Leemans R, Robson BJ, van de Leemput IA, Mooij WM (2015) How models can support ecosystem-based management of coral reefs. Progress in Oceanography 138: 559–570.

White L (1967) The historical roots of our ecological crisis. Science 10: 1203–1207.

Wilen JE, Smith MD, Lockwood D, Botsford LW (2002) Avoiding surprises: incorporating fisherman behaviour in to management models. Bulletin of Marine Science 70(2): 553–575.

Willis KS (2015) Remote sensing change detection for ecological monitoring in United States protected areas. Biological Conservation 182: 233–242.

Wilson WG (1998) Resolving discrepancies between deterministic population models and individual-based simulations. American Naturalist 151: 116–134.

Wilson J (2006) Matching social and ecological systems in complex ocean fisheries. Ecology and Society 11(1).

Witt MJ, Godley BJ (2007) A step towards seascape scale conservation: using vessel monitoring systems (VMS) to map fishing activity. Public Library of Science One 2(10): e1111.

Wu J (2004) Effects of changing scale on landscape pattern analysis: scaling relations. Landscape Ecology 19(2): 125–138.

Wu J (2008) Making the case for landscape ecology an effective approach to urban sustainability. Landscape Journal 27(1): 41–50.

Wu J (2013) Key concepts and research topics in landscape ecology revisited: 30 years after the Allerton Park workshop. Landscape Ecology 28: 1–11.

Yu R, Pan M, Railsback SF, Leung P (2009) A prototype agent based fishery management model of Hawaii's longline fishery. 18th World IMACS/MODSIM Congress, Cairns, Australia.

14

Applying Landscape Ecology for the Design and Evaluation of Marine Protected Area Networks

Mary A. Young, Lisa M. Wedding and Mark H. Carr

14.1   Introduction

The past three decades have seen a marked increase in the application of marine protected areas (MPAs) as ecosystem-based tools for both conservation and fisheries management (Gaines, Lester *et al.* 2010; Lubchenco & Grorud-Colvert 2015). Marine protected areas are often designed with the goal of reducing impacts of fishing pressure on targeted species, habitats and ecosystems (Gaines *et al.* 2003; Lester *et al.* 2009; Halpern *et al.* 2010), but MPAs also attempt to mitigate the cumulative effects from other anthropogenic impacts (*e.g.,* land-based sources of pollution, invasive species, shoreline armouring, effects of climate change) by preserving more resilient communities (Hughes *et al.* 2003). Moreover, MPAs are now distributed in networks such that larval production generated within MPAs can replenish populations across the network to achieve conservation and fisheries management goals (Hastings & Botsford 2003; Botsford *et al.* 2009; Gaines *et al.* 2010; Berglund *et al.* 2012). MPAs replicated across networks provide an additional safeguard against uncertainty over single, isolated MPAs (Allison *et al.* 2003; Gaines *et al.* 2010) and the efficacy of MPA networks can be greater than the sum of the effects of individual MPAs (Grorud-Colvert *et al.* 2014).

The science of MPA network design uses a multitude of criteria including: incorporation of biogeographic regions; habitat and ecosystem representation and diversity; human threats; natural catastrophes; size and number of MPAs; distributions of species; shape of the protected area; boundary : area ratios; location; depth range; regulated activities allowed to occur within their boundaries; and the spacing between MPAs accounting for ocean currents and the movement of organisms (*i.e.,* connectedness; Agardy 2000; Claudet *et al.* 2010; Saarman *et al.* 2013; Botsford *et al.* 2014; White *et al.* 2014). Theoretical guidelines for designing protected areas such as the equilibrium theory of island biogeography (MacArthur & Wilson 1967) and species-area relationships have been used in the design of terrestrial reserves, but there is some debate on the transferability of these theories to marine systems (McNeill & Fairweather 1993). One of the most contested arguments derived from these theories is the SLOSS (single large or several small) debate, which focuses on whether a single large reserve or several small reserves will contain more species (McNeill & Fairweather 1993). Unlike terrestrial islands, evidence in marine systems has shown that sometimes relatively small habitat patches contain more species than larger patches, including: studies on

Seascape Ecology, First Edition. Edited by Simon J. Pittman.
© 2018 John Wiley & Sons Ltd. Published 2018 by John Wiley & Sons Ltd.

corals (Scheltema 1986) and mussel beds (Paine & Levin 1981; Sousa 1984) with mixed results in other systems such as seagrass beds (McNeill & Fairweather 1993). Variable species-area relationships, dispersal patterns and species habitat requirements result in varying requirements for size, spacing and habitat representation in the design of MPAs (Roberts *et al.* 2003; Palumbi 2004).

Evaluating the effectiveness of MPA design at achieving intended management goals is imperative due to the complexity in design and the incipient state of MPA networks around the world (Sale *et al.* 2005; Carr *et al.* 2011; Grorud-Colvert *et al.* 2011). Marine protected area evaluation can inform adaptive management of MPA network design and help avoid unanticipated detrimental consequences of networks (*e.g.*, concentrating fishing mortality on source populations or vulnerable ecosystems when fishing effort is relocated due to no-take MPAs; Hilborn *et al.* 2004; Agardy *et al.* 2011). Moreover, information collected for MPA evaluation can simultaneously be applied to other management goals including fisheries management, documenting ecological consequences of climate change and identifying unanticipated environmental and anthropogenic perturbations (Schroeter *et al.* 2001; Ling *et al.* 2009; Babcock & MacCall 2011; Carr *et al.* 2011; McGilliard *et al.* 2011).

14.2   Applying Landscape Ecology Principles in the Marine Environment

By combining concepts from geography and ecology, landscape ecologists have generated a growing number of spatial pattern metrics to characterize the configuration and heterogeneity of environmental features across multiple spatial scales (Table 14.1). These metrics are then used to determine the effect of pattern (arrangement and composition of patches that compose a landscape) on the processes that drive the distribution and abundance of organisms (Turner *et al.* 2001). The use of spatial pattern metrics has enabled landscape ecologists to understand how habitat configuration and heterogeneity can influence relationships between species, communities and the environment (Turner *et al.* 2001; Turner 2005; Wiens 2009; Wu 2013). A protected area often encompasses heterogeneous areas of habitat and, as a result, a consideration of landscape structure, function and connectivity are important for effective conservation (Hansson & Angelstam 1991; Margules & Pressey 2000).

Landscape ecology often examines spatial scales that are much broader than traditional ecology (Turner *et al.* 2001) and this broad-scale approach is critical to consider when evaluating MPA networks across a broad geographic scale. Proof of the importance of landscape ecology in conservation is in its successful application to the design and evaluation of terrestrial protected areas and by the many terrestrial management organizations that recognize the need for a landscape perspective in order to achieve sound resource management (Leitão & Ahern 2002). The understanding that landscape pattern influences ecological processes (McGarigal & Marks 1995) has helped with land-use planning including preserving landscape diversity, reducing fragmentation and halting the spread of disturbance (Leitão & Ahern 2002). Additionally, there are many examples of the application of landscape ecology in terrestrial conservation to designing and / or evaluating terrestrial protected areas (*e.g.*, Armenteras *et al.* 2003;

Table 14.1 Landscape metrics used to conduct landscape ecology in terrestrial and marine studies. The first column provides the landscape metric broken into sub categories of landscape composition (2D), landscape composition (3D), spatial configuration (patch based) and spatial configuration (contagion). The second column provides a description of each metric followed by several publications that used each metric in terrestrial (third column) and marine (fourth column) studies.

Landscape metric	Description	Terrestrial examples	Marine examples
Landscape composition (2D)			
Habitat area	Measure of habitat coverage (*i.e.*, area, proportion, *etc.*)	Murphy & Jenkins 2010; Flick *et al.* 2012; Hurley *et al.* 2012; Vasques *et al.* 2012; Abdel Moniem & Holland 2013	Bell *et al.* 2008; Appeldoorn *et al.* 2011; Chatfield *et al.* 2010; Huntington *et al.* 2010; Kendall & Miller 2010; Knudby *et al.* 2010a; Knudby *et al.* 2010b; Kendall *et al.* 2011; Zajac *et al.* 2013
Habitat diversity	Calculated using Shannon's Diversity Index (Equals minus the sum of the proportional abundance of each habitat type multiplied by the ln of that proportion)	Honnay *et al.* 1999; Ricklefs & Lovette 1999; Luoto *et al.* 2002	Grober-Dunsmore *et al.* 2008; Kendall *et al.* 2011; Meager *et al.* 2011
Patch richness	Number of patch types	McGarigal *et al.* 2009; Flick *et al.* 2012; Schindler *et al.* 2013	Teixidó *et al.* 2007; Grober-Dunsmore *et al.* 2008; Moore *et al.* 2011
Habitat richness	Total number of habitat types	Lucherini & Lovari 1996; Tremblay *et al.* 2003; Lovari *et al.* 2013	Prada *et al.* 2008
Habitat evenness	A measure of how close the numbers of each habitat type are to the total number of habitats.		Teixidó *et al.* 2007; Prada *et al.* 2008
Elevation or depth	The elevation of the landscape or the depth of the seascape	McGarigal *et al.* 2009; Murphy & Jenkins 2010; Vasques *et al.* 2012; Abdel Moniem & Holland 2013	Chatfield *et al.* 2010; Kendall & Miller 2010; Knudby *et al.* 2010a; Knudby *et al.* 2010b; Monk *et al.* 2010; Giménez-Casalduero *et al.* 2011; Moore *et al.* 2011; Pittman & Brown 2011; Zajac *et al.* 2013
Patch diversity	Calculated using Shannon's Diversity Index (Equals minus the sum of the proportional abundance of each patch multiplied by the ln of that proportion)	Rossi & van Halder 2010; Flick *et al.* 2012; Schindler *et al.* 2013	Teixidó *et al.* 2007; Grober-Dunsmore *et al.* 2008; Moore *et al.* 2011
Patch density	Number of patches per area	Rossi & van Halder 2010; Flick *et al.* 2012; Schindler *et al.* 2013; Uuemaa *et al.* 2013	Meynecke *et al.* 2008; Mizerek *et al.* 2011

(Continued)

Table 14.1 (Continued)

Landscape metric	Description	Terrestrial examples	Marine examples
Landscape composition (3D)			
Slope	Measure of the slope of the terrain in either degrees or radians	Murphy & Jenkins 2010; Heinänen et al. 2012; Vasques et al. 2012; Abdel Moniem & Holland 2013	Monk et al. 2010; Giménez-Casalduero et al. 2011; Pittman & Brown 2011
Measure of habitat complexity (e.g., rugosity, slope of slope, variation in elevation, topographic roughness etc.)	Roughness of the physical structure of the seafloor	Aragón & Morales 2003; Williams et al. 2009; Murphy & Jenkins 2010	Pittman et al. 2007b; Rattray et al. 2009; Knudby et al. 2010b; Kendall et al. 2011; Meager et al. 2011b; Pittman & Brown 2011
Aspect	Measure of the direction the slope is facing	Aragón & Morales 2003; Carroll & Miquelle 2006; Williams et al. 2009	MacLeod et al. 2008; Monk et al. 2010, 2011; Ierodiaconou et al. 2011
Topographic or bathymetric position index	Relative elevation of a feature compared to its surrounding landscape	Carroll & Miquelle 2006; McGarigal et al. 2009; Murphy & Jenkins 2010	Rattray et al. 2009; Monk et al. 2010, 2011; Ierodiaconou et al. 2011
Maximum curvature	Curvature is a measure of the greatest curve relative to its neighbours of either the profile or the plan	Aragón & Morales 2003; Heinänen et al. 2012; Hurley et al. 2012; Abdel Moniem & Holland 2013	Monk et al. 2010, 2011; Ierodiaconou et al. 2011; Pittman & Brown 2011
Spatial configuration (patch based)			
Number of patches	The total number of patches	Lasanta et al. 2006; Goetz et al. 2009; Lechner et al. 2012	Anderson et al. 2009; Giménez-Casalduero et al. 2011; Moore et al. 2011
Largest patch index	Area of the largest patch in a landscape	Armenteras et al. 2003; Lasanta et al. 2006; McGarigal et al. 2009; Rossi & van Halder 2010; Schindler et al. 2013	Pittman et al. 2004; Yang & Liu 2005
Nearest neighbour distance	Distance between patch and the patch closest to it	Armenteras et al. 2003; Lechner et al. 2012	Meynecke et al. 2008; Huntington et al. 2010; Moore et al. 2011
Patch area	Size of the patches	Narumalani et al. 2004; Crist et al. 2005; Minor & Urban 2007; Lechner et al. 2012; Crouzeilles et al. 2013; Uuemaa et al. 2013	Prada et al. 2008; Huntington et al. 2010; Giménez-Casalduero et al. 2011; Claisse et al. 2012; Pinsky et al. 2012; Carroll & Peterson 2013; D'Aloia et al. 2013; Zajac et al. 2013

Landscape metric	Description	Terrestrial examples	Marine examples
Radius of gyration	Mean distance for each cell of one patch to the patch centroid	McGarigal et al. 2009; Schindler et al. 2013	Moore et al. 2011
Patch shape	Measure of patch shape (various metric used)	Sisk et al. 1997	Teixidó et al. 2002; Teixidó et al. 2007; Carroll & Peterson 2013
Landscape shape index	Ratio of the total edge to the minimum total edge	Armenteras et al. 2003; Schindler et al. 2013	Teixidó et al. 2002; Teixidó et al. 2007
Intra-patch connectivity	The number of functional joins between patches of the same type	Uezu et al. 2005; Minor & Urban 2007; Crouzeilles et al. 2013	Kendall et al. 2011; Moore et al. 2011; Treml et al. 2012
Graph structure	A landscape of habitat patches connected to some degree by edges that join pairs of nodes functionally	Minor & Urban 2007; Goetz et al. 2009; Crouzeilles et al. 2013	Treml et al. 2008
Perimeter to area ratio	Measure of patch shape complexity that equals the perimeter per area	Lechner et al. 2012; Crouzeilles et al. 2013; Schindler et al. 2013	Grober-Dunsmore et al. 2008; Meynecke et al. 2008; Huntington et al. 2010
Hydrodynamic Aperture	The sum of the width of the sections along the atoll rim that connect the internal 'wet' areas of the atoll with the open ocean	n/a	Andréfouët et al. 2003
Patch evenness index	Measure of patch diversity that considers the evenness of the patch sizes and not the number of patches	Rossi & van Halder 2010; Schindler et al. 2013	Teixidó et al. 2002; Pittman et al. 2004; Teixidó et al. 2007
Total edges	Total length of edge habitat	Trzcinski et al. 1999; Erickson & West 2002; Lasanta et al. 2006	Teixidó et al. 2002; Giménez-Casalduero et al. 2011; Kendall et al. 2011; Moore et al. 2011; Carroll & Peterson 2013
Edge density	Total length of edge per unit area	Hurley et al. 2012; Lechner et al. 2012; Schindler et al. 2013; Uuemaa et al. 2013	Pittman et al. 2004; Giménez-Casalduero et al. 2011
Fractal dimension	Patch shape complexity measure that approaches 1 for simple shapes and 2 for complex shapes	Narumalani et al. 2004; Schindler et al. 2013	Meager et al. 2011a; Moore et al. 2011; Santos et al. 2011
Shape index	Equals 1 when all patches are circular and increases as patches become more complex	Schindler et al. 2013; Uuemaa et al. 2013	Yang & Liu 2005; Giménez-Casalduero et al. 2011; Moore et al. 2011

(Continued)

Table 14.1 (Continued)

Landscape metric	Description	Terrestrial examples	Marine examples
Distance between patches	Measure of the distance between patches (either edge to edge or centroid to centroid)	Minor & Urban 2007; Bauerfeind et al. 2009	Hovel 2003; Kendall et al. 2003; Mizerek et al. 2011; Pinsky et al. 2012; D'Aloia et al. 2013
Mean nearest neighbour distance	Minimum edge to edge distance between two patches of the same type	McGarigal et al. 2009; Schindler et al. 2013	Pittman et al. 2004
Distance to features	Distance to feature of interest (i.e., coastline, mangrove prop root, shelf edge, habitat patch, etc.)	Zharikov et al. 2006; Heinänen et al. 2012; Hurley et al. 2012	Huntington et al. 2010; Kendall & Miller 2010; Monk et al. 2010; 2011; Pittman & Brown 2011
Current land use	Classification of current land use (i.e., agriculture)	Falcucci et al. 2007; Bauerfeind et al. 2009; Goetz et al. 2009	Oliver et al. 2011
Cohesion	Aggregation of the patches	Rossi & van Halder 2010; Lechner et al. 2012; Mairota et al. 2013; Uuemaa et al. 2013	Prada et al. 2008; Moore et al. 2011
Spatial configuration (contagion)			
Interspersion and juxtaposition index	Measure of evenness of patch adjacencies, equals 100 for even and approaches 0 for uneven adjacencies	Lasanta et al. 2006; Hurley et al. 2012; Schindler et al. 2013	Teixidó et al. 2002; Teixidó et al. 2007; Moore et al. 2011
Contagion	Measure of the aggregation of habitat coverage classes	McGarigal et al. 2009; Schindler et al. 2013; Uuemaa et al. 2013	Robbins & Bell 2000; Pittman et al. 2004; Moore et al. 2011

Rouget *et al.* 2003; Rayfield *et al.* 2008; Goetz *et al.* 2009; Svancara *et al.* 2009; Townsend *et al.* 2009; Mairota *et al.* 2013; Table 14.2).

Despite the fact that there is substantial overlap between the goals of landscape ecologists and marine conservation biologists, these two disciplines have not united, but future unification could advance conservation efforts (Olds *et al.* 2016). Landscape ecology techniques allow for the incorporation of important dynamic broad-scale processes such as the occurrence of natural disturbances (Pickett & White 1985; Baker 1992; Turner 2005) and the potential for ecological resilience (Nyström & Folke 2001; Cumming 2011), resulting in a strengthened ecological basis for marine spatial planning. However, the procedures have yet to be developed and integrated into the planning process (Ndubisi 2002; Termorshuizen *et al.* 2007, but see Foley *et al.* 2010). In order to more effectively evaluate and design MPAs, a landscape ecology approach is essential to characterize ecosystem patterns and processes at relevant spatial scales. Marine ecologists have applied principles of landscape ecology in the marine environment over the past few decades (reviewed in Grober-Dunsmore *et al.* 2009; Boström *et al.* 2011; see Tables 14.1 and 14.2), but these methods have yet to be extensively applied to marine resource management and protection strategies (Böstrom *et al.* 2011; Olds *et al.* 2016). In the last couple of decades, there has been a move towards the application of landscape ecology principles in MPA planning, with the incorporation of aspects of the seascape in programs such as Marxan (Ball & Possingham 2000; Possingham *et al.* 2005). Marxan uses information about the composition and configuration of seascapes to help managers design optimal MPA networks that maximize ecological gains while minimizing costs. These tools can be used to assess the representation, replication and diversity of habitats that support biodiversity, which is often a major consideration when designing networks of MPAs (Roberts *et al.* 2003; Gaines *et al.* 2010; Halpern *et al.* 2010); however, these tools are only as good as the information provided to them. More broad scale information on patterns and resulting processes across seascapes is required for landscape ecology to benefit marine spatial planning. Spatial pattern metrics created by landscape ecologists have great potential for both MPA design and evaluation of existing MPAs (Wedding *et al.* 2008; Huntington *et al.* 2010).

When designing networks of MPAs, connectivity between MPAs (Beger *et al.* 2010), between ecosystems within individual MPAs (Olds *et al.* 2014) and between MPAs and fished ecosystems (Harrison *et al.* 2012) are other major factors to consider (Olds *et al.* 2012a) and some that are greatly benefitted through principles of landscape ecology (Kool *et al.* 2013). Connectivity science has now become an important field for understanding important processes including population dynamics, evolution, dispersal, population genetics, source-sink dynamics, potential impacts of climate change (Kool *et al.* 2013) and spatial ecological resilience (Nyström & Folke 2001; Cumming 2011; Olds *et al.* 2012b). Recent explorations of how connectivity influences protected area performance have found that connectivity can improve fish abundances inside MPA boundaries (Olds *et al.* 2012a, 2012b). Additionally, global scale analyses have found that seascape connectivity inside protected areas can lead to improved conservation outcomes such as increases in productivity and biodiversity (Olds *et al.* 2016). Methods from landscape ecology provide tools for modelling connectivity across seascapes at a variety of scales for application to marine spatial planning.

Table 14.2 Examples of studies using metrics of landscape ecology to evaluate the effectiveness of protected areas in both terrestrial and marine systems. The goal of each of the studies is stated along with the question used to evaluate that goal, the landscape metrics used and the source for the study or studies

Environment	Goal	Evaluation question	Landscape metric	Source
Terrestrial	Habitat representation	Are the habitats within the reserves representative?	Habitat area	Armenteras *et al.* 2003; Rouget *et al.* 2003; Svancara *et al.* 2009; Townsend *et al.* 2009
	Habitat fragmentation	How fragmented are PAs compared to the region?	Number of Patches, Largest Patch Index, Mean Patch Size, Nearest Neighbour Distance, Landscape Shape Index	Armenteras *et al.* 2003; Townsend *et al.* 2009
		Using habitat fragmentation as a proxy for human use, how fragmented are the habitats within and around PAs?	Cohesion, Patch Density, Fractal Dimension, Edge Density, Contagion, Habitat Diversity	Mairota *et al.* 2013
	Habitat connectedness	Do the PAs have adequate connection to core habitat and are these core habitats protected?	Graph Theoretic Approach	Rayfield *et al.* 2008; Goetz *et al.* 2009; Townsend *et al.* 2009
	Landscape structure	How well is the landscape structure captured within the PAs?	Many landscape level and class level metrics used (Table 14.1).	Schindler *et al.* 2008
		How does the landscape structure affect distribution of native herbs and deer foraging around PAs?	Spatial distribution of habitat types	Hurley *et al.* 2012
	Landscape functionality	Is landscape functionality higher in protected areas than in nonprotected areas?	Landscape connectivity and habitat area.	Skokanová & Eremiášová 2013
	Protected area dimensions	Do larger parks with less edge habitat support larger populations than smaller parks?	Size and shape of PA	Hurley *et al.* 2012
	Land use change	How has the vegetation coverage changed within the Effigy Mounds National Monument (EFMO) and what are the anthropogenic impacts causing these changes?	Habitat Area, Number of patches, mean patch size, fractal dimension	Narumalani *et al.* 2004
		How has the habitat coverage and fragmentation changed inside and outside protected areas?	Habitat coverage	Wang & Moskovits 2002

Environment	Goal	Evaluation question	Landscape metric	Source
Marine	Roadless areas adjacent to PAs	Do roadless areas adjacent to PAs increase the coverage of certain habitat types and the connectivity between habitat types?	Habitat coverage, Habitat Patch Size, Distance between patches, Spatial Arrangement of Patches	Crist et al. 2005
	Reserve size and boundary length	How does the reserve size and boundary length affect the density of fish within marine reserves?	Perimeter to Area Ratio	Bartholomew et al. 2008
	Reserve boundary permeability	How does the amount of reserve boundary that intersects suitable reef habitat affect the densities of fish within the reserve?	Ratio between reserve boundary that intersects reef habitat (HI) and reef habitat area within the reserve (HA) (HI/HA)	Bartholomew et al. 2008
	Increase biodiversity and biomass of fishes	Does habitat have a greater effect on the spatial distribution of species or protection status?	Substrate type, bottom complexity, depth, protection status	Claudet et al. 2010
		Is species richness, biomass and diversity higher in protected areas when variations in habitat type are taken into consideration?	Habitat coverage, protected status	Friedlander et al. 2007
		What is the effect of a reserve on the biodiversity and biomass of fishes when the seascape variability is taken into account?	Multiple metrics for configuration, composition and patch structure of the seascape (Appendix 1 in Huntington et al. 2010)	Huntington et al. 2010
	Spillover from MPAs	Using fish–habitat associations as a controlling factor, are spillover effects closer to the MPA boundaries and decrease as you move further from the boundary?	Rugosity, benthic habitat cover, depth, slope of slope (complexity), protection status	Stamoulis & Friedlander 2013
	Habitat representation	How well are the habitat types represented within a marine protected area compared to the habitat types in the region?	Habitat area	Stevens & Connolly 2005
	Connectivity	How does seascape connectivity affect marine reserve importance?	Physical connectedness of patches in the seascape	Olds et al. 2014

In this chapter, we focus primarily on the application of landscape ecology to the evaluation of MPA network designs and how well these designs meet criteria thought to make MPAs successful. However, as we discuss below, landscape metrics also play a direct role in assessing whether MPAs are meeting conservation and management goals by informing our estimates of species distributions and the structure and function of communities and ecosystems. The effectiveness of networks of protected areas depends on the association of species with habitat patches and the overall seascape inside and outside of protected areas. The conceptual and operational framework of landscape ecology allows marine researchers to understand the consequences of environmental heterogeneity on MPAs (García-Charton & Pérez-Ruzafa 1999). In this chapter, we propose a methodological framework for MPA network evaluation based on a landscape ecology approach that informs both the criteria used to design these networks and the assessment of their effectiveness in achieving goals of conservation and restoration. We do this by highlighting a case study where we have applied our methodological framework to the Central Coast MPA Network in California, which demonstrates the utility of this approach. In this case study we: (i) consider the goals identified in the establishment of a network of MPAs and the design criteria implemented to achieve these goals; (ii) articulate hypotheses to test the effectiveness of the design process in achieving these criteria; (iii) identify the relevant landscape metrics to test the hypotheses; (iv) extract those metrics from seafloor maps (or other sources of landscape data); and (v) conduct analyses to apply landscape metrics for testing hypotheses of reserve and network effectiveness. To distil lessons learned and advance the application of a landscape ecology approach in marine conservation we reflect on the challenges and opportunities from our case study application to the network of MPAs established off the coast of California and discuss next steps in the field of seascape ecology.

14.3   Case Study: Applying Landscape Ecology to Evaluate a Network of MPAs in California

In 1999, the state of California passed the Marine Life Protection Act (MLPA), which mandated a redesign of California's current MPAs and the designation of a full network of MPAs along the entire coast of California within state waters (Carr *et al.* 2010; Gleason *et al.* 2013; Kirlin *et al.* 2013; Saarman *et al.* 2013). The MLPA Initiative and the planning process divided the coast of California into four regions (North Coast, North Central Coast, Central Coast and South Coast) and between 2004 and 2011 conducted separate planning processes for each region to designate a complete statewide network of MPAs (Gleason *et al.* 2013; Kirlin *et al.* 2013). A science advisory team was established to generate the science-based design criteria for both the individual MPAs and the network. California is the first state in the United States to develop a scientifically based network of MPAs within state waters (Kirlin *et al.* 2013) and it now hosts one of the largest, scientifically based networks of MPAs in the world. A total of 124 MPAs, covering approximately 2200 km^2 now make up the network of MPAs in California, which were created based on a specific set of scientific, ecological and socioeconomic goals that influenced the MPA network design and the future evaluation of their effectiveness (Box14.1; California Marine Life Protection Act 1999; Gleason *et al.* 2013).

Box 14.1

The six goals, paraphrased here, for which individual MPA and network design criteria were developed through the MLPA initiative. Goals 1, 2, 4 and 6 are ecologically based goals. The goals in bolded text are those goals addressed in the case studies presented.

Goals outlined in California's Marine Life Protection Act 1999

1) **To protect the natural diversity and abundance of marine life and the structure, function and integrity of marine ecosystems.**
2) **To help sustain, conserve and protect marine life populations, including those of economic value and rebuild those that are depleted.**
3) To improve recreational, educational and study opportunities provided by marine ecosystems that are subject to minimal human disturbance and to manage these uses in a manner consistent with protecting biodiversity.
4) **To protect marine natural heritage, including protection of representative and unique marine life habitats in California waters for their intrinsic value.**
5) To ensure that California's MPAs have clearly defined objectives, effective management measures and adequate enforcement and are based on sound scientific guidelines.
6) To ensure that the state's MPAs are designed and managed, as far as possible, as a network.

We focused our MPA network evaluation on the Central Coast region ranging from Pigeon Point in the north (37.1817°N, 122.3939°W) to Point Conception in the south (34.4481°N, 120.4714°W) and extending 3 NM from the coastline, covering approximately 3000 km^2 (Figure 14.1). The Central Coast region was the first to be designated on 21 September 2007 and contains a total of 28 new or modified MPAs, which cover about 18% of the region (https://www.wildlife.ca.gov/Conservation/Marine/MPAs, accessed 25 May 2017). We evaluated the design of the Central Coast network using the design criteria that were developed for that region (see Appendix R of the California Marine Life Protection Act: Master Plan for Marine Protected Areas 2015) and specifically addressed the overall goals of the MLPA using the following hypotheses: (H1) habitats within the region are adequately replicated across the Central Coast MPA network (MLPA Goal 4; Box 14.1), (H2) the Central Coast MPA network captures the diversity of habitats within the region (MLPA Goal 1; Box 14.1) and (H3) MPAs in the Central Coast region are configured to reduce movement of organisms from within their boundaries to fished waters (spillover) (MLPA Goal 2; Box 14.1).

14.3.1 California Seafloor Data Sets

To collect data on the seafloor features within the state waters, the state of California designed and implemented the California State Mapping Program, which was a collaborative effort among state and federal organizations, academia and industry to map seafloor features across the entire state waters of California. The main purpose of this program was to create a high-resolution geologic basemap of the California State waters (Kvitek & Iampietro 2010). Seafloor mapping along the California mainland was completed in 2012 and these data are now publicly available (http://seafloor.otterlabs.org/csmp/csmp.html, accessed 25 May 2017).

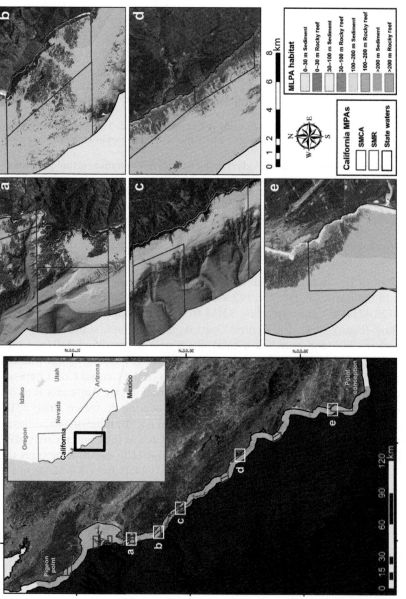

Figure 14.1 The image on the left shows the California Central Coast Marine Life Protection Act (MLPA) region from Pigeon Point to Point Conception with the marine protected areas (MPAs) outlined in blue (SMCA: State Marine Conservation Area) and red (SMR: State Marine Reserve) and the California State waters boundary outlined in black. Boxes a–e provide zoomed in views of the MLPA habitat along the central coast with the different colours representing each habitat specified in the MLPA based on substrate type and depth overlaid on shaded relief imagery of the seafloor. *Source:* Copyright 2013 Esri, DeLorme, NAVTEQ.

14.3.2   MPA Goal: Habitat Replication and Representativeness

Representation of the diversity of habitats and their associated ecosystems and biodiversity has been a primary goal for spatial conservation planning (Margules & Pressey 2000) and is a major criterion for the design of MPAs (Agardy 1995; Stevens & Connolly 2005). By protecting a variety of habitats, it is more likely that the network of MPAs will protect a greater diversity of species and communities (Saarman *et al.* 2013). Previously, landscape approaches used to assess the representation of habitat across the Central Coast MPA network revealed that the MLPA-defined habitats are proportionately represented across the network (Young & Carr 2015a). In addition to habitat representation, the science advisory team also developed criteria for replication of habitats across the region. For replication, they determined a minimum percentage area of habitat within an MPA to constitute a replicate. The habitat maps allow us to classify the seafloor into the MLPA habitat classes and evaluate representation and replication of habitat types across the network relative to their availability (proportionate area) throughout the entire region.

Landscape ecology metrics have proven useful when evaluating the distribution of habitats in both terrestrial and marine protected areas (Table 14.2). The most common type of metrics used to map distribution of habitats are for quantifying landscape composition. Landscape composition is a type of landscape metric used to provide information on the abundance and variety of habitat (or patch types) within a landscape (Dunning *et al.* 1992; Wedding *et al.* 2011). The most frequently used two-dimensional measures of landscape composition include patch area, diversity, richness and evenness of habitat types (Wedding *et al.* 2011; Table 14.1). Habitat area and elevation, or depth classified into zones, are the most commonly used two-dimensional metrics applied in seascape ecology (Table 14.1) and they have been found to have significant correlations with faunal community metrics such as fish diversity and abundance (*e.g.*, depth class and area metrics: Stoner *et al.* 2001; Chatfield *et al.* 2010; area metrics: Pittman *et al.* 2004; Bell *et al.* 2008; Grober-Dunsmore *et al.* 2008; Kendall & Miller 2010).

To address the first hypothesis in this case study regarding *adequate replication of habitats throughout the network*, we used those habitats defined by the MLPA. Benthic habitats outlined in the California Marine Life Protection Act: Master Plan for Marine Protected Areas 2015 are associated with eight combinations of two substrate types (hard rock versus soft sediment) and four depth zones (0–30 m, 30–100 m, 100–200 m, >200 m) important to many different species. To be included as a replicate for a specified habitat class (*i.e.*, the MLPA habitat classes outlined above), the habitat class had to represent 'more than a critical fraction of the entire MPA' and these critical fractions varied by habitat: 20% for 0–30 m sediment, 15% for 30–100 m sediment and 0–30 m rocky reef and 10% for >100 m sediment, >100 m rocky reef and 30–100 m rocky reef. In addition, the California Marine Life Protection Act: Master Plan for Marine Protected Areas 2015 specifies that there needs to be a minimum of three, ideally five, replicate MPAs for each habitat class. The percentage of each of these habitats within the MPAs was calculated from the habitat areas derived for the mosaicked MLPA habitat maps. These replicates were then compared to the entire Central Coast by creating 'potential MPAs' along the coastline. These potential MPAs were created in ArcGIS as ~7 km (average coastal length of Central Coast MPAs) by ~5.5 km (average distance from shore to the California state waters limit) extending from the northern (Pigeon Point) to southern

(Point Conception) boundary to represent potential MPAs across the Central Coast region. These potential MPAs were used for the inside / outside comparison because they divide the coast into areas of comparable size to the actual MPAs. Therefore, we can compare habitat composition of the actual MPAs to similarly sized areas along the coast. The percentage of each of the habitat types were calculated within these 60 potential MPAs as well. These percentages were then compared to the replicate guidelines to define if MPA or potential MPA served as a replicate for those habitat classes. The replication analysis showed that six of the eight habitat classes were adequately replicated within the MPAs as specified in the MLPA guidelines (Table 14.3). The two habitat classes that are not replicated (100–200 m rocky reef and >200 m rocky reef) are present in negligible amounts throughout the MLPA region and are essentially not available to replicate within this region. The 0–30 m rocky reef habitat class was observed more within the network of MPAs than outside with over 50% of the MPAs containing greater than the amount of habitat necessary to serve as a replicate, whereas only 4.2% of the potential MPAs within the region contain replicates of this habitat.

14.3.3 MPA Goal: Protect Diversity and Abundance of Marine Life

The majority of protected areas throughout the world are established to preserve the biodiversity of the region under protection (Agardy 2000). The first goal of the MLPA is directly related to preserving marine biodiversity off the California coast by stating that the MPAs should 'protect the natural diversity and abundance of marine life, and the structure, function, and integrity of marine ecosystems' (Box 14.1). Now that the perception of the oceans as a homogenous environment is disappearing, people are realizing that more complex structural patterns support the high biodiversity within the oceans (Agardy 1995). In order to protect this biodiversity, managers need to consider heterogeneity when establishing networks of MPAs (Hockey & Branch 1997; Roberts *et al.* 2003). One measure of the heterogeneity of habitats is habitat diversity. Increases in habitat diversity are believed to correlate with increases in species richness because of the variety of different habitats that different species associate with (Hart & Horwitz 1991; Barbault 1995; Ricklefs & Lovette 1999). In terrestrial studies, habitat diversity has been shown to be an important attribute affecting the distributions of many species including species presence (Honnay *et al.* 1999) and species richness (Ricklefs & Lovette 1999; Douglas & Lake 1994). These results, however, are not limited to the terrestrial realm. For example, Purkis *et al.* (2007) has demonstrated that there is a positive correlation between species diversity and habitat diversity in coral reefs. In contrast, some studies have found no significant relationship with habitat diversity, suggesting that seascape type and spatial scale may have important influences on the results (Pittman *et al.* 2007; Kendall & Miller 2010).

To test the second hypothesis that *MPAs within the Central Coast network contain a representative diversity of habitats*, habitat diversity, a metric of landscape composition, was calculated for each of the MPAs and compared to habitat diversity throughout the region using the potential MPAs derived for the habitat replication analysis. To calculate the diversity in habitat across the Central Coast region, the seafloor data were divided into slope, rugosity and broad (50 m) and fine scale (20 m) bathymetric position index (BPI) classes, which were calculated using the Benthic Terrain Modeler tool (Wright *et al.* 2012) in ArcGIS Desktop: Release 10. The slope and rugosity classes were based on

Table 14.3 Percentage of the MLPA habitat classes within each of the Central Coast MPAs. Those percentages that meet the criteria to be considered a replicate as specified in the MLPA are in **bold** typeface. The final row in the table contains the percentage of blocks from the 'potential MPAs' in the region that contain replicates of the each of the habitat classes

MPA Name	Sediment 0–30 m (%)	Rocky Reef 0–30 m (%)	Sediment 30–100 m (%)	Rocky Reef 30–100 m (%)	Sediment 100–200 m (%)	Rocky Reef 100–200 m (%)	Sediment >200 m (%)	Rocky Reef >200 m (%)
Año Nuevo	**43.5**	**28.3**	**19.1**	**9.1**	0.0	0.0	0.0	0.0
Asilomar	**37.5**	**53.2**	2.6	6.7	0.0	0.0	0.0	0.0
Big Creek	7.7	1.3	**19.5**	0.3	7.5	0.1	**63.6**	0.0
Cambria	**68.0**	**29.0**	2.9	0.0	0.0	0.0	0.0	0.0
Carmel Bay	**41.0**	**24.3**	**21.7**	**7.6**	3.0	1.3	1.1	0.0
Carmel Pinnacles	3.0	**19.9**	**21.1**	**55.9**	0.1	0.0	0.0	0.0
Edward F. Ricketts	**82.0**	**18.0**	0.0	0.0	0.0	0.0	0.0	0.0
Greyhound Rock	10.4	10.2	**79.0**	0.3	0.0	0.0	0.0	0.0
Lovers Point	**72.9**	**27.1**	0.0	0.0	0.0	0.0	0.0	0.0
P.G. Marine Gardens	**30.2**	**41.6**	**14.6**	**13.6**	0.0	0.0	0.0	0.0
Piedras Blancas	20.7	14.3	**48.6**	**16.4**	0.0	0.0	0.0	0.0
Point Buchon	1.4	4.7	**71.9**	4.5	**17.2**	0.2	0.0	0.0
Point Lobos	1.6	4.9	**16.3**	**12.2**	**28.8**	**2.1**	**34.0**	0.0
Point Sur	13.5	11.5	**62.7**	**11.1**	1.0	0.0	0.1	0.0
Portuguese Ledge	0.0	0.0	**15.2**	1.2	**52.0**	0.7	**31.0**	0.0
Soquel Canyon	0.0	0.0	**64.5**	0.6	**11.6**	0.4	**23.0**	0.0
Vandenberg	**59.3**	5.3	**35.1**	0.3	0.0	0.0	0.0	0.0
White Rock (Cambria)	**32.3**	**43.7**	**19.3**	4.7	0.0	0.0	0.0	0.0
Percentage of Regional Blocks Containing Habitat Replicates	31.9	4.2	94.4	19.4	22.2	0.0	22.2	0.0

the classifications in Greene *et al.* (2007) and the BPI classes were based on those defined in Iampietro *et al.* (2005). Once the rocky habitat was classified into the distinct habitat classes, the area of each of those habitat classes was tabulated for all of the designated MPAs as well as the potential MPAs created for the previous analysis. The areas of each of these habitat classes were then used to calculate the Shannon–Wiener diversity index for each MPA and potential MPA. For the MPAs within the Central Coast region, the diversity indices varied from 4.9 in the Vandenberg MPA to 15.8 in the Point Lobos MPA (Figure 14.2) with a mean diversity index of 9.6. Viewing the seafloor data for these MPAs shows that habitat diversity is visually very different. Vandenberg is a large MPA that is mostly covered in flat sand with a few rocky reef patches while Point Lobos contains rocky reef habitat with varying levels of relief, only a small portion of sediment and spans a wide range of depths. Compared to the habitat diversity seen across the region, habitats within the MPAs are more diverse. Although the range in habitat diversity of the 60 potential MPAs across the entire region is similar to that observed for the actual MPAs (4.0 to 13.3), the mean diversity is only 6.8 because the distribution of habitat diversity throughout the region is strongly skewed to the left (lower diversity; Figure 14.2). As a result, the MPAs are generally protecting higher diversity habitat than is available across the seascape throughout the region.

14.3.4   MPA Goal: Reduce Movement across Boundaries

To address the third hypothesis that *MPAs in the Central Coast region are configured to reduce organism movement out from the MPA into surrounding fished waters (spillover)*, we looked at the configuration of MPAs relative to the surrounding seascape. Landscape approaches were used to determine the ecological permeability of the MPA boundaries. One way in which MPAs can differ in the effect that they have on populations under protection is their placement within the surrounding seascape. In fact, Edgar *et al.* (2014) states that isolation of habitat inside an MPA was one of five main factors that promotes MPA performance. Many species have affinities for specific habitat types and their willingness or ability to leave that habitat patch depends on the surrounding habitat and the behaviour of the organism under study (Stamps *et al.* 1987). Permeability of the edge of a habitat patch can either be 'hard', where species tend to not disperse across the edge or 'soft' where there are no barriers to dispersal (Stamps *et al.* 1987). When setting up protected area boundaries, a new 'edge' to the seascape was introduced where the species on one side of that boundary are protected from exploitation while those on the other side are not. If the MPA boundary coincides with a physical habitat edge, the MPA is more likely to retain species, however, if the MPA boundary intersects (*i.e.*, divides) contiguous habitat, species will spillover to those areas outside the protected area (Chapman & Kramer 2000; Bartholomew *et al.* 2008). Bartholomew *et al.* (2008) found that there was a significant negative relationship between the rate of density change and the ratio between the protected area boundary that intersects reef habitat (HI) and the total area of reef habitat within the protected area (HA). The metric HI/HA provides support for the idea that MPA boundaries that follow natural habitat boundaries may have higher recovery rates than those ecosystems in protected areas that have boundaries intersecting contiguous coral reef habitat (Bartholomew *et al.* 2008). The location of MPA boundaries within the Central Coast region was evaluated to determine how well the MPAs are likely to retain populations and meet MLPA goal three. Because the MPA

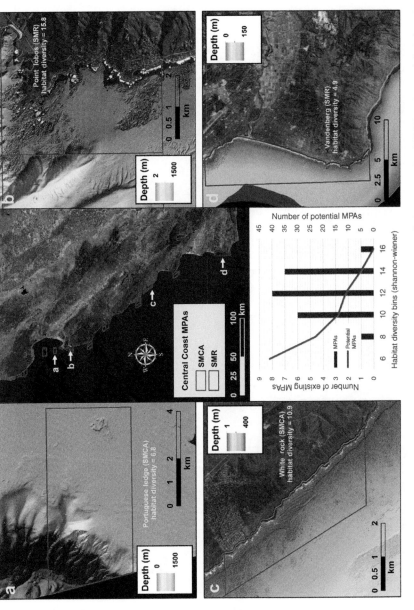

Figure 14.2 Examples of MPAs with varying habitat diversity within the Central Coast Marine Life Protection Act (MLPA) region and their locations shown on the map in the centre: (a) Portuguese Ledge State Marine Conservation Area (SMCA), (b) Point Lobos State Marine Reserve (SMR), (c) White Rock SMCA and (d) Vandenberg SMR. The shaded relief imagery of the seafloor is coloured by depth (note: depth range varies between images). The graph provides the distribution of diversity indices in the marine protected areas (MPAs, blue bars) and in the potential MPAs across the region (red line). *Source*: Copyright 2013 Esri, DeLorme, NAVTEQ.

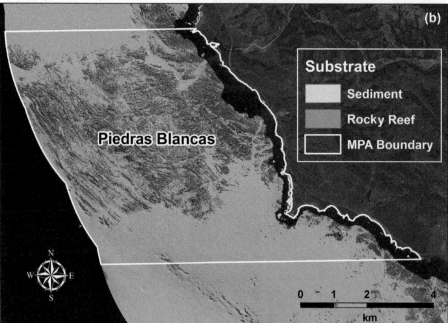

Figure 14.3 Example of how the seafloor data can be used to determine the amount of rocky reef habitat the boundaries of marine protected areas (MPAs) intersect. Rocky reef is brown, sediment is light tan and the MPA boundary is outlined in white. The Carmel Pinnacles MPA has a large area of rocky reef habitat within it but the MPA boundary intersects a large portion of the rocky reef (a) while the Piedras Blancas MPA has a large area of rocky reef habitat but the MPA boundary intersects very little rocky reef habitat (b). *Source*: Copyright 2013 Esri, DeLorme, NAVTEQ.

boundaries were placed prior to the availability of comprehensive seafloor maps, we used the newly available seafloor maps to quantify the amount of permeable habitat the MPA boundaries intersect (HI) in comparison to the total area of reef habitat within the MPA (HA).

From these seafloor maps, we calculated the HI/HA ratio for rocky reef habitat since most of the fishery targeted species are strongly associated with rocky reef. The total length of rocky reef substrate that the MPA boundary intersected was calculated (HI) by first converting the rocky reef class within the substrate raster to polygons using ArcGIS Desktop: Release 10. Then, using the Intersect tool within the Analysis Toolbox, the boundaries of the MPAs that intersected the rocky reef polygons were extracted and the total length of boundary intersecting rocky reef substrate for each MPA was summed. To calculate the total area of rocky reef substrate within the MPAs (HA), the Tabulate Area tool was applied within the Spatial Analyst toolbox. The HI/HA ratio was then calculated for each of the Central Coast MPAs. For rocky reef habitat, the HI/HA ratio varied for the Central Coast MPAs from the lowest value (0.113) in the Piedras Blancas MPA to the highest value (2.888) in the Carmel Pinnacles MPA (Figure 14.3; Table 14.4).

Table 14.4 The boundary analysis for the Central Coast MPAs. Each MPA is listed along with length of the MPA boundary intersecting reef given in kilometres (HI) and the total area of reef within the MPA (HA) in square kilometres. The final column is a measure of boundary permeability given as HI/HA, which has been multiplied by 1000 for easier comparison between MPAs

MPA name	Length of MPA boundary intersecting Reef (km) – HI	Area of reef within MPA (km²) – HA	HI/HA
Año Nuevo	4.150	9.069	0.458
Asilomar	3.775	1.642	2.300
Big Creek	0.200	1.088	0.184
Cambria	0.875	4.018	0.218
Carmel Bay	1.300	1.697	0.766
Carmel Pinnacles	3.325	1.151	2.888
Edward F. Ricketts	0.500	0.210	2.384
Greyhound Rock	1.550	3.236	0.479
Lovers Point	0.400	0.274	1.460
Pacific Grove Marine Gardens	1.800	1.283	1.403
Piedras Blancas	1.725	15.269	0.113
Point Buchon	2.550	4.262	0.598
Point Lobos	2.000	6.980	0.287
Point Sur	1.350	11.708	0.115
Portuguese Ledge	0.250	0.543	0.461
Soquel Canyon	0.150	0.564	0.266
Vandenberg	1.700	4.610	0.369
White Rock	2.275	2.819	0.807

Visually, the Piedras Blancas MPA captures a large area of rocky reef habitat within its boundaries but very little of the MPA boundary intersects that rocky reef. In contrast, the Carmel Pinnacles MPA was placed over an area of rocky reef habitat that extends beyond the MPA boundary, thus a large portion of the MPA boundary intersects the rocky reef habitat. The HI/HA ratio provides a measure of MPA boundary permeability that can be used to assess the likelihood of retaining species within each MPA. These results can help assess how well the Central Coast MPAs are meeting their design criteria both in increasing species abundances within MPAs (MLPA goals 1 & 2; Box 14.1; lower HI/HA) and contributing to fisheries through spillover (MLPA goal 3; Box 14.1; higher HI/HA).

14.4   Synthesis

Overall, incorporation of landscape ecology into marine conservation planning represents an important step forward in MPA design theory that will allow scientists and managers to view the seascape through a new lens. In this chapter, we presented a methodological approach involving the application of seafloor maps and landscape ecology metrics to quantify the replication of habitat, habitat diversity and boundary placement across and within a network of MPAs along the California coast to evaluate how well the MPA network met design criteria to achieve its ecological and management goals. These results show that, when designing networks that encompass broad geographic areas, assessment of these networks needs to incorporate the entire seascape throughout the region. Below we discuss current progress and shortcomings in the application of landscape ecology to designing and evaluating MPAs.

14.4.1   Mapping Technologies

The application of landscape ecology to the understanding of terrestrial systems has been aided by the ability to map broad geographic areas of habitat using optical remote sensing technologies. Unfortunately, these technologies are limited when used to map the ocean floor because light cannot penetrate far through the water column (Brown *et al.* 2011). The growth in acoustic technology, however, has allowed for increasing the depth range and resolution of seafloor mapping (Clarke *et al.* 1996; Lurton 2002; Mayer 2006; Brown *et al.* 2011), but only 5–10% of the seafloor is currently mapped at comparable resolution to terrestrial maps (Wright & Heyman 2008; Brown *et al.* 2011). From the case study presented in this chapter, it is clear that accurate and precise maps of marine habitats are critical to evaluating how well individual and networks of MPAs are likely to meet their conservation and management goals.

The landscape ecology approach to evaluating MPA network design presented in this chapter provides an example of how metrics of landscape ecology are derived from seafloor data and used to determine if marine spatial management goals are being met. However, we only evaluated those landscape metrics based on 2D features. The fields of digital terrain modelling, known as geomorphometry (Pike 2000), have developed and applied a wide range of morphometrics (first, second and third derivatives of DEMs) for investigating surface geometry across a landscape. Seascape topography has consequences for ecological functioning and the continued application and exploration of morphometrics in the third dimension that are relevant to MPA design and evaluation is important. For example, for patch-mosaic metrics that quantify seascape composition such as patch density, percentage of landscape and largest patch index, the analogous 3D metrics are peak density, surface volume and maximum peak height. For quantifying landscape configuration such as edge density, nearest neighbour index and fractal dimension index, the analogous 3D metrics are mean slope, mean nearest maximum index and surface fractal dimension (McGarigal *et al.* 2012). Morphometrics applied to DEMs can provide measurements that can be scaled to allow for extraction of information at spatial extents that are more appropriate for broad-scale ecosystem studies and facilitate a multiscaled approach to marine conservation planning.

14.4.2 MPA Effects on Biodiversity and Populations

In this chapter, we focused on the application of landscape ecology to informing and evaluating the design of MPA networks, however, the approach plays an equally important role in informing evaluations of the performance of networks in achieving their management goals. For example, the most common approach to evaluating MPA performance is to compare the trajectory of populations and communities inside and outside of MPAs, respectively, and test the hypothesis that these trajectories diverge over time. The relative rates of change are likely to be influenced by the seascape context and this spatial variability is rarely taken into account. Habitat maps and knowledge of how species respond to structural heterogeneity in seascape quality can inform interpretations of these comparisons. Similarly, if comparisons of fished species inside and outside of MPAs will be used to identify how species respond to the separate and combined effects of fishing and climate change (Carr *et al.* 2011), knowledge of both the within-patch variables and seascape context is essential. Taking methods from terrestrial ecology, species distribution models are increasingly applied to the marine environment to help understand the distribution and abundance of marine species (*e.g.*, Kendall *et al.* 2003; Pittman *et al.* 2007b, 2011; Young & Carr 2015b). Once these species-habitat associations are determined, the resulting habitat maps can be used as proxies for the distribution and abundance of species outside those areas where the biological observations exist (Ward *et al.* 1999; Anderson *et al.* 2009; Lucieer *et al.* 2013). Seafloor mapping data provides an opportunity to derive a number of variables that are important to fishes and these variables can be used to extrapolate biological observation data over the entire network of MPAs (Leathwick *et al.* 2008; Young & Carr 2015b).

14.4.3 Scale of Interaction between Species and Environment

The landscape ecology approach, which quantifies seascape structure across a range of scales, is appropriate largely because organisms respond to structure at a range of scales (Wiens & Milne 1989). Rather than protecting single species, MPAs spatially protect communities of species occupying a single area and these species tend to interact with their environment at differing scales (*e.g.*, sessile versus motile species). Landscape ecology allows you to analyse patterns and processes across multiple scales and, therefore, increases our understanding of how species respond to different aspects of the seascape (Boström *et al.* 2011). Species / habitat interactions not only vary across species but within species depending on their age, size or motility (Wiens & Milne 1989). Designing MPA networks across mosaics of habitat patches that take into consideration the patch types that organisms require to complete their life cycle can help to improve their success (Olds *et al.* 2013; Nagelkerken *et al.* 2015). The distribution of suitable habitat patches and distance between them can affect the likelihood of different areas supporting populations by allowing for the complementary, supplementary or neighbourhood effects of multiple habitats (Dunning *et al.* 1992). Pittman *et al.* (2007a) found that seascape structure (*i.e.,* coral reefs, seagrass beds) surrounding mangroves affected the assemblage structure and abundance of juvenile fish, showing that the seascape surrounding critical habitat can affect the composition of species using that habitat. Examples of ontogenetic habitat shifts in which species require different habitats during successive life stages are common (Love *et al.* 1991; Beck *et al.* 2001; Baskin *et al.* 2003). If ontogenetic shifts in habitat use are not taken into account when designing MPAs, species may only be protected at certain stages of their life cycle or when they are occupying one habitat (Pittman *et al.* 2007a; Saarman *et al.* 2013). Moreover, species may require refuge habitats or shift distributions in response to changing climatic conditions (*e.g.,* Pinsky *et al.* 2013; Smith *et al.* 2014). Individual and networks of MPAs that accommodate these spatial and temporal responses are key to creating networks that enhance resiliency of species to climate change (Carr *et al.* 2010; Fuller *et al.* 2010).

14.4.4   Across-System Interactions

In this chapter, we only explored the application of seascape ecology to the marine environment without considering effects from adjacent terrestrial areas. However, not only do management programs need to take into account the mosaic of habitats within the ecosystem of interest, but future conservation plans should also consider the interactions across terrestrial, marine and freshwater systems (Stoms *et al.* 2005; Lagabrielle *et al.* 2009; Beger *et al.* 2010), especially those processes that connect these systems. Both biological connectivity (*e.g.,* movement of species between systems) and transfer of energy or matter between systems (*e.g.,* pollution moving from one system to another) should be considered when making decisions on placement of protected areas. For instance, salmon that inhabit the northeastern Pacific Ocean are anadromous species, which means that they are born in fresh water, migrate to the ocean and return to freshwater to spawn (Scholz *et al.* 1976). This movement results in the transfer of both individuals and energy / nutrients from riparian ecosystems to the oceans and then from oceans back to riparian ecosystems (Gende *et al.* 2002; Merz & Moyle 2006; Moore *et al.* 2011). Pollution is another form of matter that can be transferred from one system to another (see Chapter 11 in this book). Understanding how pollution is transferred across systems could help with placing MPAs in areas where anthropogenic threats are more limited.

14.4.5   Population Connectivity

Connectivity of populations is another important characteristic considered in the design of terrestrial networks of protected areas (Kool *et al.* 2013) and marine protected areas (reviewed in Kool *et al.* 2013; Botsford *et al.* 2014; see also Chapter 9 in this book). Connectivity has been incorporated into spatial conservation for many decades through the use of critical maximum dispersal distances and minimum patch size requirements (Sarkar *et al.* 2006). Challenges remain; however, in the incorporation of connectivity into the design of protected areas, especially when considering entire communities rather than single species (Kool *et al.* 2013). The close ties between metapopulation ecology and landscape ecology have helped with the development of methods to better understand how populations are connected across broad spatial scales. Many marine species exhibit characteristics of metapopulations (Hanski & Simberloff 1997) where local populations occupying habitat patches are spatially isolated from each other by a matrix of less optimal habitats. As noted by Wiens (1997), landscape ecology can provide valuable insights into metapopulations and more recently for metacommunities (Holyoak *et al.* 2005).

Temporal and spatial variation in dispersal results in connectivity being dynamic, patchy and complex (Baguette *et al.* 2013). Landscape ecology, however, can be used to help understand connectivity by combining structural connectivity (spatial arrangement of suitable habitat patches) with functional connectivity (flow of individuals among habitat patches, Calabrese & Fagan 2004). In marine systems, the tridimensional dynamics resulting from highly variable oceanographic processes such as water currents, including coastal upwelling, can cause spatiotemporal variation in larval transport, as well as oceanographic conditions such as seawater temperature, salinity and light levels, further complicating mechanisms of connectivity (*e.g.,* Calabrese & Fagan 2004; Woodson *et al.* 2012). Despite these complexities, understanding connectivity in marine systems can provide valuable information to the design and evaluation of MPA networks (Edwards *et al.* 2010; Beger *et al.* 2015). Most studies so far have looked at structural connectivity (Cowen *et al.* 2006; Cowen & Sponaugle 2009; Sundblad *et al.* 2011), but the application of 'seascape genetics' is showing promise in understanding the functional connectivity across seascapes (Bay *et al.* 2008; Christie *et al.* 2010; Berumen *et al.* 2012; Harrison *et al.* 2012; Johansson *et al.* 2015). Networks of marine protected areas can serve as the sources in source-sink systems where individuals in exploited areas are replenished with individuals provided by protected populations (Costello *et al.* 2010; Gaines, White, *et al.* 2010; Botsford *et al.* 2014). As methods for both direct tracking of organisms and population genetics continue to improve, so too will our ability to identify spatial structure of populations, barriers to dispersal and source-sink populations (Kool *et al.* 2013). Consequently, the application of seascape ecology to understanding connectivity in marine systems is imperative to the successful design and evaluation of MPA networks (Cerdeira *et al.* 2010).

14.5   Conclusions and Future Research Priorities

The landscape ecology approach to evaluating MPA network design presented in this chapter provides an example of how metrics of landscape ecology can be used to determine whether or not marine spatial management goals are being met. Although we only focused on attributes of the seabed and their application to the development of landscape metrics and ignored the dynamics associated with the water column (*e.g.,* temperature, currents, upwelling, connectivity), this study represents an effective way to evaluate several of the fundamental design criteria for networks of MPAs, both within the marine protected area and with respect to the surrounding seascape. Future applications of landscape ecology to marine protected area evaluation would benefit from consideration of those seascape attributes that are important to the supply of food, nutrients and reproductive propagules within the ecosystem. To improve MPA science, we recommend advancements in seascape ecology that focus on the following research areas: (i) further advances in 3D terrain morphometrics for measuring topographic complexity, (ii) developing methods for the incorporation of oceanographic variables both spatially and temporally into assessments of MPA performance, (iii) understanding link between structural connectivity metrics and functional connectivity, (iv) predicting how environmental changes will affect spatial patterning, (v) across system interactions and (vi) quantification of seascape context when comparing MPA performance between MPAs and inside versus outside.

References

Abdel Moniem HEM, Holland JD (2013) Habitat connectivity for pollinator beetles using surface metrics. Landscape Ecology 28: 1251–1267.

Agardy T (1995) The Science of Conservation in the Coastal Zone: New Insights on How to Design, Implement and Monitor Marine Protected Areas. IUCN, Gland, Switzerland.

Agardy T (2000) Information needs for marine protected areas: Scientific and societal. Bulletin of Marine Science 66: 875–888.

Agardy T, di Sciara GN, Christie P (2011) Mind the gap: Addressing the shortcomings of marine protected areas through large scale marine spatial planning. Marine Policy 35: 226–232.

Allison GW, Gaines SD, Lubchenco J, Possingham HP (2003) Ensuring persistence of marine reserves: Catastrophes require adopting an insurance factor. Ecological Applications 13: 8–24.

Anderson TJ, Syms C, Roberts DA, Howard DF (2009) Multi-scale fish-habitat associations and the use of habitat surrogates to predict the organization and abundance of deep-water fish assemblages. Journal of Experimental Marine Biology and Ecology 379: 34–42.

Andréfouët S, Robinson JA, Hu C, Feldman GC, Salvat B, Payri C, Muller-Karger FE (2003) Influence of the spatial resolution of SeaWiFS, Landsat-7, SPOT, and International Space Station data on estimates of landscape parameters of Pacific Ocean atolls. Canadian Journal of Remote Sensing 29: 210–218.

Appeldoorn RS, Ruíz I, Pagan FE (2011) From habitat mapping to ecological function: incorporating habitat into coral reef fisheries management. Proceedings of the 63rd Gulf and Caribbean Fisheries Institute 63: 10–17.

Aragón R, Morales JM (2003) Species composition and invasions in NW Argentinian secondary forests, Effects of land use history, environment and landscape. Journal of Vegetation Science 14: 195–204.

Armenteras D, Gast F, Villareal H (2003) Andean forest fragmentation and the representativeness of protected natural areas in the eastern Andes, Colombia. Biological Conservation 113: 245–256.

Babcock E, MacCall AD (2011) How useful is the ratio of fish density outside versus inside no-take marine reserves as a metric for fishery management control rules? Canadian Journal of Fisheries and Aquatic Sciences 68: 343–359.

Baguette M, Blanchet S, Legrand D, Stevens VM, Turlure C (2013) Individual dispersal, landscape connectivity and ecological networks. Biological Reviews 88: 310–326.

Baker WL (1992) The landcape ecology of large distrubances in the design and management of nature reserves. Landscape Ecology 7: 181–194.

Ball IR, Possingham HP (2000) MARXAN (V1. 8.2). Marine Reserve Design using Spatially Explicit Annealing, A Manual. http://marxan.net/downloads/documents/marxan_manual_1_8_2.pdf (accessed 19 June 2017).

Barbault R (1995) Biodiversity dynamics: from population and community ecology approaches to a landscape ecology point of view. Landscape and Urban Planning 31: 89–98.

Bartholomew A, Bohnsack JA, Smith SG, Ault JS, Harper DE, McClellan DB (2008) Influence of marine reserve size and boundary length on the initial response of exploited reef fishes in the Florida Keys National Marine Sanctuary, USA. Landscape Ecology 23: 55–65.

Baskin Y, Beck MW, Heck KL, Able KW, Childers DL, Eggleston DB, Gillanders BS, Halpern BS, Hays CG, Hoshino K, Minello TJ, Orth RJ, Sheridan PF, Weinstein PM (2003) The role of nearshore ecosystems as fish and shellfish nurseries. Issues in Ecology 11: 1–12.

Bauerfeind SS, Theisen A, Fischer K (2009) Patch occupancy in the endangered butterfly Lycaena helle in a fragmented landscape: Effects of habitat quality, patch size and isolation. Journal of Insect Conservation 13: 271–277.

Bay LK, Caley MJM, Crozier RH (2008) Meta-population structure in a coral reef fish demonstrated by genetic data on patterns of migration, extinction and re-colonisation. BMC Evolutionary Biology 8: 248.

Beck MW, Heck KL, Able KW, Childers DL, Eggleston DB, Gillanders BM, Halpern B, Hays CG, Hoshino K, Minello TJ, Orth RJ, Sheridan PF, Weinstein MP (2001) The identification, conservation, and management of estuarine and marine nurseries for fish and invertebrates. Bioscience 51: 633.

Beger M, Grantham HS, Pressey RL, Wilson KA, Peterson EL, Dorfman D, Mumby PJ, Lourival R, Brumbaugh DR, Possingham HP (2010) Conservation planning for connectivity across marine, freshwater, and terrestrial realms. Biological Conservation 143: 565–575.

Beger M, Linke S, Watts M, Game E, Treml E, Ball I, Possingham HP (2010) Incorporating asymmetric connectivity into spatial decision making for conservation. Conservation Letters 3: 359–368.

Beger M, McGowan J, Treml EA, Green AL, White AT, Wolff NH, Klein CJ, Mumby PJ, Possingham HP (2015) Integrating regional conservation priorities for multiple objectives into national policy. Nature Communications 6: 8208.

Bell SS, Fonseca MS, Kenworthy WJ (2008) Dynamics of a subtropical seagrass landscape: Links between disturbance and mobile seed banks. Landscape Ecology 23: 67–74.

Berglund M, Nilsson Jacobi M, Jonsson PR (2012) Optimal selection of marine protected areas based on connectivity and habitat quality. Ecological Modelling 240: 105–112.

Berumen ML, Almany GR, Planes S, Jones GP, Saenz-Agudelo P, Thorrold SR (2012) Persistence of self-recruitment and patterns of larval connectivity in a marine protected area network. Ecology and Evolution 2(2): 444–452.

Boström C, Pittman SJ, Simenstad C, Kneib RT (2011) Seascape ecology of coastal biogenic habitats: Advances, gaps, and challenges. Marine Ecology Progress Series 427: 191–217.

Botsford LW, Brumbaugh DR, Grimes C, Kellner JB, Largier J, O'Farrell MR, Ralston S, Soulanille E, Wespestad V (2009) Connectivity, sustainability, and yield: Bridging the gap between conventional fisheries management and marine protected areas. Reviews in Fish Biology and Fisheries 19: 69–95.

Botsford LW, White JW, Carr MH, Caselle JE (2014) Marine protected area networks in California, USA. Advances in Marine Biology 69: 205–251.

Brown CJ, Smith SJ, Lawton P, Anderson JT (2011) Benthic habitat mapping: A review of progress towards improved understanding of the spatial ecology of the seafloor using acoustic techniques. Estuarine Coastal Shelf Sciences 92: 502–520.

Calabrese JM, Fagan WF (2004) A comparison-shopper's guide to connectivity metrics. Frontiers in Ecology and Environment 2: 529–536.

Carr MMH, Saarman E, Caldwell MMR (2010) The role of 'rules of thumb' in science-based environmental policy: California's Marine Life Protection Act as a case study. Stanford Journal of Law, Science Policy 2: 1–17.

Carr MH, Woodson CB, Cheriton OM, Malone D, McManus MA, Raimondi PT (2011) Knowledge through partnerships: Integrating marine protected area monitoring and ocean observing systems. Frontiers in Ecology and Environment 9: 342–350.

Carroll C, Miquelle DG (2006) Spatial viability analysis of Amur tiger Panthera tigris altaica in the Russian Far East: the role of protected areas and landscape matrix in population persistence. Journal Applied Ecology 43: 1056–1068.

Carroll JM, Peterson BJ (2013) Ecological trade-offs in seascape ecology: Bay scallop survival and growth across a seagrass seascape. Landscape Ecology 28: 1401–1413.

Cerdeira JO, Pinto LS, Cabeza M, Gaston KJ (2010) Species specific connectivity in reserve-network design using graphs. Biological Conservation 143: 408–415.

Chapman MR, Kramer DL (2000) Movements of fishes within and among finging coral reefs in Barbados. Environmental Biology of Fishes 57: 11–24.

Chatfield BS, Niel KP Van, Kendrick GA, Harvey ES (2010) Combining environmental gradients to explain and predict the structure of demersal fish distributions. Journal of Biogeography 37: 593–605.

Christie MR, Tissot BN, Albins MA, Beets JP, Jia Y, Ortiz DM, Thompson SE, Hixon MA (2010) Larval connectivity in an effective network of marine protected areas. PLoS One 5: 1–8.

Claisse JT, Pondella DJ, Williams JP, Sadd J (2012) Using GIS mapping of the extent of nearshore rocky reefs to estimate the abundance and reproductive output of important fishery species. PLoS One 7: 23–29.

Clarke JEH, Mayer LA, Wells DE (1996) Shallow-water imaging multibeam sonars: A new tool for investigating seafloor processes in the coastal zone and on the continental shelf. Marine Geophysical Research 18: 607–629.

Claudet J, García-Charton JA, Lenfant P (2010) Combined Effects of Levels of Protection and Environmental Variables at Different Spatial Resolutions on Fish Assemblages in a Marine Protected Area. Conservation Biology 25: 105–114.

Costello C, Rassweiler A, Siegel D, Leo G De, Micheli F, Rosenberg A (2010) The value of spatial information in MPA network design. Proceedings National Academy of Sciences 107: 18294–18299.

Cowen RK, Paris CB, Srinivasan A (2006) Scaling of connectivity in marine populations. Science 311(5760): 522–527.

Cowen RK, Sponaugle S (2009) Larval dispersal and marine population connectivity. Annual Review of Marine Science 1: 443–466.

Crist MR, Wilmer B, Aplet GH (2005) Assessing the value of roadless areas in a conservation reserve strategy: biodiversity and landscape connectivity in the northern Rockies. Journal Applied Ecology 42: 181–191.

Crouzeilles R, Lorini ML, Grelle CEV (2013) The importance of using sustainable use protected areas for functional connectivity. Biological Conservation 159: 450–457.

Cumming GS (2011) Spatial resilience: integrating landscape ecology, resilience, and sustainability. Landscape Ecology 26: 899–909.

D'Aloia CC, Bogdanowicz SM, Majoris JE, Harrison RG, Buston PM (2013) Self-recruitment in a Caribbean reef fish: A method for approximating dispersal kernels accounting for seascape. Molecular Ecology 22: 2563–2572.

Douglas M, Lake PS (1994) Species richness of stream stones : An investigation of the mechanisms generating the species-area relationship. Oikos 69(3): 387–396.

Dunning JB, Danielson BJ, Pulliam HR (1992) Ecological processes that affect populations in complex landscapes. Oikos 65: 169–175.

Edgar GJ, Stuart-Smith RD, Willis TJ, Kininmonth S, Baker SC, Banks S, Barrett NS, Becerro M a, Bernard ATF, Berkhout J, Buxton CD, Campbell SJ, Cooper AT, Davey M, Edgar SC, Förstera G, Galván DE, Irigoyen AJ, Kushner DJ, Moura R, Parnell PE, Shears NT, Soler G, Strain EM a, Thomson RJ (2014) Global conservation outcomes depend on marine protected areas with five key features. Nature 506: 216–220.

Edwards HJ, Elliott IA, Pressey RL, Mumby PJ (2010) Incorporating ontogenetic dispersal, ecological processes and conservation zoning into reserve design. Biological Conservation 143: 457–470.

Erickson JL, West SD (2002) Associations of bats with local structure and landscape features of forested stands in western Oregon and Washington. Biological Conservation 109: 95–102.

Falcucci A, Maiorano L, Boitani L (2007) Changes in land-use/land-cover patterns in Italy and their implications for biodiversity conservation. Landscape Ecology 22: 617–631.

Flick T, Feagan S, Fahrig L (2012) Effects of landscape structure on butterfly species richness and abundance in agricultural landscapes in eastern Ontario, Canada. Agriculture, Ecosystems and Environment 156: 123–133.

Foley MM, Halpern BS, Micheli F, Armsby MH, Caldwell MR, Crain CM, Prahler E, Rohr N, Sivas D, Beck MW, Carr MH, Crowder LB, Emmett Duffy J, Hacker SD, McLeod KL, Palumbi SR, Peterson CH, Regan HM, Ruckelshaus MH, Sandifer PA, Steneck RS (2010) Guiding ecological principles for marine spatial planning. Marine Policy 34: 955–966.

Friedlander AM, Brown EK, Monaco ME (2007) Coupling ecology and GIS to evaluate efficacy of marine protected areas in Hawaii. Ecological Applications 17: 715–730.

Fuller R a, McDonald-Madden E, Wilson K a, Carwardine J, Grantham HS, Watson JEM, Klein CJ, Green DC, Possingham HP (2010) Replacing underperforming protected areas achieves better conservation outcomes. Nature 466: 365–367.

Gaines SD, Gaylord B, Largier JL (2003) Avoiding current oversights in marine reserve design. Ecological Applications 13: 32–46.

Gaines SD, Lester SE, Grorud-Colvert K, Costello C, Pollnac R (2010) Evolving science of marine reserves: new developments and emerging research frontiers. Proceedings National Academy of Science USA 107: 18251–18255.

Gaines SD, White C, Carr MH, Palumbi SR (2010) Designing marine reserve networks for both conservation and fisheries management. Proceedings National Academy of Science 107: 18286–18293.

García-Charton JA, Pérez-Ruzafa Á (1999) Ecological heterogeneity and the evaluation of the effects of marine reserves. Fisheries Research 42: 1–20.

Gende SM, Edwards RT, Willson MF, Wipfli MS (2002) Pacific salmon in aquatic and terrestrial ecosystems. Bioscience 52: 917–928.

Giménez-Casalduero F, Gomariz-Castillo FJ, Calvín JC (2011) Hierarchical classification of marine rocky landscape as management tool at southeast Mediterranean coast. Ocean and Coastal Management 54: 497–506.

Gleason M, Fox E, Ashcraft S, Vasques J, Whiteman E, Serpa P, Saarman E, Caldwell M, Frimodig A, Miller-Henson M, Kirlin J, Ota B, Pope E, Weber M, Wiseman K (2013) Designing a network of marine protected areas in California: Achievements, costs, lessons learned, and challenges ahead. Ocean and Coastal Management 74: 90–101.

Goetz SJ, Jantz P, Jantz CA (2009) Connectivity of core habitat in the Northeastern United States: Parks and protected areas in a landscape context. Remote Sensing of Environment 113: 1421–1429.

Greene HG, Bizzarro JJ, O'Connell VM, Brylinsky CK (2007) Construction of digital potential marine benthic habitat maps using a coded classification scheme and its application. In Todd BJ, Greene HG (eds) Mapping the seafloor for habitat characterization. Special Paper 47. Geological Association of Canada, Memorial University of Newfoundland, St. John's, Canada, pp. 141–155.

Grober-Dunsmore R, Frazer TK, Beets JP, Lindberg WJ, Zwick P, Funicelli NA (2008) Influence of landscape structure on reef fish assemblages. Landscape Ecology 23: 37–53.

Grober-Dunsmore R, Pittman SJ, Caldow C, Kendall MS, Frazer TK (2009) A landscape ecology approach for the study of ecological connectivity across tropical marine seascapes. In Nagelkerken I (ed.) Ecological Connectivity among Tropical Coastal Ecosystems. Springer, Dordrecht, pp. 493–530.

Grorud-Colvert K, Claudet J, Carr M, Caselle JE, Day J, Friedlander AM, Lester SE, Loma TL De, Tissot BN, Malone DP (2011) The assessment of marine reserve networks: guidelines for ecological evaluation. In Claudet J (ed.) Marine Protected Areas: A Multidisciplinary Approach. Cambridge University Press, Cambridge, pp. 293–321.

Grorud-Colvert K, Claudet J, Tissot BN, Caselle JE, Carr MH, Day JC, Friedlander AM, Lester SE, Loma TL De, Malone D, Walsh WJ (2014) Marine protected area networks: Assessing whether the whole is greater than the sum of its parts. PLoS One 9: 1–7.

Halpern BS, Lester SE, McLeod KL (2010) Placing marine protected areas onto the ecosystem-based management seascape. Proceedings National Academy Science 107: 18312–18317.

Hanski I, Simberloff D (1997) The metapopulation approach, its history, conceptual domain, and application to conservation. In Hanski I, Gilpin M (eds) Metapopulation Biology: Ecology, Genetics, and Evolution. Academic Press, San Diego, CA, pp. 5–26.

Hansson L, Angelstam P (1991) Landscape ecology as a theoretical basis for nature conservation. Landscape Ecology 5: 191–201.

Hart DD, Horwitz RJ (1991) Habitat diversity and the species-area relationship: alternative models and tests. In Bell SS, McCoy ED, Mushinsky HR (eds) Habitat Structure. 8th edition. Springer Netherlands, Dordrecht, pp. 47–68.

Harrison HB, Williamson DH, Russ GR, Feldheim KA, van Herwerden L, Planes S, Srinivasan M, Berumen ML, Jones GP (2012) Larval export from marine reserves and the recruitment benefit for fish and fisheries. Current Biology 22: 1023–1028.

Hastings A, Botsford LW (2003) Comparing designs of marine reserves for fisheries and for biodiversity. Ecological Applications 13: S65–S70.

Heinänen S, Erola J, Numers M von (2012) High resolution species distribution models of two nesting water bird species: a study of transferability and predictive performance. Landscape Ecology 27: 545–555.

Hilborn R, Stokes K, Maguire JJ, Smith T, Botsford LW, Mangel M, Orensanz J, Parma A, Rice J, Bell J, Cochrane KL, Garcia S, Hall SJ, Kirkwood GP, Sainsbury K, Stefansson G, Walters C (2004) When can marine reserves improve fisheries management? Ocean Coastal Managment 47: 197–205.

Hockey PAR, Branch GM (1997) Criteria, objectives and methodology for evaluating marine protected areas in South Africa. South African Journal of Marine Science 18: 369–383.

Holyoak M, Leibold MA, Holt RD (eds) (2005) Metacommunities: Spatial Dynamics and Ecological Communities. University of Chicago Press, Chicago, IL.

Honnay AO, Endels P, Vereecken H, Hermy M (1999) The role of patch area and habitat diversity in explaining native plant species richness in disturbed suburban forest patches in northen Belgium. Diversity and Distributions 5: 129–141.

Hovel KA (2003) Habitat fragmentation in marine landscapes: Relative effects of habitat cover and configuration on juvenile crab survival in California and North Carolina seagrass beds. Biological Conservation 110: 401–412.

Hughes ATP, Baird AH, Bellwood DR, Card M, Connolly SR, Folke C, Jackson JBC, Kleypas J, Lough JM, Marshall P, Palumbi SR, Pandolfi JM, Rosen B, Roughgarden J (2003) Climate change, human impacts, and the resilience of coral reefs. Science 301: 929–933.

Huntington BE, Karnauskas M, Babcock EA, Lirman D (2010) Untangling natural seascape variation from marine reserve effects using a landscape approach. PLoS One 5: e12327.

Hurley PM, Webster CR, Flahspohler DJ, Parker GR (2012) Untangling the landscape of deer overabundance: reserve size versus landscape context in the agricultural midwest. Biological Conservation 146: 62–71.

Iampietro PJ, Kvitek RG, Morris E (2005) Recent advances in automated genus-specific marine habitat mapping enabled by high-resolution multibeam bathymetry. Marine Technology Society 39: 89–93.

Ierodiaconou D, Monk J, Rattray A, Laurenson L, Versace VL (2011) Comparison of automated classification techniques for predicting benthic biological communities using hydroacoustics and video observations. Continental Shelf Research 31: 28–38.

Johansson ML, Alberto F, Reed DC, Raimondi PT, Coelho NC, Young MA, Drake PT, Edwards CA, Cavanaugh K, Assis J, Ladah LB, Bell TW, Coyer JA, Siegel DA, Serrão EA (2015) Seascape drivers of Macrocystis pyrifera population genetic structure in the northeast Pacific. Molecular Ecology 24: 4866–4885.

Kendall MS, Christensen JD, Hillis-Starr Z (2003) Multi-scale data used to analyze the spatial distribution of French grunts, Haemulon flavolineatum, relative to hard and soft bottom in a benthic landscape. Environmental Biology of Fishes 66: 19–26.

Kendall MS, Miller TJ (2010) Relationships among map resolution, fish assemblages, and habitat variables in a coral reef ecosystem. Hydrobiologia 637: 101–119.

Kendall MS, Miller TJ, Pittman SJ (2011) Patterns of scale-dependency and the influence of map resolution on the seascape ecology of reef fish. Marine Ecology Progress Series 427: 259–274.

Kirlin J, Caldwell M, Gleason M, Weber M, Ugoretz J, Fox E, Miller-Henson M (2013) California's Marine Life Protection Act Initiative: Supporting implementation of legislation establishing a statewide network of marine protected areas. Ocean and Coastal Management 74: 3–13.

Knudby A, Brenning A, LeDrew E (2010a) New approaches to modelling fish-habitat relationships. Ecological Modelling 221: 503–511.

Knudby A, LeDrew E, Brenning A (2010b) Predictive mapping of reef fish species richness, diversity and biomass in Zanzibar using IKONOS imagery and machine-learning techniques. Remote Sensing of Environment 114: 1230–1241.

Kool JT, Moilanen A, Treml EA (2013) Population connectivity: Recent advances and new perspectives. Landscape Ecology 28: 165–185.

Kvitek R, Iampietro P (2010) California's seafloor mapping project. In Breman J (ed.) Ocean Globe. ESRI, Redlands, CA, pp. 75–85.

Lagabrielle E, Rouget M, Payet K, Wistebaar N, Durieux S, Baret S, Lombard A, Strasberg D (2009) Identifying and mapping biodiversity processes for conservation planning in islands: a case study in Reunion Island (Western Indian Ocean). Biological Conservation 142: 1523–1535.

Lasanta T, González-Hidalgo JC, Vicente-Serrano SM, Sferi E (2006) Using landscape ecology to evaluate an alternative management scenario in abandoned Mediterranean mountain areas. Landscape and Urban Planning 78: 101–114.

Leathwick J, Moilanen A, Francis M, Elith J, Taylor P, Julian K, Hastie T, Duffy C (2008) Novel methods for the design and evaluation of marine protected areas in offshore waters. Conservation Letters 1: 91–102.

Lechner AM, Langford WT, Bekessy SA, Jones SD (2012) Are landscape ecologists addressing uncertainty in their remote sensing data? Landscape Ecology 27: 1249–1261.

Leitão AB, Ahern J (2002) Applying landscape ecological concepts and metrics in sustainable landscape planning. Landscape and Urban Planning 59: 65–93.

Lester SE, Halpern BS, Grorud-Colvert K, Lubchenco J, Ruttenberg BI, Gaines SD, Airamé S, Warner RR (2009) Biological effects within no-take marine reserves: A global synthesis. Marine Ecology Progress Series 384: 33–46.

Ling SD, Johnson CR, Frusher SD, Ridgway KR (2009) Overfishing reduces resilience of kelp beds to climate-driven catastrophic phase shift. Proceedings National Academy Science USA 106: 22341–22345.

Lovari S, Sforzi A, Mori E (2013) Habitat richness affects home range size in a monogamous large rodent. Behavioural Processes 99: 42–46.

Love MS, Carr MH, Haldorson LJ (1991) The ecology of substrate-associated juveniles of the genus Sebastes. Environmental Biology of Fishes 30: 225–243.

Lubchenco J, Grorud-Colvert K (2015) Making waves: The science and politics of ocean protection. Science 350(6259): 382–383.

Lucherini M, Lovari S (1996) Habitat richness affects home range size in the red fox Vulpes vulpes. Behavioural Processes 36: 103–106.

Lucieer V, Hill NA, Barrett NS, Nichol S (2013) Do marine substrates 'look' and 'sound' the same? Supervised classification of multibeam acoustic data using autonomous underwater vehicle images. Estuarine Coastal and Shelf Sciences 117: 94–106.

Luoto M, Toivonen T, Heikkinen R (2002) Prediction of total and rare plant species richness in agricultural landscapes from satellite images and topographic data. Landscape Ecology 17: 195–217.

Lurton X (2002) An Introduction to Underwater Acoustics: Principles and Applications. Springer-Praxis, Chichester.

MacArthur RH, Wilson EO (1967) The Theory of Island Biogeography. Princeton University Press, Princeton, NJ.

MacLeod CD, Mandleberg L, Schweder C, Bannon SM, Pierce GJ (2008) A comparison of approaches for modelling the occurrence of marine animals. Hydrobiologia 612: 21–32.

Mairota P, Cafarelli B, Boccaccio L, Leronni V, Labadessa R, Kosmidou V, Nagendra H (2013) Using landscape structure to develop quantitative baselines for protected area monitoring. Ecological Indicators 33: 82–95.

Margules CR, Pressey RL (2000) Systematic conservation planning. Nature 405: 243–253.

Mayer LA (2006) Frontiers in seafloor mapping and visualization. Marine Geophysical Research 27: 7–17.

McGarigal K, Cushman SA, Ene E (2012) FRAGSTATS v4: spatial pattern analysis program for categorical and continuous maps.

McGarigal K, Marks BJ (1995) FRAGSTATS: spatial pattern analysis program for quantifying landscape structure.

McGarigal K, Tagil S, Cushman SA (2009) Surface metrics: An alternative to patch metrics for the quantification of landscape structure. Landscape Ecology 24: 433–450.

McGilliard CR, Hilborn R, MacCall A, Punt AE, Field JC (2011) Can we use information from marine protected areas to inform management of small-scale, data-poor stocks? ICES Journal of Marine Science 68: 201–211.

McNeill S, Fairweather P. (1993) Single large or several small marine reserves? An experimental approach with seagrass fauna. Journal of Biogeography 20: 429–440.

Meager JJ, Schlacher TA, Green M (2011) Topographic complexity and landscape temperature patterns create a dynamic habitat structure on a rocky intertidal shore. Marine Ecology Progress Series 428: 1–12.

Merz JE, Moyle PB (2006) Salmon, wildlife, and wine: marine-derived nutrients in human dominated ecosystems of Central California. Ecological Applications 16: 999–1009.

Meynecke JO, Lee SY, Duke NC (2008) Linking spatial metrics and fish catch reveals the importance of coastal wetland connectivity to inshore fisheries in Queensland, Australia. Biological Conservation 141: 981–996.

Minor ES, Urban DL (2007) Graph theory as a proxy for spatially explicit population models in conservation planning. Ecological Applications 17: 1771–1782.

Mizerek T, Regan HM, Hovel KA (2011) Seagrass habitat loss and fragmentation influence management strategies for a blue crab Callinectes sapidus fishery. Marine Ecology Progress Series 427: 247–257.

Monk J, Ierodiaconou D, Bellgrove A, Harvey E, Laurenson L (2011) Remotely sensed hydroacoustics and observation data for predicting fish habitat suitability. Continental Shelf Research 31: 17–27.

Monk J, Ierodiaconou D, Versace VL, Bellgrove A, Harvey E, Rattray A, Laurenson L, Quinn GP (2010) Habitat suitability for marine fishes using presence-only modelling and multibeam sonar. Marine Ecology Progress Series 420: 157–174.

Moore JW, Hayes SA, Duffy W, Gallagher S, Michel CJ, Wright D (2011) Nutrient fluxes and the recent collapse of coastal California salmon populations. Canadian Journal of Fisheries and Aquatic Sciences 68: 1161–1170.

Murphy HM, Jenkins GP (2010) Observational methods used in marine spatial monitoring of fishes and associated habitats: A review. Marine and Freshwater Research 61: 236–252.

Nagelkerken I, Sheaves M, Baker R, Connolly RM (2015) The seascape nursery: A novel spatial approach to identify and manage nurseries for coastal marine fauna. Fish and Fisheries 16: 362–371.

Narumalani S, Mishra DR, Rothwell RG (2004) Change detection and landscape metrics for inferring anthropogenic processes in the greater EFMO area. Remote Sensing of Environment 91: 478–489.

Ndubisi F (2002) Ecological Planning: A historical and comparitive synthesis. JHU Press, Baltimore, MD.

Nyström M, Folke C (2001) Spatial resilience of coral reefs. Ecosystems 4: 406–417.

Olds AD, Albert S, Maxwell PS, Pitt KA, Connolly RM (2013) Mangrove-reef connectivity promotes the effectiveness of marine reserves across the western Pacific. Global Ecology and Biogeography 22: 1040–1049.

Olds AD, Connolly RM, Pitt KA, Maxwell PS (2012a) Habitat connectivity improves reserve performance. Conservation Letters 5: 56–63.

Olds AD, Connolly RM, Pitt KA, Maxwell PS, Aswani S, Albert S (2014) Incorporating surrogate species and seascape connectivity to improve marine conservation outcomes. Conservation Biology 28(4): 982–991.

Olds AD, Connolly RM, Pitt KA, Pittman SJ, Maxwell PS, Huijbers CM, Moore BR, Albert S, Rissik D, Babcock RC, Schlacher TA (2016) Quantifying the conservation value of seascape connectivity: A global synthesis. Global Ecology and Biogeography 25: 3–15.

Olds AD, Pitt KA, Maxwell PS, Connolly RM (2012b) Synergistic effects of reserves and connectivity on ecological resilience. Journal of Applied Ecology 49: 1195–1203.

Oliver LM, Lehrter JC, Fisher WS (2011) Relating landscape development intensity to coral reef condition in the watersheds of St Croix, US Virgin Islands. Marine Ecology Progress Series 427: 293–302.

Paine RT, Levin SA (1981) Intertidal landscapes: Disturbance and the dynamics of pattern. Ecological Monographs 51: 145–178.

Palumbi SR (2004) Marine reserves and ocean neighbourhoods: The spatial scale of marine populations and their management. Annual Review of Environment and Resources 29: 31–68.

Pickett STA, White PS (1985) The ecology of natural disturbances and patch dynamics. In Pickett STA, White PS (eds) The Ecology of Natural Disturbance and Patch Dynamics. Academic Press, San Diego, CA, pp. 3–13.

Pike RJ (2000) Geomorphometry – diversity in quantitative surface analysis. Progress in Physical Geography 24: 1–20.

Pinsky ML, Palumbi SR, Andréfouët S, Purkis SJ (2012) Open and closed seascapes: Where does habitat patchiness create populations with high fractions of self-recruitment? Ecological Applications 22: 1257–1267.

Pinsky ML, Worm B, Fogarty MJ, Sarmiento JL, Levin SA (2013) Marine taxa track local climate velocities. Science 341(6151): 1239–1242.

Pittman SJ, Brown KA (2011) Multi-scale approach for predicting fish species distributions across coral reef seascapes. PLoS One 6: e20583.

Pittman SJ, Caldow C, Hile SD, Monaco ME (2007a) Using seascape types to explain the spatial patterns of fish in the mangroves of SW Puerto Rico. Marine Ecology Progress Series 348: 273–284.

Pittman SJ, Christensen JD, Caldow C, Menza C, Monaco ME (2007b) Predictive mapping of fish species richness across shallow-water seascapes in the Caribbean. Ecological Modelling 204: 9–21.

Pittman SJ, Kneib RT, Simenstad CA (2011) Practicing coastal seascape ecology. Marine Ecology Progress Series 427: 187–190.

Pittman SJ, McAlpine CA, Pittman KM (2004) Linking fish and prawns to their environment: a hierarchical landscape approach. Marine Ecology Progress Series 283: 233–254.

Possingham HP, Franklin J, Wilson K, Regan T (2005) The roles of spatial heterogeneity and ecological processes in conservation planning. Ecology: 389–406.

Prada MC, Appeldoorn RS, Rivera JA (2008) The effects of minimum map unit in coral reefs maps generated from high resolution side scan sonar mosaics. Coral Reefs 27: 297–310.

Purkis SJ, Graham NAJ, Riegl BM (2007) Predictability of reef fish diversity and abundance using remote sensing data in Diego Garcia (Chagos Archipelago). Coral Reefs 27: 167–178.

Rattray A, Ierodiaconou D, Laurenson L, Burq S, Reston M (2009) Hydro-acoustic remote sensing of benthic biological communities on the shallow south east Australian continental shelf. Estuarine Coastal Shelf Sciences 84: 237–245.

Rayfield B, James PMA, Fall A, Fortin MJ (2008) Comparing static versus dynamic protected areas in the Québec boreal forest. Biological Conservation 141: 438–449.

Ricklefs RE, Lovette IJ (1999) The roles of island area per se and habitat diversity in the species-area relationships of four Lesser Antillean faunal groups. Journal of Animal Ecology 68: 1142–1160.

Robbins BD, Bell SS (2000) Dynamics of a subtidal seagrass landscape: seasonal and annual change in relation to water depth. Ecology 81: 1193–1205.

Roberts CM, Andelman S, Branch G, Bustamante RH, Castilla JC, Dugan J, Halpern BS, Lafferty KD, Leslie H, Lubchenco J, McArdle D, Possingham HP, Ruckelshaus M, Warner RR (2003) Ecological criteria for evaluating candidate sites for marine reserves. Ecological Applications 13: 199–214.

Rossi JP, Halder I van (2010) Towards indicators of butterfly biodiversity based on a multiscale landscape description. Ecological Indicators 10: 452–458.

Rouget M, Richardson DM, Cowling RM (2003) The current configuration of protected areas in the Cape Floristic Region, South Africa – Reservation bias and representation of biodiversity patterns and processes. Biological Conservervation 112: 129–145.

Saarman E, Gleason M, Ugoretz J, Airamé S, Carr M, Fox E, Frimodig A, Mason T, Vasques J (2013) The role of science in supporting marine protected area network planning and design in California. Ocean and Coastal Management 74: 45–56.

Sale PF, Cowen RK, Danilowicz BS, Jones GP, Kritzer JP, Lindeman KC, Planes S, Polunin NVC, Russ GR, Sadovy YJ, Steneck RS (2005) Critical science gaps impede use of no-take fishery reserves. Trends in Ecology Evolution 20: 74–80.

Santos RO, Lirman D, Serafy JE (2011) Quantifying freshwater-induced fragmentation of submerged aquatic vegetation communities using a multi-scale landscape ecology approach. Marine Ecology Progress Series 427: 233–246.

Sarkar S, Pressey R, Faith D, Margules C, Fuller T, Stoms D, Moffett A, Wilson K, Williams K, Williams P, Andelman S (2006) Biodiversity conservation planning tools: present status and challenges for the future. Annual Review of Environment and Resources 31: 123–159.

Scheltema RS (1986) Long-distance dispersal by planktonic larvae of shoal-water benthic invertebrates among central Pacific islands. Bulletin Marine Science 39: 241–256.

Schindler S, Poirazidis K, Wrbka T (2008) Towards a core set of landscape metrics for biodiversity assessments: A case study from Dadia National Park, Greece. Ecological Indicators 8: 502–514.

Schindler S, Wehrden H Von, Poirazidis K, Wrbka T, Kati V (2013) Multiscale performance of landscape metrics as indicators of species richness of plants, insects and vertebrates. Ecological Indicators 31: 41–48.

Scholz AT, Horrall RM, Cooper JC, Hasler AD (1976) Imprinting to chemical cues: the basis for home stream selection in salmon. Science 192(4245): 1247–1249.

Schroeter SC, Reed DC, Kushner DJ, Estes JA, Ono DS (2001) The use of marine reserves in evaluating the dive fishery for the warty cucumber (*Parastichopus parvimensis*) in California, USA Canadian Journal of Fisheries and Aquatic Science 58: 1773–1781.

Sisk TD, Haddad NM, Ehrlich PR (1997) Bird assemblages in patchy woodlands: modelling the effects of edge and matrix habitats. Ecological Applications 7: 1170–1180.

Skokanová H, Eremiášová R (2013) Landscape functionality in protected and unprotected areas: Case studies from the Czech Republic. Ecological Informatics 14: 71–74.

Smith TB, Glynn PW, Maté JL, Toth LT, Gyory J (2014) A depth refugium from catastrophic coral bleaching prevents regional extinction. Ecology 95(6): 1663–1673.

Sousa WP (1984) Intertidal mosaics: Patch size, propagule availability, and spatially variable patterns of succession. Ecology 65: 1918–1935.

Stamoulis KA, Friedlander AM (2013) A seascape approach to investigating fish spillover across a marine protected area boundary in Hawai'i. Fisheries Research 144: 2–14.

Stamps JA, Buechner M, Krishnan VV (1987) The effects of edge permeability and habitat geometry on emigration from patches of habitat. American Naturalist 129: 533–552.

Stevens T, Connolly RM (2005) Local-scale mapping of benthic habitats to assess representation in a marine protected area. Marine and Freshwater Research 56: 111–123.

Stoms DM, Davis FW, Andelman SJ, Carr MH, Gaines SD, Halpern BS, Hoenicke R, Leibowitz SG, Leydecker A, Madin EMP, Tallis H, Warner RR (2005) Integrated coastal reserve planning: making the land-sea connection. Frontiers in Ecology and the Environment 3: 429–436.

Stoner AW, Manderson JP, Pessutti JP (2001) Spatially explicit analysis of estuarine habitat for juvenile winter flounder: Combining generalized additive models and geographic information systems. Marine Ecology Progress Series 213: 253–271.

Sundblad G, Bergstrom U, Sandstrom A (2011) Ecological coherence of marine protected area networks: a spatial assessment using species distribution models. Journal of Animal Ecology 48: 112–120.

Svancara LK, Scott JM, Loveland TR, Pidgorna AB (2009) Assessing the landscape context and conversion risk of protected areas using satellite data products. Remote Sensing of Environment 113: 1357–1369.

Teixidó N, Garrabou J, Arntz WE (2002) Spatial pattern quantification of Antarctic benthic communities using landscape indices. Marine Ecology Progress Series 242: 1–14.

Teixidó N, Garrabou J, Gutt J, Arntz WE (2007) Iceberg disturbance and successional spatial patterns: The case of the shelf Antarctic benthic communities. Ecosystems 10: 142–157.

Termorshuizen JW, Opdam P, Brink A van den (2007) Incorporating ecological sustainability into landscape planning. Landscape Urban Planning 79: 374–384.

Townsend PA, Lookingbill TR, Kingdon CC, Gardner RH (2009) Spatial pattern analysis for monitoring protected areas. Remote Sensing of Environment 113: 1410–1420.

Tremblay I, Thomas DW, Lambrechts MM, Blondel J, Perret P (2003) Variation in blue tit breeding performance across gradients in habitat richness. Ecology 84: 3033–3043.

Treml EA, Halpin PN, Urban DL, Pratson LF (2008) Modelling population connectivity by ocean currents, a graph-theoretic approach for marine conservation. Landscape Ecology 23: 19–36.

Treml EA, Roberts JJ, Chao Y, Halpin PN, Possingham HP, Riginos C (2012) Reproductive output and duration of the pelagic larval stage determine seascape-wide connectivity of marine populations. Integrated Computational Biology 52: 525–537.

Trzcinski MK, Fahrig L, Merriam G (1999) Independent Effects of Forest Cover and Fragmentation on the Distribution of Forest Breeding Birds. Ecological Applications 9: 586–593.

Turner MG (2005) What Is the State of the Science? Landscape Ecologist 36: 319–344.

Turner MG, Gardner RH, O'Neill R V (2001) Landscape Ecology in Theory and Practice. 1st edition. Springer, New York, NY.

Uezu A, Metzger JP, Vielliard JME (2005) Effects of structural and functional connectivity and patch size on the abundance of seven Atlantic Forest bird species. Biological Conservation 123: 507–519.

Uuemaa E, Mander Ü, Marja R (2013) Trends in the use of landscape spatial metrics as landscape indicators: a review. Ecological Indicators 28: 100–106.

Vasques GM, Grunwald S, Myers DB (2012) Associations between soil carbon and ecological landscape variables at escalating spatial scales in Florida, USA. Landscape Ecology 27: 355–367.

Wang Y, Moskovits DK (2002) Tracking fragmentation of natural communities and changes in land cover: applications of Landsat data for conservation in an urban landscape (Chicago Wilderness). Conservation Biologist 15: 835–843.

Ward TJ, Vanderklift MA, Nicholls AO, Kenchington RA (1999) Selecting Marine Reserves Using Habitats and Species Assemblages as Surrogates for Biological Diversity. Ecological Applications 9: 691–698.

Wedding LM, Christopher LA, Pittman SJ, Friedlander AM, Jorgensen S (2011) Quantifying seascape structure: Extending terrestrial spatial pattern metrics to the marine realm. Marine Ecology Progress Series 427: 219–232.

Wedding LM, Friedlander AM, McGranaghan M, Yost RS, Monaco ME (2008) Using bathymetric lidar to define nearshore benthic habitat complexity: Implications for management of reef fish assemblages in Hawaii. Remote Sensing of Environment 112: 4159–4165.

White JW, Botsford LW, Moffitt EA, Fischer DT (2014) Decision analysis for designing marine protected areas for multiple species with uncertain fishery status. Ecological Applications 20: 1523–1541.

Wiens JA (1997) The emerging role of patchiness in conservation biology. In Pickett STA, Ostfeld RS, Shachak M, Likens GE (eds) The Ecological Basis of Conservation: Heterogeneity, Ecosystems, and Biodiversity. Chapman & Hall, New York, NY, pp. 93–107.

Wiens JA (2009) Landscape ecology as a foundation for sustainable conservation. Landscape Ecology 24: 1053–1065.

Wiens JA, Milne BT (1989) Scaling of 'landscapes' in landscape ecology, or, landscape ecology from a beetle's perspective. Landscape Ecology 3: 87–96.

Williams JN, Seo C, Thorne J, Nelson JK, Erwin S, O'Brien JM, Schwartz MW (2009) Using species distribution models to predict new occurrences for rare plants. Diversity and Distributions 15: 565–576.

Woodson CB, McManus MA, Tyburczy JA, Barth JA, Washburn L, Caselle JE, Carr MH, Malone DP, Raimondi PT, Menge BA, Palumbi SR (2012) Coastal fronts set recruitment and connectivity patterns across multiple taxa. Limnology and Oceanography 57: 582–596.

Wright DJ, Heyman WD (2008) Introduction to the special issue: marine and coastal GIS for geomorphology, habitat mapping, and marine reserves. Marine Geodesy 31: 223–230.

Wright DJ, Pendleton M, Boulware J, Walbridge S, Gerlt B, Eslinger D, Sampson D, Huntley E (2012) ArcGIS Benthic Terrain Modeler (BTM) v. 3.0, Environmental Systems Research Institute, NOAA Coastal Services Center, Massachusetts Office of Coastal Zone Management, at http://esriurl.com/5754. (accessed 25 May 2017).

Wu J (2013) Key concepts and research topics in landscape ecology revisited: 30 years after the Allerton Park workshop. Landscape Ecology 28: 1–11.

Yang X, Liu Z (2005) Quantifying landscape pattern and its change in an estuarine watershed using satellite imagery and landscape metrics. International Journal of Remote Sensing 26: 5297–5323.

Young M, Carr M (2015a) Assessment of habitat representation across a network of marine protected areas with implications for the spatial design of monitoring. PLoS One 10: e0116200.

Young M, Carr MH (2015b) Application of species distribution models to explain and predict the distribution, abundance and assemblage structure of nearshore temperate reef fishes. Diversity and Distributions 21: 1428–1440.

Zajac RN, Vozarik JM, Gibbons BR (2013) Spatial and temporal patterns in macrofaunal diversity components relative to sea floor landscape structure. PLoS One 8: e65823.

Zharikov Y, Lank DB, Huettmann F, Bradley RW, Parker N, Yen PPW, Mcfarlane-Tranquilla LA, Cooke F (2006) Habitat selection and breeding success in a forest-nesting Alcid, the marbled murrelet, in two landscapes with different degrees of forest fragmentation. Landscape Ecology 21: 107–120.

15

Seascape Economics: Valuing Ecosystem Services across the Seascape

Edward B. Barbier

15.1   Introduction

The term *seascape* is now widely used to refer to spatial mosaics of interconnected coastal and near-shore marine habitat types, such as mangroves, saltmarsh, seagrasses and coral reefs (Moberg & Rönnbäck 2003; Harborne *et al.* 2006; Mumby 2006; Pittman *et al.* 2011; Berkström *et al.* 2012; Olds *et al.* 2015). As a result of nutrient fluxes and material exchange including movements of marine fauna, these habitats provide important goods and services both individually and through functional linkages across the seascape (Harborne *et al.* 2006; Olds *et al.* 2015) (Figure 15.1). Both the seascape concept and its focus on connectivity have built upon key concepts in landscape ecology that have proven useful for policy and management. Landscape ecology has demonstrated that spatial aspects of landscape composition and configuration play a major role in maintaining biodiversity and ecosystem services and consequently influence human wellbeing. As pointed out by Grober-Dunsmore *et al.* (2009) and Wedding *et al.* (2011), connectivity is an important concept inherited from landscape ecology and now applied to define a seascape.

An emerging area of research is the development of economic models for integrated coastal-marine ecosystems that focus on the consequences of spatial patterning across the seascape (Barbier & Lee 2014). A particular emphasis has been the need to understand how connectivity influences the provision of various benefits (Barbier 2007, 2012; Sanchirico & Mumby 2009; Sanchirico & Springborn 2011; Smith & Crowder 2011; Plummer *et al.* 2013; Barbier & Lee 2014). There are several ways in which connectivity is taken into account (Barbier 2011). One approach is to assess the multiple benefits arising from interconnected habitats, such as estuaries (*e.g.*, Johnston *et al.* 2002). Another method is to consider how ecosystem goods and services are affected by the biological connectivity of habitats, food webs and migration and lifecycle patterns across specific seascapes, such as mangrove-seagrass-reef systems (*e.g.*, Sanchirico and Mumby 2009; Sanchirico and Springborn 2011; Plummer *et al.* 2013). Finally, some models examine how specific ecosystem services, such as habitat-fishery linkages, storm protection and water quality, are influenced by their spatial variation across a specific seascape habitat or habitats (Aburto-Oropeza *et al.* 2008; Barbier *et al.* 2008, 2013; Smith & Crowder 2011; Barbier 2012; Barbier & Lee 2014).

Seascape Ecology, First Edition. Edited by Simon J. Pittman.

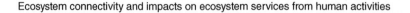

Ecosystem connectivity and impacts on ecosystem services from human activities

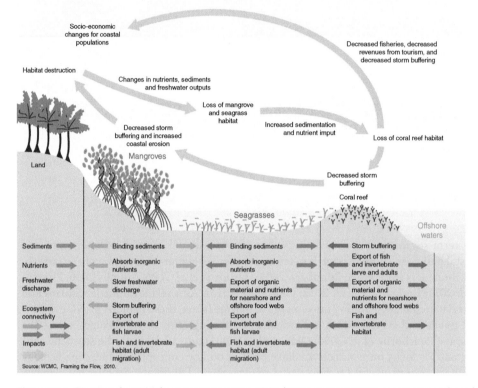

Figure 15.1 Diagram showing the ecosystem connectivity between mangroves, seagrasses and coral reefs. Ecological and physical connectivity between ecosystems is depicted for each ecosystem: terrestrial (brown arrows); mangroves (green arrows); seagrasses (blue arrows); and coral reefs (red arrows). Potential feedbacks across ecosystems from the impacts of human activities on ecosystem services are shown with yellow arrows. *Source*: Silvestri & Kershaw (2010), reproduced with permission from UNEP World Conservation Monitoring Centre.

As these approaches illustrate, valuing ecosystem goods and services across a seascape is providing new analytical, management and policy insights. The purpose of the following chapter is to review this growing literature and to illustrate with a case study how connectivity among coastal and marine habitats matters to the valuation of goods and services. Such a review is also important for identifying future valuation challenges and areas of research. The next section briefly overviews and discusses the ways in which habitat connectivity may influence the provision of ecosystem goods and services in a marine seascape. It is followed by a review of selected studies that have incorporated some aspect of this connectivity in order to value marine and coastal ecosystem goods and services. The chapter then summarizes the case of a mangrove-coral reef seascape explored by Barbier & Lee (2014) to show more explicitly the insights of applying the seascape concept to valuing ecosystem services. The chapter concludes by discussing some of the management and policy implications of valuing the various benefits of seascapes, as well as the challenges and key areas for future research.

15.2   Habitat Connectivity and Seascape Goods and Services

The influence of habitat connectivity on various ecosystem goods and services depends on whether that connectivity is *unidirectional* or *reciprocal* in the seascape. Unidirectional connectivity describes a one-way flow of biological, physical and chemical processes, such as from near-shore marine to coastal habitats, or vice versa. In contrast, reciprocal connectivity implies that these processes may interact back and forth between the marine and coastal habitats. Table 15.1 provides examples of how various types of connectivity between different patch types influence ecosystem services. Storm and flood protection usually involves unidirectional connectivity across the marine seascape, from the near-shore marine to coastal habitats. For example, coral reefs often shelter coastal habitats by buffering oceanic waves and currents and slowing down periodic storm surge (Ferrario *et al.* 2014) and thus may enhance the ability of seagrass beds, marshes, mangroves and other coastal habitats to attenuate waves and buffer winds (Moberg & Rönnbäck 2003; Harborne *et al.* 2006; Spalding *et al.* 2014) (Figure 15.1). Another unidirectional connectivity, from coastal to near-shore habitats, could arise through water and pollution control by coastal systems that not only benefits nearby human populations and activities, but also protects near-shore marine habitats, thus enhancing their goods and services. For example, mangroves, seagrass meadows, salt marsh and other coastal habitat also protect and enhance offshore coral reefs by interrupting freshwater discharge and acting as sinks for pollutants and organic materials (Mitsch & Gosselink 2000; Bowen & Valiela 2004; Shapiro *et al.* 2010; Nelson & Zavaleta 2012). Finally, coastal habitats serve as breeding and nursery grounds for marine fisheries, while at the same time, reefs and other marine habitats protect coastal systems from wave action and their ability to act as reproductive habitats. This represents reciprocal connectivity between the near-shore marine and coastal habitats of a seascape. For example, many fish and shellfish species utilize mangroves and seagrass beds as nursery grounds, then migrate to coral reefs as adults, with some species returning to the mangroves and seagrasses to spawn (Pittman & McAlpine 2003; Sheaves *et al.* 2006; Nagelkerken *et al.* 2015). Mumby *et al.* (2004) found that the biomass of adults of several commercially valuable fish species was higher where mangroves were adjacent to the adult coral reef habitats. The result is a close relationship between reef fishery densities and the proximity of mangrove and seagrass nurseries in the seascape (Dorenbosch *et al.* 2004; Mumby 2006; Kimirei *et al.* 2011). Similar synergies exist for salt marshes, seagrass beds and near-shore marine systems in estuaries (Ray 2005; Pittman *et al.* 2007; Rountree & Able 2007; Boström *et al.* 2011; Nagelkerken *et al.* 2015).

Taking into account the possible unidirectional or reciprocal connectivity of habitats comprising the seascape clearly has implications for both the valuation of benefits, such as storm and flood protection, habitat-fishery linkages and pollution control and the management of coastal and near-shore marine habitats. For example, because of seascape connectivity, a decision to develop, exploit or protect one part of the seascape, such as a coastal wetland, could have important implications for the habitat in the rest of the seascape, such as a coral reef and thus the goods and services they provide. The rest of this chapter explores these implications further, by reviewing specific case studies that have incorporated seascape connectivity into the valuation of ecosystem goods and services.

Table 15.1 Seascape connectivity of interconnected habitat types and the consequences for ecosystem services.

Ecosystem structure/function	Ecosystem services	Examples of connectivity between coastal and near-shore marine habitats in the seascape
Attenuates and/or dissipates waves, buffers wind	Storm and flood protection	*Unidirectional connectivity* (from near-shore marine to coastal habitats): reefs, seagrass beds, mangroves, marsh and sand dunes attenuate storm surge waves and slow winds, thus having a direct impact on protecting coastal communities and their property. However, reefs also shelter the coastal shore and its habitats, resulting in healthier coastal habitats and enhancing their storm and flood protection.
Provides nutrient and pollution uptake, as well as retention, particle deposition and clean water	Water pollution and sediment control	*Unidirectional connectivity* (from the coastal to near-shore marine habitats): water pollution and sediment control by mangroves, seagrass meadows, salt marsh and other coastal habitats leads to cleaner water supplies for nearby human populations and economic activities. By interrupting freshwater discharge and acting as sinks for pollutants and organic materials, these coastal habitats also protect near-shore marine habitats, thus enhancing their goods and services.
Provides suitable reproductive habitat and nursery grounds, sheltered living space	Maintains fishing, hunting and foraging activities	*Reciprocal connectivity* (between near-shore marine and coastal habitats): many fish and shellfish species utilize mangroves, marshes and seagrass beds as nursery grounds and eventually migrate to near-shore habitats as adults, returning to the coastal habitats to spawn. While coastal habitats are supporting marine fisheries, reefs may shelter seagrass beds, marshes, mangroves and other coastal habitats, thus enhancing their ability to serve as fish breeding and nursery grounds.

15.3 Valuing Seascape Goods and Services

Increasingly, economists are interested in estimating the multiple benefits arising from interconnected coastal and marine habitats. For example, Johnston *et al.* (2002) estimate the benefits arising from a wide range of ecosystem services provided by the Peconic Estuary on Long Island, New York. The tidal mudflats, salt marshes and seagrass (eelgrass) beds of the estuary support the shellfish and demersal finfish fisheries. In addition, bird watching and waterfowl hunting are popular activities. The authors incorporate production function methods to simulate the biological and food web interactions of the ecosystems in order to assess the marginal value per acre in terms of gains in commercial value for fish and shellfish, bird watching and waterfowl hunting. The aggregate annual benefits are estimated to be $67 per acre for intertidal mud flats, $338 for salt marsh and $1065 for seagrass across the estuary system. Using these estimates, the authors calculate the asset value of protecting existing habits to be $12 412 per acre for seagrass, $4291 for salt marsh and $786 for mud flats, in comparison, the asset value of restored habitats is $9996 per acre for seagrass, $3454 for marsh and $626 for mudflats.

Some economic assessments also consider how ecosystem goods and services are affected by modelling the biological connectivity of habitats, food webs and migration and lifecycle patterns across specific seascapes, such as mangrove-seagrass-reef systems (Barbier & Lee 2014). For example, Plummer *et al.* (2013) use dynamic simulations in a food web model of central Puget Sound, Washington, United States, to determine how eel-grass beds provide valuable refuge, foraging and spawning habitat for many commercially and recreationally fished species, including Pacific salmon, Pacific herring and Dungeness crab. The authors found that a 20% increase in the area of eel-grass beds would lead to increases of $3/km^2 (of eel-grass area) in the value of herring, $15/km^2 for crabs and $316/km^2 for salmon. However, the authors also examine the potential tradeoffs of enhanced eel-grass beds with other coastal ecosystem benefits, notably bird and marine mammal watching. The latter activities were not valued, but the quantity of the service was estimated in terms of days of the recreational activity. The tradeoff analysis reveals that the positive values to recreational and commercial fishing from increased eelgrass coverage is also associated with a slight rise in marine mammal watching, but does lead to a slight decline in bird watching.

Sanchirico & Mumby (2009) developed an integrated seascape model to illustrate that the presence of mangroves and seagrasses considerably enhances the biomass of coral reef fish communities. A key finding is that mangroves become more important as nursery habitat when excessive fishing effort levels are applied to the reef, because the mangroves can directly offset the negative impacts of fishing effort. The results of this study support the development of ecosystem-based fishery management and the design of integrated coastal-marine reserves that emphasize four key priorities: (i) the relative importance of mangrove nursery sites, (ii) the connectivity of individual reefs to mangrove nurseries, (iii) areas of nursery habitat that have an unusually large importance to specific reefs and (iv) priority sites for mangrove restoration projects (Mumby 2006). A further extension of this work maps the lifecycle of fish through coral reef-mangrove-seagrass systems to determine the contribution of each habitat to the biological growth and productivity of marine fisheries and to the protection of coastal properties from storm damage (Sanchirico & Springborn 2011). The authors demonstrate how payments for ecosystem services for a particular habitat with multiple services are interdependent, change over time and can be greater than the profits associated with fishing activities that degrade the coral reef-mangrove-seagrass system.

Stål *et al.* (2008) attempt to determine how three coastal habitats (soft sediment bottoms, seagrass beds and rocky bottoms with macroalgae) influence the value of marine commercial, recreational and fisheries off the west coast of Sweden. The five species studied are cod, plaice, eel, mackerel and sea trout. The distributional pattern and habitat dependence varied among fish species. For example, plaice were more highly associated with shallow soft bottoms as a nursery habitat for juveniles but migrate to offshore fisheries as adults. Eel mainly inhabit seagrass beds and rocky bottoms, which is where they are frequently harvested. Cod are primarily caught in offshore fisheries, but their juveniles depend on rocky shores with algae and seagrass beds on soft sediment bottoms. As a consequence, decrease in the seagrass beds had a significant negative impact on the catch, working hours and profits of eel fisherman. Macroalgae coverage of 30–50% of rocky bottoms reduces significantly recruitment in the plaice fishery and substantially lowers its net present value. Finally, loss of seagrass habitats on the Swedish west coast may have accounted for around 3% of the total annual cod recruitment in the entire North Sea and Swedish Skagerrak archipelago system.

Some marine ecosystem services, such as habitat-fishery linkages, storm protection and water quality, are influenced by their spatial variation across the seascape. For example, evidence suggests that, for mangroves and salt marshes, wave attenuation and nursery fish diversity are greatest on the seaward edge of these systems, but tend to diminish with the distance inshore from the seaward fringe (Peterson & Turner 1994; Rountree & Able 2007; Aburto-Oropeza *et al.* 2008; Aguilara-Perera & Appeldorn 2008; Barbier *et al.* 2008; Loneragan *et al.* 2005; Meynecke *et al.* 2008; Koch *et al.* 2009; Gedan *et al.* 2011; Ysebaert *et al.* 2011; Shephard *et al.* 2012). Increasingly, valuation studies of coastal and marine ecosystem services are taking into account the spatial variability in the ecological production functions underlying these services.

In the Gulf of California, Mexico the mangrove fringe with a width of 5–10 m has the most influence on the productivity of near-shore fisheries, with a median value of $37 500 per hectare. Fishery landings also increased positively with the length of the mangrove fringe in a given location (Aburto-Oropeza *et al.* 2008). Barbier *et al.* (2008) show that how much of a mangrove forest coastline should be converted to shrimp aquaculture may depend critically on the spatial variability of coastal storm protection across the mangrove landscape. Barbier (2012) also includes the spatial variability of the support for offshore fisheries in the decision to convert a mangrove landscape to shrimp farms.

In the absence of any spatially declining production of storm protection and habitat-linkage benefits, Barbier (2012) estimated that the total value of all mangrove ecosystem services ($18 978/ ha) exceeded the commercial net returns from shrimp farming ($9632/ha) and the economic net returns ($1220/ha), which are the commercial returns adjusted for the subsidies for shrimp farm operations. If the benefits of mangrove ecosystem services are constant across the seascape, then the entire ecosystem should be preserved. However, spatial variation in the production of storm protection and habitat-fishery linkages changes the land use outcome significantly. When these two mangrove services decline spatially, it causes the total net benefit of all ecosystem services to diminish significantly from the seaward edge to the landward boundary. When the value of preserving the mangroves is compared to the commercial net returns to shrimp farming, it is optimal to conserve only the first 118 m of mangroves from the seaward edge and to convert the rest to aquaculture. When compared to the net economic returns to shrimp farming, mangroves up to 746 m from the seaward edge should be preserved and the remaining mangroves inland converted. Thus, accounting for the spatial production and distribution of ecosystem services across a coastal seascape influences not only the amount of land converted but also where the conversion activity takes place.

Recent studies determining the economic value of coastal wetlands in protecting against damages and loss of life from storm surges are also accounting for the effects of varying wetland presence and vegetation across coastal landscapes in slowing down storm surge. In addition, mangrove trees also have the capacity to buffer winds (Das & Crépin 2013). The study by Das & Vincent (2009) of casualties from the 1999 Cyclone Orissa in India allows for topographical factors, such as low elevation, the availability of dikes and the presence of estuarine rivers, but does not include the varying wave attenuation properties of mangrove vegetation in reducing storm surge. Laso Bayas *et al.* (2011) do take into account land cover roughness (including vegetation friction) of mangroves, plantations and other coastal land uses in mitigating storm surge caused by the 2004 Indian Ocean tsunami in west Aceh, Indonesia. By combining hydrodynamic analysis of simulated hurricane storm surges and economic valuation of expected property damages, Barbier *et al.* (2013) and Barbier & Enchelmeyer (2014) showed that the presence of coastal marshes and their vegetation has a demonstrable effect on reducing storm surge levels, thus generating significant values in terms of protecting property in southeast Louisiana, United States. Simulations for four storms along a sea to land transect show that surge levels decline with wetland continuity (the ratio of wetlands to water) and vegetation roughness along each 6 km segment from open sea to the inland shore. A 0.1 increase in wetland continuity per meter reduces property damages for the average affected area analysed in southeast Louisiana, which includes New Orleans, by $99–$133 and a 0.001 increase in vegetation roughness decreases damages by $24–$43. These reduced damages are equivalent to saving three to five and one to two properties per storm for the average subplanning unit area, respectively.

Water and pollution control by mangroves, seagrass beds, marsh and other coastal habitats not only benefit nearby human populations, activities and property but also protect near-shore marine habitats, thus enhancing their goods and services (Mitsch & Gosselink 2000; Moberg & Rönnbäck 2003; Bowen & Valiela 2004; Harborne *et al.* 2006; Shapiro *et al.* 2010; Nelson & Zavaleta 2012; Berkström *et al.* 2012). Figure 15.1 shows a schematic diagram of ecosystem connectivity across interconnected coastal seascape comprising mangroves, seagrasses and coral reefs. The diagram highlights the flow of ecological processes and resultant ecosystem services across the seascape and shows some of the consequences that may occur if the integrity of one or more of the component patch types is disrupted. In this way, the interconnected system can be said to provide greater quantity and quality of services than can be estimated from the sum of its parts.

Increasingly, economists are estimating the benefits associated with this service, in terms of improved human health, residential property value and support for marine fisheries (Breaux *et al.* 1995; Leggett & Bockstael 2000; Massey *et al.* 2006; Smith 2007; Silvestri & Kershaw 2010; Smith & Crowder 2011). In some cases, the benefit attributed to water and pollution control by estuarine and coastal habitats has to be inferred (Massey *et al.* 2006). However, in an explicitly spatial predator-prey model of the commercial blue crab fishery in North Carolina, United States, Smith & Crowder (2011) determined how reductions in nitrogen pollution and hypoxia (low dissolved oxygen) in the Neuse River Estuary benefit the fishery. One of the key assumptions underlying the model is the spatial nature of nutrient loads and flow from the estuary to the open sound, that is, 'nutrient levels affect oxygen in the estuary, but there is enough mixing in the Sound to eliminate the nutrient-induced oxygen depletion, at least in the range of current loadings' (Smith & Crowder 2011, p. 2257). The authors found that a 30% reduction in nitrogen loadings in the estuary increases the discounted present value of fishery rent by $2.56 million, although this estimate could vary from $195 000 to $7.51 million.

15.4 Example of a Mangrove-Coral Reef Seascape

To provide further insight into the management implications of valuing ecosystem services across a seascape, Barbier & Lee (2014) developed a simple model of a two-habitat marine system. The model illustrates how the connectivity of two habitats (a near-shore marine ecosystem and a coastal mangrove habitat) comprising the seascape influences the provision of certain ecosystem services. For example, the ecological function of the coastal habitat as a fish nursery and breeding ground enhances the biological productivity of the broader seascape through lifecycle movements. In addition, the presence of coral reefs in the near-shore marine environment attenuates waves thus enhancing the storm protection service of the coastal habitat. The model incorporates spatial variation in the seascape

Table 15.2 summarizes the key results of the seascape model developed by Barbier & Lee (2014) for the cases of unidirectional connectivity from the marine to coastal habitat (*e.g.*, storm protection), unidirectional connectivity from the coastal to marine habitat (*e.g.*, water pollution and sediment control and reciprocal connectivity between the two habitats (*e.g.*, coastal habitat-marine fishery linkage). As indicated in Table 15.2, these scenarios were applied to a representative mangrove-coral reef system, in which the mangroves faced irreversible conversion to commercial shrimp farms. For each scenario, the outcome for when seascape connectivity was taken into account was compared to the outcome for optimal conversion of the mangrove when its ecosystem services were considered in isolation from the rest of the seascape (*i.e.*, the coral reef).

The first scenario considers the storm protection service provided by mangroves. For this service, the relevant connectivity is unidirectional across the seascape, from the coral reef of the near-shore marine habitat to the coastal mangroves. The result was the direct and indirect benefit of coral reefs in terms of storm protection. First, coral reefs attenuate storm surge waves, thus having a direct impact on protecting coastal communities and their property from flood damages caused by storm surges. Second, more area of coral reef also leads to greater sheltering of the coastal shore and the seagrass beds, marshes, mangroves and other habitats. This sheltering and protection effect is especially relevant to the seaward edge of coastal habitats, as damaging storms and surges approach the coastline from the open sea. The result is healthier coastal habitats and the vegetation they contain resulting in greater attenuation of storm surge waves and thus additional protection of coastal communities and property.

Without considering this connectivity between coral reef and mangrove storm protection, the decision to convert mangroves to commercial shrimp farms was very similar to the outcome in Barbier (2012). It was considered optimal to conserve the first 515 m from the seaward edge and convert the rest to shrimp farms. However, if the enhancement of mangrove storm protection by coral reefs is taken into account, then conservation of mangroves should extend further to 563 m inland. For water pollution

Table 15.2 Summary of seascape connectivity implications for management of mangrove-coral reef system.

Scenario	Description	Interpretation	Outcome
Unidirectional connectivity from marine to coastal habitat (e.g., storm protection)	Determining the optimal width of a coastal landscape and the amount of coral reef area affected by development when unidirectional connectivity between the habitats influences the ecosystem service	The storm protection service provided by the optimal coastal landscape width must take into account any sheltering of coastal habitats by the optimal coral reef area (10 km²). The marginal rents of any development that affects the coral reef must offset any resulting loss to the direct and indirect storm protection benefits provided by the reef.	Without connectivity, it is optimal to conserve the first 515 m of mangroves from the seaward edge, with connectivity, conservation of mangroves should extend further to 563 m inland.
Unidirectional connectivity from coastal to marine habitat (e.g., water pollution and sediment control)	Determining the optimal width of a coastal landscape and the amount of coral reef area affected by optimal development when unidirectional connectivity between the habitats influences the ecosystem service	The gains from preserving the optimal coastal landscape include not only the direct benefits of producing cleaner water in coastal areas but also the indirect benefits of water pollution and sediment control in terms of protecting marine ecosystem services. The marginal rents from any development that impact the coral reef must equal any direct loss to marine ecosystem services.	Without connectivity, it is optimal to convert up to 419 m inland from the seaward boundary and to conserve the remaining mangroves, with connectivity, only the first 385 m inland should be converted to shrimp farms and the rest of the landscape is preserved.
Reciprocal connectivity between coastal and marine habitat (e.g., coastal habitat-marine fishery linkage)	Determining the optimal width of a coastal landscape and the amount of coral reef area affected by optimal development when reciprocal connectivity between the habitats influences the ecosystem service	The support for marine fisheries provided by the optimal coastal landscape width must take into account any sheltering of coastal habitats by the optimal coral reef area (10 km²). The marginal rents of any development that affects the coral reef must offset any resulting loss to the protection of coral habitat provided by the reef.	Without connectivity, mangroves should be preserved as a fishery breeding and nursery habitat for the first 50 m inland from the seaward boundary and the remaining inland area of mangrove landscape could be converted, with connectivity, the development decision does not change, but value of preserving the first 50 m of the mangrove seaward fringe for coastal-habitat–fishery linkage is considerably greater.

(Continued)

Table 15.2 (Continued)

Scenario	Description	Interpretation	Outcome
Multiple connectivity and ecosystem services	Determining the optimal width of a coastal landscape and the amount of coral reef area affected by optimal development when unidirectional and reciprocal connectivity between the habitats influence a range of ecosystem services	The seascape benefits associated with optimal coastal landscape include both a direct benefit in terms of additional coastal ecosystem services and an indirect benefit in terms of supporting additional marine ecosystem services. The seascape benefits impacted by coral reef development include both a direct loss to marine ecosystem services and an indirect loss to coastal services.	Without connectivity, the aggregation of all ecosystem benefits always exceeds the economic returns to conversion and thus the entire landscape should be preserved, with connectivity, the decision to preserve the coastal landscape is not changed, but the value of the preserved mangroves are considerably higher.

Source: Based on Barbier & Lee (2014).

and sediment control, the relevant seascape connectivity is also unidirectional, but from the coastal landscape to the near-shore marine habitat. More coastal landscape leads to greater control of water pollution and sediments that enter the seascape from its inland boundary. In addition, less pollution and sediment will flow into the marine environment, which suggests more protection of near-shore marine habitat and coral reef. The result is a direct and indirect benefit of this service across the seascape.

Barbier & Lee (2014) compared the scenario whereby the influence of seascape connectivity on water pollution and sediment control by mangroves was included compared with a scenario when this function was ignored (see Table 15.2). In both cases, it was optimal to convert some of the seaward edge of the mangroves to shrimp farming. However, where only the direct benefits of improving water quality were considered, the optimal solution was to convert up to 419 m of mangrove inland from the seaward boundary, whilst conserving the remaining mangroves. When the indirect benefit of supporting marine ecosystem services was included, the model indicated that the first 385 m inland from the terrestrial boundary should be converted to shrimp farms and the rest preserved (Figure 15.2). The model suggests that shrimp farming should only occur in the interior of the mangrove and never within the first 50 m of the seaward fringe. If seascape connectivity is ignored, this development should take place from 50 m to 461 m inland from the seaward boundary, but if seascape connectivity is considered, then it should be restricted to an interior location between 50 m and 385 m inland.

Without the seascape connectivity considered, the optimal solution was to preserve a 50 m wide seaward fringe of mangroves as a breeding and nursery site for fishery species. The remaining inland area of mangroves could be converted to mangroves. The inclusion of the sheltering effect of coral reefs does not change this development decision. However, the value of preserving the first 50 m of the mangrove seaward fringe for coastal-habitat fishery linkage is considerably greater. The final simulation indicated in

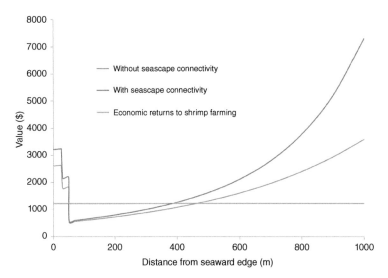

Figure 15.2 Simulation of seascape connectivity, water pollution / sediment control and habitat-fishery linkage in a mangrove-coral reef system.

Table 15.2 aggregates all three benefits: storm protection, water pollution and sediment control and coastal habitat-marine fishery linkage. The aggregation of these benefits always exceeds the economic returns of shrimp farming for all locations across the coastline. Thus, the most efficient solution would be to preserve the entire mangrove area. This outcome is not affected by seascape connectivity, although its inclusion raises substantially the value of the mangroves. The same outcome of preserving all mangroves occurs if the aggregate benefits consist only of water pollution / sediment control and storm protection. In addition, Barbier & Lee (2014) showed that the aggregate benefits of mangroves consisting of both storm protection and coastal habitat-marine fishery linkages was not significantly different if only storm protection was considered. Although the combined benefits of habitat-fishery and storm protection raise significantly the value of the seaward fringe of the mangroves, the decision to convert mangroves to shrimp farms and where to locate that development does not change compared to the case when only storm protection is considered (Table 15.2). Without connectivity, it is optimal to preserve the first 515 m of mangroves from the seaward edge. In contrast, with connectivity, preservation of mangroves should extend farther landward to 563 m from the seaward edge. However, if the aggregate benefits comprise just water pollution / sediment control and habitat-marine fishery linkages, then a very different result emerges, which is illustrated in Figure 15.2. Regardless of whether or not the effects of seascape connectivity on coastal benefits are considered, the first 50 m of the mangrove are always preserved because of their high value as nursery and breeding grounds. On the other hand, because water pollution and sediment control increases the value of mangroves on the landward edge, development should not occur on this side of the mangrove landscape either. In addition, the interior boundary of mangrove development is affected by seascape connectivity. Thus, shrimp farming should only occur in the interior of the mangrove forests. If seascape connectivity is ignored, this development should take place from 50 m to 461 m inland from the seaward boundary of the landscape, but if seascape connectivity is considered, then it should be restricted to an interior location between 50 m and 385 m inland.

Two important insights emerge from the seascape model constructed by Barbier & Lee (2014) and the various simulations based on it. First, incorporating seascape connectivity in models of ecosystem services can guide tradeoffs in coastal development and habitat conservation, specifically, quantitative thresholds can be derived to guide spatial planning decisions. As demonstrated here, taking into account the connectivity between mangrove forest and the surrounding seascape, such as a coral reef can help estimate some of the economic impacts of coastal development and where the development should be located in the landscape. Second, the unique ecological production of each service and especially how it varies spatially across a habitat, may also determine the way in which seascape connectivity may matter. In the case of storm protection, this benefit is generated mainly by the seaward part of the mangroves and more coral reef leads to greater sheltering of the coastal shore, therefore seascape connectivity extends coastal habitat conservation further inland from its seaward edge. Similarly, preserving coastal wetlands such as mangroves leads to greater control of water pollution and sediments entering the seascape from its landward boundary and this function is enhanced by preservation of vegetation at the landward boundary. However, in the case of coastal habitat-fishery linkage, because the value of this service is associated only with the extreme seaward edge of the mangroves, taking into account seascape connectivity may not affect development outcomes. Sheltering of the shoreline by coral reefs clearly increases the value of the seaward fringe for habitat-fishery linkage, but the high value of this service may already suggest protecting the mangrove fringe from development activities, even before the sheltering effect of coral reefs is considered.

This illustrative model of a mangrove-coral reef seascape focuses on a limited range of structural connectivity relationships that are determined by the spatial proximity of the mangrove and coral reef habitats (Calabrese & Fagan 2004). A more complete model of the connectivity of these two habitats within the seascape may involve more direct measurement of the biological, physical and chemical processes that create functional linkages within the seascape. In addition, economic valuation of storm protection, water pollution / sediment control and habitat-marine fishery linkages of either coral reef or mangrove habitats is relatively new and only recently have appropriate valuation methods have been developed to assess these benefits (Barbier 2007). Further methodological advances are likely necessary, if these important benefits are to be assessed for the entire connected seascape rather than individual habitat level.

15.5   Conclusions and Future Research Priorities

Ecological research spanning mosaics of habitat types have revealed that key functional linkages across the seascape influence the provisioning of ecosystem goods and services (Harborne *et al.* 2006; Olds *et al.* 2015). Mapping, measuring and valuing these benefits across the seascape is providing new bioeconomic, management and policy insights. For example, recent economic studies have focused on the multiple benefits arising from interconnected habitats (*e.g.,* Johnston *et al.* 2002), how ecosystem goods and services are affected by the biological connectivity of habitats, food webs and migration and lifecycle patterns across specific seascapes (*e.g.,* Sanchirico & Mumby 2009; Sanchirico & Springborn 2011; Plummer *et al.* 2013) and how specific ecosystem services are influenced by their spatial variation across a specific seascape habitat or habitats (Aburto-Oropeza *et al.* 2008; Barbier *et al.* 2008, 2013; Smith & Crowder 2011; Barbier 2012; Barbier & Lee 2014). Nevertheless, some important challenges for future research remain.

First, connectivity may also be a function of the distance between the various habitats comprising the seascape. For example, as emphasized by Berkstrom *et al.* (2012, p. 7): 'In some cases seagrass beds and mangroves are located close to coral reefs making different habitat types readily accessible to fish, hence creating a more connected seascape. In other cases, however, habitats are more separated in space leading to a more fragmented seascape. Studies from the Caribbean have shown that presence of seagrass and mangroves within 100–1000 m of a reef affects the abundance and species richness of coral reef fishes, especially mobile invertebrate feeders' (Pittman *et al.* 2007; Grober-Dunsmore *et al.* 2009). Taking into account distance between habitats and its impact on the connectivity of these habitats within a seascape should be an important focus of the next wave of research on the economics of the seascape.

Second, current economic studies of seascapes tend to focus on a limited range of structural connectivity relationships that are determined by the spatial proximity of the interconnected habitats (Calabrese & Fagan 2004). Further advances in valuing the additional benefits accruing from the connectivity among the mosaic of near-shore and coastal habitats may involve more direct measurement of the biological, physical and chemical processes that create functional linkages within the seascape. Determining which processes are important to measure and value seascapes will require more interdisciplinary collaboration between marine ecologists and economists. Development of spatial map-based frameworks for analyses and communication of ecosystem service values across spatially heterogeneous seascapes will help integrate economic data into the spatial planning process such as design and placement of marine protected areas and marine spatial plans.

Third, as is clear from this review, most economic models have been developed for specific types of seascapes, such as tropical coral reef-mangrove systems (Aburto-Oropeza *et al.* 2008; Barbier *et al.* 2008, 2013; Sanchirico & Mumby 2009; Sanchirico & Springborn 2011; Barbier 2012; Barbier & Lee 2014) or temperate estuaries (Johnston *et al.* 2002; Plummer *et al.* 2013). Clearly, the range of tropical and temperate seascapes considered needs to be expanded. In addition, only a limited number of ecosystem services are examined from a seascape perspective – usually habitat-fishery linkages, storm protection and water quality. Although much more work is still needed to determine how seascape connectivity may affect the value of these important benefits, other coastal and marine ecosystem goods and services need also to be assessed from a seascape perspective. Finally, economic valuation of the ecosystem services arising from the regulatory and habitat functions of coastal and marine systems is relatively new and only recently have appropriate valuation methods been developed to assess these benefits (Barbier 2007). Further methodological advances are likely necessary, if these important benefits are to be assessed at the seascape rather than individual habitat level. As this review indicates, progress in this area of research will continue to require interdisciplinary collaboration between marine ecologists and economists.

References

Aburto-Oropeza O, Ezcurra E, Danemann G, Valdez V, Murray J, E Sala (2008) Mangroves in the Gulf of California increase fishery yields. Proceedings of the National Academy of Sciences 105: 10456–10459.

Aguilar-Perera A, Appeldoorn R (2008) Spatial distribution of marine fishes along a cross-shelf gradient containing a continuum of mangrove-seagrass-coral reefs off southwestern Puerto Rico. Estuarine Coastal and Shelf Science 76: 378–394.

Barbier EB (2007) Valuing ecosystems as productive inputs. Economic Policy 22: 177–229.

Barbier EB (2011) Progress and challenges in valuing coastal and marine ecosystem services. Review of Environmental Economics and Policy 6(1): 1–19.

Barbier EB (2012) A spatial model of coastal ecosystem services. Ecological Economics 78: 70–79.

Barbier EB, Enchelmeyer B (2014) Valuing the storm surge protection service of US Gulf Coast Wetlands. Journal of Environmental Economics and Policy 3(2): 167–185.

Barbier EB, Georgiou IY, Enchelmeyer B, Reed DJ (2013) The value of wetlands in protecting southeast Louisiana from hurricane storm surges. PLOS One 8(3): e58715.

Barbier EB, Koch EW, Silliman BR, Hacker SD, Wolanski E, Primavera JH, Granek E, Polasky S, Aswani S, Cramer LA, Stoms DM, Kennedy CJ, Bael D, Kappel CV, Perillo GM, Reed DJ (2008) Coastal ecosystem-based management with nonlinear ecological functions and values. Science 319: 321–323.

Barbier EB, Lee KD (2014) Economics of the marine seascape. International Review of Environmental and Resource Economics 7: 35–65.

Berkström C, Gullström M, Lindborg R, Mwandya AW, Yahya SAS, Kautsky N, Nyström M (2012) Exploring 'knowns' and 'unknowns' in tropical seascape connectivity with insights from east African coral reefs. Estuarine Coastal and Shelf Science 107: 1–21.

Bowen JL, Valiela I (2004) Nitrogen loads to estuaries: using loading models to assess the effectiveness of management options to restore estuarine water quality. Estuaries 27: 482–500.

Boström C, Pittman SJ, Simenstad C, Kneib RT (2011) Seascape ecology of coastal biogenic habitats: advances, gaps and challenges. Marine Ecology Progress Series 427: 101–217.

Breaux A, Farber S, Day J (1995) Using natural coastal wetlands systems for wastewater treatment: an economic benefit analysis. Journal of Environmental Management 44(3): 285–291.

Calabrese JM, Fagan WF (2004) A comparison-shopper's guide to connectivity metrics. Frontiers in Ecology and the Environment 2(10): 529–536.

Carleton RG (2005) Connectivities of estuarine fishes to the coastal realm. Estuarine, Coastal and Shelf Science 64: 18–32.

Das S, Crépin A (2013) Mangroves can provide against wind damage during storms. Estuarine Coastal and Shelf Science 134: 98–107.

Das S, Vincent JR (2009) Mangroves protected villages and reduced death toll during Indian super cyclone. Proceedings of the National Academy of Sciences 106: 7357–7360.

Dorenbosch M, van Riel MC, Nagelkerken I, van der Velde G (2004) The relationship of reef fish densities to the proximity of mangrove and seagrass nurseries. Estuarine Coastal and Shelf Science 60: 37–48.

Ferrario F, Beck MW, Storlazzi CD, Micheli F, Shepard CC, Airoldi L (2014) The effectiveness of coral reefs for coastal hazard risk reduction and adaptation. Nature Communications 5(3794).

Johnston RJ, Grigalunas TA, Opaluch JJ, Mazzotta M, Diamantedes J (2002) Valuing estuarine resource services using economic and ecological models: the Peconic Estuary System. Coastal Management 30(1): 47–65.

Gedan KB, Kirwan ML, Wolanski E, Barbier EB, Silliman BR (2011) The present and future role of coastal wetland vegetation in protecting shorelines: answering recent challenges to the paradigm. Climatic Change 106: 7–29.

Grober-Dunsmore R, Pittman SJ, Caldow C, Kendall MS, Frazer TK (2009) A landscape ecology approach for the study of ecological connectivity across tropical marine seascapes. In Nagelkerken I (ed.) Ecological Connectivity among Tropical Coastal Ecosystems. Springer, New York, NY, pp. 493–530.

Harborne AR, Mumby PJ, Micheli F, Perry CT, Dahlgren CP, Holmes KE, Burmbaugh DR (2006) The functional value of Caribbean coral reef, seagrass and mangrove habitats to ecosystem processes. Advances in Marine Biology 50: 59–189.

Kimirei IA, Nagelkerken I, Giffioen B, Wagner C, Mgaya YD (2011) Ontogenetic habitat use by mangrove / seagrass-associated coral reef fisheries shows flexibility in time and space. Estuarine Coastal and Shelf Science 92: 47–58.

Koch EW, Barbier EB, Silliman BR, Reed DJ, Perillo GME, Hacker SD, Granek EF, Primavera JH, Muthiga N, Polasky S, Halpern BS, Kennedy CJ, Kappel CV, Wolanski E (2009) Non-linearity in ecosystem services: temporal and spatial variability in coastal protection. Frontiers in Ecology and the Environment 7: 29–37.

Leggett CG, Bockstael NE (2000) Evidence of the effects of water quality on residential land prices. Journal of Environmental Economics and Management 39(2): 121–144.

Laso Bayas, Carlos J, Marohn C, Dercon G, Dewi S, Piepho HP, Joshi L, van Noordwijk M, Cadisch G (2011) Influence of coastal vegetation on the 2004 tsunami wave impact Aceh. Proceedings of the National Academy of Sciences 108: 18612–18617.

Loneragan NR, Adnan NA, Connolly RM, Manson FJ (2005) Prawn landings and their relationship with the extent of mangroves and shallow waters in western peninsular Malaysia. Estuarine Coastal and Shelf Science 63(1): 187–200.

Massey DM, Newbold SC, Gentner B (2006) Valuing water quality changes using a bioeconomic model of a coastal recreational fishery. Journal of Environmental Economics and Management 52(1): 482–500.

Meynecke JO, Lee SY, Duke NC (2008) Linking spatial metrics and fish catch reveals the importance of coastal wetland connectivity to inshore fisheries in Queensland, Australia. Biological Conservation 141: 981–996.

Mitsch WJ, Gosselink JG (2000) The value of wetlands: importance of scale and landscape setting. Ecological Economics 35: 25–33.

Moberg F, Rönnbäck P (2003) Ecosystem services of the tropical seascape: interactions, substitutions and restoration. Ocean and Coastal Management 46: 27–46.

Mumby PJ (2006) Connectivity of reef fish between mangroves and coral reefs: algorithms for the design of marine reserves at seascape scales. Biological Conservation 128: 215–222.

Mumby PJ, Edwards AJ, Arias-Gonzalez JE, Lindeman KC, Blackwell PG, Gall A, Gorczynska MI, Harborne AR, Pescod CL, Renken H, Wabnitz CCC, Llewellyn G (2004) Mangroves enhance the biomass of reef fisheries in the Caribbean. Nature 427: 533–536.

Nagelkerken I, Sheaves M, Baker R, Connolly RM (2015) The seascape nursery: a novel spatial approach to identify and manage nurseries for coastal marine fauna. Fish and Fisheries 16: 362–371.

Nelson JL, Zavaleta ES (2012) Salt marsh as a coastal filter for the oceans: changes in function with experimental increases in nitrogen loading and sea-level rise. PLoS One 7(8): e38558.

Olds AD, Connolly RM, Pitt KA, Pittman SJ, Maxwell PS, Huijbers CM, Moore BR, Albert S, Rissk D, Babcock RC, Schlacher TA (2015) Quantifying the conservation value of seascape connectivity: a global synthesis. Global Ecology and Biogeography 25(1): 3–15.

Peterson GW, Turner RE (1994) The value of salt marsh edge versus interior as habitat for fish and decapods crustaceans in a Louisiana tidal marsh. Estuaries 17: 235–262.

Pittman SJ, Caldow C, Hile SD, Monaco ME (2007) Using seascape types to explain the spatial patterns of fish in the mangroves of SW Puerto Rico. Marine Ecology Progress Series 348: 273–284.

Pittman SJ, Kneib RT and Simenstad CA (2011) Practicing coastal seascape ecology. Marine Ecology Progress Series 427: 187–190.

Pittman SJ, McAlpine CA (2003) Movements of marine fish and decapod crustaceans: process, theory and application. Advances in Marine Biology 44: 205–294.

Plummer ML, Harvery CJ, Anderson LE, Guerry AD, Ruckelshaus MH (2013) The role of eelgrass in marine community interactions and ecosystem services: Results from an ecosystem-scale food web model. Ecosystems 16: 237–251.

Ray GC (2005) Connectivities of estuarine fishes to the coastal realm. Estuarine, Coastal and Shelf Science 64(1): 18–32.

Rilov G, Schiel DR (2006) Seascape-dependent subtidal-intertidal trophic linkages. Ecology 87(3): 731–744.

Rountree RA, Able KW (2007) Spatial and temporal habitat use patterns for salt marsh nekton: Implications for ecological functions. Aquatic Ecology 41: 25–45.

Sanchirico JN, Mumby PJ (2009) Mapping ecosystem functions to the valuation of ecosystem services: implications of species-habitat associations for coastal land-use decisions. Theoretical Ecology 2: 67–77.

Sanchirico JN, Springborn M (2011) How to get there from here: Ecological and economic dynamics of ecosystem service provision. Environmental and Resource Economics 48: 243–267.

Shapiro K, Conrad PA, Mazel JAK, Wallender WW, Miller WA, Largier J (2010) Effect of estuarine wetland degradation on transport of *Toxoplasma gondii* surrogates from land to sea. Applied and Environmental Microbiology 76: 6821–6828.

Sheaves M, Baker R, Johnston R (2006) Marine nurseries and effective juvenile habitats: an alternative view. Marine Ecology Progress Series 318: 303–306.

Shephard CC, Crain CM and Beck MW (2012) The protective role of coastal marshes: a systematic review and meta-analysis. PLoS One 6: e27374.

Silvestri S, Kershaw F. (eds) (2010) Framing the Flow: Innovative Approaches to Understand, Protect and Value Ecosystem Services across Linked Habitats. UNEP World Conservation Monitoring Centre, Cambridge.

Smith MD (2007) Generating value in habitat-dependent fisheries: the importance of fishery management institutions. Land Economics 83: 59–73.

Smith MD, Crowder LB (2011) Valuing ecosystem services with fishery rents: a lumped-parameter approach to hypoxia in the Neuse river estuary. Sustainability 3(11): 2229–2267.

Spalding MD, Ruffo S, Lacambra C, Meliane I, Hale LZ, Shepard CC, Beck MW (2014) The role of ecosystems in coastal protection: adapting to climate change and coastal hazards. Ocean and Coastal Management 90: 50–57.

Stål J, Paulsen S, Pihl L, Rönnbäck P, Söderqvist T, Wennhage H (2008) Coastal habitat support to fish and fisheries in Sweden: Integrating ecosystem functions into fisheries management. Ocean and Coastal Management 51: 594–600.

Wedding LM, Lepczyk CA, Pittman SJ, Friedlander AM, Jorgensen S (2011) Quantifying seascape structure: extending terrestrial spatial pattern metrics to the marine realm. Marine Ecology Progress Series 427: 219–232.

Ysebaert T, Yang S, Zhang L, He Q, Bouma TJ, Herman PMJ (2011) Wave attenuation by two contrasting ecosystem engineering salt marsh macrophytes in the intertidal pioneer zone. Wetlands 31: 1043–1054.

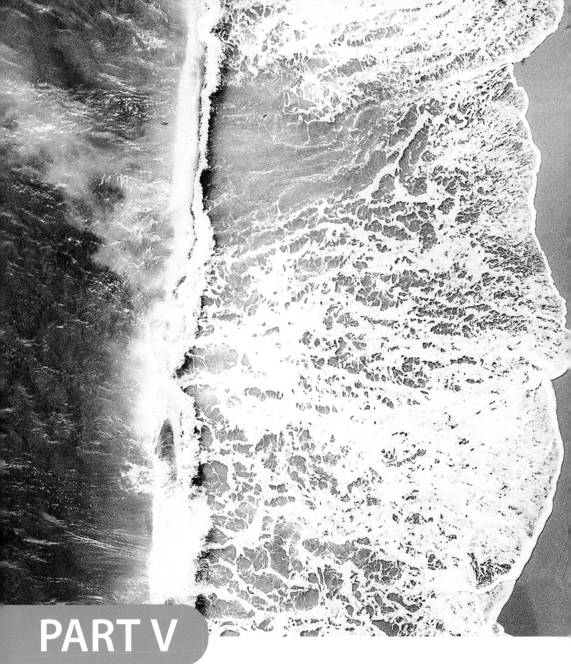

PART V

Epilogue

16

Landscape Ecologists' Perspectives on Seascape Ecology

Simon J. Pittman, John A. Wiens, Jianguo Wu and Dean L. Urban

16.1   Introduction

Here we present some thoughts on the emergence of seascape ecology from the perspective of leading ecologists who have been working at the forefront of modern terrestrial landscape ecology in North America since its inception in the 1970s and who continue to play an instrumental role in guiding the discipline into the future. This chapter consists of opinion pieces from three eminent landscape ecologists – John Wiens, Dean Urban and Jianguo Wu – who have profoundly influenced the development of landscape ecology through their prolific and generous written contributions, together with a high level of participation in the US branch of the International Association for Landscape Ecology (USIALE) and the flagship journal, *Landscape Ecology*, established in 1987 (Barrett *et al.* 2015). All three ecologists were influenced, directly or indirectly, by a landmark three-day workshop at Allerton Park, Illinois, in 1983, which is widely regarded as an important catalyst in the formation of landscape ecology in the United States (Wiens 2008; Wu 2013a; Barrett *et al.* 2015). John Wiens and Jianguo Wu are both past recipients (1996 and 2010 respectively) of the award of Distinguished Landscape Ecologist presented by the US-IALE. Dean Urban served as president of USIALE from 2010 to 2012.

This chapter emerged from the recognition that the fledgling discipline of seascape ecology has much to benefit from the experience and lessons learned in terrestrial landscape ecology. Although biophysical differences exist between marine and terrestrial systems, there are sufficient generalities in ecological pattern-process relationships that make landscape ecology applicable. What guidance, concerns and opportunities do landscape ecologists perceive for the future development of the sister discipline of seascape ecology? Are there lessons learned, pitfalls to be aware of and specific focal topics that should receive priority attention?

16.2   From Landscapes to Seascapes (and Back Again)
By John A. Wiens

Terrestrial ecologists, such as myself, sometimes view oceans as undifferentiated masses of water – places where the land ends and we pass the baton to marine ecologists.

This may be particularly true of landscape ecologists, for whom 'land' is part of their professional identity. Perhaps it is because we are used to looking at the land surface as the template on which ecological dynamics are played out and the ocean's surface seems to us homogeneous and, well, unfathomable.

All it takes is one day at sea to dispel such notions. Mine came on a trip to Cordell Bank, a rocky seamount on the edge of the continental shelf west of San Francisco. Travelling there we saw only an occasional gull but then we were suddenly surrounded by a frenzy of shearwaters, albatrosses, fulmars, auklets and scores of humpback and blue whales. All were feeding on the abundance of krill and fish concentrated about the seamount by the confluence of oceanographic conditions and undersea habitat diversity. Such spatial concentrations of marine life are not unique to Cordell Bank. For millennia, fishermen have realized that some places in the sea support masses of fish while other places are barren. Foraging aggregations of seabirds and whales have been used as indicators for designing marine reserves (Nur *et al.* 2011). On the deeper ocean floor, hydrothermal vents are (literally) hotspots of productivity and biodiversity.

These examples illustrate how the central themes of landscape ecology apply to seascapes as well. Landscapes and seascapes both have spatial *structure*: not every place is the same. Landscapes and seascapes are interconnected mosaics of patches with different environments and inhabitants. The structure of landscapes and seascapes affects how they *function*. The physical structure of Cordell Bank or patches of water of different temperatures produces spatial variations in productivity, feeding relationships and nutrient dynamics. These spatial relationships are not constant; landscapes and seascapes *change*. Feeding aggregations vary in time according to the dynamics of prey populations, which respond to such things as daily tidal cycles or seasonal patterns of upwelling. And the dynamics and patterns of landscapes and seascapes occur over multiple *scale*s. Seasonal productivity of Cordell Bank is driven by local factors but the seabirds and whales that feed there may move over thousands of kilometres during a year.

These are general parallels between landscapes and seascapes. How closely they match depends on where one looks. In intertidal and nearshore areas, ecological dynamics are strongly influenced by the relatively stable spatial structure of the substrate close to the surface. In deeper pelagic areas dynamics are largely determined by the water mass, which is fluid on multiple scales of space and time. Boundaries among water masses may shift or disappear over hours to decades. The distributions of organisms and nutrients may also change, but not in the same ways. Consequently, landscape ecologists may be able to get away with giving little attention to the air above the land, but unless they restrict attention to shallow areas, seascape ecologists must focus on spatial patterns and dynamics in the three-dimensional mass of the ocean itself.

The topics to which landscape ecology has made important contributions are connectivity – fragmentation, boundary dynamics, scale, spatial modelling and statistics, conservation planning, resource management and others (Wiens & Moss 2005) – are also ripe for development in seascape ecology, as the previous chapters illustrate. Applying the approaches and insights of landscape ecology to topographic features such as seamounts or intertidal zones may be (relatively) straightforward. It will be more difficult and require innovative approaches to translate landscape ecology to the seascapes of open pelagic waters where spatial patterns and relationships in the water change with the tides and currents over a vast range of scales (El Niño events are a good example).

The promise of seascape ecology, however, goes beyond adapting the concepts and tools of landscape ecology to the ocean setting. Terrestrial landscapes and oceans are inexorably linked. Even distant uses of the land can affect the ocean environment well out to sea. Overgrazing by cattle in interior Queensland contributes to sedimentation of the Great Barrier Reef and agricultural practices in the corn belt of Iowa promote hypoxia in the Gulf of Mexico (Nassauer *et al.* 2007). The linkages also go in the opposite direction; foraging seabirds or marine mammals can enhance the productivity of terrestrial ecosystems by importing nutrients from marine ecosystems when the animals come ashore (Anderson & Polis 1999). Placing these linkages in a spatially integrated landscape-seascape context will enhance the management of both terrestrial and marine ecosystems, although they will require different approaches. Understanding land-sea linkages rests on knowing how water runs from fields to streams to rivers and thence to and within the sea. Understanding the sea-land linkages rests instead on knowledge of how organisms move between marine areas and the land. Understanding how it all fits together over multiple scales of space and time will require the collaborative efforts of landscape ecologists, hydrologists, oceanographers, animal behaviourists and a new cadre of seascape ecologists. An exciting challenge!

Biography

John Wiens grew up in Oklahoma as an avid birdwatcher. Following degrees from the University of Oklahoma and the University of Wisconsin-Madison (MS, PhD), he joined the faculty of Oregon State University and, subsequently, the University of New Mexico and Colorado State University, where he was a professor of ecology and university distinguished professor. His work has emphasized landscape ecology, conservation and the ecology of birds, leading to over 260 scientific papers and ten books on ecology and landscape ecology. John has chaired several USIALE symposia and in 1996 received the Distinguished Landscape Ecologist award. John left academia in 2002 to join the Nature Conservancy as lead scientist, with the challenge of putting years of classroom teaching and research into conservation practice in the real world. In 2005, John was the recipient of the Cooper Ornithological Society's Loye and Alden Miller Research Award, which is given in recognition of lifetime achievement in ornithological research. In 2008, he joined PRBO Conservation Science as Chief Scientist and since retirement he now divides his time between his home in Corvallis, Oregon and the University of Western Australia where he is a Winthrop Research Professor.

16.3   Seascape Ecology and Landscape Ecology: Distinct, Related and Synergistic
By Jianguo Wu

Most ecological theories have been based on terrestrial systems, despite the fact that about 71% of the Earth's surface is covered by water (nearly 96.5% of which is contained in the oceans). With rare exceptions, terrestrial and marine systems were studied separately with little scholarly communication until the 1980s when scientists began to compare and connect them in order to understand the earth as a whole ecosystem (*e.g.,*

Steele 1985, 1991a; Levin *et al.* 1993; Okubo & Levin 2001). The past few decades have witnessed a wave of new research fronts that cut across marine and terrestrial systems. One of these exciting and emerging cross-system fields is seascape ecology, the topic in this book. Here I compare and contrast this new field with landscape ecology and discuss how they can benefit each other.

16.3.1 Landscape Ecology

While the term landscape ecology was coined in 1939, initially as the study of the relationship between biotic communities and their environment in a regional landscape mosaic, modern landscape ecology since the 1980s has become a highly interdisciplinary and comprehensive scientific enterprise, with multiple definitions and interpretations (Forman 1995; Wiens & Moss 2005; Wu 2006; Wu & Hobbs 2007; Turner & Gardner 2015). It is widely accepted that landscape ecology focuses on the relationship between landscape pattern and ecological processes. Landscape pattern refers to spatial heterogeneity, encompassing patchiness and gradients, which is usually neither random nor uniform in reality. Heterogeneity is almost always scale dependent. Thus, landscape ecology is inevitably and fundamentally a science of heterogeneity and scaling. Conceptually, scale multiplicity in pattern and process begets hierarchical thinking.

Landscape ecology is both a research field (or a body of knowledge) of how landscape composition and configuration interact with ecological processes on broad scales and a new ecological paradigm that explicitly integrates geographical patterns, ecological processes and spatiotemporal scales. Modern landscape ecology covers a wide range of topics (Wu 2013a): (i) pattern-process-scale relationships of landscapes; (ii) landscape connectivity and fragmentation; (iii) scale and scaling; (iv) spatial analysis and landscape modelling; (v) land use and land cover change; (vi) landscape history and legacy effects; (vii) landscape and climate-change interactions; (viii) ecosystem services in changing landscapes; (ix) landscape sustainability and (x) accuracy assessment and uncertainty analysis. As such, landscape ecology is really an interdisciplinary integration of science and art for studying and improving the relationship between spatial pattern and ecological processes on multiple scales, with landscape sustainability as its ultimate goal (Wu 2006, 2013b).

16.3.2 Seascape Ecology

Seascape ecology is the study of the relationship between spatial pattern and ecological processes in marine environments on a range of spatiotemporal scales. The emergence of seascape ecology was apparently inspired by the rapid development of landscape ecology in recent decades (Pittman *et al.* 2004, 2011; Boström *et al.* 2011; Kavanaugh *et al.* 2016). The current literature indicates that there are different views on seascape ecology in terms of its relationship to landscape ecology. The first view promotes seascape ecology as the application of landscape ecology principles and methods in the study of coastal marine systems (Boström *et al.* 2011; Pittman *et al.* 2011; Olds *et al.* 2016). The second view also acknowledges the relevance and usefulness of landscape ecology but places more emphasis on the open and dynamic oceanographic features of marine environments (*e.g.,* Kavanaugh *et al.* 2014, 2016). The first view is focused more on coastal seascapes whereas the second more on pelagic seascapes. Thus, these two views complement each other, together making seascape ecology more comprehensive in scope and more challenging intellectually.

Fundamentally different from the first two, the third view asserts that, because the 'properties and dynamics of the ocean fluid' differ so much from those of terrestrial landscapes, seascape ecology can benefit little from landscape ecology and that such 'terrestrial analogies' should be 'avoided' (Manderson 2016). While it is true that marine and terrestrial systems are fundamentally different in many ways, both geophysically and biologically (Steele 1985), this fact itself does not suffice the rejection of landscape ecological principles and methods in seascape ecology. On the contrary, interdisciplinary comparisons and fertilization across land and water have been necessary, fruitful and quite promising (Steele 1985, 1989, 1991b; Levin *et al.* 1993; Okubo & Levin 2001). As I discuss below briefly, landscape ecology as a body of knowledge may be of limited use to seascape ecology, but it can be quite relevant as a new ecological paradigm that focuses on pattern-process-scale relations.

16.3.3 How can Landscape and Seascape Ecology Interact with Each Other?

The conceptual similarity between landscape ecology and seascape ecology is apparent although fundamental biophysical differences exist between the two 'scapes'. I see three general ways that seascape ecology can benefit from landscape ecology. The degree of relevance or applicability of the three uses varies with the locations and spatial extents of marine environments, generally decreasing from coastal marine zones to open oceans.

First, the findings of pattern-process-scale relations in terrestrial landscapes should be heuristically useful for seascape ecology, such as the effects of the kinds and amounts of habitat, geometry and connectivity of habitat patches, edges and corridors, matrix (or context) and natural and human disturbances on biodiversity and ecological processes, as well as their scaling relations in space and time. This heuristic value, however, may be quite limited especially for pelagic systems. Second, many spatial analysis and modelling methods used in landscape ecology, such as spatial statistics, categorical and surface pattern metrics and individual-based models, can be used in seascape ecology. Indeed, some of them (*e.g.,* power spectral analysis) were used in marine studies before being introduced into landscape ecology. The Stommel diagram originated in oceanography has had profound influences on the study of scaling and hierarchy in landscape ecology. Of course, remote sensing, GIS and GPS are now frequently used in almost all field-based studies way beyond landscape ecology and geography. The third and most general way is to use landscape ecology as a spatially explicit ecological paradigm that emphasizes spatial heterogeneity, pattern-process relations, scale multiplicity, transient dynamics and holistic human-environmental interactions.

Several key principles that characterize landscape ecology may also become prominent in seascape ecology, including patch dynamics, scaling, matrix / context, connectivity / fragmentation, ecotones / gradients, ecosystem / landscape services and landscape resilience / sustainability. Patch dynamics and scaling are two science themes transcending the boundaries between physical systems and between academic disciplines, both of which had been explored in the water and on the land before the term seascape ecology existed (Levin & Paine 1974; Steele 1978, 1989; Levin *et al.* 1993). Patch dynamics had its original conceptual roots in terrestrial community ecology in the 1940s (Watt 1947), saw its first mathematical theory developed from intertidal systems in the 1970s (Levin & Paine 1974) and became a widely applied perspective in both terrestrial and marine ecology in the 1980s and 1990s (Pickett & White 1985; Levin *et al.* 1993; Wu & Loucks 1995), epitomizing modern landscape ecology as a unifying framework. Conceptually, landscapes are hierarchically structured land mosaics (Forman 1995) in which patch dynamics take place constantly on multiple scales – *i.e.*, 'hierarchical patch dynamics' in operation (Wu & Loucks 1995).

Pelagic marine environments are open, diffusive and dynamic, with less obvious physical boundaries than terrestrial systems but they also exhibit spatial patchiness and scaling relations in both their physical environment (from eddies to gyres) and ecological organization (from phytoplankton to zooplankton and higher tropic levels). In a seminal paper published in the journal *Landscape Ecology*, the eminent oceanographer John H. Steele (1989) discussed the spatial patterning and scaling of 'ocean landscapes':

> The ocean has a complex physical structure at all scales in space and time, with 'peaks' at certain wave numbers and frequencies. Pelagic ecosystems show regular progressions in size of organisms, life cycle, spatial ambit and tropic status.

These observations remain as relevant and inspiring today as they were when the article was published. Recent studies in seascape ecology have taken the hierarchical patch dynamics and scaling perspectives to a new level, conceptualizing the marine environment as 'a mosaic of distinct seascapes, with unique combinations of biological, chemical, geological and physical processes that define habitats which change over time' and integrating oceanographic and ecological paradigms in studying, managing and protecting marine systems (Kavanaugh *et al.* 2016).

Other key ideas in landscape ecology including matrix / context; connectivity / fragmentation; ecotones / gradients; ecosystem / landscape services; and landscape resilience / sustainability are also relevant, but yet to be fully explored in the context of seascapes. Some pioneering seascape ecological studies utilizing these ideas already exist (Pittman *et al.* 2004 and examples throughout this book). Such studies are crucial to the marine biodiversity conservation, marine resource management and seascape sustainability. By focusing on ecosystems services and human wellbeing in changing climates and marine environments, a seascape sustainability science is expected to occur, in parallel to landscape sustainability science (Wu 2013b). Through integrating ecological studies across land and water, a spatial ecology of landscapes and seascapes is in the making.

Biography

Jianguo (Jingle) Wu is the Dean's Distinguished Professor of Sustainability Science at the School of Life Sciences and School of Sustainability at Arizona State University, Tempe, Arizona, United States. He was awarded his PhD (1991) in ecology from Miami University, Oxford, Ohio, United States and then as a National Science Foundation (NSF) postdoctoral fellow at Cornell University (1991–1992) and Princeton University (1992–1993) working alongside Simon A. Levin. His research focuses on landscape ecology, urban ecology and sustainability science, subjects on which he has authored 14 books and more than 300 journal articles and book chapters. Jianguo has served as editor-in-chief of *Landscape Ecology* since 2005 and serves on the editorial boards of several international journals on ecology and interdisciplinary research. He is the founding director of the Sino-US Center of Conservation, Energy and Sustainability Science (2007–) and the Center for Human-Environment System Sustainability (CHESS), Beijing Normal University (2012). His contributions to landscape ecology have been recognized through several major awards and honours including the American Association for the Advancement of Science (AAAS) Award for International Scientific Cooperation (2006); Leopold Leadership Fellow (2009); Distinguished Landscape Ecologist Award and an Outstanding Scientific Achievements Award and Distinguished Service Award from United States Association for Landscape Ecology (2010, 2011 and 2012).

16.4   Seascape Ecology
By Dean L. Urban

The emergence of landscape ecology reflected a growing interest in spatial heterogeneity and scale in ecology (*e.g.*, Levin 1992). As the discipline evolved, it has become a multithreaded enterprise with branches to and from a broad range of disciplines. But a few themes are recognizable as hallmarks of the field – how we organize textbooks, how we think about our work. Landscape ecologists think a lot about where pattern comes from, how it scales in space and time and why it matters to populations, communities and ecosystem processes. Marine systems invite the same perspective. (In the landscape ecology laboratory at Duke, we deliberately decided to broach seascape ecology as a new focus and to take advantage of our marine lab. Pat Halpin led that charge. Some later beneficiaries of that decision are contributors to this book.) In general, the principles of landscape ecology translate readily to seascapes.

Agents of pattern include the physical template, biotic processes and disturbance regimes. These apply readily to seascapes, with the notable complication that the physical template is three dimensional and layered and it is dynamic over timescales much faster than most landscapes. For example, landscape ecologists might be concerned with temperature and moisture gradients that reflect the interaction of climate with landform. In the seas, we expect gradients in temperature and chemistry (*e.g.*, salinity) that reflect the interaction of the ocean climate (currents) with bathymetry and the currents change rapidly enough that applications often require data that are resolved on monthly or even finer time scales.

In landscape ecology, patchiness that arises from interacting agents of pattern have often been summarized in 'space-time' diagrams (*e.g.*, Delcourt *et al.* 1983). These diagrams emphasize the characteristic spatiotemporal scaling of various levels of patchiness. Somewhat ironically, these Stommel diagrams came to landscape ecology from marine systems (for an intriguing perspective on these diagrams, see http://rs.resalliance .org/2010/02/24/a-history-of-stommel-diagrams, accessed 6 June 2017). In particular, Steele's (1974) seminal work on the interplay between biophysical dynamics and biotic processes (food web interactions) set a precedent for how to visualize patchiness that has been foundational to landscape ecology.

One implication of pattern is the potential fragmentation of populations into metapopulations. Many landscape ecologists have embraced graph-theoretic models of metapopulations. Ted Ames, a fisherman and historian of fisheries, essentially reinvented graph theory as a conceptual model for the antique cod fisheries of New England because the system just made sense that way (Ames 2004). This underscores the ready translation of terrestrial models to seascapes – again, with the caveat that the model is temporally dynamic while landscape ecologists typically assume the habitat dynamics are very slow relative to the species inhabiting those habitats.

A central challenge in landscape ecology is inferring the relative control of spatial structure (autocorrelation) in species distributions: partitioning beta diversity (Legendre *et al.* 2005). Autocorrelation might be caused by spatially structured environmental constraints such as patterns in temperature or soil moisture; or it might be due to spatial processes such as dispersal or contagious disturbances. The difficulty lies in the reality that spatial processes are often unmeasured and so we would like to make inferences about process from spatial structure that is residual after accounting for environmental constraints. But residual spatial structure might also be due to an unmeasured environmental constraint and so this inference is a challenge. In marine systems, this challenge is further complicated by the fact that the environment moves around with currents. For example, if we observe a marine mammal at a particular location and time, we might wonder whether it is there because it is responding to bathymetry (as distance to shore, or location and experiencing whatever temperature or salinity happens to be there), or responding to temperature (at whatever location that temperature happens to be), or responding to prey species (which might be responding to bathymetry or temperature) (Schick *et al.* 2011). These are complicated questions and the answers likely will require new analytic techniques as well as new data (*e.g.*, from genomics). This work will continue in landscape and seascape ecology, with marine systems perhaps having an advantage precisely because their temporal dynamics might provide more leverage to separate 'environment' from 'location'.

In summary, seascapes might be seen as particularly fast-moving versions of landscapes. The central questions in landscape and seascape ecology are very similar and the analytic approaches overlap substantially. More crossfertilization and interdisciplinary collaboration would seem to be to the benefit of all.

Biography

Dean Urban is professor of landscape ecology and senior associate dean for academic initiatives in the Nicholas School of the Environment at Duke University. He received his PhD in ecology from the University of Tennessee (1986), working at Oak Ridge National Laboratory in the formative years of landscape ecology. A hallmark of his work is integrated studies that extrapolate our fine-scale empirical understanding of environmental issues to the larger space and time scales of management and policy. Specific research interests include the implications of climate change for forest ecosystems and the consequences of land use pattern on forest habitat connectivity and watershed function in developed landscapes. He has been named a Distinguished Landscape Ecologist but is especially proud to have placed more than a hundred masters students into positions in governmental and nongovernmental environmental organizations, where they are having a direct impact on landscape management.

References

Ames EP (2004) Atlantic cod stock structure in the Gulf of Maine. Fisheries 29: 10–28.

Anderson W, Polis G (1999) Nutrient fluxes from water to land: seabirds affect plant nutrient status on Gulf of California islands. Oecologia 118: 324–332.

Barrett GW, Barrett TL, Wu J (eds) (2015) History of landscape ecology in the United States. Springer.

Boström C, Pittman SJ, Simenstad C, Kneib RT (2011) Seascape ecology of coastal biogenic habitats: advances, gaps, and challenges. Marine Ecology Progress Series 427: 191–217.

Delcourt HR, Delcourt PA, Webb T (1983) Dynamic plant ecology: the spectrum of vegetation change in space and time. Quaternary Science Reviews 1: 153–175.

Forman RTT (1995) Land Mosaics: The ecology of landscapes and regions. Cambridge University Press, Cambridge.

Kavanaugh MT, Hales B, Saraceno M, Spitz YH, White AE, Letelier RM (2014) Hierarchical and dynamic seascapes: A quantitative framework for scaling pelagic biogeochemistry and ecology. Progress in Oceanography 120: 291–304.

Kavanaugh MT, Oliver MJ, Chavez FP, Letelier RM, Muller-Karger FE, Doney SC (2016) Seascapes as a new vernacular for pelagic ocean monitoring, management and conservation. ICES Journal of Marine Science 73: 1839–1850.

Legendre P, Borcard D, Peres-Neto PR (2005) Analyzing beta diversity: partitioning the spatial variation of community composition data. Ecological Monographs 75: 435–450.

Levin SA (1992) The problem of pattern and scale in ecology. Ecology 73: 1943–1967.

Levin SA, Paine RT (1974) Disturbance, patch formation and community structure. Proceedings of the National Academy of Sciences (USA) 71: 2744–2747.

Levin SA, Powell TM, Steele JH (eds) (1993) Patch dynamics. Springer-Verlag, Berlin.

Manderson JP (2016) Seascapes are not landscapes: an analysis performed using Bernhard Riemann's rules. ICES Journal of Marine Science 73: 1831–1838.

Nassauer JI, Santelmann MV, Scavia D (eds) (2007) From the corn belt to the Gulf. Societal and environmental implications of alternative agricultural futures. Resources for the Future, Washington, DC.

Nur N, Jahncke J, Herzog MP, Howar J, Hyrenbach KD, Zamon JE, Ainley DG, Wiens JA, Morgan K, Ballance LT, Stralberg D (2011) Where the wild things are: predicting hotspots of seabird aggregations in the California Current System. Ecological Applications 21(6): 2241–2257.

Okubo A, Levin SA (eds) (2001) Diffusion and ecological problems: Modern perspectives. 2nd edition. Springer, New York, NY.

Olds AD, Connolly RM, Pitt KA, Pittman SJ, Maxwell PS, Huijbers CM, Moore BR, Albert S, Rissik D, Babcock RC, Schlacher TA (2016) Quantifying the conservation value of seascape connectivity: A global synthesis. Global Ecology and Biogeography 25: 3–15.

Pickett STA, White PS (eds) (1985) The ecology of natural disturbance and patch dynamics. Academic Press, Orlando.

Pittman SJ, Kneib RT, Simenstad CA (2011) Practicing coastal seascape ecology. Marine Ecology Progress Series 427: 187–190.

Pittman SJ, McAlpine CA, Pittman KM (2004) Linking fish and prawns to their environment: a hierarchical landscape approach. Marine Ecology Progress Series 283: 233–254.

Schick RS, Halpin PN, Read AJ, Urban DL, Best BD, Good CP, Roberts JJ, LaBrecque EA, Dunn C, Garrison LP, Hyrenbach KD, McLellan WA, Pabst DA, Stevick S (2011) Community structure in pelagic marine mammals at large spatial scales. Marine Ecology Progress Series 434: 165–181.

Steele JH (1974) The structure of marine ecosystems. Harvard University Press, Cambridge.

Steele JH (ed.) (1978) Spatial pattern in plankton communities. Plenum Press, New York, NY.

Steele JH (1985) A comparison of terrestrial and marine ecological systems. Nature 313: 355–358.

Steele JH (1989) The ocean 'landscape'. Landscape Ecology 3: 185–192.

Steele JH (1991a) Can ecological theory cross the land-sea boundary? Journal of Theoretical Biology 153: 425–436.

Steele JH (1991b) Marine functional diversity. BioScience 41: 470–474.

Turner MG, Gardner RH (2015) Landscape ecology in theory and practice: pattern and process. 2nd edition. Springer, New York, NY.

Watt AS (1947) Pattern and process in the plant community. Journal of Ecology 35: 1–22.

Wiens J (2008) Allerton Park 1983: the beginnings of a paradigm for landscape ecology? Landscape Ecology 23(2): 125.

Wiens J, Moss M (eds) (2005) Issues and perspectives in landscape ecology. Cambridge University Press, Cambridge.

Wu JG (2006) Landscape ecology, cross-disciplinarity, and sustainability science. Landscape Ecology 21: 1–4.

Wu JG (2013a) Key concepts and research topics in landscape ecology revisited: 30 years after the Allerton Park workshop. Landscape Ecology 28: 1–11.

Wu JG (2013b) Landscape sustainability science: ecosystem services and human well-being in changing landscapes. Landscape Ecology 28: 999–1023.

Wu JG, Loucks OL (1995) From balance of nature to hierarchical patch dynamics: A paradigm shift in ecology. Quarterly Review of Biology 70: 439–466.

Wu JG, Hobbs RJ (eds) (2007) Key topics in landscape ecology. Cambridge University Press, Cambridge.

Index

Seascape Ecology, First Edition. Edited by Simon J. Pittman.
© 2018 John Wiley & Sons Ltd. Published 2018 by John Wiley & Sons Ltd.